AF669405

BASIC MECHANICAL ENGINEERING

SECOND EDITION

BASIC MECHANICAL ENGINEERING

SECOND EDITION

A.R. Israni
Professor in Mechanical Engineering (retd.),
LD Engineering College, Ahmedabad.
Professor in Mechanical Engineering,
Narayan-Swami Institute of Technology,
Jetalpur, Ahmedabad.

P.K. Shah
Principal,
Silver Oak College of Engineering & Technology,
Ahmedabad.

N. C. Nayak
Reader in Mechanical Engineering,
Indira Gandhi Institute of Technology, Sarang
(An Autonomous Institute of Government of Odisha)
Odisha.

A unit of **BSP Books Pvt. Ltd.**

4-4-309/316, Giriraj Lane, Sultan Bazar,
Hyderabad - 500 095 - T.S.
Phone : 040 - 23445605, 23445688

Basic Mechanical Engineering, ***Second Edition by A.R. Israni, P.K. Shah and N.C. Nayak***

Second Edition 2018

Published by :

4-4-309/316, Giriraj Lane, Sultan Bazar,
Hyderabad - 500 095 - T.S.
Phone : 040 - 23445605, 23445688
e-mail : info@bspbooks.net
Website : www.bspbooks.net

ISBN: 978-93-86819-69-7 (Hardback)

Dedication

My mother late Puriben Ramchand Israni and uncle Nahichal Das

Wife Late Kumariben A. Israni and daughters Sonia Tekchandani, Kalpanan Khubchandani

- Prof. A. R. Israni

Wife Minaxi and sons Viren and Milan

- Prof. P. K. Shah

My beloved Parents, Wife Suman and Daughters Deeksha & Ruchira

- Dr. N. C. Nayak

Preface to Second Edition

Basic Mechanical Engineering is the basic course which includes the fundamental aspects of various subjects in Mechanical Engineering. All students admitted to first year of engineering must undergo these basic topics of Mechanical Engineering. This subject is the base for further study in the fields of Thermal Engineering, Theory of Machines and Engineering Materials.

Mechanical power is very much essential for transportation, industry and electricity generation. Prime movers generate mechanical power by utilizing heat energy of fuels. The prime movers and the physical quantities related with heat and work are explained in introductory chapter.

Chapter 2 deals with the fuels and their combustion. Properties of gases and steam are dealt in next two chapters.

Chapter 5 introduces heat engines and various cycles. Steam generators i.e., boilers are explained in next chapter.

Chapter 7 explains details of internal combustion engines. These engines are most widely used as prime movers and are very versatile. Speed controls of engines are treated in chapter 8.

Chapter 9 deals with various pumps used for lifting fluids.

Compressed air is used in many applications. Chapter 10 explains the air compressors used to compress the air.

Refrigeration and air conditioning systems are included in Chapter 11.

Chapter 12 discusses in detail about the various production processes.

Chapter 13 dealt with couplings, clutches, brakes. Chapter 14 discusses about transmission of power.

The last chapter explains various engineering materials and their properties.

This book is in a textbook form. The text is explained with enough number of illustrations and figures. Large number of solved problems are given in each chapter. At the end of chapters, objective type questions, short and long questions and exercises are given. These will help the students for the examination.

We are thankful to our senior professors for inspiration and help. We specially thank Dr. L. N. Patel, Director, S. P College of Engineering, Visnagar for his guidance and for acknowledgement of this book. We are thankful to B.S. Publications for timely publishing this book.

We hope that this book will satisfy the demand of first year engineering students. We welcome the suggestions from the readers.

This book is dedicated to our beloved parents and children who have provided driving force to us to write this book.

-Authors

Contents

Chapter 2

Fuel and Combustion

Chapter 3

Properties of Gases

Chapter 4

Properties of Steam

Chapter 5

Heat Engines

Chapter 6

Steam Boiler

Chapter 7

Internal Combustion Engines

Chapter 8

Speed Control

Chapter 9

Pumps

Chapter 13

Couplings, Clutches and Brakes

Chapter 14

Power Transmission System

Chapter 15

Important Engineering Materials

CHAPTER 1

Introduction

1.1 Prime Movers

Prime mover is a device which converts natural source of energy into mechanical work to drive machines for various applications.

In olden days, man had to depend for power on his own physical strength and on animals like horse, bullock etc. As the horse was most important, the unit of power was named 'horse power' for the power developed by prime movers.

Petrol engine, diesel engine, steam turbine, gas turbine are few examples of prime movers.

1.2 Sources of Energy

Conventional sources of energy derived from nature for driving prime movers are coal, petroleum, natural gas, flow of river water, atom etc. Today some non-conventional sources of energy like solar, wind, biogas, biomass, tide, ocean waves, geothermal etc. have also been developed.

Coal, petroleum and natural gas are fuels. Petrol, diesel and kerosene are derived from petroleum. All these fuels are most important sources of energy. Heat energy produced by burning of these fuels may be converted to mechanical work by heat engines.

Another useful source of energy is flow of river water. Dam may be constructed across the river. The potential energy of water collected on upstream side of dam, is used to drive hydraulic turbine that generates the power.

Energy released from atom is nuclear energy. Heat energy produced by fission or fusion of atoms may be converted to mechanical work by steam turbines.

The non-conventional sources of energy mentioned above are on the way of development. These will be needed more in future when the conventional sources would be depleted. Solar energy is available freely. It is renewable and non-polluting source of energy. It may be used to produce steam to run steam turbine. Solar photovoltaic cells may be used to produce electricity.

1.3 Types of Prime Movers

Prime movers may be classified according to the sources of energy utilized by them. The main two types are:

1. Thermal prime movers
2. Hydraulic prime movers

Some other types are:

3. Wind mill
4. Tidal
5. Wave

1. A prime mover which converts heat energy available from different sources into mechanical work is known as thermal prime mover. They are further classified as (a) heat engines (b) nuclear power plants.

   Prime movers using heat energy produced by burning of fuel are called heat engines. Petrol engine, diesel engine, steam engine, steam turbine, gas turbine etc., are examples of heat engines. These engines may be either external combustion engines or internal combustion (I.C) engines.

   Nuclear power plant uses nuclear heat energy obtained by fission or fusion of atom.

   Solar power plant, biogas plant and geothermal plant are also thermal prime movers as they use heat energy to generate power.

2. Hydraulic prime movers utilize potential energy of river water and convert it into mechanical work. All hydraulic turbines are hydraulic prime movers.

   Other non-conventional prime movers include wind mill, tidal and wave power plants.

   Thus, types of prime movers may be represented as follows:

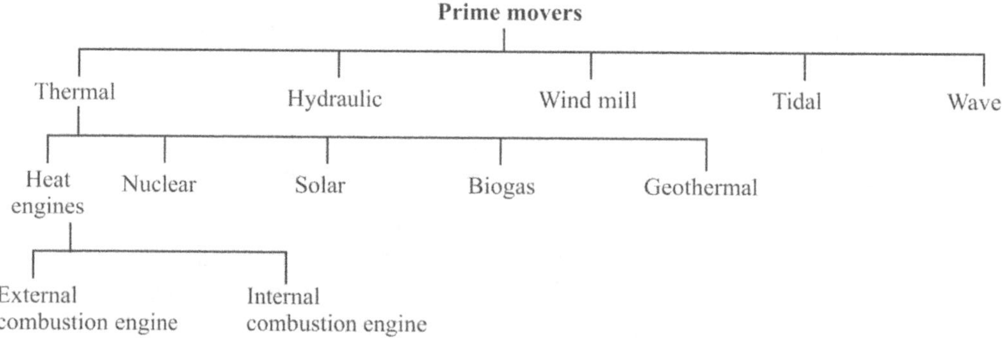

1.4 Force and Mass

We will follow SI (system international) system of units.

Mass of the body is defined as the amount of matter contained in it. Its unit is kilogramme (kg). It is denoted by m. Mass of the body always remains constant.

Force is the product of mass and acceleration of the body upon which it is applied.

Thus, force = mass × acceleration

i.e. $$F = ma \qquad .....(1.1)$$

Unit of mass is kg and unit of acceleration is m/s^2. So, the unit of force is kg-m/s^2. This unit is known as Newton (N). Thus, 1N = 1 kg-m/s^2.

One Newton is that force which when applied to a body of mass 1 kg produces an acceleration of 1 m/s^2 in it. This is a derived unit. The larger units are kN and MN.

The weight (W) of a body is the gravitational force of earth acting on the body. It is the product of mass of the body and local gravitational acceleration g.

i.e. $$W = mg \text{ N} \qquad .....(1.2)$$

The value of g at sea level is 9.80665 m/s^2 (approximately 9.81)

The weight of a body of mass 1 kg is considered 1 kgf.

$\therefore$ weight 1 kgf = 1 × 9.81 N

$\therefore$ 1 kgf = 9.81 N

1.5 Pressure

Pressure is defined as normal force per unit area of the surface.

i.e. $$\text{Pressure} = \frac{\text{force}}{\text{area}}$$

$$\therefore \quad P = \frac{F}{A} \qquad .....(1.3)$$

SI unit of pressure will be N/m^2. This unit is also known as Pascal.

i.e. $$\text{Pa} = \frac{\text{N}}{\text{m}^2}$$

One Pascal equals to force of 1 N acting on area of 1m^2. Pascal is a small unit. The larger units are kPa and MPa.

The other unit of pressure is bar.

$$1 \text{ bar} = 10^5 \text{ Pa}$$

Pressure of air at mean sea level is 1.01325 bar. This is known as 1 atmosphere, denoted by 1 atm.

i.e. $$1 \text{ atm} = 1.01325 \text{ bar}$$

Atmospheric pressure is also equal to the pressure developed by a mercury column of 760 mm. Atmospheric pressure in kgf/cm^2 unit is 1.033 kgf/cm^2.

Some of the pressure measuring instruments are Bourdon pressure gauge, U-tube manometer, barometer etc. Most of the instruments indicate pressure relative to atmospheric pressure, i.e. pressure above the atmospheric pressure. This pressure is called the 'gauge pressure'. The actual pressure is known as absolute pressure.

So, absolute pressure = gauge pressure + atmospheric pressure. .....(1.4)

When system pressure is less than atmospheric, the gauge pressure becomes negative. This negative gauge pressure is called vacuum. Thus, vacuum is the absolute pressure less than atmospheric.

If h = difference in heights of fluid in two limbs of U-tube manometer, m

g = gravitational acceleration, m/s^2

ρ = density of manometric fluid, kg/m^3

then $$p = \rho g h \; \frac{N}{m^2} \qquad .....(1.5)$$

1.6 Work

Work is said to be done when a body, under the action of force, is moving in the direction of force. The magnitude of work is the product of force and the distance moved in the direction of force.

Thus, $$\text{work} = \text{force} \times \text{distance}$$

i.e. $$W = F \times l \qquad .....(1.6)$$

This work is the mechanical work mentioned above. The SI unit of work will be Newton-meter (Nm), also known as Joule (J). One Joule is the amount of work done when a force of one Newton acts through a distance of one meter. The larger units are kJ and MJ.

When work is done by a system, it is taken positive, while when it is done on a system, it is taken negative.

When a constant pressure p is exerted on the piston of area A inside the cylinder, work is done due to piston displacement l. This work is

$$W = F \times l = (p \times A) \times l = p \times (A \times l)$$
$$= pV \qquad .....(1.7)$$

where V = volume swept by the piston

When a fluid having initial volume V_1 expands to a final volume V_2, work done

$$W = p\,(V_2 - V_1) \qquad .....(1.8)$$

Pressure-volume diagram and work:

The graph drawn by taking pressure on y-axis and volume on x-axis is called pressure-volume or P-V diagram. Fig 1.1 shows a P-V diagram.

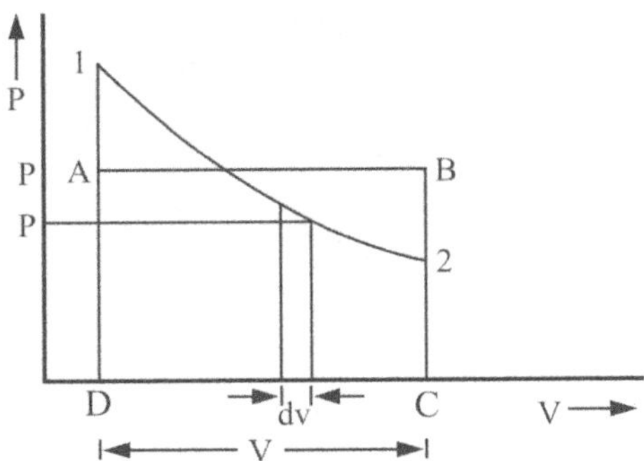

Fig. 1.1 P-V diagram.

If pressure p is constant, area of p-V diagram is area ABCD which is equal to p × V.

Work done is also equal to P × V.

So, area of P-V diagram represents amount of work done.

Now consider the case where pressure varies from 1 to 2 as shown in Fig. 1.1. Assume that the pressure P remains constant for a small change of volume dV. The small amount of work done dW will be represented by area pdV.

So, $$dW = pdV$$

Therefore, the total amount of work done from 1 to 2 will be the sum of all small areas pdv.

i.e. $$W = \int_1^2 PdV \qquad .....(1.9)$$

This is the total area under curve 1-2.

Thus in all cases, area of P-V diagram represents the amount of work done.

Flow work: The flow work is the energy required to make the fluid flow across the boundary of control volume or open system (Fig. 1.2).

If a fluid of volume V flows into a cylinder against pressure p, it has to do work equal to pV to move the piston. In other words, the fluid possesses the energy equal to pV. This energy is flow work or flow energy.

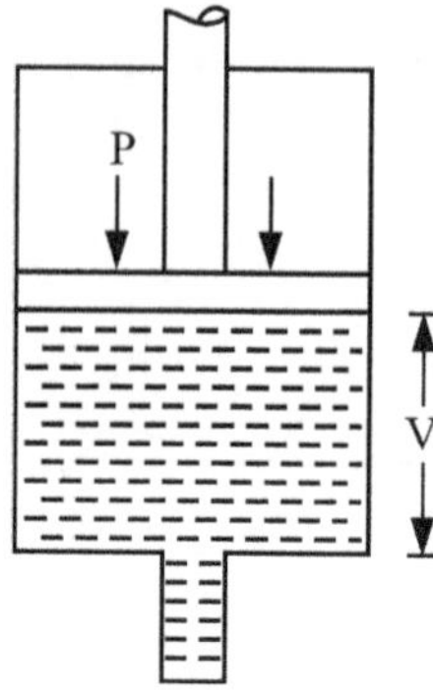

Fig. 1.2 Flow work.

Shaft work: When a shaft is rotated by motor as shown in Fig. 1.3, work is said to be done on shaft.

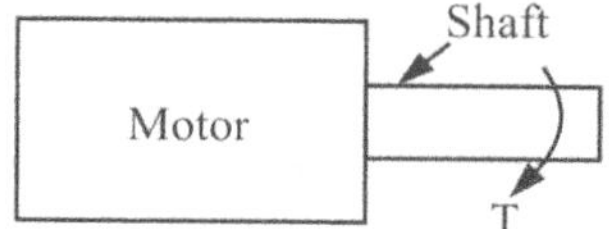

Fig. 1.3 Shaft work.

If T is the torque applied to shaft and dθ is the small angular displacement of the shaft, the shaft work will be

$$W = \int_1^2 T d\theta \quad \text{.....(1.10)}$$

1.7 Power

Power is defined as the rate of doing work or work done per unit time.

i.e. $$\text{Power} = \frac{\text{Work done}}{\text{time}}$$

So, $$P = \frac{W}{t}$$

Unit of power will be Joule/second (J/s). It is called Watt (W). The larger units are kW and MW.

The unit of power in MKS system is horse power (hp). The work done at the rate of 75 kgf-m per sec is equal to 1 hp.

Also, $$1 \text{ hp} = 735.5 \text{ W or } 0.7355 \text{ kW}$$

The power developed by an engine is termed as 'indicated power'. Power developed at the shaft of the engine is termed as 'brake power'.

The power lost due to friction in the engine is termed as 'friction power'. It is the difference between indicated power and brake power.

1.8 Energy

The energy of a body is its capacity for doing work. So, the unit of energy is same as that of work i.e. Joule. The larger units are kJ and MJ.

Another important unit of energy is kilowatt-hour (kWh). Energy consumed in one hour by 1 kW power supply is called 1 kWh.

$$1 \text{ kWh} = 1000 \text{ Wh} = 1000 \frac{J}{s} \times 3600 \text{ s} = 36 \times 10^5 \text{ J}$$

This unit is important in electrical engineering.

There are different forms of energy such as 1. mechanical energy; 2. thermal energy; 3. electrical energy; 4. chemical energy; 5. nuclear energy etc. One form of energy can be readily converted into another form, but the total amount of energy remains constant. This is known as law of conservation of energy.

There are two kinds of mechanical energy :

(a) Potential energy, (b) Kinetic energy

(a) ***Potential energy:*** The energy possessed by a body by virtue of its position is called its potential energy. If a body is raised to a height z metre, its potential energy is

$$PE = mgz \quad J \qquad .....(1.12)$$

(b) ***Kinetic energy:*** The energy possessed by a body by virtue of its motion is called its kinetic energy. If a body is moving at a velocity v m/s, its kinetic energy is

$$KE = \frac{1}{2} mV^2 \quad J \qquad .....(1.13)$$

1.9 Heat

Heat is one form of energy. So, its unit is Joule. It may be converted into work or other forms of energy. Heat can be transferred from one body to the other when temperature difference exists between them. If heat is added to the system, it is taken as positive. If heat is rejected by the system, it is taken as negative.

1.10 Temperature

Temperature is a measure of degree of hotness or coldness of a body. Its unit in SI system is Kelvin (K). The melting point of ice at atmospheric pressure is taken as 273 K, and boiling point of water is taken as 373 K.

For practical purposes, the Celsius scale for measuring temperature is more common. Formerly it was known as centigrade scale. It is denoted by degree Celsius (°C). In this scale, at atmospheric pressure, freezing point of water is taken as 0 °C and boiling point of water is taken as 100 °C. One division of celsius scale is equal to one division of kelvin scale. The relation between kelvin and celsius scales is

$$K = {}^{\circ}C + 273 \qquad .....(1.14)$$

One more unit of temperature is degree Fahrenheit (°F). Its relation with celsius scale is

$${}^{\circ}F = \frac{9}{5} \times {}^{\circ}C + 32 \qquad .....(1.15)$$

Temperature is measured by the instruments known as thermometer and pyrometer.

1.11 Units of Heat

As heat is one form of energy, its unit is Joule. The larger units are kJ and MJ.

In metric system, calorie (cal) and kcal are taken as the units of heat. The quantity of heat required to raise the temperature of 1 kg of water through 1 °C is known as 1 kcal.

$$1 \text{ kcal} = 4186 \text{ J} = 4.186 \text{ kJ}$$

1.12 Specific Heat Capacity

The specific heat capacity (also called specific heat) of a substance is defined as the amount of heat required to change the temperature of unit mass of a substance through one degree. Its value is different for different substances. It is denoted by c. It Q J is the amount of heat required to raise the temperature of m kg of a substance by Δt K, then specific heat of this substance.

$$C = \frac{Q}{m\,\Delta t} \text{ J/kg-K} \qquad \because Q = mc\Delta t \qquad .....(1.16)$$

The SI unit of specific heat is J/kg-K. The larger unit is kJ/kg-K.

If the specific heat does not remain constant during heating, it may be assumed that for a small change of temperature dt, specific heat c remains constant. So, the small amount of heat transfer is

$$dQ = mCdt$$

$$\therefore \quad Q = m\int_1^2 Cdt \qquad \text{.....(1.17)}$$

In case of gaseous substance, volume of gas changes with the change in pressure. So heat added to the gas by keeping its volume constant and heat added by keeping the pressure constant would be different. So there will be two different values of specific heat for the gas, one specific heat at constant volume c_v and the other specific heat at constant pressure c_p.

As solids and liquids are generally incompressible, they have single value of specific heat.

The product of mass and specific heat is called the heat capacity of the substance. It is denoted by C.

1.13 Interchange of Heat

When a body at higher temperature comes in contact with another body at lower temperature, there will be an interchange of heat. Heat flows from the body at higher temperature to the body at lower temperature. Heat lost by hot body is equal to heat gained by cold body. There are three modes of heat transfer – conduction, convection and radiation. There will be no interchange of heat if two bodies in contact have equal temperatures.

1.14 Change of State

Matter can exist in three phases – solid, liquid and gas (or vapour). State of the substance can be changed by addition or rejection of heat.

The constant temperature at which solid substance changes to liquid state is called melting point.

When heat is added to solid, first of all, its temperature increases. After melting point is reached, if heat is added further, solid gets converted into liquid at this constant temperature.

Amount of heat required to convert 1 kg of solid into liquid or heat to be rejected to freeze 1 kg of liquid into solid at constant temperature is called latent heat of fusion.

The constant temperature at which liquid changes to vapour is called boiling point or saturation temperature.

If heat is added to liquid, its temperature increases until saturation temperature is reached. At this constant temperature, liquid is converted to vapour by addition of heat.

Amount of heat required to convert 1 kg of liquid into vapour or heat to be rejected to condense 1 kg of vapour into liquid at its saturation temperature is called latent heat of vaporization.

If heat is further supplied to vapour, there will be an increase in its temperature.

1.15 Mechanical Equivalent of Heat

Law of conservation of energy states that energy can neither be created nor destroyed, it can be transformed from one form to another. So, heat energy can be converted into mechanical work.

Thermodynamics is a branch of science which deals with the relation between heat energy and mechanical work. Joule established that heat and work are mutually convertible. This is the statement of First Law of Thermodynamics.

Thus, W = Q when the units of work and heat are same (J) and W = JQ when units of work and heat are different.

Where J (after the name of Joule) is constant and known as mechanical equivalent of heat or Joule's equivalent.

$$J = \frac{W}{Q} \qquad .....(1.18)$$

When W is in Joule and Q in calorie, J = 4.186 J/cal

1.16 Internal Energy

The amount of heat transferred to a body is not completely converted to work. It is found that when a quantity of heat Q is supplied to a body, some amount of external work W is done due to the expansion and the remaining amount is stored in the body. This stored energy is called internal energy (I.E) of the body, denoted by U. It is neither heat nor work. As it is energy, its unit is Joule.

Thus, $$Q = W + \Delta U \qquad .....(1.19)$$

Where ΔU is increase (or change) in I.E. of the body. (The body may already have some I.E., so by interaction of Q & W, the change in I.E. is considered).

This relation applies to closed system or non-flow process, where PE and KE are zero.

For a flowing fluid $$Q = W + \Delta E \qquad .....(1.20)$$

where ΔE is change in total energy of the system, including I.E., PE and KE.

i.e. $$E = U + PE + KE \quad .....(1.21)$$

The internal energy per unit mass is denoted by u. Its unit is J/kg or kJ/kg.

Internal energy is the sum of all microscopic forms of energy of a system. It is associated with the molecular structure and activity of the constituent particles of the system. The average velocity and activity of the molecules are proportional to the temperature of the system. Higher the temperature of the body, higher is its I.E. This energy is also associated with intermolecular forces.

1.17 Enthalpy

We have seen that internal energy U is a form of energy stored by a fluid due to supply of heat. The flow work PV is another form of energy possessed by a fluid which is necessary for maintaining the fluid flow. Sum of these two quantities is known as enthalpy of fluid. It is denoted by H. Its unit is Joule.

Thus, $$H = U + PV \quad .....(1.22)$$

The enthalpy per unit mass is denoted by h. Unit of h will be J/kg or kJ/kg.

We know that $$Q = dU + W$$
$$= dU + pdV$$

When gas is heated at constant pressure from an initial condition 1 to a final condition 2, change in I.E.

$$dU = U_2 - U_1$$

and work done by gas

$$W = PdV$$
$$= P(V_2 - V_1)$$

$$\therefore \quad Q = (U_2 - U_1) + P(V_2 - V_1)$$
$$= (U_2 + PV_2) - (U_1 + PV_1)$$
$$= H_2 - H_1 \quad .....(1.23)$$

For a unit mass, $$q = h_2 - h_1 \quad .....(1.24)$$

Thus, for a constant pressure process, heat supplied to the gas is equal to the change of enthalpy.

1.18 Efficiency

Engine performance is indicated by the term efficiency. Efficiency is denoted by η. The amount of energy supplied to the engine is not completely converted into work. The ratio of work done to the energy supplied is called efficiency.

Thus,

$$\text{efficiency } \eta = \frac{\text{work done by the engine}}{\text{energy supplied}} \quad \text{.....(1.25)}$$

Some of the important engine efficiencies are defined below.

Thermal efficiency: Thermal efficiency of a heat engine is the ratio of heat energy converted to mechanical work to the heat energy supplied to the engine.

$$\text{Thermal efficiency} = \frac{\text{mechanical work done}}{\text{heat energy supplied}} \quad \text{.....(1.26)}$$

Air standard efficiency: It is the thermal efficiency of the engine working with air as a working fluid. This is the ideal efficiency.

Indicated power (ip) of an engine: Power developed in the cylinder of an engine is known as indicated power.

Brake power (bp): Power available for work at the shaft of an engine is known as brake power.

Calorific value of fuel (cv): It is defined as the amount of heat generated by burning of 1 kg of fuel. Its unit is KJ/kg

Thermal efficiency can be based either on ip or bp as shown below.

Indicated thermal efficiency: It is defined as the ratio of ip to heat supplied by fuel.

$$\eta_i = \frac{\text{indicated power, W}}{\text{mass of fuel, kg/s} \times \text{calorific value of fuel, J/kg}}$$

$$= \frac{\text{ip}}{m_f \times CV} \quad \text{.....(1.27)}$$

Brake thermal efficiency: It is defined as the ratio of brake power to heat supplied by fuel.

$$\eta_b = \frac{\text{brake power, KW}}{\text{mass of fuel, kg/s} \times \text{calorific value of fuel, KJ/kg}}$$

$$= \frac{\text{bp}}{m_f \times CV} \quad \text{.....(1.28)}$$

Mechanical efficiency: It is the ratio of brake power to indicated power

$$\eta_m = \frac{\text{brake power}}{\text{indicated power}}$$

$$= \frac{\text{bp}}{\text{ip}} \quad \text{.....(1.29)}$$

Relative efficiency: It is the ratio of thermal efficiency of an actual cycle to that of an ideal cycle.

$$\eta_r = \frac{\text{actual thermal efficiency}}{\text{air standard efficiency}} \qquad .....(1.30)$$

1.19 Statement of First Law of Thermodynamics

This law may be stated as follows:

(a) "The heat and mechanical work are mutually convertible". According to this law, when a closed system undergoes a thermodynamic cycle, the net heat transfer is equal to net work transfer. In other words, cyclic integral of heat transfer is equal to cyclic integral of work transfer.

$$\oint \delta Q = \oint \delta W \qquad .....(1.31)$$

(b) "The energy can neither be created nor destroyed. It can be transferred from one form to another". Thus, 1st law of thermodynamics is the statement of law of conservation of energy. According to this law, when a system undergoes a change of state (or a thermodynamics process), both heat transfer and work transfer take place, the net energy is stored in the system. If Q is the amount of heat transferred and W is the amount of work transferred during the process, the net energy Q – W is stored in the system. This stored energy is known as total energy of the system which includes I.E., PE and KE

Thus $\quad Q - W = \Delta E$

Where $\quad \Delta E = \Delta U + \Delta PE + \Delta KE \qquad .....(1.32)$

For a closed system, change in PE and KE are zero.

So $\quad Q - W = \Delta U \qquad .....(1.33)$

Steady Flow Energy Equation (SFEE)

An open system undergoes a steady flow process as shown in Fig. 1.4. The fluid enters the system at point 1 and leaves the system at point 2. At point 1, various quantities per unit mass are

P_1 = pressure of the fluid

V_1 = volume

v_1 = velocity

z_1 = height from datum

u_1 – internal energy

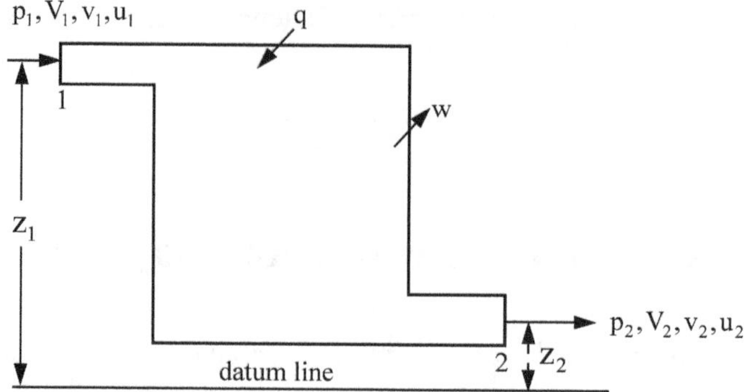

Fig. 1.4 Steady flow process.

P_2, V_2, v_2, z_2 and u_2 are the corresponding quantities at point 2.

q = heat transferred to the system

W = work done by the system

At 1, Flow work = $p_1 V_1$

Kinetic energy = $\frac{1}{2}v_1^2$

Potential energy = gz_1

So, the total energy entering the system at 1

$$= u_1 + P_1V_1 + \frac{1}{2}v_1^2 + gz_1 + q$$

Total energy leaving the system from 2

$$= u_2 + P_2V_2 + \frac{1}{2}v_2^2 + gz_2 + w$$

According to law of conservation of energy,

Total energy entering the system = total energy leaving the system

$$\therefore \quad u_1 + p_1V_1 + \frac{1}{2}v_1^2 + gz_1 + q = u_2 + p_2V_2 + \frac{1}{2}v_2^2 + gz_2 + w$$

$$\therefore \quad h_1 + \frac{1}{2}v_1^2 + gz_1 + q = h_2 + \frac{1}{2}v_2^2 + gz_2 + w \qquad .....(1.34)$$

This equation is known as SFEE. Further, it may be simplified to

$$q = w + (h_2 - h_1) + \frac{1}{2}(v_2^2 - v_1^2) + g(z_2 - z_1) \qquad .....(1.35)$$

1.20 Statement of Second Law of Thermodynamics

The 1st law states that heat and work are mutually convertible. According to this law, in a cycle, work is completely converted into heat and vice versa. But we know that heat is not completely converted into work

We will study two statements of second law of thermodynamics.

1. Kelvin-Planck statement 2. Clausius statement.

1. ***Kelvin-Planck statement***

It is impossible to construct an engine working on a cyclic process, whose sole effect is to convert heat energy from a single reservoir into an equivalent amount of work. In other words, no actual heat engine can convert whole of the heat supplied to it into mechanical work. It means that to produce net work in a cycle, certain amount of heat has to be rejected to a lower temperature body. See Fig. 1.5. The body at higher temperature T_1 from which heat is extracted is known as source. The body at lower temperature T_2 to which heat is rejected is called sink.

$$W < Q_1, \quad T_1 > T_2, \; Q_2 > 0$$

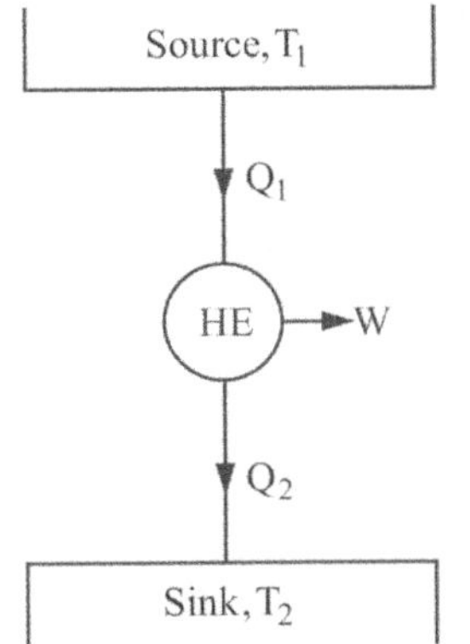

Fig. 1.5 Heat engine operating between two temperatures.

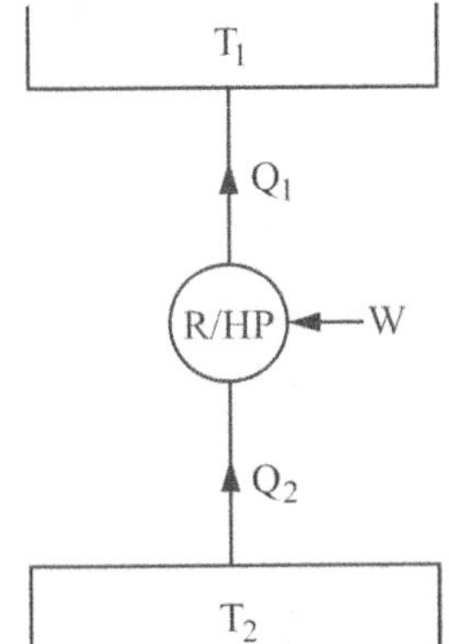

Fig. 1.6 Work has to supply to transfer heat from a cold to hot body.

2. ***Clausius statement***

It is impossible to construct a device, working in a cyclic process, to transfer heat from a body at lower temperature to a body at higher temperature without the aid of an external agency. It means that heat cannot flow itself from a cold body to a hot body without expenditure of mechanical work. Refrigerator and a heat pump, while operating in a cyclic process, can transfer heat from a cold body to hot body. They require an input work for this purpose as shown in Fig. 1.6.

The two statements of 2nd law of thermodynamics mentioned above, are equivalent.

Solved Problems

1. The pressure of steam inside a boiler is recorded by a pressure gauge which shows 1.2×10^6 N/m^2. If the barometer reads the atmospheric pressure as 770 mm of Hg, find the absolute pressure of steam inside the boiler in N/m^2, kPa and bar.

Atmospheric pressure = 760 mm of Hg = 1.01325×10^5 N/m^2

Solution: So, 770 mm of Hg $= \dfrac{1.01325 \times 10^5 \times 770}{760}$ N/m^2

$= 1.02641 \times 10^5$ N/m^2

$= 0.1026 \times 10^6$ N/m^2

This is atmospheric pressure measured by barometer.

Absolute pressure = gauge pressure + atmospheric pressure

$= 1.2 \times 10^6 + 0.1026 \times 10^6$ N/m^2

$= 1.3026 \times 10^6$ N/m^2

$= 1.3026 \times 10^6$ Pa

= 1302.6 kPa

= 13.026 bar

2. The density of mercury is 13600 kg/m^3. Calculate the pressure due to 600 mm of Hg in kPa and bar.

Solution: Pressure $P = \rho g h$

$= 13600 \times 9.81 \times \dfrac{600}{1000}$ N/m^2

$= 80.05 \times 10^3$ N/m^2

$= 80.05 \times 10^3$ Pa

= 80.05 kPa

= 0.8005 bar

3. The temperature of 4.5 kg of water is raised from 15 °C to 100 °C at constant atmospheric pressure. Determine the heat transferred. Specific heat of water = 4.186 kJ/kg-K.

Solution: Heat transferred $= mc\,(t_2 - t_1)$

$= 4.5 \times 4.186 \times (100 - 15)$

= 1601 kJ

4. In a thermal power plant 220000 kg of coal is burnt per hour. The calorific value of coal is 30000 kJ/kg. The efficiency of conversion is 30%. Determine the power developed.

Solution:

$$\eta = \frac{\text{Work done}}{\text{Heat supplied}} = \frac{W}{Q}$$

$\therefore$ $$\eta = \frac{W}{m_f \times CV}$$ m_f should be taken in kg/s

$\therefore$ $$W = \eta \times m_f \times CV$$

$$= 0.3 \times \frac{220000}{3600} \times 30000 \ \frac{kJ}{s}$$

$$= 550 \times 10^3 \text{ kW}$$

$$\mathbf{= 550\ MW}$$

5. An engine has thermal efficiency of 30%. The engine rejects 1000 MJ/hr. Calorific value of fuel is 30000 kJ/kg. Determine power output and fuel consumed by the engine.

Solution:

$$\eta = \frac{W}{Q_1}$$ Work done = heat supplied – heat rejected

$= Q_1 - Q_2$

$\therefore$ $$\eta = \frac{Q_1 - Q_2}{Q_1} = 1 - \frac{Q_2}{Q_1}$$

$\therefore$ $$0.3 = 1 - \frac{1000}{Q_1}$$

$\therefore$ $$\frac{1000}{Q_1} = 0.7 \qquad \therefore Q_1 = \frac{1000}{0.7} = 1428.6 \text{ MJ/hr}$$

Now $$W = Q_1 - Q_2$$

$$= 1428.6 - 1000$$

$$= 428.6 \text{ MJ/hr}$$

$$= \frac{428.6 \times 10^3}{3600} \text{ kJ/s}$$

$$\mathbf{= 119\ kW}$$

$$Q_1 = m_f \times CV$$

$\therefore$ $$1428.6\ \text{MJ/hr} = m_f \frac{\text{kg}}{\text{hr}} \times 30 \frac{\text{MJ}}{\text{kg}}$$

$\therefore$ $$\mathbf{m_f = 47.62\ kg/hr}$$

6. A power station has an output of 800 MW and thermal efficiency is 28%. Determine the coal consumed in tonne/hr. Calorific value of coal is 31 MJ/kg.

Solution:

$$\eta = \frac{W}{Q} = \frac{W}{m_f \times CV}$$

$\therefore$ $$m_f = \frac{W}{\eta \times CV} = \frac{800\ \text{MW}}{0.28 \times 31\ \text{MJ/kg}}$$

$$= 92.166\ \text{kg/s}$$

$$= \frac{92.166 \times 3600}{1000}\ \text{tonne/hr}$$

$$\mathbf{= 331.8\ tonne/hr}$$

7. The work produced in an engine cylinder in 6 minutes is 2870 kJ and the work done against the engine brake in 10 minutes is 3800 kJ. Find (i) Power produced in the cylinder (ii) Power delivered to the brake.

Solution:

$$\text{Power produced} = \frac{2870}{6 \times 60}\ \text{kJ/s}$$

$$\mathbf{= 7.97\ kW}$$

$$\text{Power delivered to brake} = \frac{3800}{10 \times 60}\ \text{kJ/s}$$

$$\mathbf{= 6.33\ kW}$$

8. A hydraulic engine receives 25000 kg of water per minute and develops 80 kW. Find the effective pressure exerted on the piston of the engine.

Solution:

$$1\ \text{kg of water} = 1\ \text{litre of water} = 10^{-3}\ \text{m}^3$$

$\therefore$ $$25000\ \text{kg of water} = 25000 \times 10^{-3}\ \text{m}^3$$

$$= 25\ \text{m}^3\ \text{of water}$$

$$\text{The volume flow rate of water} = 25\ \text{m}^3/\text{min}$$

$$= \frac{25}{60}\ \text{m}^3/\text{s}$$

$$\text{Work developed/sec} = P \times V$$

$$\therefore \quad 80\text{kW} = P\frac{\text{kN}}{\text{m}^2} \times \frac{25}{60}\frac{\text{m}^3}{\text{s}}$$

$$\therefore \quad P = \frac{80 \times 60}{25}\frac{\text{kN}}{\text{m}^2}$$

$$= \mathbf{192\ kN/m^2}$$

9. In a non-flow process there is a heat loss of 1055 kJ and an internal energy increase 210 kJ. How much work is done?

For a non-flow process, $Q = W + \Delta U$

Solution: Here, it is heat loss, so Q is negative

$$\therefore \quad -1055 = W + 210$$

$$\therefore \quad \mathbf{W = -1265\ kJ}$$

Work is negative, so work is spent here.

10. A cycle consists of four processes and the heat transfer during the processes were +84 kJ, –8.4 kJ, –21 kJ and +4.2 kJ. Calculate the net work done during the cycle.

Solution: Heat transfer during the cycle is

$$\oint dQ = 84 - 8.4 - 21 + 4.2\ \text{kJ}$$

$$= 58.8\ \text{kJ}$$

Now

$$\oint dW = \oint dQ$$

$$\therefore \quad \oint \mathbf{dW = 58.8\ kJ}$$

11. In a system 45 kg/s of working fluid enters with 10 m/s velocity and leaves with 19 m/s velocity. The system receive 46 kJ/s of heat and does 15 kW of work. The outlet is 30 m above the inlet. Find change in specific enthalpy of working fluid.

Solution: SFEE for unit mass flow rate is

$$h_1 + \frac{1}{2}v_1^2 + gz_1 + \frac{Q}{m} = h_2 + \frac{1}{2}v_2^2 + gz_2 + \frac{W}{m}$$

$$\therefore \quad h_2 - h_1 = \frac{1}{2}\left(v_1^2 - v_2^2\right) + g\left(z_1 - z_2\right) + \frac{Q}{m} - \frac{W}{m}$$

$$= \frac{1}{2}\frac{(10^2 - 19^2)}{1000} + 9.8 \times \left(\frac{-30}{1000}\right) + \frac{46}{45} - \frac{15}{45}$$

$$= -0.130 - 0.294 + 1.022 - 0.333\ \text{kJ/kg}$$

$$= 1.022 - 0.757$$

$$= \mathbf{0.265\ kJ/kg}$$

Objective Type Questions

1. Which of the following is a renewable source of energy?

   (a) Solar (b) Natural gas (c) Coal (d) Petroleum

2. Which of the following is not the unit of pressure?

   (a) Bar (b) Newton (c) Pascal (d) mm of mercury

3. Which is the unit of pressure?

   (a) Newton (b) Watt (c) Joule (d) N/m^2

4. Barometer is used to measure

   (a) Atmospheric pressure (b) Temperature

   (c) Work done (d) Power

5. 1 bar is equal to

   (a) 1 N/m^2 (b) 760 mm of Hg (c) 10^5 Pascal (d) 100 Pascal

6. Unit of work is

   (a) Newton (b) N-m (c) Bar (d) Watt

7. 1 hp is equal to

   (a) 750 W (b) 50 W (c) 735.5 W (d) 735.5 kW

8. kWh is the unit of

   (a) Power (b) Force (c) Pressure (d) Energy

9. 1 kWh is equal to

   (a) 36×10^5 J (b) 36×10^5 kJ (c) 10^5 Joule (d) 1 Joule

10. Relation between K and celsius scales of temperature is

    (a) $1\ ^\circ C = 1\ K$ (b) $K = ^\circ C + 273$ (c) $^\circ C = K + 273$ (d) $1K = 100\ ^\circ C$

11. 1 kcal is equal to

    (a) 4.186 J (b) 4186 kJ (c) 4186 J/s (d) 4186 J

12. 1st law of thermodynamics deals with

    (a) Conservation of momentum (b) Conservation of heat

    (c) Conservation of energy (d) Conservation of mass

13. 1st law of thermodynamics defines

    (a) Work (b) Internal energy (c) Heat (d) Entropy

Answers to Objective Type Questions

1. (a)	2 (b)	3. (d)	4. (a)	5. (c)
6. (b)	7. (c)	8. (d)	9. (a)	10. (b)
11. (d)	12. (c)	13. (b)		

Short Questions

1. What is a prime mover?
2. Which are the different sources of energy?
3. Define

   (a) Force (b) Weight (c) Pressure (d) Work (e) Power (f) ip (g) bp (h) Energy (i) PE (j) KE (k) Temperature (l) Specific heat (m) Heat capacity (n) Melting point (o) Latent heat of fusion (p) Saturation temperature (q) Latent heat of vaporization (r) Mechanical equivalent of heat (s) Internal energy (t) Enthalpy (u) Thermal efficiency (v) Air standard efficiency (w) Indicated thermal efficiency (x) Brake thermal efficiency (y) Mechanical efficiency (z) Relative efficiency
4. Give SI units of following quantities

   (a) Force (b) Pressure (c) Work (d) Power (e) Energy (f) Temperature (g) Heat (h) Specific heat (i) Latent heat (j) Internal energy (k) Enthalpy (l) Efficiency (m) Calorific value
5. Give different units of pressure.
6. What is the magnitude of atmospheric pressure?

7. Give names of pressure measuring instruments
8. Give relation between absolute pressure and gauge pressure.
9. What is flow work?
10. What is the difference between heat and temperature?
11. Give names of different units of heat. Write relation between them.
12. What is the old name of Celsius scale?
13. Why has gas two specific heats?
14. Which are the different modes of heat transfer?
15. What is the law of conservation of energy?
16. What is Thermodynamics?
17. What is the value of J when units of heat and work are same?
18. Why efficiency of an engine is less than 100%?
19. What is calorific value of fuel?
20. Can heat be completely converted to work?
21. Write SFEE.

Long Questions

1. Write short notes on

   (a) Sources of energy (b) Types of prime movers (c) Temperature (d) Specific heat (e) Internal energy (f) 1st law of Thermodynamics
2. Show that area of P.V diagram represents work done.
3. Explain change of state of a substance.
4. Show that for a constant pressure process, heat supplied to gas is equal to change of enthalpy.
5. Explain Kelvin-Planck statement of 2nd law of Thermodynamics.
6. Explain Clausius statement of 2nd law of Thermodynamics.

Exercises

1. The pressure of steam inside the boiler, as measured by pressure gauge is 1×10^6 N/m^2. The barometric pressure of the Atmosphere is 765 mm of mercury. Find the absolute pressure of steam in N/m^2, kPa, bar and N/mm^2.

   [1.102×10^6 N/m^2, 1102 kPa, 11.02 bar, 1.102 N/mm^2]

2. Certain quantity of air undergoes a change of state in such a way that 50 kJ of heat is transferred to the system. For this change of state, work produced is 40 kJ. Calculate the change in its internal energy. Assume that the process is non-flow.

   [10 kJ]

3. A steam engine piston having an area of 160 cm^2 moves a distance of 200 mm inside the cylinder. Find the amount of work done if the pressure exerted on the piston is 70 kN/m^2

   [224 J]

4. A fluid in a cylinder is at a pressure of 7 bar. It is expanded at constant pressure from 0.3 m^3 to 1.5 m^3. Calculate the work done.

   [840 kJ]

5. Find the theoretical power developed by an water wheel if it receives 500 kg of water per sec at a pressure of 1.5 bar.

   [75 kW]

6. 10 kg of steel, having a specific heat of 450 J/kg.K is heated from 20 °C to 100 °C. How much heat is required?

   [360 kJ]

7. A steam plant uses 3 tonne of coal in one hour. The steam is fed to turbine. Output of turbine is 4 MW. Calorific value of coal is 30 MJ/kg. Calculate the thermal efficiency of the plant.

   [16 %]

8. 2000 m^3 of water is to be delivered in 30 minutes to a vertical distance of 30 m. Determine the power required.

   [327 kW]

9. An engine rejects 1260 MJ/hr when running at a efficiency of 22%. The calorific value of fuel used is 42 MJ/kg. Determine the power output of the engine and mass of fuel used in one hour.

[98.72 kW, 38.5 kg/hr]

10. A gas enters a system at an initial pressure of 0.5 MPa and flow rate of 0.15 m^3/s and leaves at a pressure of 0.95 MPa and flow rate of 0.09 m^3/s. During the passage through the system, the increase in internal energy is 22 kJ/s. Find the change in enthalpy of the medium.

[32.5 kJ]

11. A system undergoes a cycle composed of four processes. Heat transfer in these processes are 400 kJ, –365 kJ, –200 kJ and 250 kJ. The respective work transfers are 140 kJ, 0, –55 kJ and 0. Is the data consistent with 1st law of Thermodynamics?

[Yes]

CHAPTER 2

Fuel and Combustion

2.1 Introduction

Fuels may be defined as substance which produces heat when burns in presence of air (oxygen). Carbon and hydrogen are the main constituents of fuel which produces heat.

As the principal constituents of fuel are carbon and hydrogen therefore fuels are known as hydrocarbon fuels. Most of the solid and liquid fuels contain a small percentage of sulphur also (1 to 3%).

2.2 Classification

Fuels may be classified according to

1. The physical state in which they exist in nature
   - (i) solid
   - (ii) liquid
   - (iii) gaseous
2. Mode of their procurement
   - (i) Natural – They occur in nature and also known as primary fuels
   - (ii) Synthetic – The artificially prepared fuels known as secondary fuel.
3. Classification as per its calorific valve.

The detailed classification of fuels is given in Table 2.1.

Table 2.1

Primary Fuels (Natural fuels)	Secondary Fuels (prepared fuels)
Solid 1. Wood 2. Peat 3. Coal 4. Lignite 5. Oil shale	**Solid** 1. Coke 2. Charcoal 3. Briquettes
Liquid Petroleum	**Liquid** 1. Gasoline 2. Naphtha 3. Diesel oil 4. Kerosene 5. Fuel oil 6. Shale oil 7. Coal tar 8. Colloidal fuel 9. Alcohols 10. Benzol
Gaseous Natural (gas from oil mines)	1. Coal gas 2. Producer gas 3. Water gas (sewer-gas) 4. Oil gas 5. Blast furnace gas 6. Coke-oven gas 7. Acetylene 8. Butane 9. Petroleum-gas

Primary fuels: These are those fuels which occur naturally, such as wood, peat, coal, lignite, petroleum and natural gas.

Secondary fuels: These are fuels produced artificially from primary fuels by some treatment processes.

Coal: Coal is the stratified rock of decaying trees and vegetation that under went changes by anaerobic bio-degradation through millions of year under heat and pressure of earth crust above them.

Coal passes through different stages during its formation. These stages are discussed below.

Plant debris – peat – lignite – brown coal – sub-bituminous coal – bituminous coal – semi bituminous coal – semi anthracite coal – anthracite coal – graphite.

How was petroleum formed: Petroleum was formed by anaerobic decay of marine plant and animal life (Fish).

Dead plants and animal of aquatic origin deposited at bottom of shallow seas were covered to thick layer of silt and set to decay under nearly anaerobic conditions. The steady accumulation of mud above the decaying the remains of organic matter to increase of temperature and pressure that causes the formation of liquid and gaseous hydrocarbons.

What is fixed carbon: It does not imply existence of uncombined carbon in coal material nor it bear any relationship with the total carbon as determined by ultimate analysis. It is obtained by subtracting from 100, the sum of the percentage of moisture, volatile matter and ash content of coal sample.

What are basic components of volatile matter?

There are three basic components of volatile matter.

1. *Gases:* Containing CO, H_2, CH_4, C_2H_6, and H_2S.
2. *Tar:* A complex mixture of hydrocarbons such as benzene, toluene, naphthalene, xylene, phenols, cresols, anthracine and free carbon.
3. *Ammonical liquor:* Aqueous condensate of nitrogen and sulphur compounds.

Volatile matter produces smoke. The percentage of volatile varies with rank of coal, it ranges as low as 3% to as high as 50%. Volatile matter is determined when finely ground coal specimen is heated in closed vessel at 1200 °F (649 °C) for 40 minutes. After cooling and drying the residue (coke) is weighed.

$\therefore$ % volatile matter = 100% – C coke and moisture) %.

2.3 Solid Fuels

The natural solid fuels are wood, peat, lignite, bituminous and anthracite. The artificial or prepared coal are charcoal, coke, and pulverised coal.

2.3.1 Wood

It is extensively used as domestic fuel for cooking in India. It is not considered as a commercial fuel except in industries where large amount of waste wood is available. It ignites easily (about 200°-250 °C). Calorific value of wood varies with its kind and moisture content. The average calorific value of wood is 19700 kJ/kg. Wood is converted in to coal when burned in the absence of air.

2.3.2 Peat

Peat is first stage in the process of transformation of buried vegetation and wood in to coal, compare to all type of coals. It is spongy, humified substance. It has a large amount of water content even after air drying (20% to 30%). Generally it is sun dried (1 to 2 months) to remove greater part of moisture. It emits a characteristic odor during burning and generates a smoky flame. Its average C.V. is 23000 kJ/kg. It is used as domestic fuel in Europe and for power generation in Russia.

2.3.3 Lignite or Brown Coal

It represents next stage of peat coal formation and it is intermediate variety between bituminous coal and peat. It is brown in colour and exhibits woody structure, so it is also called brown coal. It contains 40% moisture and 60% carbon. They break easily therefore they impose transport difficulties and cannot be stored well. It burns with smoky flame so they are suitable for local use only. Its average C.V. is 25000 kJ/kg. In India it is being mined at Nayvali in Tamil Nadu, and Rajasthan.

2.3.4 Bituminous Coal

It is shining black in appearance. It represents next stage of lignite in coal formation. It contains 4% to 6% moisture and 70% to 90% carbon. It is weather resistance and burn with yellow flame. The average C.V. of bituminous coal is 33500 kJ/kg. In India coal reserves are located in Bihar, Bengal and Orissa. There are two types of Bituminous coal (a) caking Bituminous coal (b) non caking bituminous coal.

(a) Caking bituminous coal softens and swells on heating. Its pieces adhere together, and forms pasty mass if heated at higher temperatures, which makes firing difficult. This coal is used for manufacturing of coke and gas. Its specific gravity is 1.26 to 1.36. Its average C.V. is 35000 kJ/kg.

(b) ***Non-caking bituminous:*** It contains 75 to 84% carbon and does not become soft on heating. It burns with shorter flame and gives little or no smoke. Its specific gravity is 1.22 to 1.42. It is used in power plants. Therefore it is known as steam coal, its average C.V. is 33000 kJ/kg.

The bituminous coal is non-caking if its carbon contain is 78 to 81%. If percentage of carbon is 81 to 82.5% it is slightly caking. If percentage of carbon is 82.5 to 84% it is medium caking. If percentage of carbon is 84 to 89% it makes the coal strongly caking.

2.3.5 Anthracite Coal

It is the last stage in the process of transformation of buried vegetation in to coal and contains 90% or more carbon with very little volatile matter. It is comparatively smoke less and produces shorter flame. Its c.v. is 36000 kJ/kg is therefore valuable, used in steam power plant for steam raising and power generation. It is hard, brittle and lustrous or shiny black in appearance. It contains 2 % to 6 % ash.

Bituminous coal when heated give off 20 to 30% of volatile matter, where as anthracite contains very small quantity of volatile matter so bituminous coal is easy to ignite than Anthracite coal.

2.3.6 Artificial or Prepared Fuels

The prepared solid fuels are used for different purposes are listed below.

(a) ***Wood char coal*:** It is made by heating wood with a limited supply of air below 280 °C. During process volatile matter and moisture are removed. It contains 80 to 90% carbon. It burns rapidly with clear flame, producing no smoke. It is a well prepared fuel for various metallurgical processes and as cooking and heating. C.V of dry char coal is 28000 kJ/kg.

(b) ***Coke*:** It is made from Bituminous coal by heating continuously for 42 to 48 hours in absence of air in closed chamber or vessel. It contains 85 to 95% carbon. It is hard in structure, dull black in colour, porous and smokeless.

If carbonization of coal is carried out at 500° to 700 °C, the resulting coke is called lower temperature coke or soft coke. It is used for domestic purpose. The coke produced by carbonization of coal at 900° to 1100 °C, it is known as hard coke used in gas producing plants and blast furnace for extracting pig-iron from iron core and to some extent used in cupola furnace for producing cost iron.

(c) ***Briquetted coal*:** It is made of finely grounded coal by molding under pressure with or without binding material. Binding materials are pitch, coal tar, crude oil and clay. The use of blocks or briquettes as fuel reduces loss of fuel through the grate openings. Briquettes can be given any shape. It is possible to increase heating value of low quality coal.

(d) ***Pulverised coal*:** When coal is powdered, it is known as pulverized coal. Coal is first dried and then crushed in to a fine powder by pulverizing machines. There is better contact between air and fuel hence combustion efficiency increases. We can use low grade coal with high ash content.

It is widely used in cement industries, in metallurgical processes and power plants. Using the fuel in pulverized form offers many advantages such as flexibility of control, complete combustion with less excess air and high temperature. Large quantities can be burnt in pulverized form.

***Note*:** Out of all above mentioned types of solid fuels anthracite coal is used in all type of heat engines, (power plants).

2.4 Liquid Fuels

Liquid fuels are classified according to mode of procurement as (i) natural or crude oil, (ii) artificial or manufactured oils.

Distilled natural oil is petrol, benzene, petroleum sprit, kerosene, benzol, fuel oil, diesel oil etc.

Distilled artificial oils are natural gas oil, shale oil, tar oil, coal tar etc.

Almost all commercial fuels are derived from natural petroleum (crude oil). The crude oil is obtained from bore-holes in earth's crust in certain part of the world.

The natural petroleum may be separated in to petrol (gasoline), paraffin oil or kerosene, fuel oil and lubricating oil by boiling the crude oil at different temperatures subsequent fractional distillation. It is the distillations by stages. Distillation is carried out in such a way that liquid with lowest boiling point is first evaporated and then condensed. The liquid with next higher boiling point is then evaporated and condensed and so on until all available liquid fuels are separately recovered in sequence of their boiling points.

Natural crude petroleum (crude oil) contains 83-87% carbon, 10-14% hydrogen and small percentages of sulphur, nitrogen, oxygen etc.

Useful fuels from crude oil are obtained by distillation process. The oils obtained vary in colour from light yellow to almost black, some are highly mobile while others are thick and viscous.

The distillation is used to separate different groups of fuels. Petrol (gasoline) is obtained between 65 to 220 °C, kerosene is obtained 150-350 °C and diesel oil is obtained between 350 to 470 °C during distillation process and remainder is heavy fuel oil. Heavy oils are used in oil fired boiler.

The following liquid fuels are generally considered.

2.4.1 Petrol or Gasoline

It is the lightest and most volatile liquid fuel. It is chiefly used in petrol engines. It is distilled at a temperature from 35 °C to 220 °C.

2.4.2 Kerosene

This is also called paraffin oil. It is distilled at a temperature from 220 to 345 °C. It is heavier and less volatile fuel than petrol. It is used as fuel for heating and lighting.

2.4.3 Alcohol

It is a liquid obtained from vegetable matter by fermentation process, energy contain of alcohol is less than petrol but cost is more than petrol. This is another fuel alternative to petrol but it is not yet used in engines on commercial basis. It is used in chemical processes.

2.5 Advantages and Disadvantages of Liquid Fuel over Solid Fuel

Advantages

1. Less air is required for complete combustion of fuel as uniform mixing of liquid fuel and air is possible due to liquid atomization.
2. Size of fan supplying air is reduced, fan cost is also reduced.
3. Requires less storage space and handling (about 40% than coal).
4. Requires no ash disposal system as it contains practically no ashes.
5. Uniform mixing of liquid fuel and air gives high combustion efficiency.
6. Better control of fuel burning rate by providing valves and burners.
7. More cleanliness of operation and freedom from dust.
8. Operation labour required is less and maintenance costs are also reduced.
9. Non-corrosion of boiler plates.
10. Transportation fuel losses are reduced.
11. Liquid fuel has higher calorific value than solid C.V.

Disadvantages (limitations) of liquid Fuel are listed below

1. Cost of liquid fuel is much higher than coal (4 to 5 times).
2. Danger of explosion during transport and storage is higher.
3. Special arrangement are required for their storage.
4. Its use is totally dependent on its availability in India.
5. Liquid fuel sources are exhaustible.

2.6 Gaseous Fuels

2.6.1 Gaseous Fuels are Classified into two Classes

(a) Natural gaseous fuels

(b) Manufactured (prepared) gaseous fuels.

(a) (i) ***Natural gaseous fuel*:** It is found in or near petroleum fields under the earth's surface. It is colourless, odourless gas composed of methane (marsh gas) together with small amount of other gases such as ethane (C_2H_6) carbon dioxide and carbon monoxide. They are accumulating in the upper part of oil of gas-wells suitable for domestic and industrial use. Its C.V. is 21000 kJ/m^3.

(ii) ***Liquefied petroleum gas*:** It is the primary flash distillation product of crude petroleum. Also it is produced by fractionating natural gas. It is mainly propane and butane which are easily liquefied, used as both domestic and industrial fuel.

(iii) ***Refinery oil gas:*** Composed of some light gases obtained during processing of crude petroleum. They are light hydrocarbons with small percentage of hydrogen and carbon monoxide. They cannot be economically converted into liquid, used as in-plant source of fuel.

2.6.2 (b) Manufactured Gaseous Fuels

They are

1. Coal gas
2. Producer gas
3. Water gas
4. Mond gas
5. Blast furnace gas
6. Coke oven gas
7. Sewer gas

(i) ***Coal gas*:** It is also known as town gas. It is the gaseous product of carbonization of fuel. It consists of mainly CO_2, CO, CH_4 and N_2. Quality of coal gas depends upon 1. type of coal used 2. temperature of carbonization 3. type of plant. It is very rich among combustible gases.

Carbonization	Gas composition
At low Temperature 775 to 900 K	$CH_4 = 46\%$
	$H_2 = 28\%$
	$CO = 12\%$
	$N_2 = 7\%$
	$CO_2 = 4\%$
	$H_2O = 2\%$
	$O_2 = 1\%$
At high temperature 1250–1600 K	$CH_4 = 25\%$
	$H_2 = 52\%$
	$CO = 2.5\%$
	$N_2 = 12.5\%$
	$O_2 = 0.2\%$

This gas is used as fuel in gas fired boiler. It is also used in furnaces and for running gas engines also for domestic cooking. Its C.V. is about 21000 to 25000 kJ/m^3.

(ii) ***Producer gas*:** It is obtained by partial combustion of coal or coke in limited stream of air and steam. It is produced in specially designed retorts. It is mostly used for furnaces particularly for glass melting. It is also used in steel industries for firing open heart furnaces and power generation. Its cost is low. Its C.V. varies from 5000 to 6700 kJ/m^3.

(iii) ***Water gas*:** It is a mixture of hydrogen and carbon monoxide. It is also known as blue water gas as it burns with blue flame.

It is produced by blowing steam over condescent coke or coal. Decomposition of steam takes place and free hydrogen is liberated. Oxygen in the steam combines with carbon to form carbon monoxide.

$$C + H_2O\text{ (steam)} = CO + H_2$$

It is mixed with coal gas to form town gas. Water gas is used in furnaces and for welding. Its calorific value varies 11200 to 12000 kJ/m^3. It is also used as synthesis gas in chemical processing.

(iv) ***Mond gas*:** It is produced by passing air and large amount of steam over waste coal at 650 °C. Its C.V. is about 5850 kJ/m^3. It is used for heating and steam generation purposes.

(v) ***Blast furnace gas*:** It is a by-product in production of pig-iron in the blast furnace. This gas contains high percentage of dust so it should be removed before it is used. This gas is used as fuel in steel works, steam generation and gas engines. It is also used as fuel for metallurgical-furnaces. It has low heating value of about 3750 kJ/m^3.

(vi) ***Coke oven gas*:** It is obtained during the production of coke by heating Bituminous coal. Its c.v. varies from 14500 to 18500 kJ/m^3. It is used for industrial heating and power generation. This gas must be thoroughly filtered before using it in gas engines.

(vii) ***Sewer-gas:*** It is obtained from sewage disposal waste in which fermentation and decay occurs. It consists of mainly marsh gas (CH_4) and is collected at large disposal plants. It is used as fuel for gas engines, which in turn drive the plant pumps and agitators.

2.7 Advantages and Disadvantages of Gaseous Fuels

2.7.1 Advantages of Gaseous Fuels are Listed below

(i) Requires less excess air for complete combustion.

(ii) Smoke less combustion is possible.

(iii) No ash formation. No smoke formation.

(iv) They are directly used in internal combustion engines.

(v) Higher calorific value.

(vi) More cleanliness operation.

(vii) Provides quicker furnace light-up.

(viii) Better response to boiler load variations.

(ix) No problem of storage if the supply is available from public supply line.

(x) Better control of combustion.

(xi) Distribution is easy with the help of pipe lines.

(xii) Economy in fuel and more efficient operation of furnace.

(xiii) They are free from solid and liquid impurities.

(xiv) Transport of gaseous fuel through pipe line is easier than transport of coal and oil.

Disadvantages

1. The major disadvantage of using natural gas or other gases as fuel is that the combustion system must be located near source of gas supply or production.
2. They are readily inflammable.
3. They require large store volume capacity, if public supply system in not available.
4. Liquefied gases require high pressure or low temperature insulated tanks which are expensive.

***Note*:** gaseous fuels have all the advantages of liquid fuel except storage as problem.

Table 2.2 Comparison of solid, liquid and gaseous fuels.

S.H	Basis	Solid Fuel	Liquid Fuel	Gaseous Fuel
1.	State	Solid	Liquid	Gas
2.	Cost	Cheap	Medium	Costly
3.	C.V	Low	Medium	High
4.	Storage space	Large	Medium	Least
5.	Weight	Heavy	Medium	Lightest
6.	Product of combustion	Large amount of smoke and ash	No ash but some smoke	No ash, no smoke
7.	Ignition Temperature	Highest, requires other fuel to reach ignition temperature	Medium but easily flammable	Low but creates difficulties
8.	Transportation	Does not require closed container	Requires container	Requires special container
9.	Conveyance	Cannot be piped	Can be piped	Can be piped
10.	Loss of heat	More, by radiation and convection	Less	No loss
11.	Use or application	Used as domestic fuel	Suitable for I.C engines	Suitable for gas engines

2.8 Proximate Analysis and Ultimate Analysis of Coal

2.8.1 Proximate Analysis

It gives the composition of moisture, volatile matter, ash and carbon contain of coal.

The proximate analysis of most of coals indicates the following percentage range of various constituents by mass.

Constituents	Moisture	Volatile matter	Ash	Fixed carbon
Percentage	3 to 30%	3 to 50%	2 to 30%	16 to 20%

2.8.2 Ultimate Analysis

It is chemical analysis of coal to determine the percentage of composition of coal which are basically carbon, hydrogen, nitrogen, sulphur and ash and occasionally phosphorous and chlorine.

This analysis serves as basis for calculating the amount of air required for complete combustion of coal. This analysis also enables us to determine the c.v. of the coal.

The ultimate analysis of coal indicates the following ranges of various constituents by mass.

Constituents	Carbon	Hydrogen	Oxygen	Sulphur	Nitrogen	Ash
Percentage	50-95	2-5.5	2-40	0.5-3	5-7	2-30

2.9 Calorific Value of Fuel

It is denoted by C.V. is also called heating value of fuel.

The calorific value of solid or liquid fuel is defined as the amount of heat given out by complete combustion of 1 kg of coal. It is expressed (unit) in terms of kJ/kg. However C.V. of gaseous fuel is expressed in terms of kJ/m^3 at specified temperature and pressure. Following are two types calorific value of fuel.

1. Gross or higher calorific value (H.C.V.)
2. Net or lower calorific value (L.C.V.)

2.9.1 Gross or Higher Calorific Value- (theoretically)

- It is briefly written as H.C.V.
- It's unit is kJ/kg of coal or kJ/m^3
- The amount of heat produced or obtained by complete combustion of 1 kg of fuel, when product of its combustion are cooled down to the temperature of supplied air (usually taken as 15°C) is called gross or higher calorific value.
- H.C.V of fuel can be determined by Dulong's formula,

$$\text{H.C.V} = 33800\ C + 144000\ \text{kJ}\ H_2 + 9270\ S\ \text{kJ/kg.}$$

where C, H_2 and S represent mass of carbon, hydrogen and sulphur respectively in 1 kg of fuel and the numerical values indicate their respective calorific value.

- If fuel contains 'O_2' kg of oxygen. Then, it is assumed that whole amount of O_2 will combine with hydrogen. The mass of hydrogen required will be $1/8^{th}$ of the mass of oxygen.

$$2H_2 + O_2 = 2H_2O$$

$$4 + 32 = 36$$

$$1 + 8 = 9$$

1 kg of hydrogen requires 8 kg of O_2 and will produce 9 kg of H_2O. Then net amount of H_2 available for burning is reduced. It is equal to $(H_2 - O_2/8)$.

- Then H.C.V of fuel, when fuel contains, C, H, S and O is given by Dulong's method as follows.

$$\text{H. C.V} = [33800\ C + 144000\ (H_2 - \frac{O_2}{8}) + 9270\ S]\ \text{kJ/kg.}$$

2.9.2 Net or Lower Calorific Value (Theoretically)

- In brief it is written as L.C.V.
- It's unit is kJ/kg of fuel or kJ/kg
- All fuels containing hydrogen, producing steam during process of combustion. Latent heat absorbed, is carried away by product of combustion is not recovered in practice and steam (water vapour) formed during combustion is not condensed. The actual amount of heat evolved (produced) is reduced by the amount of heat carried away by the steam. This value of heat is known as L.C.V. of fuel.

If higher calorific value is known, then L.C.V. value may be obtained by subtracting the amount of heat carried away by steam (water vapour) from H.C.L.

L.C.V. = H.C.L – Heat of steam formed during combustion.

Let m_s be the mass of steam formed in kg per kg of fuel burnt.

Since the amount of heat per kg of steam is the latent heat of vaporization of water corresponding to standard temperature of 15 °C and standard pressure of 760 mm Hg is 2466 kJ/kg.

Therefore

$$\text{L.C.V.} = \text{H.C.L.} - 9H_2 \times 2466\ \text{kJ/kg.}$$

The Calorific value of gaseous fuels is calculated by combining the amount of heat generated by each constituent, and calorific value of each constituent must be known.

The heat generated by complete combustion of one cubic meter of gaseous fuel is known as calorific value of gaseous fuel. Gaseous fuels also have H.C.V. and L.C.V. like liquid and solid fuel.

***Note*:** It is advisable to use L.C.V. of fuel for all practical purposes as this is net heat available for use.

2.9.3 Experimental Determination of Calorific Value of Liquid and Solid Fuel by using Bomb Calorimeter

Bomb Calorimeter

***Function*:** It is used to find out H.C.V. of liquid and solid fuel. But H.C.V. of gaseous fuel cannot be found by Bomb Calorimeter. It is found by Junkers Calorimeter (Boy's Calorimeter)

***Construction*:** Name of apparatus arises from shape of the vessel in which fuel is burnt. Its shape resembles that of Bomb.

It consists of following components as shown in Fig. 2.1.

1. Calorimeter
2. Stainless bomb
3. Crucible
4. Iron fuse wire
5. Oxygen supply valve
6. Release valve
7. Stirrer with motor
8. Thermometer
9. Main supply connection
10. Briquette mould apparatus (Fig. 2.2)

The bomb is made up of stainless steel which is capable of with standing high pressure (100 atm), heat and corrosion. Cover head of bomb carries valve for admitting oxygen through which, bomb is filled with oxygen and another valve for releasing exhaust gases. Two pillars are provided inside the bomb to support the crucible made of silica quartz.

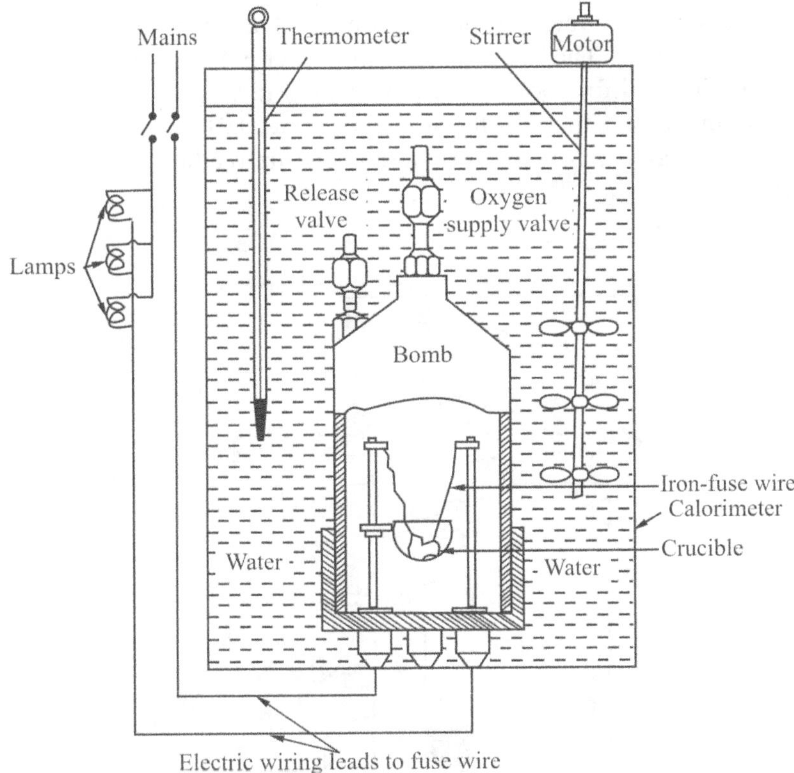

Fig. 2.1 Bomb calorimeter.

Bomb is placed in water bath and water bath itself is placed in another container (not shown in figure) which acts as heat insulator by forming an air space around the calorimeter. A thermometer and stirrer are inserted in apparatus.

Procedure

The sample of coal to be tested is reduced to powder then compressed into tablets (briquette) with help of Briquette mould apparatus. Mass of coal taken is one gram and calorimeter is filled with 2000 C.C. of water.

The known quantity of coal is placed inside the crucible in such a way that iron fuse wire touches the fuel. Iron fuse wire is connected to the lower end of two electrodes.

The bomb is then charged with oxygen to a pressure of 30 atmospheres. Then bomb is submerged in quantity of water (about 2000 C.C.) contained in large copper vessel known as calorimeter. The water in the calorimeter is stirred and when the temperature of water remains steady, the fuel is ignited by passing current through the fuse wire.

Temperature reading of water in the calorimeter is taken at one minute interval for period of 10 to 15 minutes. The stirrer is kept in motion all the time and temperature are taken at ½ minute interval for about five minutes until maximum temperature is reached. Beckmann thermometer which reads thermometer up to 0.01 °C accurately may be used. Once maximum temperature is reached then temperature reading are taken at one minute interval for 10 minutes.

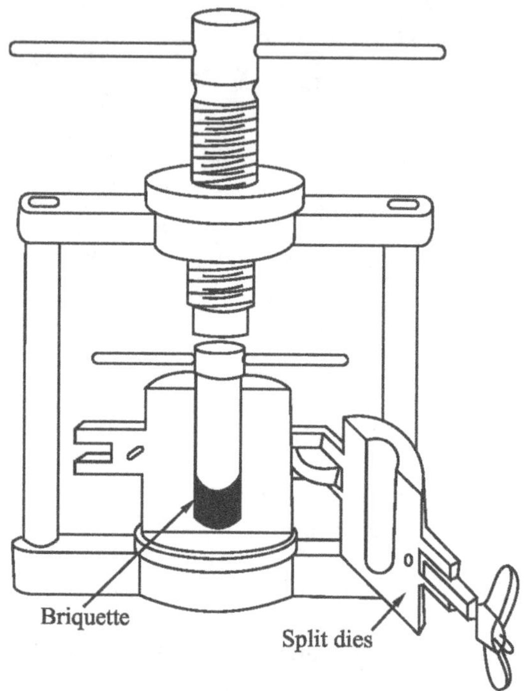

Fig. 2.2 Briquette mould apparatus.

The heat generated by combustion of sample of the coal is partly absorbed by the known quantity of the water in calorimeter and partly of metal of Bomb and calorimeter. The water equivalent of bomb and calorimeter is usually supplied by the manufactures or it can be determined experimentally.

M_f = mass of fuel sample burnt in sample kg.

M_w = mass of water filled in calorimeter in kg.

M_e = water equivalent of apparatus in kg. (Bomb and calorimeter)

t_1 = Initial (before starting the experiment) temperature of water and apparatus, in °C

t_2 = final temperature of water and apparatus, in °C

C_{pw} = specific heat of water, m_{wf} = mass of fuse wire, C.V. = calorific value of wire.

We know that heat liberated by fuel = $M_f \times$ H.C.V.

We know that heat liberated by wire = $m_{wf} \times$ C.V. of wire

Heat absorbed by water and apparatus = $(M_w + M_e)\, C_{pw}\, (t_2 - t_1)$.

∴ Heat liberated by fuel and fuse wire = Heat absorbed by water and apparatus

$$\therefore \qquad \text{H.C.V.}C_{pw} = \frac{(m_w + m_e)(t_2 - t_1) - m_{wf}CV}{M_f} \qquad .....(A)$$

The C.V. given by above equation is H.C.V. of fuel causes the water vapour formed during combustion is condensed. Finding L.C.V. its equation should be used for the very accurate determination of C.V, the corrections are made for the following.

(i) Heat may loss or heat may gain from the atmosphere by apparatus during experiment.

(ii) Combustion takes place in Oxygen, resulting in the formation of nitric acid and sulphuric acid.

(iii) Fuse wire librates some heat energy.

(iv) Product of combustion are cooled to original temperature.

The Correction for radiation during the period of burning (or cooling correction) is to be made in observed temperature rise, which will eliminate error due to reason (i).

Cooling correction is expressed in degree centigrade and should be added to observed temperature rise of water to get correct temperature rise.

The heat resulting from formation of nitric acid and due to heating of fusing wire when subtracted from heat absorbed by apparatus and water, will eliminate error due to (ii) and (iii) reasons. The error due to (iv) is only small fraction about 1%. Calorific value of liquid fuel is measured in same manner by bomb calorimeter. The sample of liquid fuel is filled is a gelatin capsule is placed in platinum crucible.

Note: To compensate for loss of heat by radiation a cooling correction is to added to observed temperature rise. Equation (A) is modified as follows

$$\therefore \qquad \text{H.C.V.} = \frac{[(M_w + M_e)C_{p_w}(t_2 - t_1 + t_c)]}{m_f}$$

t_C = cooling correction.

2.10 Combustion of Fuel

Under the heading of combustion of fuel we shall discuss, the combustion of different constituents of fuel which are frequently used in combustion of fuel.

All the initial substances that undergo the combustion process are called the reactants and substance that results from combustion process is called products of combustion.

Combustion is a chemical process in which substance combines with oxygen at temperature above ignition temperature of substance and results in evolution of heat and light. A fuel at the time of ignition must be in ignition form so as to combine with oxygen.

All substances are composed of minute particles called atoms and these atoms are arranged in group are called molecules. Elements are representing by C, H, O and S which indicates one atom each of Carbon, Hydrogen, oxygen and sulphur respectively.

The mass of one standard volume of any element at 0 °C and 760 mm of Hg pressure in form of gas is atomic weight of the element. The molecular weight of substance is the mass of two standard volumes (22.4 m^3) of that substance when in the form of gas at normal temperature and pressure. Both element and compound have molecular weight.

Atomic and molecular weights of some important elements, compounds and gases are shown in Table 2.2.

Table 2.2

Substance	Molecule		Atom	
	Symbol	Molecular weight	Symbol	Molecular weight
Hydrogen	H_2	2	H	1
Oxygen	O_2	32	O	16
Nitrogen	N_2	28	N	14
Carbon	C	12	C	12
Sulphur	S	32	S	32
Water (steam)	H_2O	18	-	-
Carbon monoxide	CO	28	-	-
Carbon dioxide	CO_2	44	-	-
Sulphur dioxide	SO_2	64	-	-
Marsh gas (Methane)	CH_4	16	-	-
Ethylene	C_2H_4	28	-	-
Butane	C_2H_6	30	-	-
Acetylene	C_2H_2	26	-	-

How to Calculate Molecular Weight of Compound Substances

The molecular weight of compound substances can be calculated by adding the atomic weight of elements concerned.

For example one molecule of water (steam) consists of 2 atom of Hydrogen + 1 atom of oxygen.

$\therefore$ Molecular weight of water (steam) $H_2O = 2H + O$

$$= 2 \times 1 + 16$$

$$= 18$$

Similarly molecular weight of SO_2

1 molecule of SO_2 consists of 1 atom of sulphur and two atom of oxygen.

$\therefore$ molecular weight of SO_2 = 1 atom of sulphur + 2 atom of oxygen

$$= 32 + 2 \times 16$$

$$= 64$$

Chemical Reaction

Chemical union of substances may be expressed by an equation or reaction. Chemical equations are valuable to engineer because they indicate, the proportion with which the elements react.

The chemical equations for combustion of different constituents of fuel:

1. ***Burning of Carbon to Carbon dioxide (complete combustion)***

$$C + O_2 = CO_2$$

$$12 + 16 \times 2 = 12 + 16 \times 2$$

$$12 + 32 = 44$$

$$1 + \frac{32}{12} = \frac{44}{12}$$

$$1 + \frac{8}{3} = \frac{11}{3}$$

i.e., 1 kg of carbon requires $\frac{8}{3}$ kg of oxygen and produces $\frac{11}{3}$ kg of carbon.

2. ***Burning of carbon to carbon monoxide (incomplete combustion)***

$$2C + O_2 = 2CO$$

$$2 \times 12 + 16 \times 2 = 2\,(12 + 16)$$

$$24 + 32 = 56$$

$$1+\frac{32}{24}=\frac{56}{24}$$

$$1+\frac{4}{3}=\frac{7}{3}$$

i.e., 1 kg of Carbon requires $\frac{4}{3}$ kg of oxygen and produces $\frac{7}{3}$ kg of carbon monoxide.

3. ***Carbon monoxide***: It combines with oxygen to produce CO_2.

$$2CO + O_2 = 2CO_2$$

$$2(12 + 16) + 16 \times 2 = 2(12 + 16 \times 2)$$

$$56 + 32 = 88$$

$$1+\frac{32}{56} = \frac{88}{56}$$

$$1+\frac{4}{7} = \frac{11}{7}$$

1 kg of CO requires $\frac{4}{7}$ kg of oxygen to produce $\frac{11}{7}$ kg of carbon dioxide.

4. ***Sulphur***: Burning of sulphur to sulphur dioxide

$$S + O_2 = SO_2$$

$$32 + 16 \times 2 = 32 + 16 \times 2$$

$$32 + 32 = 64$$

$$1 + 1 = 2$$

1 kg of sulphur needs 1 kg of O_2 and produce 2 kg of SO_2.

5. ***Burning of Hydrogen to steam***

$$2H_2 + O_2 = 2H_2o$$

$$2(1 \times 2) + 16 \times 2 = 2(1 \times 2 + 16)$$

$$4 + 32 = 36$$

$$1 + 8 = 9$$

1 kg of Hydrogen needs 8 kg of oxygen and produce 9 kg of steam.

6. ***Burning of Methane (CH_4)***

$$CH_4 + 2O_2 = CO_2 + 2H_2o$$

$$(12 + 1 \times 4) + 2(16 \times 2) = (12 + 16 \times 2) + 2(1 \times 2 + 16)$$

$$16 + 64 = 44 + 36$$

$$1 + 4 \qquad = \frac{11}{4} + \frac{9}{4}$$

1 kg of methane requires four kg of O_2 to produce $\frac{11}{4}$ kg of CO_2 and $\frac{9}{4}$ kg of H_2O.

7. ***Burning of Ethylene (C_2H_4)***

$$C_2H_4 + 3O_2 = 2CO_2 + 2H_2O$$

i.e.,

$$(12 \times 2 + 1 \times 4) + 3(16 \times 2) = 2(12 + 16 \times 2) + 2(1 \times 2 + 16)$$

$$28 + 96 = 88 + 36$$

$$1 + \frac{96}{28} = \frac{88}{28} + \frac{36}{28}$$

$$1 + \frac{24}{7} = \frac{22}{7} + \frac{9}{7}$$

i.e., 1 kg of ethylene needs $\frac{24}{7}$ kg of O_2 to produce $\frac{22}{7}$ kg of CO_2 and $\frac{9}{7}$ kg of steam.

8. ***Acetylene (C_2H_2)***

$$2C_2H_2 + 5O_2 = 4CO_2 + 2H_2O$$

$$2(12 \times 2 + 1 \times 2) + 5(16 \times 2) = 4(12 + 16 \times 2) + 2(1 \times 2 + 16)$$

$$52 + 160 = 176 + 36$$

$$1 + \frac{160}{52} = \frac{176}{52} + \frac{36}{52}$$

$$1 + \frac{40}{13} = \frac{44}{13} + \frac{9}{13}$$

i.e., 1 kg of Acetylene requires $\frac{40}{13}$ kg of O_2 to produce $\frac{44}{13}$ kg of CO_2 and $\frac{9}{13}$ of steam.

This extra amount of air supplied is known as excess air and varies with type of fuel and method of firing. In practice excess air, to the extent of 30 to 50% of theoretical (minimum) air is supplied to burn the fuel completely.

It should be noted that too great an excess air result in:

(i) lower furnace temperature,

(ii) increased heat carried away by flue gases,

(iii) increased smoke,

(iv) more power is required to handle air and flue gases, to produce artificial draught,

(v) incomplete combustion.

Minimum Weight of Air Required for Complete Combustion of Fuel

1. From the point of view of combustion it is usual to assume that air is a mixture of oxygen and nitrogen. Their proportion being:

   By weight Nitrogen = 77%
   Oxygen = 23%
   By volume Nitrogen = 79%
   Oxygen = 21%

2. Nitrogen is an inert gas and takes no part in chemical reactions. Its presence in air lowers the temperature of combustion, retards the intimate mixing of fuel with oxygen and also carries away heat up the chimney.
3. The amount of oxygen required for each of the constituents of fuel is calculated separately with help of chemical equations, or by applying the formula.
4. Then total amount of oxygen required is obtained by adding oxygen required by each constituent. If some oxygen is present already in fuel, then it must be deducted from the calculated amount of oxygen for combustion of the constituents of fuel.
5. Amount of air required for complete combustion of 1 kg of fuel can be obtained from oxygen requirements.
6. *Excess air:* In practice the quantity of excess air, in excess of the theoretical air, must be supplied to ensure complete combustion, as some of the air may pass off without any chemical reactions.

Weight of Flue Gas per kg Fuel Burnt

It can be obtained as follows.

(i) Calculate weight of carbon present in flue gases.

(ii) Weight of carbon in fuel is known.

(iii) As the analysis of the exhaust gases is volumetric, should be reduced to weight analysis.

(iv) Apply following formula and mass of flue gases can be calculated.

$$\text{Total weight of carbon in one kg of flue gas} = \left(\frac{3}{11}CO_2 + \frac{3}{7}CO\right)$$

$$\therefore \text{ weight of flue gases/ kg of flue} = \frac{\text{weight of carbon in one kg of fuel}}{\text{weight of carbon in one kg of flue gases}}$$

Solved Problems

Problem may be solved by Formula or from Fundamentally by Chemical Reactions

By Formula

Minimum or theoretical mass air required for complete combustion per kg of fuel

$$= \frac{100}{23}\left[\frac{8}{3}C + 8H + S - O\right]$$

where c = mass of carbon in one kg of fuel

H = Mass of Hydrogen in one kg of fuel

S = Mass of sulphur in one kg of fuel

O = Mass of oxygen in one kg of fuel

1. Percentage composition of fuel on mass basis is C = 87.1 H_2 = 4.4, O_2 = 1.2 and ash. Find minimum or theoretical mass and volume of air required to burn one kg of fuel.

***Solution*:**

First method by formula

In this problem sulphur in fuel is not given

Therefore $S = 0$

$C = 0.871$

$H = 0.044$

$O = 0.012$

$$\text{Minimum air required/kg of fuel} = \frac{100}{23}\left[\frac{8}{3}C + 8H + S - O\right]$$

$$\therefore \quad \text{Minimum air required/kg of fuel} = \frac{100}{23}\left[\frac{8}{3}\times 0.871 + 8\times .044 + 0 - 0.012\right]$$

$$= 11.578 \text{ kg of air/kg of fuel}$$ **Ans.**

Second method

By chemical reaction

$$C + O_2 = CO_2$$

$$12 + 32 = 44$$

$$1 + \frac{8}{3} = \frac{11}{3}$$

$$2H_2 + O_2 = 2H_2O$$

$$4 + 32 = 36$$

$$1 + 8 = 9$$

Equation shows that 1 kg of carbon requires:

(i) $\frac{8}{3}$ kg of oxygen to burn it, if mass of carbon in one kg of fuel is C kg then

oxygen required to burn 0.871 kg of carbon $= \frac{8}{3} \times 0.871$

$= 2.323$ kg

(ii) Mass of Hydrogen in 1 kg of fuel is 0.044 kg

$\therefore$ O_2 required to burn H kg $= 8 \times 0.044 = 0.352$ kg

$\therefore$ Total O_2 required $= 2.323 + 0.352$

$= 2.675$ kg

But 0.012 of O_2 is already present in one kg fuel.

$\therefore$ Net o_2 required to burn 1 kg of fuel $= 2.675 - .012 = 2.663$ kg

$\therefore$ Minimum mass of air required to burn 1 kg of fuel $= \frac{100}{23} \times 2.663$

$= 11.578$ kg **Ans.**

To Find Theoretical or Minimum Volume of Air Required

(i) Density of air is proportion to molecular weight and inversely proportion to molar volume.

(ii) Molar volume of gas at NTP i.e., 0 °C and 760 mm of Hg is 22.4 m^3.

$\therefore$ Density of O_2 at O °C and 760 mm of Hg $= \frac{\text{Molecular weight}}{\text{Molar volume}}$

$= \frac{32}{22.4} = 1.429 \frac{kg}{m^3}$

$\therefore$ Volume of O_2 required $= \frac{\text{Mass of } O_2}{\text{density}}$

$= \frac{2.663}{1.429}$

$= 1.864$ m^3 of O_2

∴ Minimum volume of air required for complete combustion of 1 kg of fuel

$$=1.864\times\frac{100}{21}$$

$= 8.876\ m^3$ of air / kg of fuel **Ans.**

2. The percentage composition of coal on mass basis is given as

$C = 90$, $H_2 = 3.5$, $O_2 = 1$, $S = 0.5$ and ash 5. Calculate minimum air required for complete combustion of 1 kg of fuel. Also calculate air supplied if excess air is 50%

Solution: Given $C = 0.9$, $H_2 = 0.035$, $O_2 = 0.01$, $S = 0.005$, ash 0.05

$$\text{Minimum air required } =\frac{100}{23}\left[\frac{8}{3}C+8H+S-O\right]$$

$$=\frac{100}{23}\left[\frac{8}{3}\times.9+8\times.035+0.005-0.01\right]$$

$= 44.55$ kg of air/kg of fuel **Ans.**

Excess air required $= 0.5 \times 44.55$

$= 22.27$ kg / kg of fuel

∴ Air supplied per kg of fuel = Minimum air required + Excess air

$= 44.55 + 22.27$

$= 66.82$ kg of air/kg of fuel **Ans.**

3. A sample of fuel on analysis is found to contain.

Carbon = 85%

Hydrogen = 10%

Sulphur = 2%

Ash = 3%

Find the H.C.V. of fuel

Solution: Applying Dulong's formula

$C = 0.85$, $H_2 = 0.1$, $S = 0.02$, $O_2 = 0$ (as 'O_2' is not given in problem)

$$\text{H.C.V.} = 33800\text{C} + 144000\,(\text{H}_2 + \frac{\text{O}_2}{8}) + 9270\ \text{S}$$

$$= 33800 \times 0.85 + 144000\,(0.1 - 0) + 9270 \times 0.02$$

$$= 28730+14400+185.4$$

$$= 43315.4$$

$$= \mathbf{44984\ kJ/kg} \qquad \textbf{Ans.}$$

4. The composition of sample of anthracite coal was found to be

C = 91%

H_2 = 3%

O_2 = 2%

N_2 = 0.8%

S = 0.8% and remainder ash.

Calculate H.C.L. and L.C.V. of the fuel.

Solution: Given

C = 0.91

H_2 = 0.03

O_2 = 0.02

N_2 = 0.008

S = 0.008

By Dulong's method

$$\text{H.C.V} = 33800\ \text{C} + 144000\,(\text{H}_2 - \frac{\text{O}_2}{8}) + 9270\ \text{S}$$

$$= 33800 \times 0.91 + 144000\,(0.03 - \frac{0.02}{8}) + 9270 \times 0.008$$

$$= 30758 + 3960 + 74.16$$

$$= 34792.2\ \text{kJ/kg}$$

L.C.V. = H.C.V. – 9H × 2257

($\because$ h_{fg} = 2257 at atmospheric pressure from steam table)

= 34792.2 – 9 × 0.03 × 2257

= 34182.77 kJ/kg **Ans.**

5. A sample of coal has the following analysis by mass

C = 80%

H_2 = 10%

O_2 = 4%

S = 2% and remaining is ash.

Determine H.C.V. and L.C.V. of the coal.

Solution: Given

C = 0.8

H_2 = 0.1

O_2 = 0.04

S = 0.02

$$\text{H.C.V.} = 33800\ C + 144000\ (H_2 - \frac{O_2}{8}) + 9270\ S$$

$$= 33800 \times 0.8 + 144000\ (0.1 - \frac{0.04}{8}) + 9270 \times 0.02$$

= 40905.4 kJ/kg

L.C.V = H.C.V. – 9 H_2 × 2257

= 40905.4 – 9 × 0.1 × 2257

[h_{fg} = 2257 at 1.0215 bar from steam table]

= 38874.1 kJ/kg **Ans.**

6. A fuel contains 86 percent carbon and 14 percent hydrogen. Determine H.C.V. if 1 kg of carbon burns to Co_2 will produce 33800 kJ/kg and hydrogen burning to produce 142000 kJ/kg.

Solution: Given C = 0.86

H_2 = 0.14

Note:

1. In the problem C.V. of carbon and C.V. of hydrogen is given therefore we can not use Dulong's method for find H.C.V.
2. In problem fuel contains only hydrogen and carbon therefore O_2 should be taken zero and sulphur also equal to zero i.e. $O_2 = 0$, $S = 0$.

$$\text{H.C.V.} = 33800\ C + 142000\ (H_2 - \frac{O_2}{8}) + 9270 \times S$$

$$= 33800 \times 0.86 + 142000 \times (0.14 - 0) + 9270 \times O$$

$$= \mathbf{48948\ kJ/kg\ of\ fuel} \qquad \textbf{Ans.}$$

7. Sample of coal having the following composition by mass.

Carbon = 75%

Hydrogen = 6%

Oxygen = 8%

Nitrogen = 1.5%

Sulphur = 1.5%

Ash = 7%

Calculate H.C.V. and L.C.V.

Solution: Given

$C = 0.75$

$O_2 = 0.08$

$N_2 = 0.025$

$S = 0.015$

$H_2 = 0.06$

$$\text{H.C.V.} = 338000\ C + 144000\ (H_2 - \frac{O_2}{8}) + 9270\ S$$

$$= 33800 \times 0.75 + 144000\ (0.06 - \frac{0.08}{8}) \times 9270 \times 0.015$$

$$= 32689\ \text{kJ/kg} \qquad \textbf{Ans.}$$

L.C.V. $= \text{H.C.V.} - 9H_2 \times 2466$

[from steam table corresponding to atmospheric pressure $h_{fg} = 2466$ kJ/kg]

$= 32689 - 9 \times 0.06 \times 2466$

= 31358.5 kJ/kg **Ans.**

8. Following data were recorded during experiment to find calorific valve sample of coal.

Mass of coal burnt = 1 gm

Mass of water in calorimeter = 2.5 kg

Water equivalent of apparatus = 0.75 kg

Initial temperature of water = 17.5 °C

Maximum observed temperature = 20 °C

Cooling correction = + 0.015 °C

If fuel contains 4% of hydrogen find lower calorific value of the fuel.

Solution:

$M_w = 2.5$ kg

$M_e = 0.75$ kg

$C_{p_w} = 4.186$ kJ/kgk

$t_1 = 17.5$ °C

$t_2 = 20$ °C

$$M_f = \frac{1}{1000} = 0.001 \text{ kg}$$

Cooling correction $= t_C = 0.015$ °C

$$\text{H.C.V.} = \frac{(M_w + M_e)C_{p_w}(t_2 - t_1 + t_C)}{M_f}$$

$$= \frac{(2.5 + 0.75) \times 4.186(20 - 17.5 + 0.015)}{0.001}$$

$= 34215.32$ kJ/kg

$\therefore$ L.C.V. = H.C.V. – 9 H_2 × 2466

[$\because$ h_{fg} = 2466 kJ/kg at standard

temperature 15 °C from steam table]

= 33996.25 – 9 × 0.04 × 2466

= 33327.56 kJ/kg. **Ans.**

9. Fuel contains 86% carbon and 14% Hydrogen. Determine HCV if 1 kg of Carbon burning to Co_2 liberates 33800 kJ and 1 kg of Hydrogen on burning evolves 142000 kJ.

Solution: Given

C = 0.86

H = 0.14

C.V of carbon = 33800 kJ/kg of fuel

C.V. of hydrogen = 142000 kJ/kg of fuel

H.C.V. of fuel = 0.86 × 33800 + 0.14 × 142000

= 48948 kJ/kg of fuel **Ans.**

Ultimate Analysis of Solid Fuel

S. No.	Fuel	Carbon C	Hydrogen H_2	Oxygen O_2	Nitrogen N_2	Sulphur S	Incombustibles	Higher C.V. kJ/kg
1.	Wood dry	48.5	6.0	43.5	0.5	-	1.5	17300
2.	Peat	58	6.3	30.8	0.9	-	4	18700
3.	Lignite Brown Coal	66	5	20	1.0	1.0	3.5	21650
4.	Bituminous coal	81	5	8	1.5	1.0	3.5	33500
5.	Anthracite	91	3	2.5	0.5	0.5	2.5	34600
6.	Coke	-	-	-	-	-	-	33600
7.	Charcoal	-	-	-	-	-	-	34300

Ultimate Analysis or Composition by Mass

So. No.	Fuel	C	H_2	S	LCV kJ/kg	H.C.V. kJ/kg	Specific gravity
1.	Gasoline	85.4 %	14.6 %	-	43700	46820	0.74
2.	Paraffin	86.3 %	13.6 %	0.1 %	43250	46250	0.79
3.	Diesel oil	86.3 %	12.8 %	0.9 %	43260	46000	0.187
4.	Heavy oil	86 %	11.8 %	2.1 %	41300	44000	0.95
	100 octane petrol				44000	47300	-
	Propane L.P.G.				45900	-	-

Objective Type Questions

1. Colorific value of anthracite is approximate

   (a) 35700 kJ/kg (b) 12600 kJ/kg (c) 10500 kJ/kg, (d) 14700 kJ/kg

2. Lower calorific value = H C V – xh_{fg}

   (a) x is hydrogen fraction in fuel and h_{fg} is enthalpy of evaporation at 25 °C saturation temperature.

   (b) x is fraction of water vapour in product of combustion and h_{fg} is enthalpy of evaporation at atmospheric pressure.

   (c) x is fraction of water vapour in product of combustion and h_{fg} is enthalpy of evaporation at standard temperature of 15 °C.

   (d) x is fraction of water vapour in product of combustion and h_{fg} is enthalpy of evaporation at ambient temperature.

3. C.V. of fuel determined at constant volume.

   (a) is equal to C.V. of fuel at constant pressure.

   (b) It is more than C.V. of fuel at constant pressure.

   (c) It is less than C.V. of fuel at constant pressure.

   (d) It may be more or may less depending upon molecular contraction or expansion of the product of combustion.

4. The C.V. of hydrogen/kg, when burnt at constant pressure in atmosphere of air as compared to that at constant volume is

   (a) Less (b) more (c) equal (d) none of above

5. Does Bomb calorimeter gives

   (a) L C V (b) H C V (c) LCV and HCV (d) non of above

6. The higher calorific value of fuel whose composition is C = 89%, S = 4% and in-combustible 7%.

   (a) will be higher than is LCV (b) will be lower than its LCV

   (c) will be same as its LCV (d) will be lower or higher than LCV.

7. Bomb calorimeter is used to determine C.V. of

   (a) solid fuel only (b) liquid fuel only

   (c) gaseous fuel only (d) solid as well as liquid fuel.

8. Bomb calorimeter determines

   (a) HCV of fuel at constant volume (b) LCV of fuel at constant volume

   (c) HCV of fuel at constant pressure (d) L.C.V. of fuel at constant pressure.

9. The O_2 percentage by volume in air is.

   (a) 21% (b) 23% (c) 30% (d) 79%

10. The O_2 percentage by weight in air is

    (a) 21 (b) 23% (c) 30% (d) 79

11. Ultimate analysis of fuel means to determine

    (a) H.C.V. of fuel

    (b) L.C.V. of fuel

    (c) Percentage of carbon, hydrogen, nitrogen, sulphur and moisture

    (d) Percentage of fixed carbon, ash, volatile matter and moisture.

12. Proximate analysis of fuel means determination of

    (a) HCV of fuel

    (b) LCV of fuel

    (c) Percentage of fixed carbon, ash, volatile matter and moisture

    (d) None of above

13. Calorific value of gaseous fuel can be found by
    (a) Bomb calorimeter
    (b) Junker's calorimeter
    (c) Fire point apparatus
    (d) Flash point apparatus

Answers

1. (a)	2. (c)	3. (d)	4. (b)
5. (b)	6. (c)	7. (d)	8. (a)
9. (a)	10. (b)	11. (c)	12. (c)
13. (b)			

Fill in the blanks

(a) Solid fuel may be classified (1) _______ and (2) _______ or_______

(b) C.V. determined by bomb calorimeter is the _______

(c) Kerosene is less _______ Petrol

(d) Calorific value of alcohol is about _______ kJ/kg.

(e) Water gas is a mixture of _______ and _______

(f) C.V. of petrol is about _______kJ//kg

(g) In _______ analysis, a complete chemical breakdown of coal into its chemical constituents is carried out.

(h) _______ gas calorimeter is used to determine c.v. of gaseous fuel.

(i) _______ analysis is the separation into its physical components.

Answers

(a) Natural fuels, artificial fuel or prepared fuels

(b) Higher

(c) Volatile

(d) 27200 kJ/kg

(e) Hydrogen and carbon monoxide

(f) 43950 kJ/kg

(g) Ultimate

(h) Junker's

(i) proximate

Questions

1. Define the term fuel. What are the principal constituents of fuel? Classify the various fuels.
2. Explain ultimate analysis and proximate analysis.
3. Differentiate between solid, liquid and gaseous fuels.
4. Name some important natural solid fuels and some important artificial solid fuels.
5. Define the term calorific value of a fuel. Distinguish between the L.C.V and H.C.V of a fuel.
6. State and explain Dulong's Formula for theoretical determination of C.V. of a fuel.

   (i) When there is no oxygen present in the fuel

   (ii) When oxygen is present in the fuel.
7. Describe with the help of sketch the working of a Bomb Calorimeter.
8. Describe different types of coals giving their properties and fields of application.
9. Discuss the advantages and disadvantages of liquid fuel over solid fuels.
10. List out the merits and demerits of gaseous fuels over solid fuels.
11. Determine from first principles, the minimum quantity of air required for complete combustion of 1 kg of (i) Carbon (ii) Hydrogen (iii) Methane gas (CH_4)
12. Explain the terms "minimum mass of air" and "excess air" with reference to combustion of fuel.
13. How you will find mass of flue gases? Give proper formula for the same.
14. What is meant by the term 'Producer gas'?

Fuel and Combustion (Some useful points)

1. The commercial sources of energy are fossil fuels, hydropower and nuclear energy.
2. The primary sources of energy are coal, oil and uranium.
3. Solar, wind and water are secondary source of energy.
4. At Chandrapur largest T.P.S. is located in India.
5. Percentage oxygen by weight in atmospheric air is 23%.
6. Percentage oxygen by volume in atmospheric air is 21%
7. High CO content in fuel gases indicate incomplete combustion of fuel.
8. The amount of heat liberated due to combustion of 1 kg of fuel is known as calorific value of fuel.
9. Principal constituents of fuel are carbon and hydrogen.

Exercises

1. Coal used has the following composition

   Carbon = 70%

   Hydrogen = 5%

   Oxygen = 10%

   Sulphur = 2%

   Ash = 13%

   Calculate H.C.V. and L.C.V

   [H C V = 2924.5 kJ/kg, L C V = 2813.5 kJ/kg]

2. From the analysis of dry coal it was found to have 80% of carbon and 4% of hydrogen. C.V. of carbon = 33700 kJ/kg and that of hydrogen is 144600 kJ/kg. Find higher calorific value of coal.

[35103 kJ/kg]

3. The following particulars refer to an experimental determination of C.V. of sample of coal containing 89% carbon and 4.1% hydrogen. Mass of coal = 0.847gm, mass of fuse wire = 0.026 gm of C.V. = 6700 J/gm. Mass of water in calorimeter = 1940 gm, observed temperature rise = 3.040 °C. Cooling correction = +0.017 °C. Find HCV and LCV of the fuel.

[34945.86 kJ/kg, 34035.9 kJ/kg]

4. The following observation were made when sample of oil fuel having 0.88 specific gravity and C = 86.5% and hydrogen = 13.5% was tested in Bomb calorimeter. Mass of oil = 0.9 gm, observed rise in temperature of water 3.475 °C correction for loss by radiation = + 0.25 °C. Determine HCV and LCV of oil fuel.

[43953 kJ/kg, 41211 kJ/kg]

CHAPTER 3

Properties of Gases

3.1 Introduction

Any substance must be available in three different forms at appropriate pressure and temperature conditions. Such as (i) solid; (ii) liquid; (iii) gas. By heating process, at particular pressure and temperature condition, solid can be converted into liquid and liquid in to gaseous form. By cooling reverse conversion is possible.

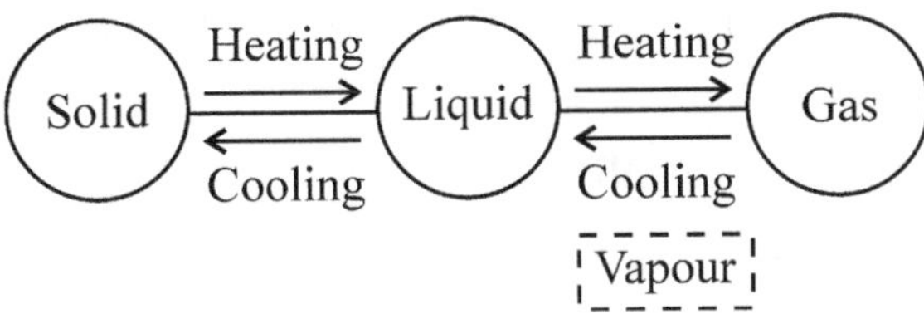

A vapour is a form lies between liquid form and gaseous form. Difference between vapour and gas is explained as below.

3.2 Difference between Vapour and Gas

(a) Vapour:

It can be defined as that state (form) of the substance in which the evaporation from its liquid state is not complete. A vapour consists of a mixture of the pure gaseous form and liquid particles in suspension.

Example: Wet steam. Vapours readily condense when expanded or cooled in any process.

Behavior of wet vapour can't be determined by Boyle's and Charle's law.

(b) A perfect Gas:

A perfect gas is one which obeys the Boyle's and Charle's laws and characteristic equation of a gas, which is obtained by above laws. No gas is perfect but many gases behave nearly as perfect gas in temperature and pressure ranges of applied thermodynamics.

Many gases like air, nitrogen, hydrogen etc can be regarded as perfect gases. They are known as real gases.

3.3 Boyle's Law

This law was discovered by Robert Boyle in 1662.

Definition: The volume of a given mass of a perfect gas varies inversely as the absolute pressure, when the temperature is kept constant.

Let P = The absolute pressure of the gas in pa

V = Volume occupied by the gas at absolute pressure in m^3.

Form above law,

$$V \alpha \frac{1}{P} \quad \text{or } V = \frac{C}{P} \qquad \text{.....(3.1)}$$

Where C = constant of proportionality

i.e.,

$$P.V = C \qquad \text{.....(3.2)}$$

During the process, if temperature is kept constant, the product of pressure and volume always remains constant.

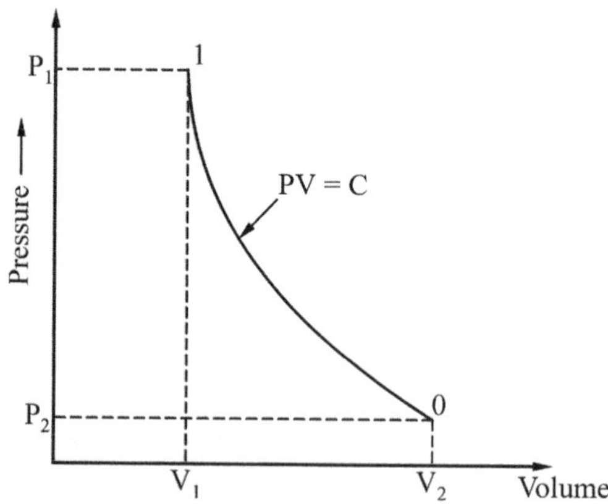

Fig. 3.1 Relation between P and V as per Boyle's law.

In Fig. 3.1 let a quantity of gas at pressure P_1 and volume V_1 changes its pressure and volume in cylinder without change in temperature. Let P_2 and V_2 are respective final pressure and volume, after end of expansion of gas.

We can say $$P_1V_1 = P_2V_2 = \ldots\ldots. = \text{constant}$$

3.4 Charle's Law

This law was discovered by Charles in 1787.

Definition: If the pressure of the given mass of a gas is kept constant, then the volume of the gas varies directly in proportion to its absolute temperature.

If V = volume occupied by a given mass of the gas in m^3

$T = 273 + t$ Absolute temperature in Kelvin or K

According to the Charle's law with constant pressure

$$V \alpha T \quad \text{or} \quad V = C.\,T$$

$$\therefore \quad \frac{V}{T} = C \text{ constant} \qquad \ldots..(3.3)$$

where C = constant of proportionality

Refer to Fig. 3.2

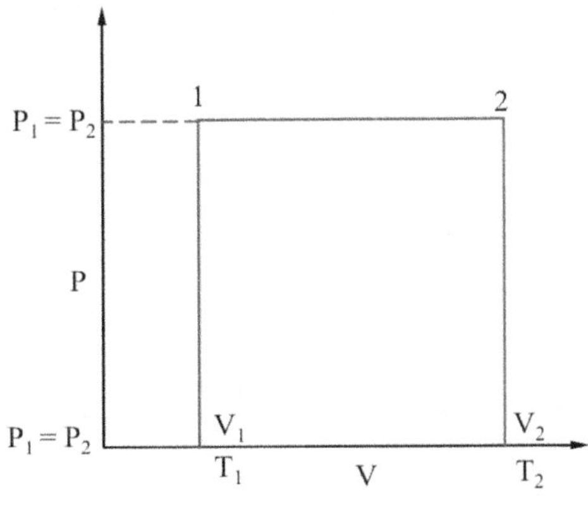

Fig. 3.2

If V_1 and T_1 are initial pressure and temperature at point 1 and V_2 and T_2 are final pressure and temperature at point 2.

By equation

$$\frac{V_1}{T_1} = \frac{V_2}{T_2} = \text{Constant}$$

i.e. $$\frac{\mathbf{V_1}}{\mathbf{T_1}} = \frac{\mathbf{V_2}}{\mathbf{T_2}} \qquad \text{.....(3.4)}$$

3.4.1 Charle's law

Charle's law can also be stated as the pressure of a gas varies with the absolute temperature, provided the volume occupied by the given mass of gas is constant.

$$P \propto T, \quad P = C \times T$$

or $$\frac{P}{T} = \text{constant},$$

or $$\frac{P_1}{T_1} = \frac{P_2}{T_2} = \text{constant}$$

where '1', '2' represent initial and final state points

3.5 Combined Gas Law

Let a given mass [m : 1 kg] of a perfect gas has the initial condition at state (1) (See Fig. 3.3), with pressure, volume and temperature are P_1, V_1 and T_1 respectively. Process (1) $\xrightarrow{\text{to}}$ (2) at point 2 pressure is kept constant. The new pressure, volume and temperature are P_2 V_2 and T_2

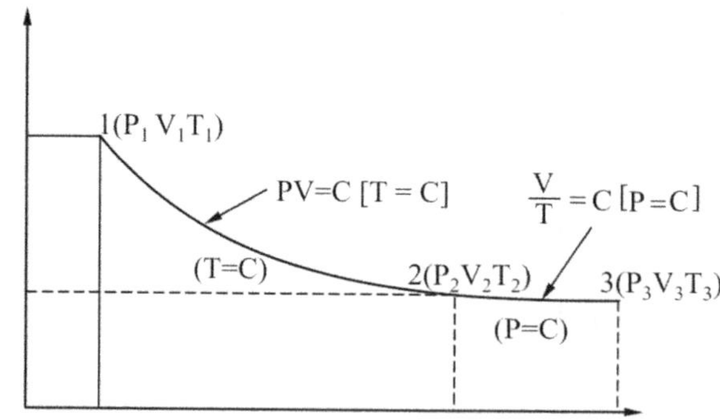

Fig. 3.3 Combination of Charle's law and Boyle's law.

Now let the gas at state point (2) have respectively pressure, volume and temperature are P_2, V_2 and T_2.

Apply Boyle's law between point 1 to point 2.

$$P_1V_1 = P_2V_2$$

$$V_2 = \frac{P_1V_1}{P_2} \qquad .....(3.6)$$

Now gas from state (2) it moves at constant pressure condition to point (3).

At the state '3', new pressure, volume and temperature are P_3, V_3 and T_3 respectively.

Applying Charle's law, we have

$$\frac{V_2}{T_2} = \frac{V_3}{T_3} \qquad \because P = C$$

$$\therefore \quad V_2 = V_3 . \frac{T_2}{T_3} \qquad .....(3.7)$$

Comparing (3.6) and (3.7)

$$\frac{P_1V_1}{P_2} = \frac{V_3.T_2}{T_3}$$

But

$$T_1 = T_2$$

$$P_2 = P_3$$

$$\therefore \quad \frac{P_3V_3}{T_3} = \frac{P_1V_1}{T_1} = \frac{P_2V_2}{T_2}$$

$$\frac{P_1V_1}{P_3} = \frac{V_3.T_1}{T_3} \qquad \text{or}$$

$$\therefore \quad \mathbf{\frac{PV}{T} = constant = mR} \qquad .....(3.8)$$

Equation **PV = mRT** is called general gas equation, equation of state or characteristic gas equation.

In eq. (3.8), 'R' is specific or characteristic gas constant of particular gas under consideration.

Eq. 3.8 can be written as

$$Pv = RT \text{ for 1 kg mass of gas} \qquad .....(3.9)$$

where v is specific volume m^3/kg.

(specific volume is the volume of unit mass)

For m kg of mass $Pv \times m = mRT$

$$\therefore \quad \mathbf{Pv = mRT} \qquad .....(3.10)$$

The above equation is called the characteristic gas equation. In equation,

V = Volume of gas in m^3

P = Absolute pressure in Pa

T = Absolute temperature given by [T = 273 + t K]

R = Characteristic gas constant J/kg K

3.6 Unit of Gas Constant 'R'

In eq. (3.9) for m kg as mass

$$PV = mRT$$

$$\therefore \quad R = \frac{\text{Work done in J (PV)}}{mT} = \text{J/kg K}$$

$$= \frac{\frac{N}{m^2} \times m^3}{kg \times {}^{o}K} = \frac{N.m}{kg\ {}^{o}K} = \text{J/kg K}$$

It is obvious that R represents the work done against external resistance when 1 kg of gas is heated by 1° temperature at constant pressure.

$$R = \frac{N.m}{1\,kg \times 1^{o}K} = J \quad \text{(work done) when } m = 1 \text{ kg } \ t = 1 \text{ K}$$

Table 3.1 Specific Gas constant

Gas	Chemical Symbol	Molecular Weight (kg / kg . mol)	Sp. Gas constant (kg/kg °K)
Air	–	28.97	0.287
Carbon dioxide	CO_2	44.01	0.189
Hydrogen	H_2	2.016	4.134
Helium	H_e	4.003	2.077
Nitrogen	N_2	28.016	0.297
Oxygen	O_2	32.0	0.2598
Methane	CH_4	16.043	0.518

3.7 Universal Gas Constant

It can be observed that the product of molecular mass of the gas and its specific gas constant (R) is same for all the gases.

This constant is known as universal gas constant which is 8.3143 kJ/kg kg.mol °K

So for SI system

$$R_o = U_o = 8.3143 \text{ kJ/kg.mol °K}$$

3.8 Joule's Law of Internal Energy

The internal energy is a property of the system and hence is depends only on temperature change. If during the process, temperature of gas increases, it means its internal energy or energy content increases. Reverse is the case, if final temperature decreases compare to initial, its internal energy also decreases.

Change in internal energy can be written as $\delta u = m\,C_v\,(T_2 - T_1)$

$$u_2 - u_1 = m\,C_v\,(T_2 - T_1) \text{ kJ}$$

When any of the system parameters like pressure, volume, temperature etc changes there may be change in internal energy because of change of state.

But Joule's law of internal energy states that **"the internal energy of a perfect gas is a function of temperature only"**.

It means that change in internal energy $(u_2 - u_1)$ when temperature changes from T_1 to T_2 will be same no matter how pressure P and volume V changes.

So from Joule's experiments it is concluded that internal energy is only function of temperature and not of pressure and volume.

3.9 Specific Heats of a Gas

The amount of heat required to raise the temperature of 1 kg of mass by 1 K, is known as its specific heat.

$$\delta Q = m\,Cv\,\Delta T$$

$$\therefore \quad C_v = \frac{\delta Q}{m \times \Delta T}$$

$$\therefore \quad \text{Unit of } C_p = \text{kJ/kg K}$$

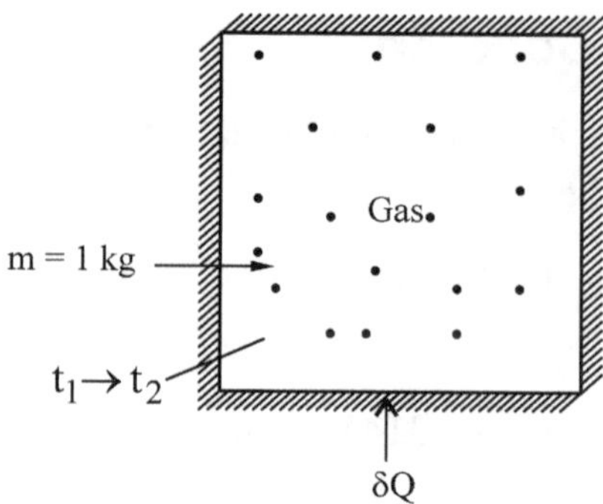

Fig. 3.4 Gas heated at constant volume.

Its unit is J/kg K or kJ/kg K. All types of liquids and solids have only one value of specific heat as they are practically incompressible while gases have two values of specific heat.

(A) Specific heat of gas at constant volume [C_v]

Let, certain quantity of gas is filled in a rigid vessel as shown in Fig. 3.4. It will be heated at

Constant volume: The whole of the heat energy is used to increase the internal energy of the gas which can be seen as increase in temperature of the gas $[\delta Q = \delta u + \delta w]$ $\because \delta w = 0$, as $dv = 0$. $\therefore \delta Q = \delta u$ where $\delta u = m\, C_v\, (T_2 - T_1)$

Thus heat required to increase the temperature of 1 kg of gas by 1 K when volume is kept constant is known as specific heat at constant volume. It is denoted by C_v.

(B) Specific Heat of a gas at constant pressure: [C_p]

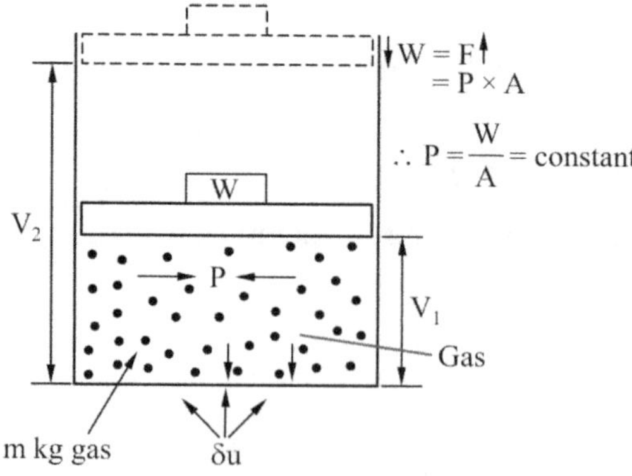

Fig. 3.5 Gas heated at constant pressure.

Now let certain quantity of gas is filled in piston cylinder arrangement as shown in Fig. 3.5 and piston carrying fixed load W acting downward.

Initial volume of gas is V_1. When heat is supplied to the gas it will be heated at constant pressure. The pressure will be net pressure on piston can be calculated as

$$p = \frac{W}{A} \qquad \text{where} \qquad A = \frac{\pi}{4} d^2 \qquad d = \text{diameter of piston}$$

Here heat is supplied will be utilized to increase internal energy of the gas and also to lift the weight under external force W acting downwards.

Thus heat supplied is used to increase the temperature and volume of the gas, volume V_1 will be increased to V_2.

∴ work done by lifting weight

$$\delta w = P \times dV$$

$$= P (V_2 - V_1)$$

Change in internal energy = $\delta u = m\, C_v\, (T_2 - T_1)$

$$\therefore \qquad \delta Q = \delta u + dw$$

$$= m\, C_v\, (T_2 - T_1) + P_1V_2 - P_1V_1 \qquad P_1 = P_2$$

$$= m\, C_v\, (T_2 - T_1) + mR\, (T_2 - T_1) \qquad \because P_2V_2 = mRT_2$$

$$= m\, (T_2 - T_1)\, (C_v + R) \qquad P_1V_1 = mRT_1$$

$$\delta Q = m\, C_p\, (T_2 - T_1) \text{ kJ/kg °K} \quad \text{as} \quad C_p - C_v = R$$

$$\therefore \qquad C_p = \frac{\delta Q}{m\, \Delta T} \qquad \therefore C_p = C_v + R$$

Thus heat required to increase the temperature of 1 kg of gas by 1 K at constant pressure is known as specific heat at constant pressure. It is denoted by C_p.

Since heat supplied at constant pressure is used to increase internal energy of the gas and also to do external work. While heat supplied at constant volume is used only to increase internal energy.

Hence C_p is always greater than C_v

3.10 Relation between two specific heats of a gas and specific gas constant

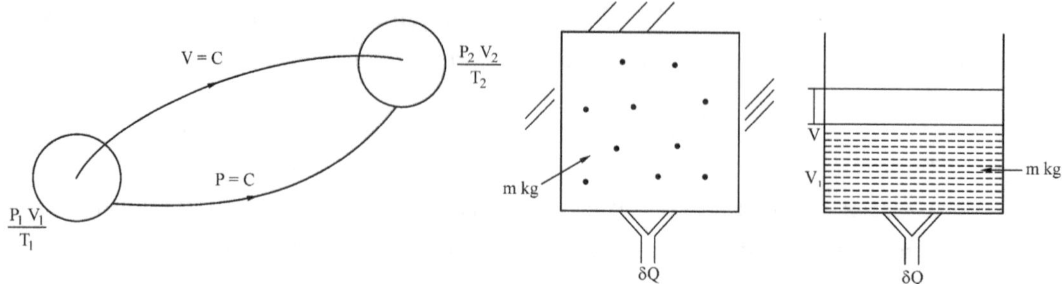

***Case* – I**: When heat is supplied at constant volume, temperature increases from initial temperature T_1 to final temperature T_2. V_1 = initial volume and V_2 = final volume.

Let 1 kg of gas is heated at constant volume and temperature increases from T_1 to T_2

Heat supplied at constant volume is given by $\delta Q = C_v (T_2 - T_1)$

Since heat supplied at constant volume increases only internal energy of steam due to increases of its temperature and no external work is done.

∴ According 1st law of thermo dynamics,

$$\delta Q = \delta u + \delta w$$

$$= \delta u + p\,(dv) \qquad \text{as} \qquad dv = 0$$

$$\delta Q = \delta u$$

$$u_2 - u_1 = C_v (T_2 - T_1) \qquad .....(1)$$

Case – II: Heat supplied at constant pressure is utilized for, state 1, state 2 P = C

(i) To increase internal energy of the gas

Increase in internal energy of gas = $C_v (T_2 - T_1)$

(ii) For overcoming external resistance for doing work

$$\text{work done} = P (V_2 - V_1)$$

$$= P_2V_2 - P_1V_1 \qquad P_2 = P_1$$

$$= RT_2 - RT_1$$

$$\delta w = R (T_2 - T_1) \qquad .....(2)$$

Net heat supplied at constant pressure is $\delta Q = C_p (T_2 - T_1)$ for same temperature rise

$$\delta Q = \delta u + \delta w$$

$$C_p (T_2 - T_1) = C_v (T_2 - T_1) + R (T_2 - T_1)$$

$\therefore \quad C_p = C_v + R$

$\therefore \quad C_p - C_v = R$

The unit of C_p, C_v and R will be the same and expressed as kJ/kg °K.

3.11 Ratio of Specific Heats

The ratio C_p/C_v is denoted by Greek letter γ. It is known as adiabatic index.

$$C_p - C_v = R$$

Divide by C_v on both sides,

$$\therefore \quad \frac{C_p}{C_v} - \frac{C_v}{C_v} = \frac{R}{C_v}$$

$$\gamma - 1 = \frac{R}{C_v}$$

$$\therefore \quad C_v = \frac{R}{(\gamma - 1)}$$

For air, value of specific heat at constant pressure is taken as 1.005 kJ/kg K, value of $C_v = 0.718$ kJ/kg K and value of $\frac{C_p}{C_v} = \gamma = 1.4$

$$C_p = 1.005 \text{ kJ/kg K}, C_v = 0.718 \text{ kJ/kg K}.$$

Solved Problems

Problem 1. A gas occupies a volume of 0.1 m^3 at a temperature of 30°C and a pressure of 1.5 bar. Find the temperature of gas, it is compressed to pressure of 7.5 bar and occupies a volume of 0.04 m^3.

$$1 \text{ bar} = \frac{10^5 P}{10^3} = 10^2 \text{ kPa}$$

Solution: Given

$$P_1 = 1.5 \text{ bar}$$
$$= 1.5 \times 10^2 \text{ kPa}$$
$$V_1 = 0.1 \text{ m}^3$$
$$T_1 = 30 + 273$$
$$= 303 \text{ k}$$
$$V_2 = 0.04 \text{ m}^3, P_2 = 7.5 \text{ bar } T_2 = ?$$

We have

$$\frac{P_1V_1}{T_1} = \frac{P_2V_2}{T_2} = C$$

During any process, characteristic equation remains same

$$\frac{1.5 \times 10^2 \times 0.1}{303} = \frac{7.5 \times 10^2 \times 0.04}{T_2}$$

$$\therefore \quad T_2 = 606 \text{ K} - 273 = 333\ ^\circ\text{C}$$

Problem 2. One kg of an ideal gas is heated from 18 °C to 98 °C. Assuming R = 0.27 kJ/kg K and γ = 1.18 for the gas, calculate (i) specific heats (ii) change in internal energy (iii) change in enthalpy.

Solution: Given $T_1 = 18 + 273 = 291$ K

$$T_2 = 98 + 273 = 371 \text{ K}$$

(i) $C_p - C_v = R = 0.27$ (given) and $\frac{C_p}{C_v} = \lambda = 1.18$ (given)

$$\frac{C_p}{C_v} - \frac{C_v}{C_v} - \frac{R}{C_v} \qquad \therefore \gamma - 1 = \frac{R}{C_v}$$

$$\therefore \quad C_v = \frac{R}{(\gamma - 1)} = \frac{0.27}{1.18 - 1} = \frac{0.27}{0.18} = 1.5 \text{ kJ/kg K}$$

$$\gamma = \frac{C_p}{C_v} = 1.18$$

$$\therefore \quad C_p = 1.18 \times 1.5$$

$$= 1.77 \text{ kJ/kg K}$$

(ii) Change in internal energy (δv), let m = 1 kg

$$\delta U = C_v (T_2 - T_1)$$

$$= 1 \times 1.5 (371 - 291)$$

$$= \mathbf{120 \text{ kJ/kg}}$$

(iii) Change in enthalpy $\delta H = \delta Q$

$$\delta H = m\, C_p (T_2 - T_1)$$

$$= 1 \times 1.77 (371 - 291)$$

$$= \mathbf{141.6 \text{ kJ/kg}} \qquad .....(3)$$

3.12 Flow and non-flow Processes

A process may be a flow process in which mass is entering and leaving through the boundary of an open system.

A non-flow process is one in which a fixed mass within a defined boundary is undergoing a change of state.

3.13 Various non-flow Processes for a Perfect Gas

As mentioned earlier a non-flow process is one in which there is no mass flow across the boundary of the system. Thus non-flow processes mean process undergone by closed systems. There are five important non-flow processes in the study of thermodynamics.

They are:

1. Constant volume process or Isochoric process
2. Constant pressure process or Isobaric process
3. Constant temperature process or Isothermal process
4. Adiabatic process (reversible) or isentropic process
5. Polytropic process.

Now we compare all processes with respect to following aspects.

(a) representation on p-V diagram

(b) work done

(c) change in internal energy

(d) heat transfer and

(e) change in enthalpy of the system during such process.

3.13.1 Constant Volume Process

A gas or vapour heated in a fixed enclosed space. It will be considered as constant volume process. Where volume will remain constant, only pressure and temperature will change during the process of heating.

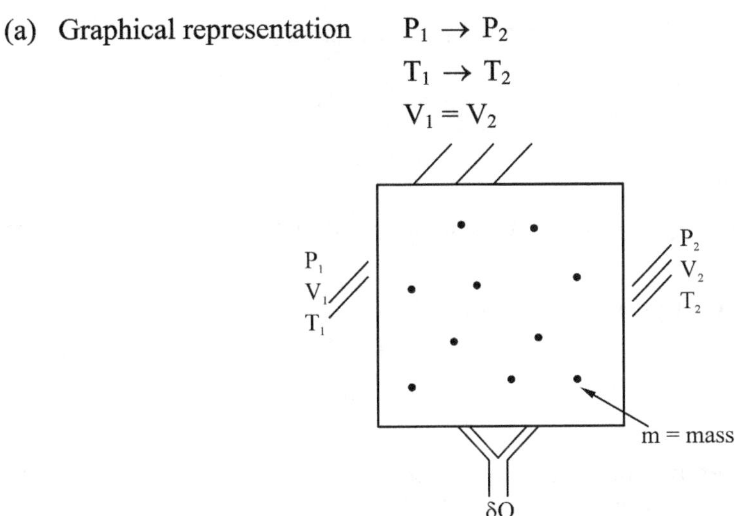

(a) Graphical representation.

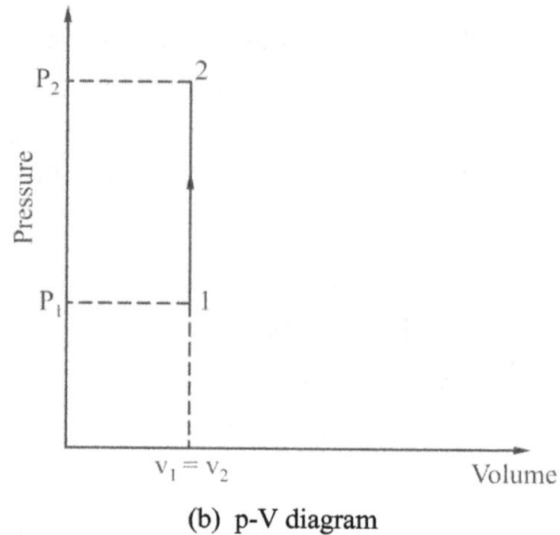

(b) p-V diagram

During heating at constant volume only pressure increases from P_1 to P_2 while volume remains constant $V_1 = V_2$. The process will be represented by vertical line from 1 to 2 as shown in above figure.

(b) Work done during the process:

$$\text{Work done} = \int_1^2 p\,dv \qquad \text{as volume is constant i.e., } dv = 0,$$

$$= 0$$

Since during constant volume, process dv = 0

Work done = 0

Relation between P_1V and T_1 between state points 1 and 2

$$\frac{P_1V_1}{T_1} = \frac{P_2V_2}{T_2} \; [V_1 = V_2] \; \therefore \; \frac{P_1}{T_1} = \frac{P_2}{T_2} \qquad \text{i.e., } \mathbf{P \alpha T}$$

(c) Change in Internal Energy

During constant volume process, due to heat supplied, temperature increases from T_1 and T_2

$\therefore$ change in internal energy $du = m\,C_v\,(T_2 - T_1)$

$$\therefore \qquad du = m\,C_v \int_{T_1}^{T_2} dT$$

It may be seen from above that change in internal energy depends on temperature change only and not on process involved. Hence the above equation is used irrespective of process being used.

(d) Heat transferred / kg

By first law of thermodynamics

$$\delta Q = \delta u + \delta w \qquad \text{In this process } \delta w = 0$$

$$\therefore \quad \delta Q = \delta u$$

$$\delta Q = C_v (T_2 - T_1) \qquad m = 1 \text{ kg}$$

(e) Change in enthalpy / kg

enthalpy can be given as

$$H = u + PV$$

For state 1. $H_1 = u_1 + P_1V_1$

For state 2. $H_2 = u_2 + P_1V_1$

$$\Delta H = \text{change in enthalpy / kg}$$

$$= H_2 - H_1$$

$$= (u_2 - u_1) + (P_2V_2 - P_1V_1)$$

$$= C_v (T_2 - T_1) + R (T_2 - T_1)$$

$$\because P_1V_1 = RT_1 \text{ and } P_2V_2 = mRT_2$$

$$\therefore \quad \Delta H = C_v (T_2 - T_1) + R (T_2 - T_1)$$

$$= (T_2 - T_1)(C_v + R)$$

But $\therefore$ $C_p - C_v = R$

$$\therefore \quad C_p = C_v + R$$

$$\therefore \quad \Delta H = C_p [T_2 - T_1]$$

3.13.2 Constant Pressure Process

When m kg gas is enclosed in a cylinder piston arrangement as shown in figure.

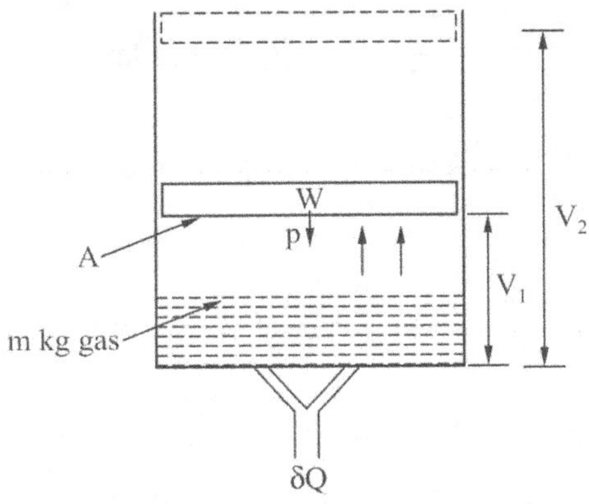

The gas will occupy initial volume V_1, pressure P_1 and temperature T_1

(a) Representation on p-V diagram

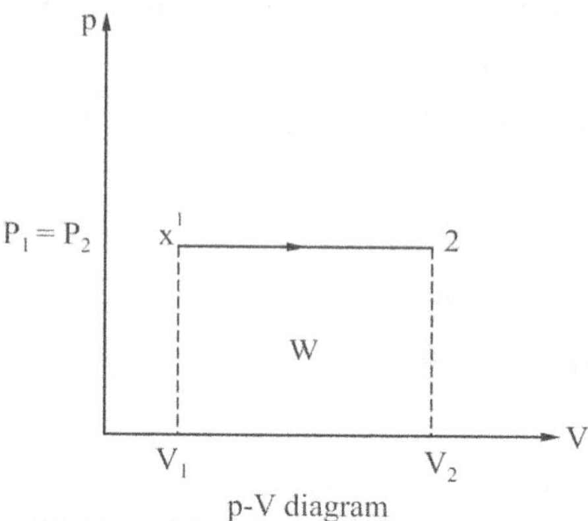

p-V diagram

When gas is heated at constant pressure its volume increases from V_1 to V_2 and temperature also increases from T_1 to T_2.

It represent horizontal line $1 \rightarrow 2$.

(b) Work done during the process

$$W.D/kg = \int_{V_1}^{V_2} pdv = P\int_{V_1}^{V_2} dv \qquad [\because P = \text{constant}]$$

$$= P(V_2 - V_1)$$

$$= P_2V_2 - P_1V_1 \qquad \ldots\ldots P_1 = P_2$$

$$= R(T_2 - T_1)$$

Relationship between P,V and T

$$\frac{P_2V_2}{T_2} = \frac{P_1V_1}{T_1} \qquad \text{(Charle's law)}$$

$$\frac{V_2}{T_2} = \frac{V_1}{T_1} \qquad \text{as } P_1 = P_2 = \text{constant}$$

(c) Change in internal energy

Change in internal energy $\delta u = C_v (T_2 - T_1)$ .... (m = 1)

(d) Heat transform/kg

$$\therefore \qquad \delta Q = \delta w + W.D$$

$$= C_v (T_2 - T_1) + P(V_2 - V_1)$$

$$= C_v (T_2 - T_1) + P_2V_2 - P_1V_1$$

$$= C_v (T_2 - T_1) + R (T_2 - T_1)$$

$$\delta Q = (T_2 - T_1)(C_v + R)$$

$$\delta Q = C_p (T_2 - T_1) \qquad \text{as } C_v + R = C_p$$

(e) Change in enthalpy / kg

$$\Delta H = C_p (T_2 - T_1)$$

Hence for constant pressure process

$$\Delta H = \delta Q.$$

3.13.3 Isothermal Process

(a) Representation on p-V diagram

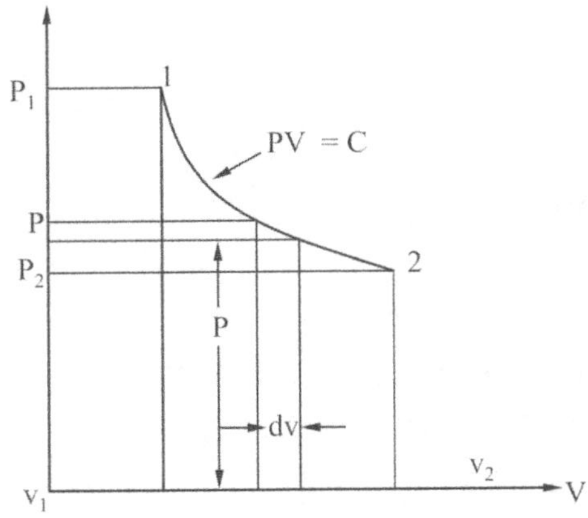

For isothermal process, from Boyle's law we can write PV = C. It means that relation between p and V will produce hyperbolic expansion curve.

(b) Work done:

$$W.D = \int_{V_1}^{V_2} p\,dv \qquad \text{.....(1)}$$

For isothermal process $PV = C = P_1V_1 = P_2V_2$

$$P = \frac{C}{V}$$

$$\therefore \quad W = C\int_{V_1}^{V_2}\frac{dV}{V}$$

$$= C\left[\log_{e^V}\right]_{V_1}^{V_2}$$

$$W_{1-2} = C\log\frac{V_2}{V_1}$$ But $C = P_1V_1$

$$\therefore \quad W_{1-2} = P_1V_1\log\frac{V_2}{V_1}$$ for isothermal process

First form of equation

$P_1V_1 = P_2V_2$

$$\therefore \quad \frac{P_1}{P_2} = \frac{V_2}{V_1}$$

$$\therefore \quad W_{1-2} = P_1V_1\log\frac{P_1}{P_2}$$ Second form of equation

$$W_{1-2} = mRT_1\log_e\frac{P_1}{P_2}$$ Third form of equation

Relation between P, V and T

For change of state from P_1, V_1, T_1 to P_2, V_2, T_2

$$\therefore \quad \frac{P_1V_1}{T_1} = \frac{P_2V_2}{T_2}$$ $\because T_1 = T_2 \quad \therefore P_1V_1 = P_2V_2 = C$

(c) Change in internal energy since change in internal energy is a function of temperature only and during isothermal process T = C.

$$\therefore \quad \Delta t = 0$$

$$\therefore \quad \Delta u = m\,C_v\,\Delta t = 0$$

Change in Internal energy will be zero due to $\Delta t = 0$

(d) Heat transfer

From 1st law of thermodynamics,

$$\delta Q = \delta w + \delta u$$ $\delta u = 0$

$$= P_1V_1\log\frac{V_2}{V_1}$$ $P_1V_1 = P_2V_2$

$$= P_1V_1 \log \frac{P_1}{P_2} \qquad \therefore \quad \frac{V_2}{V_1} = \frac{P_1}{P_2}$$

$$= mRT \log \frac{P_1}{P_2}$$

(e) Change in enthalpy

$$\Delta H = m\, C_p\, (T_2 - T_1) \qquad \text{since } T_1 = T_2$$

$$\therefore \quad \Delta H = 0$$

3.13.4 Adiabatic Process

A process in which there is no heat transfer to or from the system is known as adiabatic process.

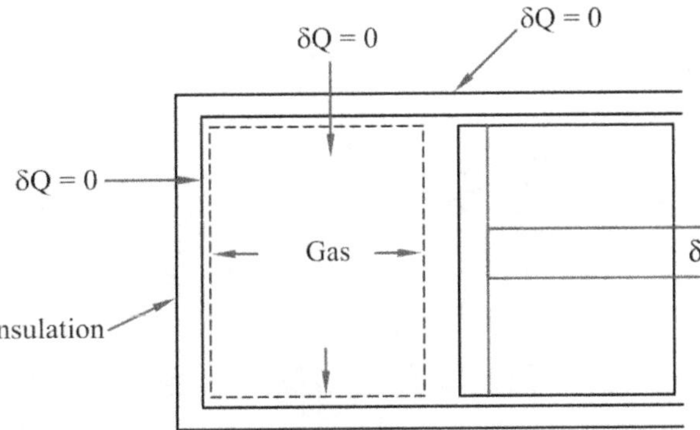

In this process the work is done at the expense of its own internal energy. Hence the temperature will fall during an adiabatic expansion and will rise during an adiabatic compression.

Exercise – 1

Derive the equation for $PV^\gamma = C$ constant for an adiabatic process.

By first law of thermodynamics for unit mass i.e. m = 1 kg

$$\delta q = \delta w + \delta u = 0$$

No heat transfer as for adiabatic process $\delta q = 0$

$$\therefore \quad P dv + C_v\, dT = 0 \qquad \text{....(i)}$$

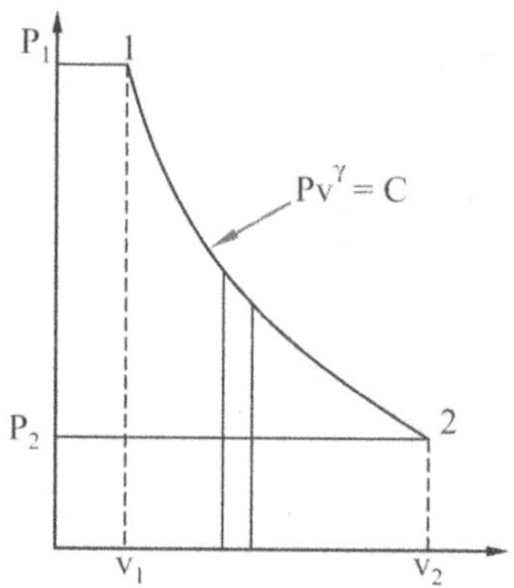

Now equation of state for m = 1 kg

$$PV = RT$$

$\therefore$ $P = \dfrac{RT}{V}$ put value of 'p' in eq (i) we get,

$$RT\,\frac{dv}{V} + C_v\,dT = 0$$

Divide by ($C_v \times T$)

$$\frac{R}{C_V}\int_1^2 \frac{dv}{v} + \int_1^2 \frac{dT}{T} = \int 0 \qquad \frac{C_p}{C_v} - \frac{C_v}{C_v} = \frac{R}{C_v} \qquad (\gamma - 1) = \frac{R}{C_V}$$

$$\frac{R}{C_v}\log V + \log T = \log C$$

$$(\gamma - 1)\log V + \log T = \log C$$

$$\therefore\ \log T\,V^{\gamma-1} = \log C$$

$$\text{i.e., } T \times V^{\gamma-1} = C \qquad \text{.....(1)}$$

Put $PV = RT$

$\therefore$ $T = \dfrac{PV}{R}$

$$\therefore \quad \frac{PV}{R} \times V^{\gamma-1} = C$$

$$\therefore \quad PV^{1+\gamma-1} = RC = \text{constant}$$

$$\therefore \quad \mathbf{PV^{\gamma} = C} \qquad \text{This is adiabatic law}$$

$$P_1V_1^{\gamma} = P_2V_2^{\gamma} = C....$$

(a) Representation on p-V diagram

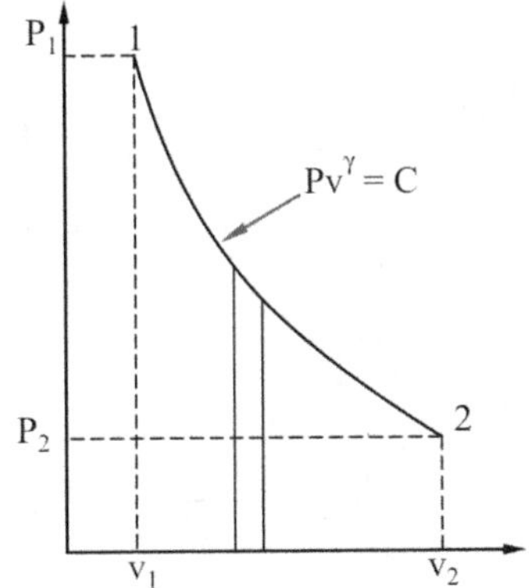

Adiabatic expression $P_1V_1^{\gamma} = P_2V_2^{\gamma} = C$

(b) Work done $= \int_{V_1}^{V_2} P dv$

where $PV^{\gamma} = C$

$$\therefore \quad P = \frac{C}{V^r} = CV^{-\gamma}$$

$$\therefore \quad W.D = C\int_{V_1}^{V_2} V^{-\gamma} dv$$

$$= C\left[\frac{V^{-\gamma+1}}{-\gamma+1}\right]_{V_1}^{V_2} = \left[\frac{V^{-\gamma+1}}{-\gamma+1}\right]_{V_2}^{V_2}$$

$$= C\left[\frac{V_2^{1-\gamma} - V_1^{1-\gamma}}{1-\gamma}\right]$$

$\therefore$ $$W.D = C\left[\frac{V_1^{1-\gamma} - V_2^{1-\gamma}}{\gamma - 1}\right]$$

let $$C = P_1V_1^{\gamma} = P_2V_2^{\gamma}$$

$\therefore$ $$W = \frac{V_1^{\gamma} \times P_1V_1^{1-\gamma} - P_2V_2^{1-\gamma} \times V_2^{\gamma}}{\gamma - 1}$$

$$W.D. = \frac{P_1V_1 - P_2V_2}{\gamma - 1}$$

$$W.D = \frac{R(T_1 - T_2)}{\gamma - 1} \qquad [\because P_1V_1 = mRT_1 : P_2V_2 = mRT_2]$$

and m = 1 kg

$$= \frac{(C_p - C_v)(T_1 - T_2)}{\gamma - 1}$$

$$= \frac{(C_p - C_v)(T_1 - T_2)}{\frac{C_p}{C_v} - 1} \qquad \text{Put } \gamma = \frac{C_p}{C_v}$$

$$= \frac{(C_p - C_V)(T_1 - T_2)C_v}{(C_p - C_v)} \qquad \delta q = \delta u + \delta w \text{ Since } \delta q = 0\ \delta w = \delta u$$

$$= C_v (T_1 - T_2)$$

$$= - C_v (T_2 - T_1)$$

$$= - \delta u \qquad \delta w = - \delta u$$

Work done W.D = decrease in I.E.

Relation between P, V and T

$$PV^{\gamma} = C$$

For two points on curve

$\therefore$ (i) $$P_1V_1^{\gamma} = P_2V_2^{\gamma}$$

$\therefore$ $$\left(\frac{P_1}{P_2}\right) = \left(\frac{V_2}{V_1}\right)^{\gamma}$$

$$\frac{V_2}{V_1} = \left(\frac{P_1}{P_2}\right)^{\frac{1}{\gamma}} \quad \text{....(1)}$$ Relation between put and V

(ii) $$\frac{P_1V_1}{T_1} = \frac{P_2V_2}{T_2}$$

$$\frac{T_2}{T_1} = \frac{P_2V_2}{P_1V_1}$$ By putting value of $\frac{V_2}{V_1}$ from eq. (i)

$$= \frac{P_2}{P_1} \cdot \left(\frac{P_1}{P_2}\right)^{\frac{1}{\gamma}}$$

$$= \frac{P_2^{1-\frac{1}{\gamma}}}{P_1^{1-\frac{1}{\gamma}}}$$

$$\frac{T_2}{T_1} = \left(\frac{P_2}{P_1}\right)^{\frac{\gamma-1}{\gamma}}$$ relation between P and T (ii)

(iii) $$\frac{P_1V_1}{T_1} = \frac{P_2V_2}{T_2}$$

$$\frac{T_2}{T_1} = \frac{P_2}{P_1} \cdot \frac{V_2}{V_1} \quad \text{.....(a)}$$

But $$P_1V_1^{\gamma} = P_2V_2^{\gamma}$$

$$\frac{P_2}{P_1} - \left(\frac{V_1}{V_2}\right)^{\gamma}$$

$\therefore$ $$\frac{T_2}{T_1} = \left(\frac{V_1}{V_2}\right)^{\gamma} \cdot \frac{V_2}{V_1}$$

$$= \frac{V_1^{\gamma} . V_1^{-1}}{V_2^{\gamma} . V_2^{-1}}$$

$$\frac{T_2}{T_1} = \left(\frac{V_1}{V_2}\right)^{\gamma-1}$$ relation between V and T ....(iii)

$$\frac{\text{Initial temperature}}{\text{Final temperature}} = \left[\frac{\text{Final volume}}{\text{Initial volume}}\right]^{\gamma-1} = \left[\frac{\text{Initial pressure}}{\text{Final pressure}}\right]^{\frac{\gamma-1}{\gamma}}$$

(c) Change in Internal Energy

By first law

$$\delta Q : \delta w + \delta u \qquad \text{since } \delta Q = 0 \text{ for adiabatic process}$$

$$\delta w = -\,\delta u$$

In adiabatic process, change in internal energy is numerically equal to the work done by the gas

OR

$$\delta u = m\, C_v\, (T_2 - T_1) \qquad \frac{C_p}{C_v} - \frac{C_v}{C_v} = \frac{R}{C_v}$$

But

$$\delta u = \frac{m\,.\,R}{(\gamma - 1)}\,(T_2 - T_1) \qquad (\gamma - 1) = \frac{R}{C_v}$$

$$= \frac{m\,RT_2 - mRT_1}{(\gamma - 1)} \qquad \therefore\ C_v = \frac{R}{(\gamma - 1)}$$

$$= \frac{P_2V_2 - P_1V_1}{\gamma - 1}$$

$$= -\left(\frac{P_1V_1 - P_2V_2}{\gamma - 1}\right)$$

$$\delta u = -\text{ work done} \qquad \delta u = -\,\delta w$$

(d) Heat transferred

During the adiabatic process, the heat transfer is zero

(e) Change in enthalpy

$$dH = m\, C_p\, (T_2 - T_1)$$

3.13.5 Polytropic process

(a) Introduction

Changes of state of working substance in thermodynamic system are often brought about by the expansion or compression of the working substance.

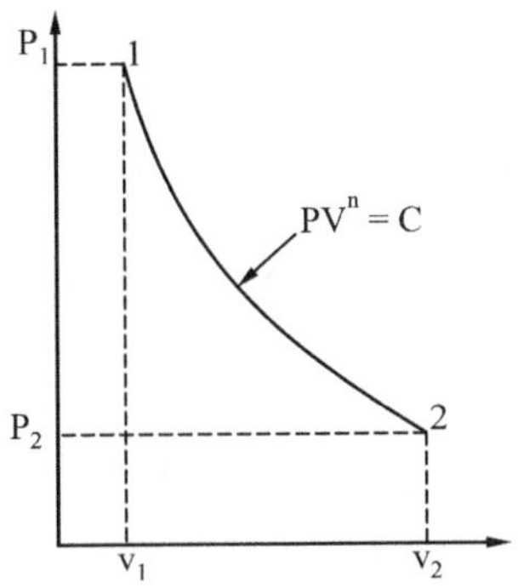

Let working substance (gas) may be expanded from state '1' to state '2' as shown in Fig. The expansion process is governed by polytrophic law i.e. $PV^n = C$.

Where 'n' is polytrophic index. During this process, heat exchange between system and surrounding is also permitted.

(b) Relation between P, V and T between two states (replace 'n' in place of 'r').

The polytrophic law is $PV^n = C$ where value of n varies between r and 1.

If we replace r of an adiabatic process by 'n' for the polytrophic process, we get the relations between pressure, volume and temperature for polytrophic process.

Thus we have

$$\frac{T_2}{T_1} = \left(\frac{P_2}{P_1}\right)^{\frac{n-1}{n}}, \quad \frac{T_2}{T_1} = \left(\frac{V_1}{V_2}\right)^{n-1}, \quad \frac{P_2}{P_1} = \left(\frac{V_1}{V_2}\right)^{n}$$

(c) Representation on p-V diagram

Many processes in actual practice follow the equation $PV^n = C$

where n = constant. Such processes are called polytrophic processes.

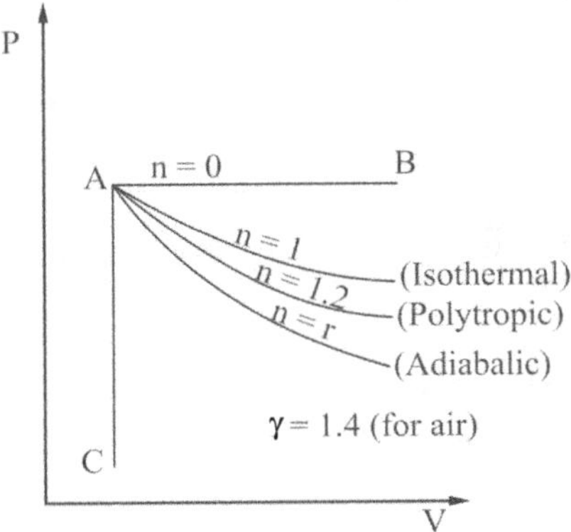

(i) when n = 0

$$\therefore \quad \frac{P_2}{P_1} = \left(\frac{V_1}{V_2}\right)^n = \left(\frac{V_1}{V_2}\right)^o = 1$$

$\therefore$ $P_2 = P_1$ ..... (AB) represents constant pressure process.

(ii) When n = 1

$$\frac{T_2}{T_1} = \left(\frac{V_1}{V_2}\right)^{n-1} = \left(\frac{V_1}{V_2}\right)^{1-1} = \left(\frac{V_1}{V_2}\right)^o = 1$$

$$T_2 = T_2$$

The process becomes isothermal process

(iii) When n = γ

$$\frac{P_2}{P_1} = \left(\frac{V_1}{V_2}\right)^n = \left(\frac{V_1}{V_2}\right)^\gamma$$

The process becomes adiabatic process

(iv) When n = ∞

$$P_1V_1^n = P_2V_2^n$$

$$\frac{P_2}{P_1} = \left(\frac{V_1}{V_2}\right)^n \quad \therefore \frac{V_1}{V_2} = \left(\frac{P_2}{P_1}\right)^{\frac{1}{n}} = \left(\frac{P_2}{P_1}\right)^{\frac{1}{\alpha}} = \left(\frac{P_2}{P_1}\right)^o = 1$$

The process becomes constant volume process and it will be represented by line AC on p-v diagram.

(d) Work done during process:

$$w = \int_1^2 pdv \qquad \text{for polytrophic process } PV^n = C$$

$$= C\int_1^2 \frac{dv}{V^n} \qquad \text{i.e., } P_1V_1^n = P_2V_2^n = \ldots\ldots = C$$

$$= C\int_1^2 V^{-n}dv \qquad \therefore P = \frac{C}{V^n}$$

$$= C\left[\frac{V^{-n+1}}{-n+1}\right]^2$$

$$= C\left[\frac{V_2^{-n+1} - V_1^{-n+1}}{-n+1}\right]$$

$$= \frac{CV_2^{-n+1} - CV_1^{-n+1}}{-n+1} \quad \text{Put } C = P_2V_2^n = P_1V_1^n \text{ appropriate value of C}$$

$$= \frac{P_2V_2^n.V_2^{-n+1} - P_1V_1^n.V_1^{-n+1}}{-n+1}$$

$$= \frac{(P_2V_2 - P_1V_1)\times -1}{(-n+1)\times -1} \qquad \text{By changing sign}$$

$$\mathbf{w_{12}} = \frac{P_1V_1 - P_2V_2}{n-1}$$

$$w_{12} = \frac{mR\,(T_1 - T_2)}{n-1}$$

(e) Change in internal energy

$$\delta u = m\,C_v\,(T_2 - T_1) \qquad \text{always for all processes}$$

(f) Heat transfer during process

$$Q = \delta u + w$$

$$= m\,C_v\,(T_2 - T_1) + \frac{P_1V_1 - P_2V_2}{n-1}$$

$$\frac{C_p}{C_v} - \frac{C_v}{C_v} = \frac{R}{C_v}$$

Now $C_v = \dfrac{R}{r-1}$

$$(\gamma - 1) = \frac{R}{C_v}$$

$$= \frac{m\,R\,(T_2 - T_1)}{\gamma\text{-}1} + \frac{P_1V_1 - P_2V_2}{n-1}$$

$$C_v = \frac{R}{(r-1)}$$

$$= \frac{P_2V_2 - P_1V_1}{\gamma - 1} + \frac{P_1V_1 - P_2V_2}{n-1}$$

$$= \frac{P_1V_1 - P_2V_2}{n-1} - \frac{P_1V_1 - P_2V_2}{\gamma - 1}$$

$$= (P_1V_1 - P_2V_2)\left[\frac{1}{n-1} - \frac{1}{r-1}\right]$$

$$= (P_1V_1 - P_2V_2)\left[\frac{\gamma-1-n+1}{(n-1)\,(r-1)}\right]$$

$$= \left(\frac{\gamma-n}{\gamma-1}\right)\left(\frac{P_1V_1 - P_2V_2}{n-1}\right)$$

$$\therefore \qquad Q = \left\{\frac{\gamma-n}{\gamma-1}\right\} \times \text{work done during process.}$$

3.14 Index of Compression or Expansions

If system changes in states from 1 to 2, the index of expansion or compression can be calculated as follows.

The polytrophic law can be written as $P_1V_1^n = P_2V_2^n$ $\quad \therefore \quad \frac{P_1}{P_2} = \left(\frac{V_2}{V_1}\right)^n$

If we take logarithm's on both sides

$$ln\left(\frac{P_1}{P_2}\right) = ln\left(\frac{V_2}{V_1}\right)^n,$$

$$\therefore \qquad ln\left(\frac{P_1}{P_2}\right) = n\, ln\left(\frac{V_2}{V_1}\right)$$

$$\therefore \qquad n = \frac{ln\left(\frac{P_1}{P_2}\right)}{ln\left(\frac{V_2}{V_1}\right)}$$

To prove that for any heat engine process using a gas as the working substance, the change in enthalpy is given by

$$\Delta H = \frac{\gamma}{\gamma-1}\,[P_2V_2 - P_1V_1] \text{ with usual notations}$$

We know, during any process, change in enthalpy is given by

$$\Delta H = m\,C_p\,(T_2 - T_1) \qquad \frac{C_p}{C_p} - \frac{C_v}{C_p} = \frac{R}{C_p}$$

$$\therefore \quad C_p = \left(\frac{\gamma}{\gamma - 1}\right) R \qquad 1 - \frac{1}{\gamma} = \frac{R}{C_p}$$

$$\therefore \quad \Delta H = \frac{\gamma\, mR\,(T_2 - T_1)}{\gamma - 1} \qquad \therefore\ \frac{\gamma - 1}{\gamma} = \frac{R}{C_p} \quad \therefore\ C_p = \frac{R.\gamma}{\gamma - 1}$$

$$\Delta H = \frac{\gamma}{\gamma - 1}\,[P_2V_2 - P_1V_1]$$

To prove that for polytrophic process with usual notation.

Heat transferred $= \dfrac{n - \gamma}{n - 1}\; m\,C_v\,(T_2 - T_1)$

By first law of thermodynamics

$$\therefore \quad \delta Q = \delta u + \delta w$$

We know that for polytrophic process

(i) $\delta u = m\,C_v\,(T_2 - T_1)$

(ii) $\delta w = \dfrac{P_1V_1 - P_2V_2}{\gamma - 1}$

$$= m\,C_v\,(T_2 - T_1) + \frac{P_1V_1 - P_2V_2}{n - 1}$$

$$= m\,C_v\,(T_2 - T_1) + \frac{mR\,(T_1 - T_2)}{n - 1} \qquad (\because PV = mRT)$$

But

$$\frac{C_p}{C_v} - \frac{C_v}{C_v} = \frac{R}{C_v}$$

$$r - 1 = \frac{R}{C_v}$$

$$\therefore \quad R = C_v\,(\gamma - 1)$$

$$\delta Q = m\,C_v\,(T_2 - T_1) + \frac{m\,C_v\,(\gamma - 1)\,(T_1 - T_2)}{n - 1}$$

$$\delta Q = m\,C_v\,(T_2 - T_1) - \frac{m\,C_v\,(\gamma-1)\,(T_2-T_1)}{n-1}$$

$$= m\,C_v\,(T_2 - T_1)\left[1-\frac{\gamma-1}{n-1}\right]$$

$$= m\,C_v\,(T_2 - T_1)\left[\frac{n-1-\gamma+1}{n-1}\right]$$

$$dQ = \left(\frac{n-\gamma}{n-1}\right) \times m\,C_v\,(T_2 - T_1)$$

Solved Problems

1. A tank contains 3 m^3 of air at 25 bar absolute pressure. This air is cooled until the pressure and temperature decreases to 15 bar and 21 °C respectively. Determine the following.

   (i) Change in internal energy

   (ii) Change in enthalpy

   (iii) Heat transferred during the process

   Take $C_p = 1.005$ kJ/kg K and $C_v = 0.718$ kJ/kg K for air

Solution: Since air is contained in a tank it is a constant volume process

Given $V_1 = V_2 = 3\ m^3$

$P_1 = 25$ bar $P_2 = 15$ bar

$T_2 = 21°C + 273 = 294$ K

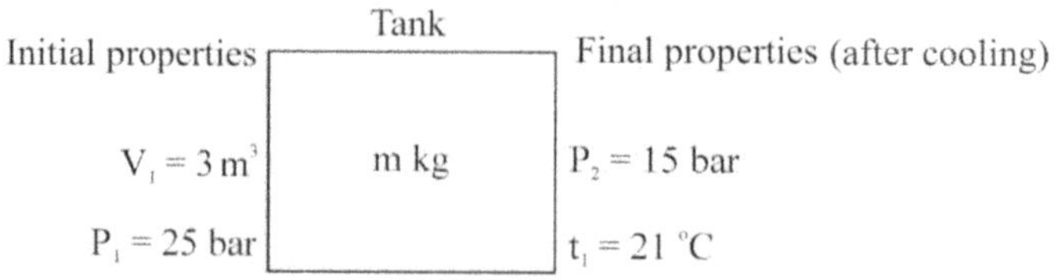

Initial properties	Tank	Final properties (after cooling)
$V_1 = 3\ m^3$	m kg	$P_2 = 15$ bar
$P_1 = 25$ bar		$t_1 = 21\ °C$

(i) change in internal energy

$$\delta u = m\,C_v\,(T_2 - T_1)$$

$\therefore$ $m = ?$ $T_1 = ?$

Apply for constant volume process

$$\frac{P_1}{T_1} = \frac{P_2}{T_2} \qquad \therefore T_1 = T_2 \times \frac{P_1}{P_2}$$

$$\frac{P_1}{P_2} = \frac{T_1}{T_2} \qquad = 294 \times \frac{25}{15}$$

$$= 490 \text{ K}$$

mass of air in the tank can be calculated using characteristic gas equation.

i.e., $PV = mRT \qquad \therefore R = C_p - C_v$

$$m = \frac{PV}{RT} \qquad = 1.005 - 0.718$$

$$= 0.287 \text{ kJ/kg K.}$$

P_1 should be converted in kN/m^2 unit

$$P_1 = 25 \text{ bar}$$

$$= \frac{25 \times 10^5}{10^3} \text{ N/m}^3$$

$$= 2500 \text{ KN/m}^2$$

$$\therefore \qquad m = \frac{P_1V_1}{RT_1} = \frac{2500 \times 3}{0.287 \times 490} = 53.33 \text{ kg}$$

Change in internal energy

$$\delta u = m\, C_v\, (T_2 - T_1)$$

$$= 53.33 \times 0.718\, (294 - 490)$$

$$= -\,7505 \text{ kJ (decreases)}$$ **Ans.**

Change in enthalpy

$$\delta H = m\, C_p\, (T_2 - T_1)$$

$$= 53.33 \times 1.005\, (294 - 490)$$

$$= -\,10504.9 \text{ kJ (decreases)}$$ **Ans.**

Since it is a constant volume process. $\therefore w = 0$

From first law of thermodynamics

Heat exchanged $\delta Q = \delta u + \delta w$

$= -7505 \qquad \delta w = 0 \qquad \because dV = 0$

$= -7505$ kJ

i.e., system rejects 7505 kJ of heat during cooling process. **Ans.**

2. A gas whose pressure, volume and temperature are 2.75 bar, 0.09 m^3 and 185 °C respectively, has the state changed at constant pressure, until its temperature becomes 15 °C. Calculate,

(i) Heat transferred

(ii) Work done during the process

(iii) Change in internal energy.

Take R = 0.29 kJ/kg K and C_p = 1.005 kJ/kg K for gas

Solution: Given

Initial		Final
$P_1 = 2.75$ bar	**constant pressure process**	$P_2 = P_1$
$V_1 = 0.09\ m^3$	⟶	$V_2 = ?$
$T_1 = 185 + 273$		$T_2 = 15 + 273$
$= 458$ K		$= 288$ K

It is a constant pressure process.

R = 0.29 kJ/kg K, Cp = 1.005 kJ/kg K

$\therefore \qquad C_p - C_v = R$

$\therefore \qquad C_v = C_p - R$

$= 1.005 - 0.29$

$= 0.715$ kJ/kg K

mass can be calculated using characteristic gas equation.

i.e., $\qquad P_1V_1 = mRT, \qquad P_1 = 2.75 \times 10^2\ KN/m^2$

$$\therefore \qquad m = \frac{P_1V_1}{RT_1} = \frac{2.75 \times 10^2 \times 0.09}{0.29 \times 458} = 0.186\ kg.$$

For constant pressure process.

$$\frac{V_1}{T_1} = \frac{V_2}{T_2}$$

$$\therefore \qquad V_2 = T_2 \times \frac{V_1}{T_1} = 288 \times \frac{0.09}{458}$$

$$= 0.057 \text{ m}^3$$

$\therefore$ work done $= P(V_2 - V_1)$

$$= 2.75 \times 10^2 (0.057 - 0.09)$$

$$= -9.075 \text{ kJ} \qquad (-\text{ve work})$$

i.e., work is done on the system.

Heat transferred during the process

$$\delta Q = m\, C_p (T_2 - T_1)$$

$$= 0.186 \times 1.005 (288 - 458)$$

$$= -31.75 \text{ kJ}$$

i.e., 31.75 kJ of heat is rejected by the system. **Ans.**

$$\delta u = m\, C_v (T_2 - T_1)$$

$$= 0.186 \times 0.715 (288 - 458)$$

$$= -22.61 \text{ kJ} \qquad \text{(decreases)}$$ **Ans.**

Also from first law of thermodynamics

$$Q = \Delta u + w$$

$$\therefore \qquad \Delta u = Q - w$$

$$= -31.75 + 9.08$$

$$= -22.67 \text{ kJ} \qquad \text{(decreases)}$$ **Ans.**

3. 0.67 kg of gas at 14 bar and 29°C is expanded to four times the original volume according to the law $PV^{1.3}$ = constant.

Calculate:

(i) The original and final volume of the gas

(ii) The final temperature of the gas

(iii) The final pressure of gas

Take R = 0.287 kJ/kg K for gas

Solution: Given

$$m = 0.67 \text{ kg} \qquad R = 0.287 \text{ kJ/kg K}$$

$$P_1 = 14 \text{ bar} \times 10^2 \text{ KN/m}^2 \; PV^{1.3} = C$$

$$T_1 = 290°C + 273 = 563 \text{ K}$$

Initial volume can be calculated by using equation of state

i.e., $P_1V_1 = m\,RT_1$

$$V_1 = \frac{m\,RT_1}{P_1} = \frac{0.67 \times 0.287 \times 563}{14 \times 10^2}$$

$$= 0.077 \text{ m}^3 \qquad \textbf{Ans.}$$

$$V_2 = 4 \times V_1 = 0.308 \text{ m}^3 \quad \textbf{Ans.}$$

For polytrophic process, we have

(i) $$\frac{P_2}{P_1} = \left(\frac{V_1}{V_2}\right)^n \qquad \therefore \; P_2 = P_1\left(\frac{V_1}{V_2}\right)^n$$

$$= 14\left(\frac{1}{4}\right)^{1.3}$$

$$= 2.31 \text{ bar}$$

(ii) $$\frac{T_2}{T_1} = \left(\frac{V_1}{V_2}\right)^{n-1} \qquad \therefore \; T_2 = T_1\left(\frac{V_1}{V_2}\right)^{n-1}$$

$$= 563\left(\frac{1}{4}\right)^{0.3}$$

$$= 371.44 \text{ K} \qquad \textbf{Ans.}$$

4. A gas initially at 5.1 bar and volume 142 litres undergoes a process. The new pressure and volume are 1.7 bar and 275 litres. The enthalpy of gas decreases by 65 kJ. Taking $C_V = 0.718$ kJ/kg K, calculate:

(i) Change in internal energy

(ii) C_p and R for gas

(iii) The final pressure of gas

Solution: Given

Initial Final $1m^3 = 1000$ litres

$P_1 = 5.1 \text{ bar} \times 10^2$ $P_2 = 1.7 \text{ bar} \times 10^2$

$V_1 = 142$ litres $V_2 = 275$ litres

$= 0.142 \text{ m}^3$ $= 0.275 \text{ m}^3$.

$$\Delta H = H_2 - H_1 = -65 \text{kJ}$$

We can write for state 1 and 2

$$H_1 = u_1 + P_1V_1$$

$$H_2 = u_2 + P_2V_2$$

$\therefore$ $(H_2 - H_1) = u_2 - u_1 + (P_2V_2 - P_1V_1)$

$$-65 = (u_2 - u_1) + \{170 \times 0.275 - 510 \times 0.142\}$$

$(u2 - u1) = -39.33$ kJ (decreased) **Ans.**

Now $\Delta H = (H_2 - H_1) = m\, C_p\, (T_2 - T_1)$ and

$$\Delta u = (u_2 - u_1) = m\, C_v\, (T_2 - T_1)$$

$$\therefore \quad \frac{\Delta H}{\Delta u} = \frac{H_2 - H_1}{u_2 - u_1} = \frac{m\, Cp\, (T_2 - T_1)}{m\, C_v\, (T_2 - T_1)}$$

$$\therefore \quad \gamma = \frac{H_2 - H_1}{u_2 - u_1} = \frac{-65}{-39.33} = 1.652$$

$\gamma = \dfrac{C_p}{C_v}$ $\therefore C_p = r \times C_v = 1.652 \times 0.718 = 1.186$ kJ/kg K **Ans.**

and $R = C_p - C_v$

$$= 1.186 - 0.718$$

$= 0.468$ kJ/kg K **Ans.**

5. 1 kg of gas expands according to the law $PV^{1.3} = C$. The initial and final temperatures of the gas are 300 °C and 200 °C respectively. The work done is 90 kJ. Calculate molecular mass of the gas if $R_o = 8.3143$ kJ/kg mole K.

Solution: Given

$m = 1$ kg $PV^{1.3} = C$ $w = 90$ kJ.

Initial Final

$T_1 = 300\ °C + 273 = 573$ K $T_2 = 200\ °C + 273 = 473$ K

Work done for polytropic process is given by

$$w = \frac{P_1V_1 - P_2V_2}{n-1} = \frac{mR\,(T_1 - T_2)}{n-1}$$

$$90 = \frac{1 \times R \times (300 - 200)}{1.3 - 1}$$

$\therefore$ $R = 0.27$ kJ/kg K **Ans.**

$R_o = M \times R$

$\therefore$ $M = \frac{R_o}{R} = \frac{8.3143}{0.27} = 30.8$ kg/kg mole. **Ans.**

6. A steel cylinder contains O_2 at a pressure of 25 bar and temperature of 27 °C. After using some quantity of gas, the pressure is found to be 5 bar and temperature of 20 °C. 700 litres of O_2 was originally put in the cylinder at NTP. Density of O_2 at NTP = 1.43 gm/litres. Find out the mass of O_2 is used.

Solution: Given

Initial	Final
$P_1 = 25$ bar	$P_2 = 5$ bar
$T_1 = 27°C + 273$	$T_2 = 20°C + 273$
$T_1 = 300$ °K	$T_2 = 293$ K
$V_1 = 700$ litres	
(at NTP) = $0.7\ m^3$	

ρ_o at NTP = 1.43 gm/litres = 1.43 kg/m^3

$$1 \text{ gm/liter} = \frac{\frac{kg}{1000}\;\frac{1}{m^3}}{1000} = 1 \text{ kg/m}^3$$

$\therefore$ Equation of state at NTP

$P_oV_o = m\,RT_o$ $P_o = 1.01325$ bar $\rho_o = \frac{m}{V_o}$

$T_o = 0$ °C = 273 K

$$R = \frac{P_o}{\rho_o T_o} = \frac{1.01325 \times 10^2}{1.43 \times 273}$$

$= 0.26$ kJ/kg K

At initial condition (at NTP)

$$P_oV_o = m_oRT_o$$

Initial mass $m_1 = \dfrac{P_oV_o}{RT_o} = \dfrac{101.325 \times 0.7}{0.26 \times 273} = 0.999 = 1 \text{ kg}$

$$P_1V_1 = m\, RT_1$$

Volume of tank $V_1 = \dfrac{mRT_1}{P_1}$

$$= \frac{1 \times 0.26 \times 300}{25 \times 10^2}$$

$$= 0.0312 \text{ m}^3$$

$$V_2 = V_1 \quad \text{constant volume}$$

Final mass, $P_2V_2 = m_2\, RT_2$

$$m_2 = \frac{P_2V_2}{RT_2} = \frac{5 \times 10^2 \times 0.0312}{0.26 \times 293}$$

$$= 0.205 \text{ kg}$$

∴ Mass of gas used

$$= m_1 - m_2$$

$$= 1 - 0.205$$

$$= 0.792 \text{ kg} \qquad \textbf{Ans.}$$

7. Initial volume of 0.18 kg of certain gas was 0.15m^3 at a temperature 15°C and pressure 1 bar. After adiabatic compression to 0.056 m^3. The pressure was found to be 4 bar. Find

(i) Gas constant R

(ii) Molecular mass

(iii) Adiabatic index 'γ'

(iv) C_P and C_V

(v) Change in internal energy

Solution: Given

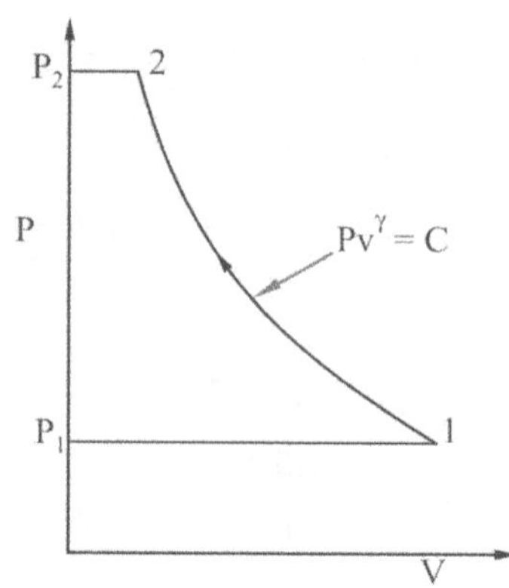

Initial	Final
$m = 0.18$ kg	$V_2 = 0.056\ m^3$
$V_1 = 0.15\ m^3$	$P_2 = 4$ bar

$$T_1 = 15 + 273$$

$$= 288\ K$$

$$P_1 = 1\ \text{bar}$$

Using equation of state at inpal point 1

$$P_1V_1 = m\,RT_1$$

$$R = \frac{P_1V_1}{mT_1} = \frac{1 \times 10^2 \times 0.15}{0.18 \times 288}$$

$$= 0.289\ \text{kJ/kg K} \qquad \textbf{Ans.}$$

Molecular mass, $R_o = M \times R$

$$M = \frac{R_o}{R} = \frac{8.3143}{0.289} = 28.77\ \text{kg/kg mole} \qquad \textbf{Ans.}$$

For adiabatic process $\gamma = ?$

$$P_1V_1^{\gamma} = P_2V_2^{\gamma}$$

$$\frac{P_2}{P_1} = \left(\frac{V_1}{V_2}\right)^{\gamma}$$

$$ln\left(\frac{P_2}{P_1}\right) = \gamma \times ln\left(\frac{V_1}{V_2}\right)$$

$$\gamma = \frac{ln\left(\frac{P_2}{P_1}\right)}{ln\left(\frac{V_1}{V_2}\right)} = \frac{ln\left(\frac{4}{1}\right)}{ln\left(\frac{0.15}{0.056}\right)} = 1.41 \qquad \textbf{Ans.}$$

$$R = C_p - C_v = 0.289$$

Also $$\frac{C_p}{C_v} = \gamma = 1.41$$

$$C_p = 1.41\ C_v \qquad C_v = 0.705 \text{ kJ/kg }^oK$$

$$1.41\ C_v - C_v = 0.289$$

$$C_v \times 0.41 = 0.289$$

$$C_p - C_v = R$$

$$C_p = C_v + R$$

$$= 0.704 + 0.289$$

$$= 0.994$$

$$= 1 \text{ kJ/kg K}$$

$$\frac{T_2}{T_1} = \left(\frac{P_2}{P_1}\right)^{\frac{\gamma-1}{\gamma}} = (4)^{\frac{0.41}{1.4}}$$

$$T_2 = 431 \text{ K}$$

$$u_2 - u_1 = m\ C_v\ (T_2 - T_1)$$

$$= 0.18 \times 0.704\ (431 - 288)$$

$$\Delta u = 18.12 \text{ kJ} \qquad \textbf{Ans.}$$

8. For a polytropic process derive

$$W = \frac{P_1V_1}{n-1}\left[1 - \frac{1}{\gamma^{n-1}}\right] \text{ where } \gamma = \frac{V_2}{V_1}$$

Solution: For a polytropic process, work done can be expressed as,

$$W = \frac{P_1V_1 - P_2V_2}{n-1}$$

$$= \frac{P_1V_1}{n-1}\left[1 - \frac{P_2}{P_1}\frac{V_2}{V_1}\right] \qquad \text{Put } \frac{P_2}{P_1} = \left(\frac{V_1}{V_2}\right)^n$$

$$= \frac{P_1V_1}{n-1}\left[1-\left(\frac{V_1}{V_2}\right)^{n}\left(\frac{V_1}{V_2}\right)^{-1}\right] \qquad \text{But given } \frac{V_2}{V_1}=\gamma$$

$$= \frac{P_1V_1}{n-1}\left[1-\left(\frac{V_1}{V_2}\right)^{n-1}\right]$$

$$= \frac{P_1V_1}{n-1}\left[1-\left(\frac{1}{\gamma^{n-1}}\right)\right]$$

9. Derive the equation of polytropic specific heat, with usual notations.

$$C_n = C_v \left\{\frac{n-\gamma}{\gamma-1}\right\}$$

Solution: From first law of thermodynamics for polytropic process

$$\delta Q = \Delta u + \delta w$$

$$= m\,C_v\,(T_2 - T_1) + \frac{m\,R\,(T_1 - T_2)}{n-1}$$

$$= m\,C_v\,(T_2 - T_1) - \frac{m\,R\,(T_2 - T_1)}{n-1}$$

But $\frac{C_p}{C_v} - \frac{C_v}{C_v} = \frac{R}{C_v}$

$$(\gamma - 1) = \frac{R}{C_v}$$

Put $R = C_v\,(\gamma - 1)$

$$\delta Q = m\,C_v\,(T_2 - T_1) - m\,C_v\,\frac{(\gamma-1)\,(T_2-T_1)}{n-1}$$

$$= m\,C_v\,(T_2 - T_1)\left\{1-\left(\frac{\gamma-1}{n-1}\right)\right\}$$

$$= m\,C_v\,(T_2 - T_1)\left\{\frac{n-1-\gamma+1}{n-1}\right\}$$

$$= m\,C_v\left(\frac{n-\gamma}{n-1}\right)(T_2 - T_1) \qquad \text{.....(1)}$$

But heat transfer to raise temperature of m kg of gas from T_1 to T_2 for polytropic process can be written as

$$Q = m\, C_n\, (T_2 - T_1) \qquad .....(2)$$

where C_n = polytropic specific heat

Comparing 1 and 2 we get,

$$C_n = C_v \left[\frac{n-\gamma}{n-1}\right] \qquad \textbf{Ans.}$$

10. Prove the following relations when gas is compressed according to the law $PV^n = C$ when n = polytropic index.

(i) $\dfrac{\text{Heat rejected}}{\text{Change in internal energy}} = \dfrac{n-\gamma}{n-1}$,

(ii) $\dfrac{\text{Work done}}{\text{Change in enthalpy}} = \dfrac{\gamma-1}{\gamma\,(1-n)}$,

(iii) Hence or otherwise prove that $\dfrac{\text{Change in enthalpy}}{\text{Change in internal energy}} = \gamma$

Solution: For polytrophic process

(i) To prove: $\dfrac{Q}{\Delta u} = \dfrac{n-\gamma}{n-1}$ .....(1)

We have $Q = \dfrac{n-\gamma}{n-1} \times$ work done

$$= \frac{n-\gamma}{n-1} \times \frac{mR\,(T_1 - T_2)}{n-1} \text{ and}$$

$$\Delta u = m\, C_v\, (T_2 - T_1)$$

$$\frac{Q}{\Delta u} = \frac{\dfrac{\gamma-n}{\gamma-1} \times \dfrac{m\,R\,(T_1 - T_2)}{n-1}}{m\,C_v\,(T_2 - T_1)}$$

$$= \frac{\gamma-n}{\gamma-1} \times \frac{R}{C_v} \times \frac{1}{(1-n)}$$

Also we have $\dfrac{C_p}{C_v} - \dfrac{C_v}{C_v} = \dfrac{R}{C_v}$

$$(\gamma - 1) = \frac{R}{C_v}$$

Putting in equation (1) we get

$$\frac{Q}{\Delta u} = \frac{\gamma - n}{\gamma - 1} \times \frac{(\gamma - 1)}{1} \times \frac{1}{(1-n)}$$

$$\mathbf{\frac{Q}{\Delta u} = \frac{n - \gamma}{n - 1}}$$

(ii) To prove $\frac{W}{\Delta H} = \frac{\gamma - 1}{\gamma\ (1-n)}$ .....(2)

We have, $W = \frac{m R (T_1 - T_2)}{n - 1}$

$$\Delta H = m\, C_p\, (T_2 - T_1)$$

$$\therefore \quad \frac{W}{\Delta H} = \frac{\frac{m R (T_1 - T_2)}{n-1}}{m C_p (T_2 - T_1)}$$

$$= \frac{R}{(1-n) C_p} \qquad \frac{C_p}{C_p} - \frac{C_v}{C_p} = \frac{R}{C_p}$$

$$= \frac{1}{(1-n)} \times \frac{\gamma - 1}{\gamma} \qquad 1 - \frac{1}{\gamma} = \frac{R}{C_p}$$

$$= \frac{(\gamma - 1)}{\gamma\ (1-n)} \qquad \frac{\gamma - 1}{\gamma} = \frac{R}{C_p}$$

(iii) We have $\Delta H = m\, C_p\, (T_2 - T_1)$ $\qquad \therefore R = \frac{C_p\ (\gamma - 1)}{\gamma}$

$$\Delta u = m\, C_v\, (T_2 - T_1)$$

$$\therefore \quad \frac{\Delta H}{\Delta u} = \frac{C_p}{C_v} = \gamma$$

Also $\frac{Q}{\Delta u} = \frac{n - \gamma}{n - 1}$

and $\frac{W}{\Delta u} = \left(\frac{\gamma - 1}{\gamma (1-n)}\right)$

or

from 1 and 2 we have,

$$\frac{\Delta H}{\Delta u} = \frac{\frac{\Delta H}{w} \times w}{\frac{\Delta u}{Q} \times Q}$$

But $$Q = \left(\frac{\gamma - n}{\gamma - 1}\right) \times w$$

$$\frac{Q}{w} = \left(\frac{\gamma - n}{\gamma - 1}\right) \qquad .....(3)$$

$$Q = \frac{\left(\frac{Q}{\Delta u}\right) \times w}{\left(\frac{w}{\Delta H}\right) \times Q}$$

$$= \frac{\left(\frac{n - \gamma}{n - 1}\right) \times w}{\frac{(\gamma - 1)}{\gamma(1 - n)} \times Q}$$

$$= \frac{(n - \gamma)}{(n - 1)} \times \frac{\gamma(1 - n)}{(\gamma - 1)} \times \left(\frac{\gamma - 1}{\gamma - n}\right)$$

$$= \frac{\gamma(\gamma - n)(n - 1)}{(n - 1)(\gamma - n)}$$

$$\frac{\Delta H}{\Delta u} = \gamma \text{ proved}$$

But we have $$Q = \left(\frac{\gamma - n}{\gamma - 1}\right) \times w$$

$$\frac{Q}{w} = \frac{\gamma - n}{\gamma - 1}$$

$$= \frac{n - \gamma}{n - 1} \times \frac{\gamma\ (1-n)}{r-1} \times \frac{\gamma - 1}{\gamma - n}$$

$$= \frac{n-\gamma}{n-\gamma} \times \frac{\gamma\ (n - 1)}{n - 1} = \gamma$$

CHAPTER 4

Properties of Steam

4.1 Introduction

The conversion of heat energy in to mechanical work needs working agent which receive the heat released from a fuel and be used to drive the engine to produce required work.

Steam is considered an excellent working agent for the purpose because of following unique properties.

(i) It can carry large quantity of heat

(ii) It is produced from water which is cheap and readily available.

(iii) It can be used for heating purpose after its duty as working agent is completed.

Water is most important vapour used as a working fluid, as it is readily available. There are some other substances which are also used. Such as Mercury and Freon are used in vapour power cycles and refrigerants like NH_3, CO_2 and Freons are used in refrigeration systems.

In this chapter, properties of water and steam will be dealt with.

The condition of vapour is liable to change either by evaporation or condensation on changing the pressure and temperature and therefore the gas laws do not apply to vapour.

***Steam*:** A steam is water vapour. It behaves like a gas only when it is perfectly dry or super heated.

Steam is generated in boiler at constant pressure and used as working fluid in steam turbines and steam engines to produce mechanical power.

The following properties of water are responsible for the wide use of steam in thermal power plants.

(i) Easily available

(ii) It is cheap

(iii) Non poisonous

(iv) Odour less

(v) Good thermal capacity during steam formation.

(vi) No decomposition at high temperature during steam formation.

***Vapour*:** It is intermediate stage between liquid and gases. So it is not a perfect gas. Therefore it does not obey gas laws unless its temperature is very high.

Although water is most important vapour used as working fluid, there are many other substances which are also used commonly such as Mercury and Freon in vapour power cycles. The refrigerants like NH_3, CO_2 and Freons are used in refrigeration systems.

Tables of properties and charts similar to steam table are also available for these fluids.

Pure Substance

(i) It has homogeneous molecular structure and invariable chemical composition.

(ii) Pure substance can exist in single phase such as solid, liquid or gaseous phase or it may exist in more than one phase. But its molecular structure and chemical composition is always remain same at all conditions.

(iii) Water is example of pure substance because its molecular structure and chemical composition does not change when it is in single pure or two phase wet steam (mixture of dry steam + moisture). Ice is solid phase of water still its molecular structure and chemical composition does not change.

Air is not a pure substance because at very low temperature gases present in air such as Nitrogen, Oxygen, Hydrogen etc., liquefies at their boiling temperature. Therefore chemical composition and molecular structure of air change.

(iv) Under certain condition all three phases may co-exist. The co-existence of two phases is very common. When ice melts there is transformation of phase from solid to liquid. This is called melting or fusion of ice. Transformation can take place in the opposite direction from liquid to solid. This is called freezing or solidification

Similarly the transformation from the liquid phase to vapour phase is called vaporization. The transformation in opposite direction i.e., from vapour to liquid phase is called condensation.

4.2 Formation of Steam (Steam Generation at constant pressure)

1. Fig. 4.1 shows a frictionless system of piston and cylinder. 1 kg of water at 0 °C is kept in it and water is heated at constant pressure (Atmospheric) Fig. 4.1 (a).

2. When heat is added, water temperature rises and continues to rise until boiling point of 100 °C is reached. This boiling point is at 760 mm of Hg, but it increases with increase in pressure. During this process the change in volume of water is negligible. The heat received by water is known as enthalpy of water or sensible heat of water Fig. 4.1 (b).
3. With further addition of heat, the water starts evaporating at boiling (saturation) temperature. The temperature remains constant and volume of the vapour increases during this process. During this process the steam is known as wet steam because it contains water particles in suspension Fig. 4.1 (c).
4. This process is continued till all water is converted into steam at constant temperature. Heat added during this process is known as latent heat or Enthalpy of evaporation. Latent heat means "Not sensible by change in temperature". At the end of evaporation, there is no water present in the steam, such a steam is known as dry-saturated steam. The amount of dry steam present in unit mass of wet steam is known as dryness fraction of wet steam. The constant temperature during evaporation process is known as the saturation temperature and corresponding pressure is known as saturation pressure. Saturation temperature increases with saturation pressure Fig. 4.1 (d).
5. If Dry-saturated steam is further heated its temperature again starts increasing at constant pressure. This steam is known as super heated steam and heat contained is known as Enthalpy of super heated steam. Temperature above saturation temperature is known as super-heated temperature. Difference between ($T_{sup} - T_{sat}$) is known as degree of super heat. This process shown in Fig. 4. 1 (e) is known as super-heating.

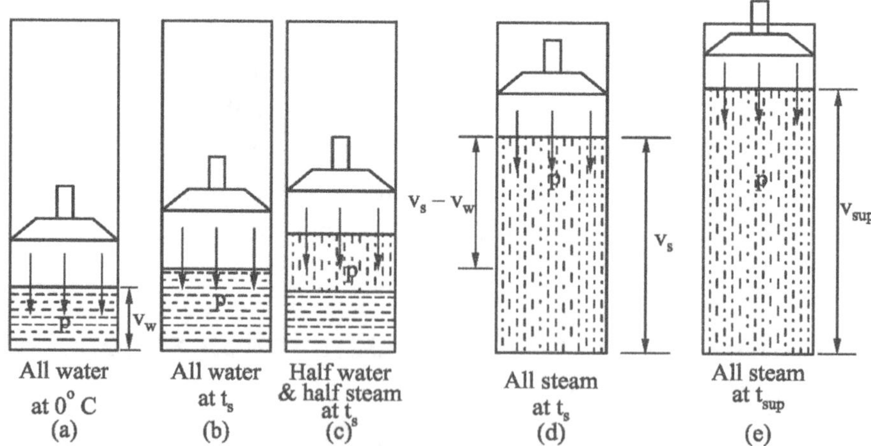

Fig. 4.1 Formation of steam at constant pressure.

4.2.1 Temperature Versus total Heat or Enthalpy Graph during Formation of Steam at °C OR Temperature versus Total Heat Graph during Formation of Steam

Shown in it (i) liquid line (ii) dry-steam line (iii) wet steam region (iv) critical point (v) super heated steam region (vi) water region.

The process of steam formation can be represented on a graph taking enthalpy (heat) on x-axis and temperature on y-axis.

The formation of super heated steam from 0 °C at atmospheric pressure is shown in Fig. 4.2.

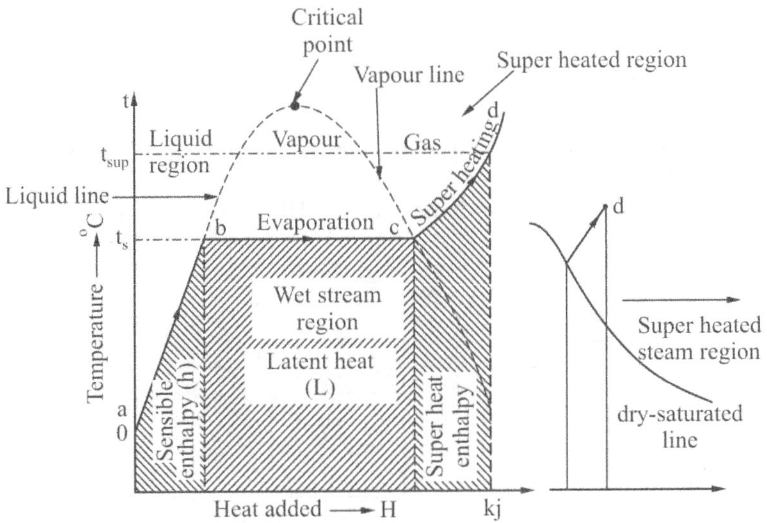

Fig. 4.2 Formation of steam at Constant Pressure.

The line (curve) a b c d shows variation of temperature as water gets converted into super heated steam. During the formation of super-heated steam from water at 0 °C and pressure p_1, heat is absorbed in three stages.

(i) Heating of water from 0 °C to its boiling temperature is shown by line a b. During this heating increases continuously. Heat absorbed by water during this heating is known as enthalpy of water or sensible heat of water and is denoted as h_f.

(ii) During further heating, at constant pressure evaporation of water takes place at constant temperature (saturation temperature), heat absorbed by water during the formation of steam is known as latent heat of steam or enthalpy of evaporation and denoted by h_{f_g} . This transformation is shown by line bc. The total heat of dry saturated steam at point C is the sum of sensible heat plus heat of evaporation. Total heat or Enthalpy of dry saturated steam is denoted as hg.

$$\therefore \qquad h_g = h_f + h_{f_g}$$

(iii) Super-heating of steam above saturation temperature is shown by line c-d. The heat absorbed by steam during this stage (c-d) of super-heating is known as super heat and steam is known as super-heated steam.

$$\text{Super heat} = c_{p_s}(t_{sup} - t_{sat})$$

where cps = specific heat of super heated steam = 2.1 KJ/Kg K

t_{sup} = super heated temperature

t_{sat} = saturation temperature

4.2.2 The formation of Super Heated Steam from Ice at –10 °C

The Formation of super heated steam from ice at – 10 °C is represented by line PQRVTS

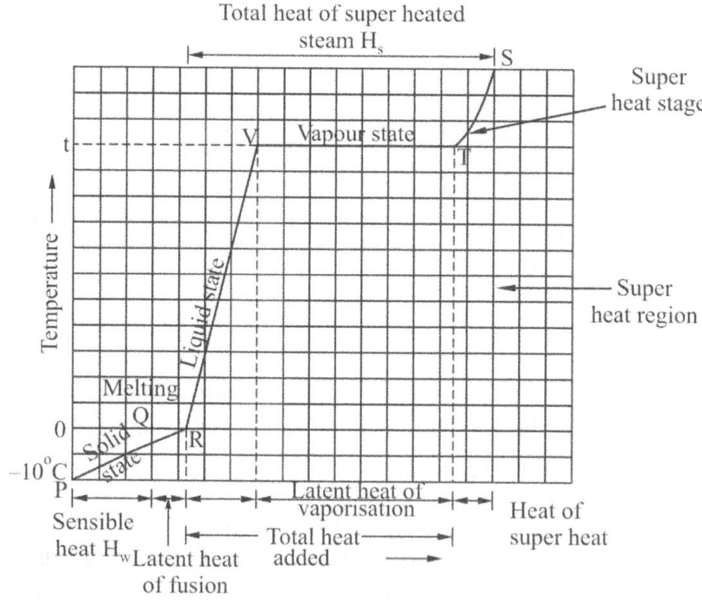

Fig. 4.3

During formation of steam from ice at –10 °C at given pressure P_1, heat is absorbed in five stages.

State I

Initial condition of ice is represented by point P. i.e., ice is at –10 °C and pressure P_1. As heat is added to ice its temperature will increases from –10 °C to 0 °C. The process of heating is represented by line PQ. During this process sensible heat is absorbed and at Q ice is at 0 °C.

Sensible heat of ice $= mC_{p_i}(T_2 - T_1)$

$= C_{p_i}\ [(0 - (-10)]$

$= 10\ C_{p_i}$ kJ/kg If m = 1kg

C_{p_i} = Specific heat of ice = 4.187 kJ/kg K

Stage II

After point Q, ice will absorb latent heat and will start melting at constant temperature of 0 °C till whole ice is melted and converted in to water at 0 °C. This process is represented by line QR.

The heat added during line QR is known as latent heat of fusion of ice (latent heat of ice is 335 kJ/kg).

Stage III

When heat is further added, temperature of water rises and continues to rise until boiling temperature related to pressure P_1 is achieved. This process is represented by line RV. During this process the change in volume of water is negligible. The heat received by water is known as Enthalpy of water or sensible heat of water and denoted as h_f, at point "V" temperature of water is known as saturation temperature and pressure is known as saturation pressure.

Stage IV

With addition of heat, water starts evaporating and temperature of steam remains constant and volume of the vapour increases during this process and steam is known as wet steam. If Addition of heat is continued, till all water is converted into steam at constant temperature as shown by line VT, at point "T" steam will be dry saturated and heat added during process VT is known as latent heat or enthalpy of evaporation.

The constant temperature during VT process is known as saturation temperature and pressure is known as saturation pressure. Saturation temperature (boiling temperature) increases with saturation pressure.

Stage V

Dry saturated at point T is further heated its temperature again starts increasing at constant pressure. This steam is known as super-heated steam. Heat added during process TS is known super-heat or enthalpy of super-heat. Temperature above saturation temperature is known as super heated temperature. Difference between (T_{sup} – T_{sat}) is known as degree of super heat. Process TS is known as super-heating process.

4.3 Properties of Steam

Melting Point: The point at which the solid is just begins to convert into liquid when heat is added is known as melting point (Fusion of ice point).

***Boiling Point*:** The point at which liquid is converted into vapour when heat is added is known as boiling point.

Saturation Temperature: It is defined as constant temperature or boiling temperature at which water is evaporated in to steam at a particular pressure.

As saturation pressure increased the saturation (boiling) temperature also increases. In steam table corresponding to given pressure, saturation temperature (also known as boiling temperature) is indicated. It is denoted by t_s or t_{sat}. Its unit is degree centigrade (°C) or Kelvin.

E.g. saturation (boiling) temperature corresponding to 10 bar from steam table = 179.9 °C.

***Dryness Fraction*:** It is related to wet steam (that is two phase mixture of water and dry steam). It is defined as ratio of mass of dry-saturated steam to mass of wet steam. It is denoted by D.F. or q. It is a constant number.

If m_s = mass of dry saturated steam in kg of wet steam

m_w = mass of water particles suspended in it in kg

∴ Mass of wet steam

= mass of dry saturated steam + Mass of water particle suspended in it

$= m_s + m_w$

$$\therefore \quad \text{Dryness fraction (D.F)} = \frac{\text{Mass of dry saturated steam}}{\text{mass of wet steam}}$$

$$= \frac{m_s}{m_s + m_w}$$

Its value lies in between 0 to 1

For wet steam D.F = $x < 1$

For Dry steam D.F = $x = 1$

For super heat steam = $x > 1$

Some of the point related to Dryness fraction

- Dryness fraction indicates condition of steam. Higher the value of D.F. less will be the water particles suspended in steam.
- Let us discuss what we mean by D.F = 0.8.

  It indicates that 1 kg of wet steam will contain 0.8 kg of dry steam and (1 – x) = (1 – 0.8) = 0.2 kg of water particles suspended in it.

  What is meant by quality of steam and wetness fraction of steam?

  ***Quality of Steam*:** The dryness fraction when expressed in percentage is called quality of steam (or vapour).

  Quality of vapour = 100 x where x = D.F of steam
- Quality of steam is say 90%. It indicates that 1 kg of wet steam contain 0.9 kg of dry saturated steam and (1 – 0.9) = 0.1 kg of water particles suspended in it.

***Wetness Fraction*:** It is defined as mass of water particles or moisture in suspension in a wet steam to its total mass.

$$\text{Wetness Fraction} = \frac{m_w}{m_s + m_w} \qquad \text{.....(i)}$$

Let us establish relation between dryness fraction and wetness fraction

(Add and subtract 1 from the above eq. (i))

$$\text{Wetness fraction} = \frac{m_w}{m_s + m_w} + 1 - 1$$

$$= 1 + \frac{m_w}{m_s + w} - 1 \qquad \text{(Rearranging)}$$

$$= 1 - \left(1 - \frac{m_w}{m_s + m_w}\right)$$

$$= 1 - \left[\frac{m_s + m_w - m_w}{m_s + m_w}\right]$$

$$= 1 - \left[\frac{m_s}{m_s + m_w}\right]$$

$$= 1 - x \qquad \left(\because x = \frac{m_s}{m_s + m_w}\right)$$

∴ Wetness fraction is equal to one minus dryness fraction.

Priming: The wetness fraction when expressed in percentage is called priming.

Wetness Fraction = (1 – Dryness Fraction) = (1 – x)

∴ Priming = 100 (1 – x)

- Suppose wetness fraction is 0.2 it indicates that 1 kg of wet steam contains 0.2 kg of water particles suspended in it and 0.8 kg of dry-saturated steam.
- Quality + Priming = 100% (Prove)

$100\,x + 100(1 - x)$ because quality = $100\,x$

$100\,x + 100 - 100\,x$ priming = $100(1 - x)$

= 100% (proved)

Volume (V): It is defined as space occupied by fluid. Its S.I unit is cubic meter.

(a) ***Specific Volume of Water***: It is defined as volume of 1 kg of water. It is expressed as m^3/kg and it is denoted by v_f. It can be seen from steam tables corresponding to given pressure. During heating change in volume of water is negligible as shown in Fig. 4.4(a).

(b) ***Specific Volume of Dry-Saturated Steam***: It is defined as the volume of one kg of dry saturated steam. It is expressed in m^3/kg and denoted by v_g. It is to be seen from steam table corresponding to given pressure. As pressure increases specific volume of dry steam decreases.

The density of dry saturated steam

$$= \frac{1}{v_g} \quad kg/m^3$$

(c) ***Specific Volume of Wet Steam***: It is always less than specific volume of dry saturated steam for given pressure.

It is defined as volume of one kg of wet steam having D. F = x

Specific volume of 1 kg of wet steam

= sum of volume of dry steam + volume of water contained in 1 kg of wet steam

If D.F of wet steam is = x

Then mass of dry steam will be x kg and

mass of water particles = (1 – x) kg

∴ ***Specific Volume of Wet Steam*** $xv_g + (1 - x)v_f$:

where v_g is volume of dry saturated steam per kg

v_f is volume of water particles suspended in wet steam.

v_g and v_f can be read from steam tables corresponding to given pressure. Value of volume of water v_f is very small compare to v_g. Therefore it is generally neglected.

∴ Volume of wet steam

$$= xv_g \qquad m^3/kg$$

∴ density of wet steam

$$= \frac{1}{\text{Specific value of wet steam}}$$

$$= \frac{1}{xv_g} \qquad kg/m^3$$

(d) ***Specific Volume of Super heated Steam*:** It is defined as volume of 1 kg of super-heated steam. It is denoted by v_{sup}. Super-heated steam obeys laws of perfect gas and superheating takes place at constant pressure. Therefore super-heated steam follows Charles Law.

If v_{sup} and v_g are specific volume super heated steam and dry-saturated steam in m^3/kg respectively and T_{sup} and T_{sat} are the absolute temperature of super-heated steam and dry saturated steam respectively.

Then by Charle's law

$$\frac{v_{sup}}{v_g} = \frac{T_{sup}}{T_{sat}}$$

$$\therefore \qquad v_{sup} = v_g \cdot \frac{T_{sup}}{T_{sat}}$$

∴ Density of super heated steam

$$= \frac{1}{v_{sup}} \qquad kg/m^3$$

***Enthalpy*:** It is defined as total heat in form of internal energy u and flow work p.v. (Product of Pressure and Volume) containing in steam (fluid). Its S.I unit is Joules or Kilo Joules (KJ).

It is denoted by "H"

$$\therefore \qquad H = U + PV$$

***Specific Enthalpy*:** It is defined as total heat contained in 1 kg of steam. Its unit in S.I is kJ/kg and denoted by "h"= U + p.v kJ/kg

(a) ***Specific Enthalpy of Water (Sensible heat of water) per kg*:** It is also known as specific enthalpy of liquid. It may be defined as amount of heat required to raise the temperature of 1kg of water from 0 °C to saturation temperature (boiling temperature). During heating pressure remain constant.

It is denoted by h_f and its unit is kJ/kg. As pressure increase the specific enthalpy of water increases. Its value can be seen from steam tables corresponding to given pressure or it can be calculated by formula as given below.

$h_f = m\, c_{pw} (t_{sat} - 0)$, if we take m = 1 kg and $cp_w = 4.1868$ kJ/kg, then

$h_f = 4.1868\, t_{sat}$ kJ/kg

h_f increases as temperature and pressure increases.

(b) ***Specific Enthalpy of Evaporation*:** Amount of heat required to convert one kg of water at saturation temperature and at given pressure in to steam at same pressure and temperature. It is also called latent heat. It is denoted by h_{fg} and it is expressed in kJ/kg. It decreases with increase in pressure. At critical point latent heat becomes zero.

(c) ***Specific Enthalpy of Dry-Saturated Steam*:** Amount of heat required to convert one kg of water at 0 °C (boiling temperature) into dry saturated steam.

It is sum of specific enthalpy of water and enthalpy of evaporation. It is represented by H or h, its unit in S.I is kJ/kg. As pressure increases the specific enthalpy, of dry steam increases.

h = Enthalpy of dry steam

$h = h_f + h_{f_g}$ (when temperature of water taken in boiler is at 0 °C.)

$h = h_f + h_{f_g} - c_{p_w} (t - o)$ when temperature of water in boiler is initially at t °C

$= h_f + h_{f_g} - c_{p_w} t$ c_{p_w} = Specific heat of water = 4.1868 kg/kgk

(d) ***Specific Enthalpy of Wet Steam*:** Total heat (Enthalpy) of wet steam in kJ/kg may be defined as quantity of heat required to convert one kg of water at 0 °C temp, into wet steam. During heating pressure remain constant.

Let mass of wet steam is one kg contains x kg of dry steam and (1 – x) kg of water particles in suspension.

The heat required to convert 1 kg of water at 0 °C (Freezing point) to boiling temperature (saturation temperature) at constant pressure = h_f

The heat required to convert x kg of water in to wet steam = x h_{f_g}

At saturation temperature at constant same pressure

∴ Total Enthalpy of wet steam = $h_f + xh_{f_g}$ when water is at 0 °C

$= h_f + xh_{fg} - c_{pw}(t - 0)$ when feed water is at t °C

Or
$$h_{wet} = (1 - x)h_f + xh_g$$
$$= (1 - x)h_f + x(h_f + h_{fg})$$
$$= h_f - xh_f + xh_f + xh_fa$$

$h_w = h_f + xh_{fg}$ when Feed water is at 0 °C

$= h_f + x_{hfg} - c_{p_w}(t - 0)$ when Feed water is t °C

(e) ***Total Heat of Super Heated Steam*:** Total heat (Enthalpy) of super heated steam may be defined as amount of heat required to convert 1 kg of water at 0°C temperature into super heated steam at constant pressure.

It can also be defined as sum of specific enthalpies of dry saturated steam and the total heat due to super heat.

It is denoted by h_{sup} and unit is kJ/kg.

$h_{sup} = h_f + h_{f_g} + c_{p_s}(t_{sup} - t_{sat})$ .... When feed water is initially at 0°C

$h_{sup} = h_f + h_{f_g} + c_{p_s}(t_{sup} - t_{sat}) - c_{p_w}(t - 0)$ .....(ii)

when water initially is at t °C

Take c_{p_s} = specific heat of super heated steam = 2.1 kJ/kgk to 2.3 kJ/kgk

(f) ***Degree of Super Heat*:** When wet steam at a given pressure having saturation temperature t_{sat} is heated at constant pressure, the wet steam gets converted into dry saturation steam. During this temperature remain constant = t_{sat}.

Further addition of heat the temperature of steam will increase above saturation temperature say t_{sup} ,which will be measured (indicated) by thermo-meter.

Degree of super heat may be defined as difference between super heated temperature and saturation temperature indicated by thermometer.

$\therefore$ Degree of super heat = $(t_{sup} - t_{sat})$

Its unit is degree centigrade (°C) as shown in Fig. 4.4(a)

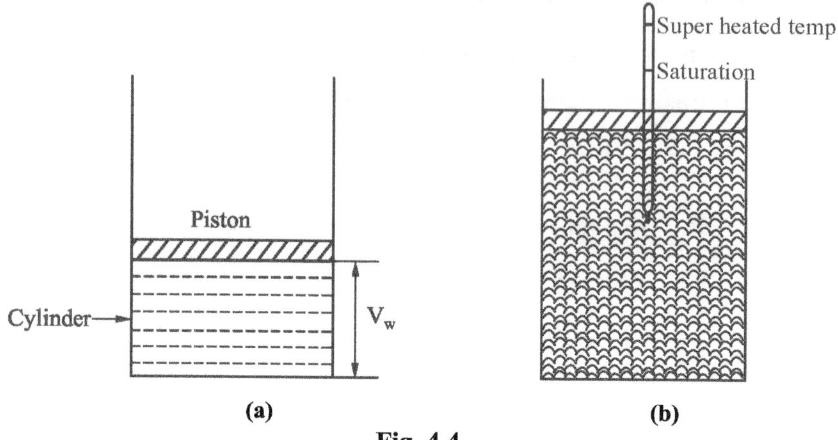

Fig. 4.4

4.3.1 Work During Steam Formation

(i) When 1 kg mass of water at 0 °C is heated up to saturation temperature (boiling temperature) at constant pressure, the volume of water and change of volume of water are negligible. The work done during this process is zero as shown in Fig 4.4(a).

(ii) During evaporation of water at saturation temperature and constant pressure, water gets converted into steam and there is considerable increase in volume, work is done for this increase in volume. This is known as work of evaporation. In other words, the energy required for performing this work is obtained during absorption of latent heat as shown in Fig. 4.4(b).

If P = Absolute pressure at which evaporation takes place, in bar

V_g = specific volume of steam corresponding to "p" from steam tables

External work of evaporation = w

$$w = p \text{ in bar} \times v_g \text{ in } \frac{m^3}{kg}$$

$$= p \times 10^5 \text{ N/m}^2 \times v_g \text{ m}^3\text{/kg}$$

$$= pv \times 10^5 \text{ N–m/kg}$$

$$= pv_g \times 10^5 \text{ J/kg}$$

$$= 10^2 pv_g \text{ KJ/kg} \qquad \text{..... (i)}$$

W_{wet} = For wet steam work of evaporation = 10^2 p.x.v_g kJ/kg .....(a)

W_{dry} = For Dry steam work of evaporation = 10^2 p. v_g kJ/kg ..... (b)

W_{sup} = For super heated steam work of evaporation $= 10^2 p.v_g . \frac{T_{sup}}{T_{sat}}$ kJ/kg..... (c)

Internal Energy of Steam

Internal Energy of Steam may define as the actual heat energy stored in steam above freezing point of water.

Out of total heat (or enthalpy) supplied to the fluid (water), some energy is used for work done and remaining energy is stored in fluid as internal energy.

It is the difference between total heat of steam and external work of evaporation.

$\therefore$ $U = H - 10^2$ p. v kJ (U = Internal energy)

where H = Enthalpy of steam kJ

P = Pressure of steam in bar

v = volume of steam based on condition of steam

U = Specific internal energy in kJ/kg

(a) ***Specific Internal Energy***

It is defined as internal energy per unit mass of fluid. It is denoted by u and its unit is kJ/kg. It is the difference between the specific enthalpy (h) and work done w per kg.

For wet steam

$$u_{wet} = h_f + \mathbf{x}\ h_{f_g} - 10^2\ p\ (x\ v_g)$$

(When initial feed water temperature is zero)

$$= \left(h_f + xh_{fg} - c_{p_w} t\right) - 10^2\ p\ (x\ v_g)$$

(When feed water temp is t° C)

For Dry Saturated Steam

$$u_g = h_f + h_{f_g} - 10^2\ p\ v_g$$

(When feed water temperature is 0 °C)

$$u_g = \left(h_f + h_{f_g} - c_{p_w} t\right) - 10^2 p v_g$$

(When feed water temp is t °C)

For Super Heated Steam

$$U_{sup} = \left(h_f + h_{f_g} - c_{p_s}(t_{sup} - t_{sat}\right) - 10^2 pv_g . \frac{T_{sup}}{T_{sat}}$$

(When Feed water temp is 0 °C)

$$U_{sup} = \left(h_f + h_{f_g} + c_{p_s}(t_{sup} - t_{sat} - c_{p_w} t\right) - 10^2 pv_g . \frac{T_{sup}}{T_{sat}}$$

(When feed water temp is t °C)

C_{ps} = spec: fie heat of supper heated steam

Note: "U" is total internal energy in kJ

"u" is total internal energy of 1 kg of steam in kJ/kg

4.3.2 Internal Latent Heat

Latent heat consists of true latent heat and work of evaporation. This true internal latent heat is called the internal latent heat, may be calculated as below.

Internal latent heat $= h_{f_g} - p.v_g$ for dry steam

$= x\,[h_{f_g} - p.v_g]$ for wet steam

Entropy of Steam

(a) ***Entropy of Water***: The entropy of water is always reckoned above the freezing point of water i.e., 0 °C or 273 °k.

$\therefore$ Entropy of water $= c_{p_w} \log_e \frac{T}{273}$

T = absolute temperature of formation of steam

$$\therefore \quad s_f = \log_e \frac{T}{273}$$

S_f can also be seen from steam table corresponding to given pressure or temperature directly.

(b) ***Entropy of Evaporating*** $\left(s_{f_g}\right)$

Now change in entropy is $= \int \frac{dq}{T}$

Since temperature during evaporation remain constant and q is heat absorbed during evaporation. Then from above integration we get

Change in entropy = $\frac{q}{T}$

During the evaporation of water into dry steam Q = L

$$\therefore \quad s_{f_g} = \frac{L}{T} \quad \text{(for dry)}$$

$$\therefore \quad s_{f_g} = \frac{xL}{T} \quad \text{(for wet steam having D.F = x)}$$

(c) Entropy of Wet Steam = Entropy of water + change of entropy during evaporation

$$= s_f + \frac{xL}{T} \qquad \text{but} \quad \frac{xL}{T} = x(s_g - s_f)$$

or $$= s_f + x(s_g - s_f) \qquad \text{(If we want to use steam table)}$$

(d) Entropy of Dry Steam $= s_f + \frac{L}{T}$ $\qquad (\because x = D.F = 1)$

or $$= s_f + (s_g - s_f) \qquad \text{But} \quad \frac{L}{T} = (s_g - s_f)$$

$= s_g$ $\qquad$ (s_g can be seen from steam table corresponding to given pressure)

(e) Entropy of super heated steam

$$= \log e \frac{T}{273} + \frac{L}{T} + c_{p_s} \log_e \frac{T_{sup}}{T_{sat}}$$

$$= s_f + s_{f_g} + c_{p_s} \log_e \frac{T_{sup}}{T_{sat}}$$

$$= s_g + c_{p_s} \log_e \frac{T_{sup}}{T_{sat}} \qquad \because s_g = (s_f + s_{fg}) \qquad \text{..... (3)}$$

4.4 Advantages of Super heated Steam

1. The temperature of super heated steam being higher, it gives higher thermal efficiency when used in prime-movers.

2. The super-heated steam has an increased capacity of doing work. So use of super heated steam result in an economy in steam consumption, in turbine or in steam engine.
3. The super-heated steam content more heat per kg of steam as a result, for given power turbine size is reduced.
4. Super-heated steam can be considerably expanded in an engine or turbine before its temperature falls so low to cause condensation. So heat lost as a result of condensation of steam on cylinder walls is greatly reduced.
5. By using super-heated steam erosion and corrosion in turbines is greatly reduced in comparison of using dry-saturated steam or wet steam.
6. The super heating is done in super-heater, which obtain heat from gases. These gases would have otherwise passed uselessly through the chimney.

4.5 Difference between Dry Steam and Super Heated Steam

Dry Steam	Super Heated Steam
1. Total heat content of dry saturated steam is less than super heated steam for same mass and pressure of steam. $h_{dry} = h_f + h_{f_g}$	1. Total heat content of super heated steam is greater than Dry-saturated steam for same mass and pressure of steam. $h_{sup} = h_f + h_{fg} + c_{ps}\,(t_{sup} - t_{sat})$
2. The saturation temperature of steam at given pressure is less than super heated temperature.	2. Super heated temperature is greater than saturation temperature at same pressure to that of dry-saturated steam.
3. Saturation temperature is denoted by t_{sat}	3. Super heated temperature is denoted by t_{sup}. Difference between super heated temperature and saturation temperature is known as degree of super heat.
4. For producing dry steam boiler is provided with steam separator. No need of super heater.	4. For producing super heated steam boiler and superheater is needed.
5. At all pressure ranging from 0.01 to 221.2 Bar h_f, h_{fg}, h_g specific volume, saturation temperature, entropy are directly available from steam table.	6. At all pressure and super heated temperature, total heat, specific volume, entropy etc are not available directly from steam table. These properties are to be calculated by using formula.

4.6 Difference between Enthalpy and Entropy of Steam

Enthalpy	Entropy
1. It is summation of internal energy and flow work.	1. The cyclic integral of $\frac{d_q}{T}$ for eversible cycle being equal to zero suggests the quantity $\frac{dQ}{T}$ is a point function and therefore it is a property of system.
2. It is denoted by 'h' or H	2. It is denoted by "S" or ϕ
3. Unit of Enthalpy is kJ and unit of specific enthalpy if kJ/kg	3. Unit of Entropy is kJ/k. But unit of specific entropy is $\frac{kJ}{kgk}$.
4. Enthalpy can be converted to an other form like work and kinetic energy	4. It can not be converted into an other form. But it can be increased or decreased by addition of heat or removal of heat respectively.
5. It is not used for deciding the direction of the process.	5. It is useful for deciding the direction of process and measure the molecular disorder.
6. Mathematically it is written as $H = U + PV$	6. Mathematically it is written as $d_s = dQ/T$
7. For mixing, it obeys additive law.	7. It does not obey additive law while mixing.

4.7 Difference between Steam and Gas

Wet Steam (Vapour)	Gas
1. Wet steam is not perfect gas. Therefore it does not obey, Boyle and Charles law. It also does not obey characteristic gas law	1. Gas is perfect. Therefore it obeys all gas laws.
2. It is state of substance in which evaporation is not completed from its liquid state.	2. Gas is complete vapouration of liquid. It is gaseous state.
3. Steam may be having three conditions, wet, dry and super heated steam.	3. There are no such three states exist in gas. O_2, N_2, Air, dry steam and super heated is taken as gases.
4. When the steam is cooled. It gets condensed.	4. It remains in gaseous state at Moderate pressure and temperature.
5. Specific volume of steam is less compare to gases.	5. Specific volume of gases is more comparative to steam.
6. It is mixture of steam and water particles are suspended in it.	6. Gases of mixture of O_2, N_2, H_2, air etc,.

4.8 Thermodynamic Process

Different processes of expansion and compression used for gases are also adopted for steam. The general energy equation applicable to perfect gases is also applicable to dry vapour. The procedure adopted for finding the change of internal energy is also same.

The different thermodynamic processes for vapours (steam) are discussed here.

1. Constant volume process. (Isochronic process)
2. Constant pressure process (Isobaric process)
3. Isothermal or constant temperature process
4. Hyperbolic process (pv = c process)
5. Reversible adiabatic process (Isentropic process or constant entropy process)
6. Polytrophic process ($pv^n = c$ process)
7. Throttling process (Constant Enthalpy process)

We shall now discuss these processes, in detail:

In all processes we will develop expression for

(i) Final condition of steam

(ii) W.D/kg

(iii) Change in Internal Energy

(iv) Heat transfer or Heat absorbed

4.8.1 Constant Volume Process (Isochronic Process)

The heating or cooling of steam in closed vessel is an example of constant volume process.

In this process, the volume or mass of steam remain constant before and after the process. In this process no work is done. Consider 1 kg of wet steam having D.F. = x_1 at pressure p_1 is heated at constant volume till its pressure become p_2 and DF = x_2.

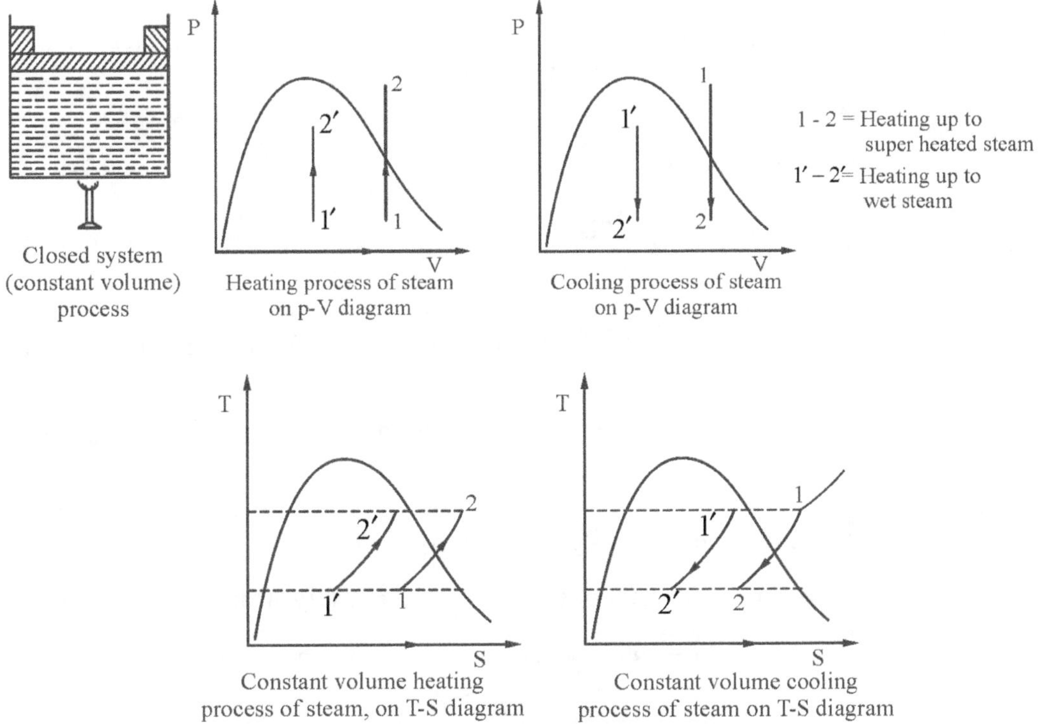

Fig. 4.5

1. Mass before Heating = Mass after heating

$$v_1 . \rho_1 = v_2 . \rho_2$$

$$x_1 v_{g1} \rho_1 = x_2 v_{g2} \rho_2 \text{ for wet steam} \qquad \text{......(i)}$$

$$x_1 \ v_{g_1} = v_{g2} \ x_2 \qquad [\because \rho_1 = \rho_2]$$

$$\therefore \qquad x_2 = x_1 . \frac{vg_1}{vg_2} \qquad \text{..... (ii)}$$

v_{g_1} and v_{g_2} are specific volumes of dry saturated steam at pressure p_1 and p_2 respectively which can be obtained from steam tables, corresponding to given pressure.

If $x_2 > 1$, it indicates that steam is super heated after heating

$$\therefore \quad x_1 v_{g_1} = v_{sup_2} \quad \text{..... (A)}$$

Let us find value of v_{sup2}

Putting this value in eq. (A)

$$v_{sup_2} = v_{g_2} \cdot \frac{T_{sup_2}}{T_{sat_2}}$$

$$\therefore \quad x_1 v_{g_1} = v_{g_2} \cdot \frac{T_{sup_2}}{T_{sat_2}} \quad [\because x_1 v_{g1} = x_2 v_{g2} \text{ ... from equation (i)}] \text{..... (ii)}$$

From equation (ii) T_{sup_2} can be calculated and degree of super heat is given by $(T_{sup} - T_{sat})$ or $(t_{sup} - t_{sat})$

2. Change in internal energy per kg of steam $(u_2 - u_1)$

$$u_1 = h_1 - 100\, p_1 v_1 \quad \text{For Dry steam} \quad \text{..... (iii)}$$

$$= h_1 - 100\, p_1 x_1 v_{g1} \quad \text{For wet steam}$$

$$= h_1 - 100\, p_1 v_{g1} \frac{T_{sup}}{T_{sat}} \quad \text{For super heated steam}$$

Final internal energy of steam, u_2 will depend upon its final condition

$$u_2 = h_2 - 100\, p_2 v_2$$

$$= h_2 - 100\, p_2 x_2 v_{g_2} \quad \text{..... For wet steam}$$

$$= h_2 - 100\, p_2 v_{g_2} \quad \text{..... For dry steam}$$

$$= h_2 - 100\, p_2 \cdot v_{g_2} \frac{T_{sup_2}}{T_{sat}} \quad \text{..... For super heated steam}$$

$\therefore$ Change in Internal energy $= d_u = (u_2 - u_1)$

If d_u is – ve, then it indicates that steam is cooled and there is a decrease in internal energy.

Note: Initial and Final enthalpy of steam i.e., h_1 and h_2 may be obtained depending upon its condition as below.

$$h_1 = h_{f_1} + x_1 h_{f_{g1}}$$ For wet steam,

$$h_1 = h_{f_1} + h_{f_{g1}}$$ For Dry steam

$$h_1 = h_{f_1} + h_{f_{g1}} + c_{p_s} (t_{sup} - t_{sat})$$ For super heated steam

$$h_2 = h_{f_2} + x_2 h_{f_{g2}}$$

$$h_2 = h_{f_2} + h_{f_{g2}}$$

$$h_2 = h_{f_2} + h_{f_{g2}} + c_{p_s} \left(t_{sup_2} - t_{sat_2}\right)$$

3. Heat absorbed or heat transferred

$$q_{1-2} = \text{change in internal energy} + \text{work done } 1-2$$

$$q_{1-2} = d_u + w_{1-2}$$

$$q_{1-2} = d_u$$

$$\because \left(W.D = \int_1^2 pdv = 0;\ \because dv = 0 \text{ for constant value}\right)$$

$$\therefore \quad q_{1-2} = u_2 - u_1 \ \ kJ/kg$$

- Therefore we can say that heat transferred or heat absorbed is equal to change in internal energy during constant volume.
- If q_{1-2} is –ve, then it indicates that steam is being cooled and heat is rejected from the steam.
- Always change in internal energy is equal to $(u_2 - u_1)$ and should not be written as $(u_1 - u_2)$.

4.8.2 Constant Pressure Process

Generation of steam in boiler is example of constant pressure process.

Let us consider 1 kg of wet steam is heated at constant pressure till it become Dry-saturated. Heating of wet steam into dry saturated steam at constant pressure is also known as constant temperature process. But when steam at constant pressure is heated up to super heated temperature then it will be no longer Isothermal process.

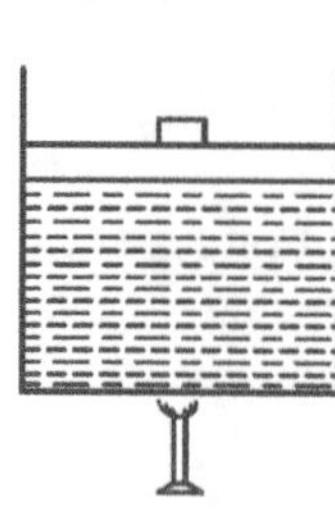

In engineering constant pressure is shown
by cylinder piston arrangement
Process is represented by piston cylinder arrangement

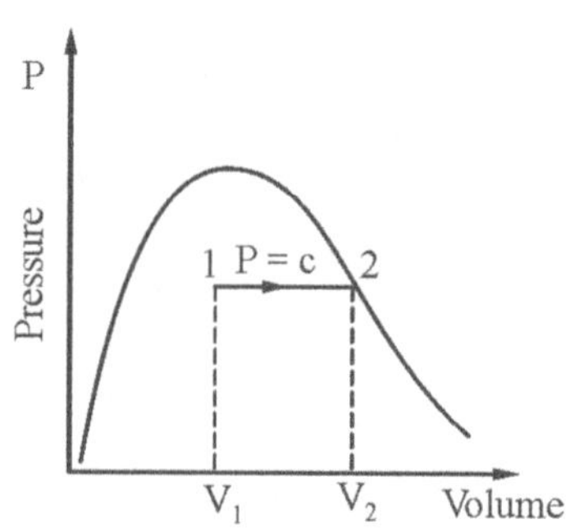

p-V diagram

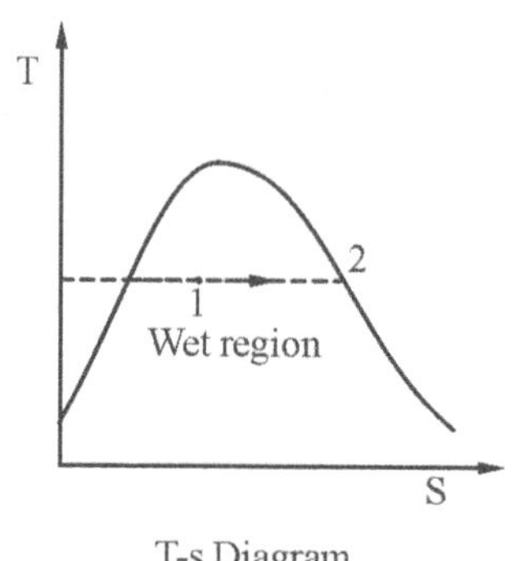

T-s Diagram

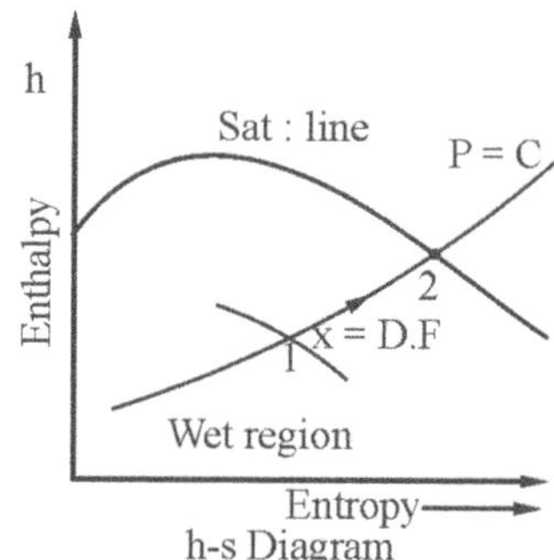

h-s Diagram

Fig. 4.6

Note: If wet steam is heated till, it comes super heated. The representation on T-s, h-s diagram will change.

1. Let 1 kg of wet steam is heated at constant pressure and its dryness fraction (D.F) changes from x_1 to x_2.

   Since pressure of steam remain constant, its specific volume of dry steam v_g corresponding to given pressure from steam tables will also remain same.

   $$\therefore \quad v_{g1} = v_{g2} = v_g \quad \text{.....(i)}$$

   If initial D.F = x_1 and final D.F = x_2

   ∴ Initial volume of

   $$v_1 = x_1 v_g \quad \text{.....(ii)}$$

Final volume of steam

$$v_2 = x_2 v_g$$

If initial D.F = x_1 and final condition of steam is dry saturated

$$\left.\begin{aligned} \text{Initial volume of steam} &= x_1\, v_g \\ \text{Final volume of steam} &= v_g \end{aligned}\right\} \quad \text{.....(iii)}$$

as x = 1 Dry saturated

If initial D.F = x_1 and final conditions of steam is super heated

$$\left.\begin{aligned} \therefore \quad \text{Initial volume of steam} &= x_1\, v_g \\ \text{Final volume of steam} &= v_g \frac{T_{sup}}{T_{sat}} \end{aligned}\right\} \quad \text{.....(iv)}$$

2. W.D during the process = $w_{1-2} = 100\, p\, (v_2 - v_1)$ kJ/kg

   where P should always be in bar

3. Change in Internal energy

$$u_1 = h_1 - 100\, p\, v_1 \qquad \text{(General expression)}$$

$$= h_{wet} - 100\, p x_1\, v_{g1} \qquad \text{(For wet steam)}$$

$$= h_{dry} - 100\, p\, v_{g_1} \qquad \text{(For dry steam)}$$

$$= h_{sup} - 100\, p\, v_{g_1} \frac{T_{sup}}{T_{sat}} \qquad \text{(For super heated)}$$

$$u_2 = h_{wet} - 100\, p\, x_2\, v_{g_1}$$

$$= h_{dry} - 100\, p\, v_{g_2}$$

$$= h_{sup} - 100\, p\, v_{g_2} \cdot \frac{T_{sup}}{T_{sat}}$$

$\therefore$ change in internal energy = $u_2 - v_1$ kJ/kg

4. Heat absorbed or Heat transferred

   q_{1-2} = Heat transferred per kg of steam (while Q = Total heat transfer)

   = work done + change in internal energy (First law of thermodynamics)

$$= W.D + (u_2 - u_1)$$

$$= (u_2 - u_1) + w_{1-2} \qquad \text{rearranging}$$

$$= (h_2 - 100p\ v_{g_2}) - (h_1 - 100\ p\ v_{g_1}) + w_{1-2}$$

$$= h_2 - 100\ p\ v_{g_2} - h_1 + 100\ p\ v_{g_1} + 100\ p(v_{g_2} - v_1) = h_2 - h_1$$

$$q_{1-2} = h_2 - h_1$$

i.e., During constant pressure heating process the heat absorbed or transferred by steam is equal to change in enthalpy or total heat of steam.

4.8.3 Constant Temperature Process

We know that wet steam has only one temperature i.e., saturation temperature corresponding to given pressure.

The heating of wet steam at constant temperature till it becomes dry-saturated is similar to that of heating at constant pressure. The same expression for work done, change in internal energy and heat absorbed may be used as for constant pressure. As soon as steam becomes super heated, it behaves like a perfect gas and will follow hyperbolic law (Pv = constant).

Therefore the isothermal process may be regarded as constant pressure process for wet steam and hyperbolic process for super heated steam.

Let us represent isothermal expansion of steam on p – v, T – s, and h – s diagram as shown below:

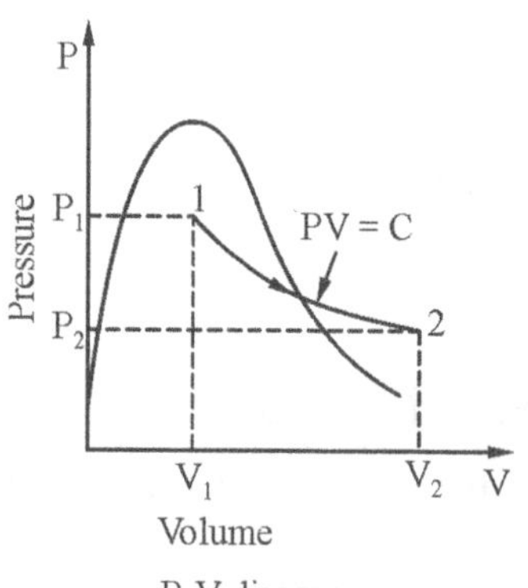

P-V diagram

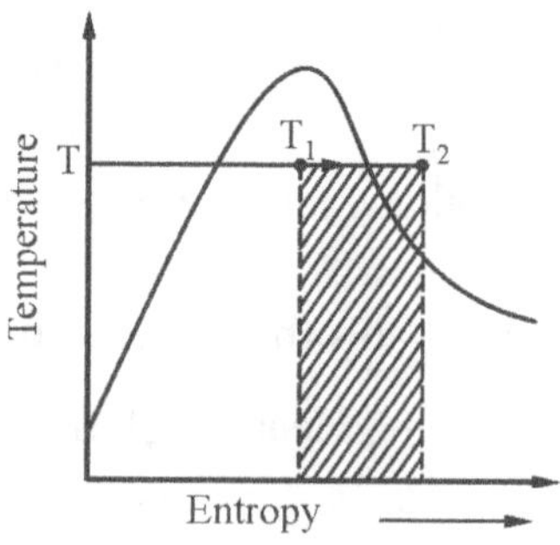

T-s diagram

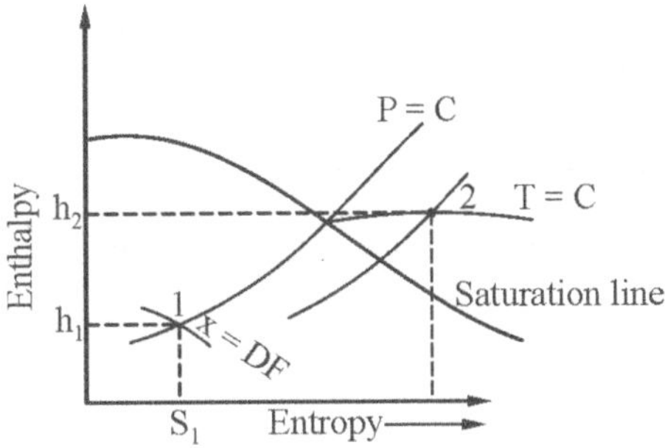

h-s diagram

Fig. 4.7

1. Work done = $P(v_2 - v_1)$ as in wet region constant temperature process is also constant pressure process

   $v_2 = v_g$ at saturation pressure corresponding to given temperature from steam table.

2. Heat absorbed or heat transferred

   $$q = h_2 - h_1 \quad \text{same as constant pressure}$$

3. Change in Internal energy

   First law of thermodynamics

   $$q = \Delta_u + W.D/kg$$

   $$\therefore \quad \Delta u/kg = q - W.D/kg \text{ same as constant pressure}$$

4.8.4 Hyperbolic Process

In case of perfect gases we can apply or pv = CT if T is constant then C × T will be constant i.e., we can write pv = C i.e., Isothermal process (in perfect gases) is regarded hyperbolic process. But pv = C (hyperbolic process) is not equal to Isothermal process in wet steam because characteristic gas constant pv = RT or pv/T = C cannot be applied.

Super heated steam behaves like perfect gas, and the processing of super heating of steam at constant temperature i.e., isothermal process in super heated region is regarded as hyperbolic process.

Representation of hyperbolic process on pv, T–s and h–s diagram (Fig. 4.8).

$p_1 =$ Initial pressure of steam

$x_1 =$ Dryness fraction (D.F)

v_{g_1} = Specific volume of dry steam corresponding to p_1 from steam table in m^3/kg.

p_2, x_2, v_{g_2} = Corresponding values for final condition of steam when it is expanded according to law pv = C till the pressure becomes P_2.

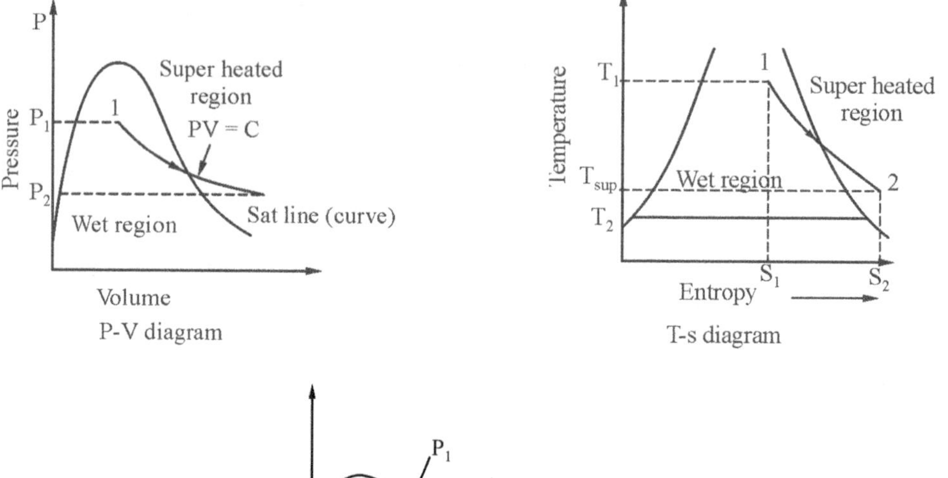

P-V diagram

T-s diagram

h-s diagram

Fig. 4.8

Applying the given law pv = C to expansion process

$$p_1 v_1 = p_2 v_2$$

$$p_1 x_1 v_{g1} = p_2 x_2 v_{g2}$$

$$\therefore \quad x_2 = \frac{p_1 x_1 v_{g1}}{p_2 v_{g2}} \quad \text{For wet steam} \qquad .....(1)$$

The value of x_2 gives us important information about final condition of steam as below.

1. If $x_2 < 1$, the final condition of steam is wet

2. If $x_2 = 1$, the final condition of steam is dry saturated
3. If $x_2 > 1$ final condition of steam is super heated.

In case of Dry saturated steam we shall find saturation temperature and volume directly from steam table corresponding to pressure P_2.

But in case of super heated steam we will find out volume and temperature as below.

$$P_1 x_1 v_1 = P_2 v_{sup}$$

$$v_{sup} = \frac{P_1 x v_1}{P_2}$$

Now T_{sup} can be found by Charles law

$$\frac{v_{sup}}{T_{sup}} = \frac{v_{g_2}}{T_{sat_2}}$$

1. Work done during process

$$W.D/kg = \int_{v1}^{v2} Pdv \qquad \text{.....(i)}$$

$$Pv = C$$

$$\therefore \quad P = \frac{C}{v} \qquad \text{Put value of P in eq (i)}$$

$$W.D/kg = \int_1^2 \frac{C}{v} dv$$

$$= C\left[\log_e^v\right]_{v1}^{v2}$$

$$= C(\log_e v_2 - \log v_1)$$

$$= P_1 v_1 \text{ loge } v_2/v_1 \qquad \because P_1 v_1 = C = P_2 v_2$$

$$\therefore \quad W.D/kg = 100 P_1 v_1 \log_e v_2/v_1 \qquad \text{where P must be in bar}$$

$$P_1 v_1 = P_2 v_2$$

$$\therefore \quad \frac{v_2}{v_1} = \frac{P_1}{P_2}$$

$$\therefore \quad W.D/kg = 100 P_1 v_1 \log_e P_1/P_2$$

2. Change in internal energy

$$u_1 = h_1 - P_1 v_1$$

$$u_2 = h_2 - P_2 v_2$$

$$\therefore \quad (u_2 - u_1) = (h_2 - P_2 v_2 - h_1 + P_1 v_1)$$

$$(u_2 - u_1) = h_2 - h_1 \qquad \because P_1 v_1 = P_2 v_2$$

3. Heat absorbed = change in internal energy + W.D/kg

$$= (h_2 - h_1) + 100\, P_1 v_1 \log_e v_2/v_1$$

4.8.5 Isentropic Process (Non-flow process)

Isentropic Process is also known as reversible adiabatic process or constant entropy process.

During isentropic process the heat transfer is zero and change in entropy is also zero.

If one kg steam at pressure P_1 and dryness fraction x_1 expands isentropically until the pressure falls to P_2 then the entropy before expansion is equal to entropy after expansion.

Initial entropy of steam before expansion = Final entropy of steam after expansion

$$S_{f1} + x_1 S_{fg1} = S_{f2} + x_2 S_{fg2} \qquad .....(i)$$

From this Dryness fraction x_2 can be calculated

where S_{f1} = entropy of water corresponding to pressure P_1 from steam table.

$S_{f_{g1}}$ = entropy of evaporation corresponding to P_1

S_{f_2} = entropy of water corresponding to pressure P_2 from steam tables.

$S_{f_{g2}}$ = entropy of evaporation corresponding to pressure P_2 from steam tables.

The value of x_2 gives us important information about final condition of steam as below:

1. If value of $x_2 < 1$, then final condition of steam is wet and equation (i) give the value of D.F x_2 of wet steam.
2. If value of $x_2 = 1$ then final condition of steam is Dry-saturated.

$$S_{f1} + x_1 S_{f_{g1}} = S_{f_2} + S_{f_{g2}}$$

3. If value of $x_2 > 1$ then final condition of steam is super heated

$$S_{f_1} + x_1 S_{f_{g1}} = s_{g2} + c_{ps} \log_e \frac{T_{sup_2}}{T_{sat_2}}$$

From this T_{sup} can be calculated

By applying charles law we can find v_{sup}

$$\frac{v_{sup}}{T_{sup}} = \frac{v_{g_2}}{T_{sat_2}}$$

2. Change in internal energy

$$u_1 = h_1 - 100\ P_1\ v_1 \qquad \text{(general equation)}$$

$$= h_1 - 100\ P_1\ x_1\ v_{g_1} \qquad \text{for wet steam}$$

$$= h_1 - 100\ P_1\ v_{g_1} \qquad \text{for Dry steam}$$

$$= h_1 - 100\ P_1\ v_{g_1}\ \frac{T_{sup}}{T_{sat}} \qquad \text{for super heated steam}$$

$$u_2 = h_2 - 100\ P_2\ x_2\ v_{g_2} \qquad \text{for wet steam}$$

$$= h_2 - 100\ P_2\ v_{g_2} \qquad \text{for Dry steam}$$

$$= h_2 - 100\ P_2 V_2\ \frac{T_{sup}}{T_{sat}} \qquad \text{for super heated steam}$$

$\therefore$ change in internal energy $= u_2 - u_1$

3. Heat absorbed or Heat transfer

$$q_{1-2} = 0$$

4. Work done during process

According to first law of thermodynamics

$$q_{1-2} = d_u + w_{1-2}$$

But $q_{1-2} = 0$

$\therefore$ $W_{1-2} = -(u_2 - u_1)$

In this process W.D equal to change in Internal energy

If steam is flowing through steam turbine isentropically then instead of first law of thermodynamic apply energy equation for this process.

$$h_1 - W.D + Q = h_2 \qquad \text{assuming PE = 0, K.E = 0}$$

$\therefore$ $W.D = h_1 - h_2$ as $Q = 0$ for isentropic process

This indicates that W.D during isentropic flow is equal to change in enthalpy

For isentropic process

$Pv^{1.13} = C$ For wet steam

$Pv^{1.3} = C$ for super heated steam

(A) Isentropic Process on P-v, T-s and h-s diagram when initially steam is wet and final condition of steam is also wet.

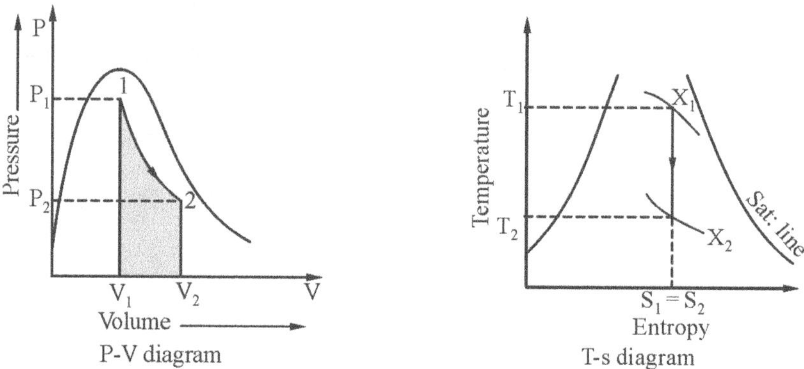

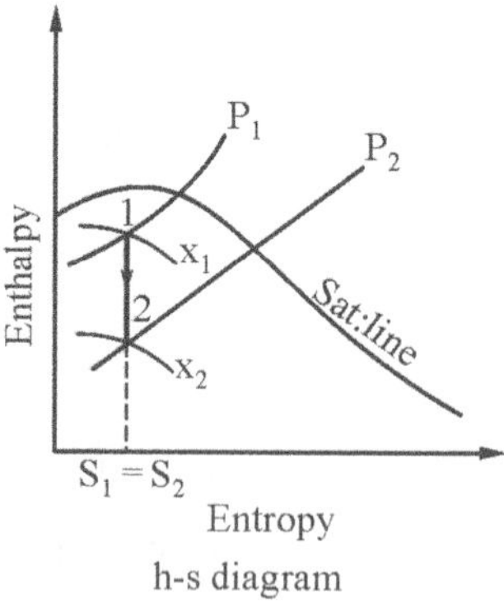

Fig. 4.9 Enthalpy – Entropy Diagram.

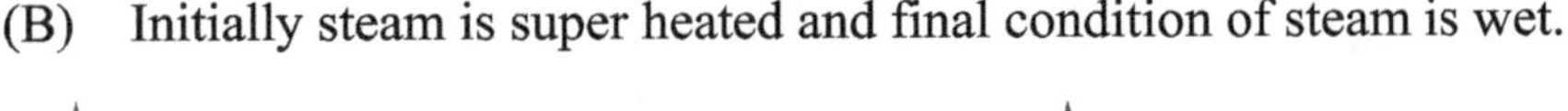

(B) Initially steam is super heated and final condition of steam is wet.

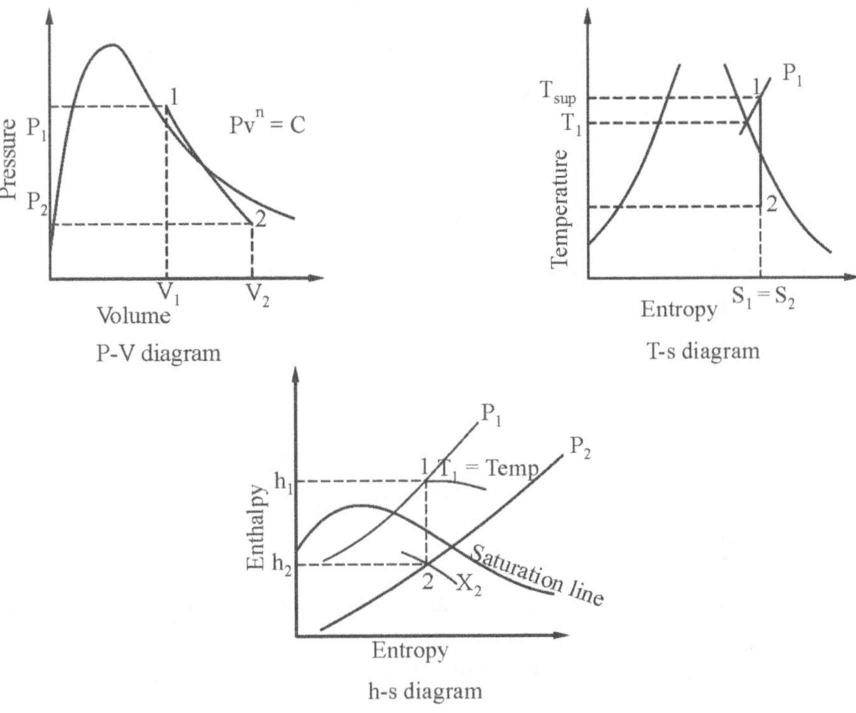

Fig. 4.10

4.8.6 Polytropic Process ($Pv^n = C$)

- In practice Isothermal process can be achieved by providing perfect cooling and process of expansion or compression should be carried out very slowly so that time available should be sufficient to transfer heat to cooling medium.
- Isentropic process in practice can be achieved by providing perfect insulation which increases the size of machine or vessel. Also process should be carried out very-very fast so that time available for heat transfer is reduced to great extent.
- Due to above limitation, in practice both isothermal and isentropic are not applicable. All process of expansion or compression follows polytrophic process, $PV^n = C$.

1. Consider 1 kg of wet steam being heated polytrophically from initial state 1 to final state 2.

Let

P_1 = Initial pressure of steam in bar

V_{g_1} = Specific volume of dry steam corresponding to pressure P_1 from steam tables.

x_1 = D.F of steam at point 1

P_2, V_{g2}, x_2 = corresponding value for final condition of steam at pressure P_2

η = Polytrophic index

$\therefore$ Initial volume of steam = $x_1 V_{g1}$

Final volume of steam = $x_2 V_{g2}$

$$P_1 v_1 = P_2^n v_2$$

$\therefore$ $P_1(x_1 V_{g1})^n = P_2 (x_2 V_{g2})n$ from this x_2 can be calculated

x_2 gives an important information about final condition of steam as follows:

1. If value of $x_2 < 1$, then final condition of steam is wet.
2. If $x_2 = 1$ then final condition of steam is dry saturated and value of $v_2 = v_{g2}$.
3. If $x_2 > 1$ then final condition of steam is super heated, in such case $v_2 = v_{sup}$.

$$\therefore \quad P_1(x_1 V_{g_1})^n = P_2(V_{sup})^n$$

$$P_2 (v_{sup})^n = P_1 (x \; v_{g_1})^n \qquad \text{(rearranging)}$$

$$\left(\frac{v_{sup}}{x_1 v_{g_1}}\right)^n = \frac{P_1}{P_2}$$

$$\therefore \quad V_{sup} = x_1 \; v_{g_1} \left(\frac{P_1}{P_2}\right)^{\frac{1}{n}}$$

T_{sup} can also be found out by charle's law

$$\frac{v_{sup}}{T_{sup}} = \frac{v_{g2}}{T_{sat}}$$

$$V_{sup} = V_{g_2} \cdot \frac{T_{sup}}{T_{sat}}$$

2. $W.D_{1-2} = 100 \dfrac{P_1 v_1 - P_2 v_2}{n-1}$

3. Change in internal energy

$$u_1 = h_1 - 100\ P_1\ v_1$$

$$u_2 = h_2 - 100\ P_2\ v_2$$

$$\therefore \quad u_2 - u_1 = (h_2 - h_1) + (100\ P_1\ v_1 - 100\ P_2\ v_2)$$

$$= h_2 - h_1 + (100\ P_1\ v_1 - 100\ P_2\ v_2)$$

4. Heat absorbed or heat transferred

$$q_{1\text{-}2} = d_{1\text{-}2} + W.D_{1-2}$$

$$= (u_2 - u_1) + W.D_{1-2}\ \ KJ/kg$$

4.8.7 Throttling Process (Constant Total heat Process)

1. It is also known as wire drawing process.
2. Steam is said to be throttled when it passes through a restricted opening. Such as partially open valve or narrow opening.
3. The leakage of fluid through crack in the vessel is example of throttling.
4. The restricted opening is so narrow that due to friction resistance between fluid and surface of aperture, the velocity of fluid is almost reduced to zero.

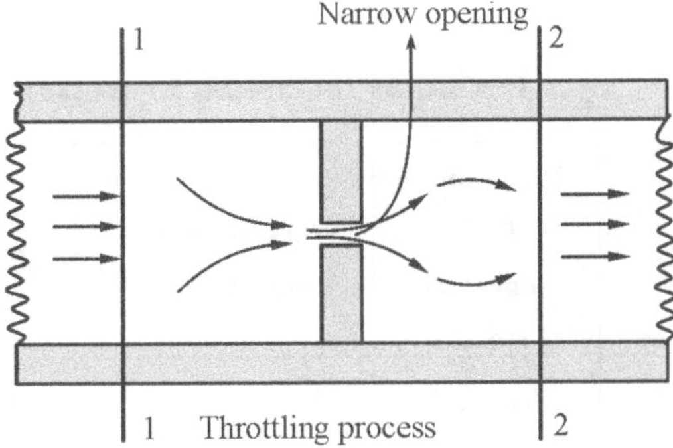

Fig. 4.11

5. K.E of fluid is converted into heat by friction. Initially if steam is wet, it will start drying up as it expands, the nearly dry steam after throttling becomes super heated.

6. During throttling no heat is supplied or rejected i.e., $q_{1-2} = 0$.
7. No work is done by expanding fluid i.e., W.D = 0
8. No change in internal energy of fluid i.e., du = 0.
9. Enthalpy before throttling = enthalpy after throttling.
10. Steam after throttling generally becomes super heated.
11. Applying S.F.E.E between section 1– 1 and 2 – 2 for unit mass of fluid

$$h_1 + \frac{v_1^2}{2} + gz_1 + q_{1-2} = h_2 + \frac{v_2^2}{2} + gz_2 + w_{1-2}$$

(i) v_1 and v_2 are small $\quad\therefore\ v_1^2/2$ and $v_2^2/2$ are negligible.

(ii) $q_{1-2} = 0$ as it is insulated

$W_{1-2} = 0$ as no work is done by fluid (Pipe in which fluid is flowing through restricted opening does not move the pipe)

(iii) $z_1 = z_2$

$\therefore\quad h_1 = h_2$

Let x_1 = D.F of steam entering the throttling

x_2 = D.F of steam after throttling

$\therefore\quad h_{f_1} + x_1\, h_{f_{g1}} = h_{f_z} + x_2\, h_{f_{g2}}$

Formulas for Steam Processes

Type of Expansion	Remarks	Equation for final condition of steam	Work done on Piston = W.D	Internal Energy	Heat absorbed q kJ/kg
Constant volume	-	$x_1V_1 = x_2 V_2$	0	$u_2 - u_1$	$u_2 - u_1$
Constant Pressure	Represents Evaporation and Condensation	-	10^2 P($V_2 - V_1$)	$(h_2 - h_1) - (P\, x_2\, v_{g_2} - P\, x_1\, V_{g_1})$	$h_{fg}(x_2 - x_1)$
Constant Temperature		-	10^2 P($V_2 - V_1$)	$(h_2 - h_1) - (P\, x_2\, V_{g_1} - Px_1\, V_{g_1})$	$h_{fg}(x_2 - x_1)$

Type of Expansion	Remarks	Equation for final condition of steam	Work done on Piston = W.D	Internal Energy	Heat absorbed q kJ/kg
Hyperbolic	PV = C	$P_1 x_1 V_{g1} = P_2 x_2 V_{g2}$	$10^2 P_1 V_1 \log_e \frac{V_2}{V_1}$	$u_2 - u_1$	$W + (u_2 - u_1)$
Reversible adiabatic	Isentropic	$S_1 = S_2$	$(u_2 - u_1)$	-	0
Throttling	Constant Total Heat	$H_1 = H_2$	0	0	0
General case	Polytrophic $PV^{\eta} = C$	$P_1(x_1 V_{g1})^n = P_2(x_2 V_{g2})^n$	$\frac{10^2(P_1V_1 - P_2V_2)}{n-1}$	$(u_2 - u_1)$	W.D+ $(u_2 - u_1)$

4.9 Mixing

When the quantity of steam generated by one boiler is not sufficient, then two or more boilers can supply steam to steam engine or turbine.

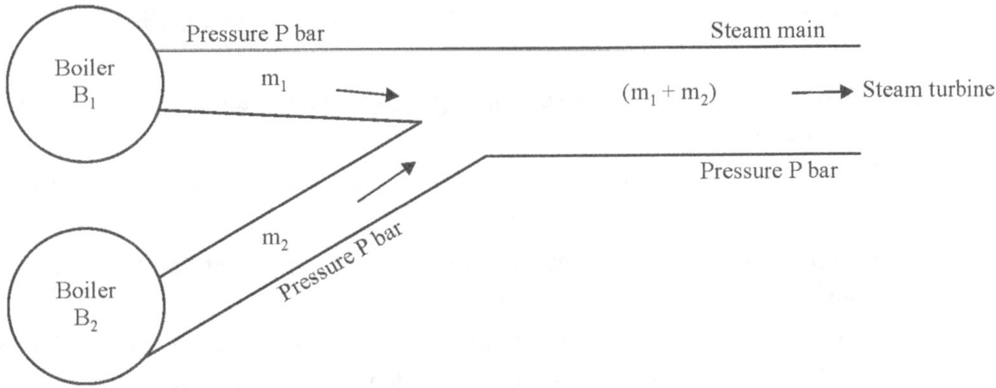

Fig. 4.12

1. Steam supplied by Boiler $B_1 = m_1$
2. Steam supplied by Boiler $B_2 = m_2$
3. Pressure of the steam supplied by both boilers is same so there is no pressure loss during mixing and pressure of steam in main pipe will be same i.e., P bar.
4. If quality supplied by each boiler is different then quality of steam in main is to be calculated.

5. If D.F of steam supplied by boiler $B_1 = x_1$ and steam supplied by Boiler B_2 is super heated?.
6. If it is assumed that steam in main is wet then we can write

$$m_1\left(h_{f_1} + x_1 h_{f_{g1}}\right) + m_2\left[h_{f_1} + h_{f_{g1}} + c_{p_s}\ (t_{sup} - t_{sat})\right] = (m_1 + m_2)\left(h_{f_1} + x_2 h_{f_{g2}}\right)$$

If we get $x_2 > 1$, it indicates that steam in main is superheated then rewrite above equation and final T_{sup} of steam in main.

$$m_1(h_{f1} + x_1\ h_{fg1}) + m_2(h_{f1} + h_{fg1} + c_{ps}\ (t_{sup1} - t_{sat}) = (m_1 + m_2)$$
$$(h_{f1} + h_{fg1} + c_{ps}\ (t_{sup2} - t_{sat})$$

t_{sup_2} can be calculated

4.10 Super heater

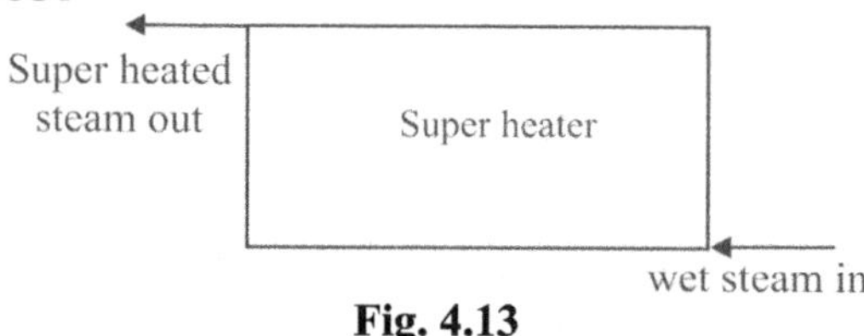

Fig. 4.13

1. Steam coming out from boiler is always wet as it remains in contact with water.
2. Steam from boiler is passed through superheater tubes. These tubes are placed in the path of hot gases.
3. Heat of hot gases is transferred to the steam passing through the superheater tubes and steam is super heated.
4. Assuming there is no pressure loss of steam passing through the super heater.
5. Suppose Dryness fraction of steam entering to the superheater is "x". Then (1 – x) kg will be the suspended water particles in it. These particles are at saturation temperature corresponding to given pressure, so they will absorb latent heat and will be converted into dry saturated steam. After that dry steam absorbs sensible heat, if temperature is increases from saturation temperature to superheated temperature.
6. Heat given to steam through super heater

= Final enthalpy of steam leaving the super heater – Initial enthalpy of wet steam entering the super heater.

$$= h_f + h_{fg} + c_{ps}(t_{sup} - t_{sat}) - (h_f + xh_{fg})$$

$$= h_{fg} - xh_{fg} + c_{ps}(t_{sup} - t_{sat})$$

$$= (1 - x)h_{fg} + c_{ps}(t_{sup} - t_{sat})$$

4.11 De-super heating of Steam

1. The temperature of steam coming out from superheater is to be controlled for safe operation of turbine or steam engine.
2. If temperature of steam leaving the superheater is more than predetermined temperature, then temperature of steam entering the turbine is controlled by injecting the water into steam in form of spray.

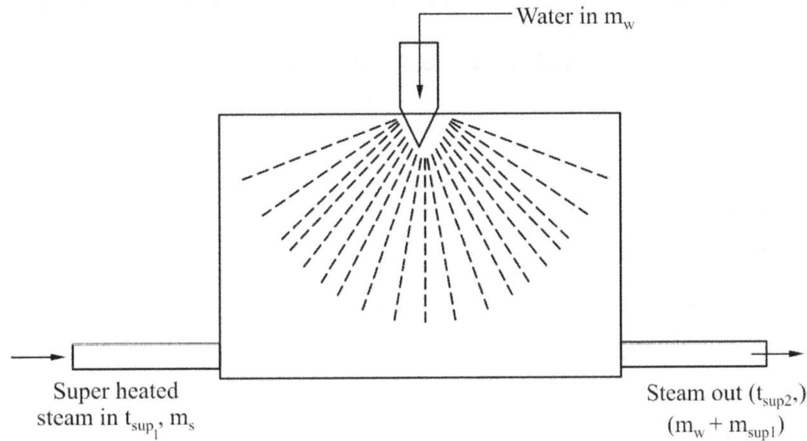

Fig. 4.14

3. Fig. 4.14 shows the arrangement of de-super-heater. Water is injected into stream of steam. Excess super-heat is reduced by giving heat to the water injected.
4. The temperature of steam entering and leaving the de-super-heater are t_{sup_1} and t_{sup_2} respectively.
5. The mass of steam flowing and mass of water injected per unit time are m_s and m_w kg respectively.
6. Assuming there is no pressure and heat loss, applying law of conversation of energy.
   Enthalpy of super heated steam + Initial heat of water sprayed entering into the de-super heater = Enthalpy of super heated steam leaving the desuper-heater.

$$m_s[h_{f_1} + h_{fg} + c_{ps}(t_{sup_1} - t_{sat})] + m_w h_f = (m_s + m_w)[h_{f_1} + h_{f_{g1}} + c_{ps}(t_{sup_2} - t_{sat})$$

where h_f = enthalpy of water; $c_{pw} t_w = 4.1868\ t_w$ kJ/kg.

4.12 Determination of Dryness Fraction of Steam

There are four methods of measuring Dryness fraction of steam

1. Barrel Calorimeter
2. Separating calorimeter
3. Throttling calorimeter
4. Combined separating and throttling Calorimeter or combined calorimeters

4.12.1 Barrel (Tank) or Bucket Calorimeter

Construction

It consists of

1. Main steam pipe (M.S.P)
2. Sampling tube
3. Control valve (c.v)
4. Pressure gauge (P.G)
5. Thermometer (T)
6. Inner vessel (calorimeter) (c.c)
7. Outer vessel (o.v)
8. Insulating material (space between internal and outer vessel with air)
9. Platform of weighing machine (P)
10. Wooden cover (Placed on calorimeter and outer vessel) (W.C)

Working

Dryness fraction can easily be calculated by barrel calorimeter.

1. A vessel (Calorimeter) contains a known mass of water.
2. The water equivalent of copper calorimeter is determined experimentally and its value stamped on it. It is surrounded by an outer vessel thus providing an air space in between, which acts as an insulator.
3. Thermometer is inserted through one of the holes in wooden cover to measure temperature of water.

4. The vessel is placed on the platform of weighing machine, sample steam pipe is connected to main steam pipe.
5. Steam from main pipe enters into sampling tube via control valve and the pipe, flows into the cold water through fine exit holes of the ring, attached at the end of steam pipe.
6. The condensation of steam takes place as it comes in contact with cold water as a result temperature of water rises.

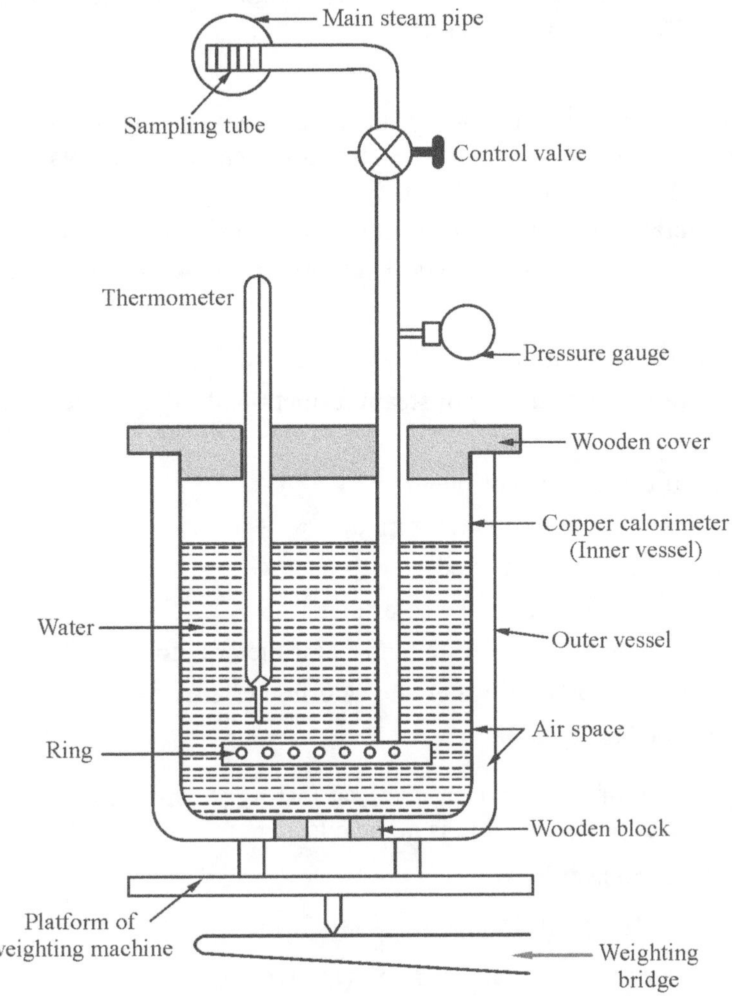

Fig. 4.14 Barrel Calorimeter

7. The mass of steam condensed can be known from the difference of reading of weighing machine before and after condensation of steam. The pressure at which steam enters the calorimeter is indicated by pressure gauge.
8. Sufficient rise in temperature be obtained by blowing large quantity of steam into calorimeter. But at the same time the final temperature of mixture (water) should not be too high. This increases heat loss by conduction and radiation.
9. Too much opening of the control valve should be avoided otherwise steam blowing into cold water will result in bubbles of steam left uncondensed escaping out of cold water thus leading to wrong result.

Limitation

1. The value of Dryness fraction by this method gives only approximate value. The value of Dryness fraction obtained by this method is always lower than actual value of Dryness fraction of steam.
2. Final temperature of water in calorimeter should not rise too much. This will cause error in D.F. of steam due to increased losses by conduction and radiation.

Calculation

P = Pressure of steam in bar

t_{sat} = Saturation temperature of steam formation at P (From steam table)

m^*_c = Mass of calorimeter

c_c = Specific heat of Calorimeter

m_e = equivalent of Calorimeter = $m_c\, c_c$

m_s = Mass of steam condensed

m_w = Mass of cold water in the calorimeter

t_1 = Initial temperature of water and Calorimeter before steam supply

t_2 = Final temperature of water and calorimeter after mixing the steam.

c_{p_w} = Specific heat of water

According the law of conversation, the heat lost by steam is equal to the heat gain by water and calorimeter provided there is no heat loss.

Provided there is no heat loss

Heat lost by steam $= ms\,[xh_{fg} + c_{p_w}\,(t_{sat} - t_2)]$ ..... (1)

Heat gain by water $= (m_w\, c_{p_w} + m_c\, c_c)\,(t_2 - t_1)$ ..... (2)

∴ Equating equation (1) and (2)

$$m_s[xh_{fg} + c_{p_w}\,(t_{sat} - t_2)] = (m_w\, c_{p_w} + m_c\, c_c)\,(t_2 - t_1)$$

From this expression, the dryness fraction "x" may be determined.

4.12.2 Separating Calorimeter

The separating Calorimeter is used to determine Dryness fraction of steam by mechanically separating the water particles from wet steam.

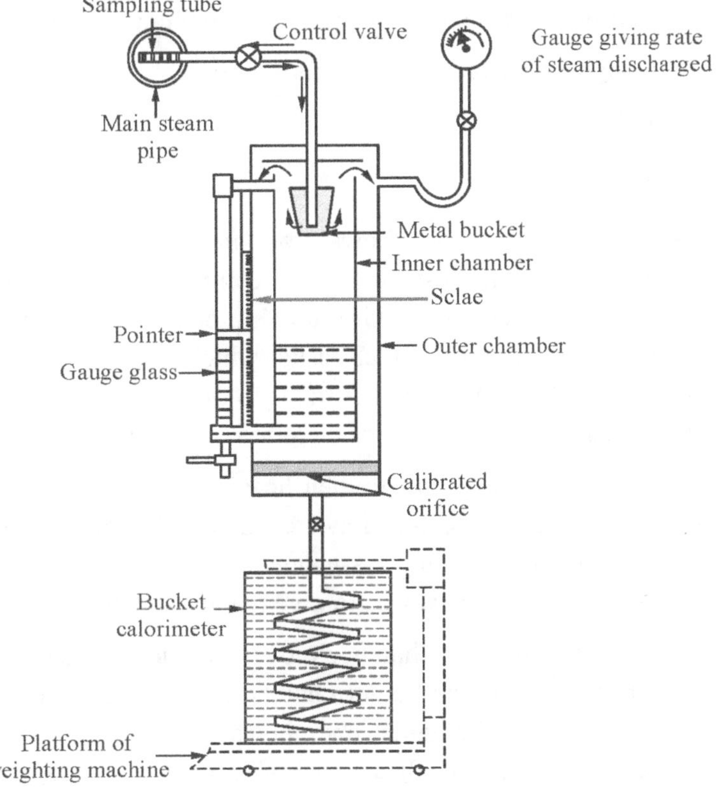

Fig. 4.15 Separating Calorimeter.

Construction

1. Main steam Pipe (M.S.P)
2. Control valve fitted with steam pipe.
3. Two concentric chambers inner and out chamber (I.C and OC)
4. Both chambers communicate with each other by the opening of the top of inner chamber.
5. Sampling tube or sampling steam pipe.
6. Metal basket (Baffle plate) (M.B)
7. Gauge glass (G.G) indicating level of water inside inner chamber

8. Calibrated orifice (C.O)
9. Scale (s)
10. Bucket calorimeter (BC)
11. Platform balance (P.B) or flatform weighing machine
12. Gauge giving rate of steam discharge
13. Pointer (P)

Working

The steam to be tested flows from main steam pipe through sample pipe (perforated) facing up steam via the control valve which must be fully opened to ensure that there is no wire drawing or throttling taking place. As steam is discharged through the metal basket which has a large number of perforations all around but not at base, it is made to move obtuse (more than 90^0) angle and undergoes a sudden reverse of direction of motion.

The water particles due to their heavier momentum get separated from steam and water is collected into inner chamber. The quantity of water collected is indicated by the level in glass-gauge and the pointer moves on scale. The scale is graduated in such a way as to give reading directly in kg of water.

Dry steam in inner chamber moves up and then downwards in the annular space between inner and outer chamber. The gauge fitted to the outer chamber has two scales. The inner one indicates the pressure while the outer one indicates the rate of discharge of dry saturated steam per 10 minutes. Afterwards the steam flows through the calibrated orifice to bucket calorimeter which is placed on the platform balance by which the mass of steam discharged can be further conformed.

m_s = Mass of dry steam condensed in kg

m_w = Mass of suspended moisture collected in inner chamber in kg

$$x = \text{Dryness fraction} = \frac{m_s}{m_s + m_w}$$

This method does not give accurate D.F of steam as the steam discharged is not absolutely dry i.e., water particles of wet steam are not separated by mechanical means completely.

4.12.3 Throttling Calorimeter

Principle: The fact that total heat of steam during throttling process remains constant is made use in throttling calorimeter.

Construction

It consists of:

1. Main steam pipe (MSP)
2. Sampling tube (S.T)
3. Throttling valve (T.V)
4. Inner chamber (I.C)
5. Outer chamber (O.C)
6. Thermometer in the pocket filled with cylinder oil
7. Exhaust (E)
8. Manometer (M)
9. Gauge pressure after throttling in mm of H_g (h)
10. Annular space (A.S)

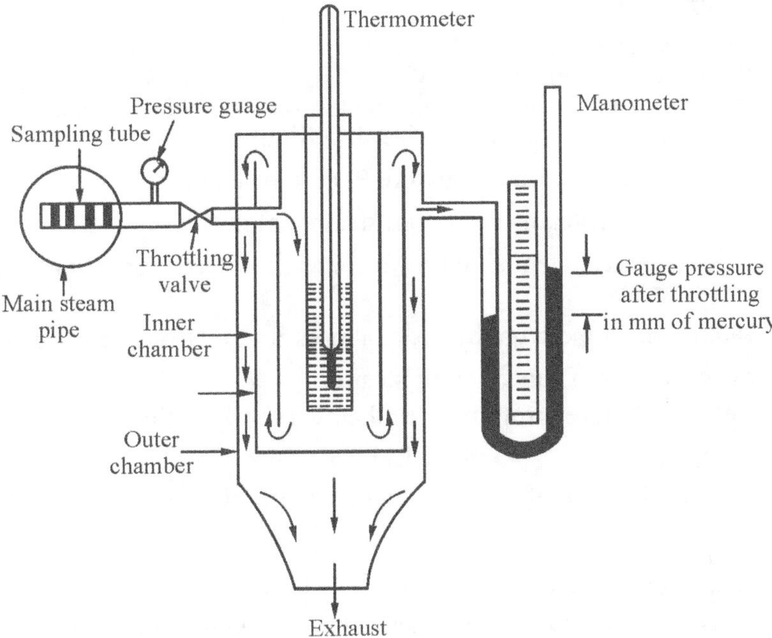

Fig. 4.16 Throttling Calorimeter.

Working

It consists of the sampling tube through which the steam from main steam pipe flows to throttling valve where steam is throttled from higher pressure to lower pressure and steam becomes super heated (5 °C super heat).

Steam then flows into inner chamber at the top, turns downwards into annular space between inner and outer chamber which is insulated to avoid heat losses. Loss of heat by radiation from inner chamber by the hot steam, around the outside of the inner chamber (i.e., in the annular space).

Before temperature reading are taken, the steam should flow long enough to obtain internal steady state i.e., long enough for all parts to be heated to a temperature that remains constant.

This method of determining dryness fraction will not give correct results unless steam issuing from throttle valve and entering the inner chamber is super- heated.

In order to ensure it, manometer is attached to measure the pressure of steam after throttling. Corresponding to this 'P_2' in bar absolute, saturation temperature of steam is read from steam table. If this value of saturation temperature is lower than that indicated by thermometer, the steam is superheated otherwise not. The steam finally escapes out through exhaust.

Let

P_1 = Initial steam pipe pressure of wet steam before throttling i.e., pressure of steam main in bar absolute.

P_2 = Pressure of steam after throttling in bar absolute (generally P_2 is measured by manometer in "h" mm of Hg. This gauge pressure should be converted into bar and by adding Barometer reading we get actual pressure OR absolute pressure of steam after throttling)

h_{fg_1} = Latent heat of steam at pressure P_1 (From steam table)

h_{fg_2} = latent heat of steam at P_2 (from steam table)

t_{sup} = super heated temperature of steam after throttling indicated by thermometer)

t_{sat} = saturation temperature of steam at pressure p_2 (from steam table)

Cps = Specific heat of super heated steam

Enthalpy before throttling = Enthalpy after throttling

$$h_{f_1} + xh_{fg_2} = h_{g_2} + c_{ps}(t_{sup} - t_{sat})$$

From this Dryness fraction of steam x can be determined.

Limitation

Condition of steam after throttling must be super-heated or $t_{sup} > t_{sat,}$ then D.F can accurately be measure by throttling calorimeter. If wet steam from main steam pipe is less then 0.9, the steam after throttling will not be super heated and throttling calorimeter will not measure D.F correctly.

4.12.4 Combined Separating and Throttling Calorimeter

Introduction

It measures accurately the D.F of steam coming out from boiler main steam pipe because limitations of only separating calorimeter and only throttling calorimeter are overcome by combined separating and throttling calorimeter.

The steam passing out from a separating calorimeter may still contain some water particles in it because it is impossible to separate out all suspended particles from steam by mechanical means. In other words steam may not be absolutely dry and hence this calorimeter gives higher value than actual dryness fraction of steam.

Again in throttling calorimeter, the steam passing through the throttling valve (Restricted valve opening) must be super heated. This limits the extent of D, F that can be measured by throttling calorimeter.

It has been observed that minimum D.F that can be reliably measured depends upon pressure of steam of in the main steam pipe. e.g, it varies from 0.97 for 4.1 bar pressure to 0.94 for 45.1 bar pressure. Thus the sample of steam which has D.F less than this value may be still wet after throttling valve i.e., it will not be super heated.

To overcome these difficulties we make use of combined separating and throttling calorimeter (or combined calorimeters).

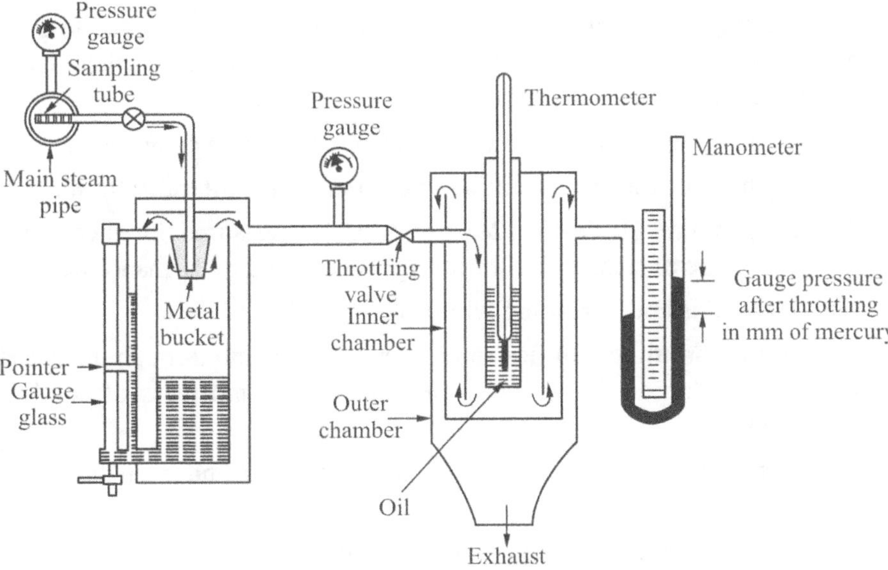

Fig. 4.18 Combined Separating and throttling calorimeter.

Construction

1. Main Steam Pipe (MSP)
2. Sampling tube (S.T)
3. Metal bucket (M.B)
4. Pointer (P)
5. Glass gauge (G.G)
6. Inner chamber (IC)
7. Outer chamber (O.C)
8. Pressure gauge (PG) Two numbers (one is attached at MSP and other before throttling)
9. Throttle valve (T.V)
10. Thermometer in pocket filled with cylinder oil (T)
11. Exhaust (E)
12. Manometer (M)
13. Gauge pressure after throttling in mm of Hg = h

Working

1. The steam to be tested flows from main steam pipe through sample pipe (sample tube) via control valve into separating calorimeter.
2. In separating calorimeter suspended water particles are removed to great extent and steam is dried up to certain extent, mass of water particle separated in separating calorimeter is measured.
3. The resulting semi dry steam in throttled into throttling calorimeter.
4. This method ensures that steam will be super heated after throttling.
5. The temperature of steam after throttling is measured by thermometer and gauge pressure of steam after throttling is measured by U-manometer.
6. Note down Barometer reading while taking readings.
7. Steam after throttling will flow into inter chamber and through outer chamber it leaves the calorimeter. Steam is then condensed and mass of condensated is measured.
8. Thermometer pocket should be half filled with cylinder oil before inserting thermometer.
9. After erection, the instrument should be thoroughly lagged to minimize heat losses.

10. The commencement of taking reading should not be hurried even when good lagging is provided.
11. To allow sufficient time before taking reading, so that proper heating of various parts of unit is established.

Calculation of Dryness Fraction

m = Mass of steam in main steam pipe in kg

m_w = Mass of suspended water particles (moisture) separated in separating calorimeter

m_s = Mass of steam entering the throttling calorimeter in kg or leaving the separating calorimeter

x = Dryness fraction of steam in the main steam pipe in kg

x_1 = Apparent D.F. of steam entering the throttling calorimeter or D.F of steam leaving the separating calorimeter.

P_1 = Pressure of steam entering the separating calorimeter in bar or pressure of steam entering the throttling calorimeter.

P_2 = Pressure of steam after throttling

m_s = mass of steam entering the throttling calorimeter which is condensed after passing through calorimeter.

If x_1 is D.F of steam from separating calorimeter

$$x_1 = \frac{m_s}{m_s + m_w}$$

Let x be the initial D.F of steam in mains and let us prove that $x = x_1 \times x_2$

x_1 = D.F of steam in separating calorimeter

x_2 = D.F. of steam passed through throttling calorimeter

Proof: Initial water particle in the main contains $(1 - x) \times$ total mass of steam in main steam pipe (not to prove in examination)

$$= (1 - x)(m_s + m_w) \quad \text{.....(i)}$$

Out of this $(1 - x_1)(m_s + m_w)$ is removed in separating calorimeter and $(1 - x_2)\,m_s$ is passed through throttling calorimeter.

Mass of initial water particles in mains contains = mass of water removed in separating calorimeter + mass of water particles passed through throttling calorimeter.

$$(1 - x)(m_s + m_w) = (1 - x_1)(m_s + m_w) + (1 - x_2)m_s$$

$$1 - x = (1 - x_1)\frac{m_s + m_w}{m_s + m_w} + (1 - x_2)\frac{m_s}{m_s + m_w}$$

$$1 - x = (1 - x_1) + (1 - x_2)\, x_1 \qquad \text{where } x_1 = \frac{m_s}{m_s + m_w}$$

$$1 - x = 1 - x_1 + x_1 - x_2 \,.\, x_1$$

Type $\quad -x = -x_2\, x_1$

$\therefore \quad x = x_1\, x_2 \quad$ Proved

Enthalpy before throttling = Enthalpy after throttling

(Before throttling steam is always wet) = After throttling steam is super heated.

$$h_{f_2} + x_2\, h_{fg_2} = h_{g_3} + C_{ps}\,(t_{sup} - t_{sat}) \qquad \text{where } h_{g3} = h_f + h_{fg}$$

4.13 Solved Problem

1. Calculate Dryness fraction of steam containing 2 kg of water suspended with 52 kg of Dry saturated steam.

Solution: Given $m_w = 2$ kg

$m_s = 52$ kg

$$\therefore \quad D.F = \frac{m_s}{m_w + m_s} = \frac{52}{2+52} = \frac{52}{54} = 0.963$$

2. Find volume and quality of steam at 15 bar, 10500 KJ of internal energy and mass of 5 kg.

Solution: From steam table for P = 15 bar

$v_f = 0.001154$ m^3/kg

$v_g = 0.13167$ m^3/kg

$h_f = 844.6$ KJ/kg

$h_{fg} = 1945.3$ KJ/kg

$$\text{Specific internal energy} = u = \frac{u}{5} = \frac{10500}{5} = 2100 \text{ KJ/kg}$$

But $\quad u = h - W.D$

$$\therefore \quad 2100 = h_f + x\, h_{fg} - 10^2 \,.\, 15 \,.\, x \,.\, v_g$$

$$= 844.6 + x \times 1945.3 - 10^2 \times 15 \times x \times 0.13167$$

$$2100 - 844.6 = 1945.3x - 197.505\, x$$

$$= 1747.795\, x$$

$\therefore \quad x = 0.7183$

$\therefore$ quality of steam = 0.7183 × 100 = 71.83% **Ans.**

Specific volume of steam = $(1 - x)\ v_f + x v_g$

$= (1 - 0.7183) \times 0.001154 + 0.7183 \times 0.13167$

$= 0.0949\ m^3/kg$

$\therefore$ Total volume of steam = 5 × 0.0949 = 0.4745 m^3 **Ans.**

3. Calculate the total heat of 4 kg of steam at pressure 8 bar having Dryness fraction of 0.8. Also calculate the heat required to convert this steam into dry saturated steam.

Solution: Given P = 8 bar

$x = D.F = 0.8$

$m_s = 4$ kg

From steam tables at 8 bar

$h_f = 720.94$ kJ/kg

$h_{fg} = 2046.5$ kJ/kg

$h_g = 2767.5$ kJ/kg

Enthalpy of wet steam/kg

$= h_f + x\ h_{fg}$

$= 720.94 + 0.8 \times 2046.5$

$= 2358.14$ kJ/kg

Enthalpy of 4 kg of steam

$= 9432.56$ kJ

Enthalpy of 1 kg of Dry-saturated steam at 8 bar from steam tables

$= 2767.5$ kJ/kg.

$\therefore$ Enthalpy of 4 kg of dry saturated steam

$= 4 \times 2767.5 = 11070.0$ kJ

Total heat required to convert 4 kg of wet steam into dry saturated steam

$= 11070 - 9432.56 = 1637.44$ kJ **Ans.**

4. What amount of heat would be required to produce 5 kg of steam at pressure 5 bar and temperature of 250 °C from water at 30 °C. Take $c_{ps} =$ 2.1 KJ/kg k

Solution: Given

$m_s = 5$ kg

$P = 5$ bar

$t_{sup} = 250$ °C

$t_w = 30$ °C

$c_{ps} = 2.1$ kJ/kg K

From steam tables corresponding to 5 bar

$h_f = 640.12$ kJ/kg

$h_{fg} = 2107.4$ kJ/kg

$h_g = 2747.5$ kJ.kg

$t_{sat} = 151.84$ °C

Enthalpy of 1 kg of super heated steam from water at 30 °C

$$h_{sup} = h_g + c_{ps}(t_{sup} - t_{sat}) - c_{pw}(t - 0)$$

$$= 2747.5 + 2.1(250 - 151.84) - 4.187 \times 30$$

$$= 2828 \text{ kJ/kg}$$

Enthalpy (Total heat) required to produce 5 kg of steam

$$= 5 \times 2828$$

$$= 14140 \text{ kJ}$$ **Ans.**

5. How much heat is required to be added to convert 4 kg of water at 20 ^{0}C to steam at 8 bar pressure and 200 ^{0}C temperature. Take c_{ps} for super heated steam as 2.1 kJ/kgk and cp_w for water as 4.187 kJ/kg k.

Solution: Given

$m_s = 4$ kg

$t_w = 20$ ^{0}C

$P = 8$ bar

$C_{ps} = 2.1$ kJ/kg K

$t_{sup} = 200$ °C

$C_{pw} = 4.187$ kJ/kg K

From steam table corresponding to pressure 8 bar

$t_{sat} = 170.41$ °C

$h_g = 2767.5$ kJ/kg

Enthalpy of super heated steam per kg of steam

$= h_g + c_{ps} (t_{sup} - t_{sat})$ when water is at 0 ^{0}C

But in this problem water is taken at 20 ^{0}C, corresponding formula will be

Enthalpy of super heated steam/kg $= h_g + c_{ps} (t_{sup} - t_{sat}) - c_{pw} (t - 0)$

$= 2767.5 + 2.1 (200 - 170.41) - 4.187 \times 20$

$= 2745.899$ kJ/kg

Enthalpy of 4 kg of steam

$= 4 \times 2745.99$

$= 10983.596$ kJ **Ans.**

6. Find the enthalpy and specific volume of steam at pressure 10 bar when steam is (i) Dry saturated (ii) wet having Dryness fraction of 0.8 (iii) super heated having temperature of 240 °C. Take C_{ps} = specific heat of super heated steam = 2.1 kJ/kgk

Solution: (i) Given

P = 10 bar

Steam Dry saturated

Find specific volume and enthalpy

From steam table corresponding to 10 bar

the enthalpy of Dry-saturated steam

$h_g = 2776.2$ kg/kg and **Ans.**

$v_g = 0.194$ m^3/kg **Ans.**

(ii) Given Pressure of steam P = 10 bar

Wet steam x = 0.8

From steam table corresponding to 10 bar

$h_f = 762.6$ kJ/kg

$h_{fg} = 2013.6$ kJ/kg

$h_g = 2776.2$ kJ/kg

$v_g = 0.194$ m^3/kg

$t_{sat} = 179.9$ kJ/kg

$v_f = 0.001127$ m^3/kg

Enthalpy of wet steam

$$h = h_f + x\,h_{fg} = 762.6 + 0.8 \times 2013.6$$

$$= 2373.48 \text{ kJ/kg} \quad \textbf{Ans.}$$

Specific volume wet steam

$$= (1 - x)\,v_f + x\,v_g$$

$$= (1 - 0.8) \times 0.001127 + 0.8 \times 0.194$$

$$= 0.0002254 + 0.1552$$

$$= 0.155425 \text{ m}^3\text{/kg} \quad \textbf{Ans.}$$

Note: Quantity $(1 - x)\,v_f$ is very small and therefore specific volume of water may be neglected therefore.

$\therefore$ Specific volume of wet steam

$$= xv_g$$

$$= 0.8 \times 0.194$$

$$= 0.1552 \text{ m}^3\text{/kg} \quad \textbf{Ans.}$$

(iii) h = Enthalpy of super heated steam

$$= h_g + cp(T_{saup} - T_{sat})$$

$$= 2776.2 + 2.1\,(240 - 179.9) = 2902.41$$

$$V_{sup} = v_g\,\frac{T_{sup}}{T_{sat}} = 0.194\,\frac{240 + 273}{179.9 + 273} = 0.22 \text{ m}^3\text{/kg}$$

7. How much quantity of heat would be required to produce 5 kg of steam at a pressure of 5 bar and temperature of 250 ^{0}C from feed water at 30 °C. Assume specific heat of super heated steam as 2.1 KJ/kgk and specific heat of water as 4.187 KJ/kg k.

Solution: Given

Mass of steam = m_s = 5 kg

$$P = 5 \text{ bar}$$

$$t_{sup} = 250\ ^0\text{C}$$

$$t_w = 30\ ^0\text{C}$$

$$C_{ps} = 2.1 \text{ KJ/kg k}$$

$$C_{pw} = 4.18 \text{ kg/kg k}$$

From steam table corresponding to 5 bar

$$h_g = 2747.5 \text{ kJ/kg}$$

$$T_{sat} = 151.8\ ^0C$$

Enthalpy of super heated steam per kg of steam

$$= h_g + c_{ps}(t_{sup} - t_{sat}) - t_w\ c_{pw}$$

$$= 2747.5 + 2.1\ (250 - 151.8) - 4.187 \times (30 - 0)$$

$$= 2953.72 - 125.66$$

$$= 2828.11 \text{ kJ/kg}$$

Total Enthalpy requires produce 5 kg of steam = 5 × 28 28.11

$$= 14140.55 \text{ kJ} \quad \textit{Ans.}$$

8. Water at 90 ^0C is injected into 10 kg of steam at a pressure of 500 KN/m^2 absolute and 180^{0}C until it becomes wet steam at same pressure with dryness fraction of 0.8. Find accuracy of water injected; assume c_p for steam as 2.23 kJ/kg k and specific heat of water as 4.1868 kJ/kg k.

Solution: Given

Water injected is at 90 °C

Mass of steam = m_s = 10 kg

Pressure of steam = P = 500 $\dfrac{kN}{m^2}$ = 5 bar

Temperature of 180°C is super heated temp because at 5 bar from steam table t_{sat} = 151.86°C

t_{sup} = 180^{0}C

Steam after injection of water becomes wet x = 0.8

Find accuracy of water injected i.e., mass of water injected

$$C_{ps} = 2.23 \text{ kJ/kg k}$$

$$C_{pw} = 4.187 \text{ kJ/kg k}$$

***Note*:**

Step I Find total heat of super heated steam at 5 bar and 180 ^0C

Step II After injection of water steam becomes wet. So calculate total heat of wet steam at 5 bar and D.F = 0.8

Step III Difference of step I and step II will give heat absorbed by injected water.

Step IV Difference of Enthalpy

$= m_w\, c_{pw}\, (t_s - 90)$ from step I and step II

Where m_w = Mass of water injected; t_s = Saturation temperature of water at 5 bar.

At 5 bar from steam table

$h_f = 640.1$ kJ/kg

$h_{fg} = 2107.4$ kJ/kg

$h_g = 2747.5$ kJ/kg

$t_s = 151.8\ ^0C$

Step I $h = 10[h_g + c_{ps}(t_{sup} - t_s)]$ kJ

$= 10[2747.5 + 2.23(180 - 151.8)] = 2810.386$ kJ

Step II $h = 10[h_f + x\, h_{fg}] = 10[640.1 + 0.8 \times 2107.4] = 2326.02$ kJ

Step III Heat absorbed by water

step IV $=$ Step I – step II $= 28103.86 - 23260.2 = 483.66$ kJ

$$484.366 = m_w\, c_{pw}(t_s - 90) \text{ i.e., } m_w = \frac{4843.66}{4.1868(151.8 - 90)}$$

$\therefore\ m_w = 1.872$ kg **Ans.**

9. Determine condition of steam in the following cases.

(i) At 10 bar and temperature 200 °C

(ii) A pressure 10 bar and volume 0.175 m^3/kg

Solution: (i) Corresponding to pressure 10 bar from steam tables

saturation temperature of dry-saturated steam is 179.9 °C. But in problem the temperature of steam is 200 °C, which is greater than dry saturated steam.

Therefore steam is super heated and degree of super heat

$= (200 - 179.9) = 20.1$ °C *Ans.*

(ii) Corresponding to pressure 10 bar from steam tables.

The volume of dry saturated steam is 0.194 m^3/kg but in this problem the volume of steam is 0.175 m^3/kg which is less than volume of dry saturated steam. Therefore in problem steam is wet and its D.F. can be calculated as below.

At 10 bar $v_g = 0.194$ m^3/kg (From steam tables)

$V_{wet} = x.v_g$ (neglecting the volume of water)

$0.175 = 0.194\ x$

$$x = \frac{0.175}{0.194}$$

$= 0.902$

$\therefore$ In given problem condition of steam is wet having D.F = 0.902 **Ans.**

10. Determine the condition of steam in each of the following cases.

(i) At a pressure of 10 bar and temperature of 200 ^{0}C

(ii) At a pressure of 8 bar and volume 0.22 m^3/kg

Solution: (i) The saturation temperature corresponding to 10 bar is 179.9 °C from steam table. But in problem temperature of steam at 10 bar is 200°, which is more than Dry-saturated temperature. Therefore steam is superheated.

The degree of super heat = 200 – 179.9

= 20.1 °C *Ans.*

(ii) At 8 Bar from steam table, the specific volume of dry-saturated steam is 0.24 m^3/kg. But specific volume of steam in this problem is given as 0.22 m^3/kg which is less than specific volume of Dry-saturated steam at 8 bar.

$\therefore$ steam is wet

$V_{wet} = (1 - x)\ v_f + x\ v_g$

$0.22 = (1 - x)\ 0.001115 + 0.24\ x$

$\therefore \quad x = 0.916$

Steam is wet having D.F = x = 0.916 **Ans.**

11. Determine condition of steam in the following cases.

(i) At 10 bar and temperature 200 °C

(ii) At pressure 12 bar and 2688 KJ/kg heat required to produce it from water at 0 °C

Solution: Note (i) From steam table corresponding to 10 bar read only saturated temperature column as in this problem 200 ^{0}C temperature is given. (Do not see other columns in steam tables corresponding to 10 bar) corresponding to 10 bar from steam table the saturation temperature of Dry saturated steam is t_{sat} = 179.9 °C.

But in problem temperature of steam is 200 ^{0}C which greater than saturation temperature therefore steam is super heated and degree of super heat = (200 – 179.9)

= 20.1 °C **Ans.**

(ii) Corresponding to 12 bar from steam table read only h_g column (not to see other column in steam tables) the enthalpy of dry-saturated steam is 2782.7 KJ/kg. But in this problem enthalpy of steam given is 2688 KJ/kg which is less than dry-saturated steam.

∴ steam is wet and D.F = x

At 12 bar from steam tables

$h_f = 798.4$ kJ/kg

$h_{fg} = 1984.3$ kJ/kg

$h = h_f + x\, h_{fg}$

$2688 = 798.4 + 1984.3\, x$

∴ $x = 0.952$

Steam is wet and D.F = 0.952 **Ans.**

12. Sample of steam having pressure of 7 bar and Enthalpy = 2550 KJ/kg.

Find

(i) Dryness fraction of steam (ii) Specific volume (iii) Internal energy

Solution: Corresponding to 7 bar from steam tables enthalpy of dry-saturated steam is 2762 kJ/kg. But in this problem enthalpy of steam at 7 bar is 2550 kJ/kg which is less than dry-saturated steam. Therefore condition of steam is wet. Now let us find its Dryness fraction.

(i) At 7 bar from steam table

$h_f = 697.1$ kJ/kg

$h_{fg} = 2064.9$ kJ/kg

$v_g = 0.273\ m^3/kg$

∴ $h = h_f + x\, h_{fg}$

$2550 = 697.1 + 2064.9\, x$

∴ x = Dryness fraction = 0.897 **Ans.**

(ii) Specific volume of wet steam

$v = x v_g$

$= 0.897 \times 0.273 = 0.245$ **Ans.**

(iii) Internal energy = Enthalpy – 10^2 Pxv_g

$= h_f + x\, h_{fg} - 10^2\, P\, xv_g$

$= 647.1 + 0.897 \times 2064.9 - 10^2 \times 7 \times 0.897 \times 0.273$

$= 2377.898$ kJ/kg

13. Calculate the state of steam i.e., it is wet, dry or super heated

(i) Steam has pressure 15.5 bar and specific volume of 0.12 m^3/kg.

(ii) Steam has pressure 9 bar and temperature 200 ^{0}C

(iii) Steam has pressure of 27 bar and if 2700 KJ/kg of heat is required to generate steam.

Solution: (i) From steam table corresponding to pressure of 15.5 bar

the specific volume of Dry-saturated steam is 0.128 m^3/kg. But specific volume of steam given in problem is 0.12 m^3/kg, which is less than specific volume of dry-saturated steam. Therefore steam is wet.

At 15.5 bar from steam tables

$v_f = 0.001156$ m^3/kg

$v_g = 0.128$ m^3/kg

The condition of steam is found as below.

$v = (1 - x)\ v_f + x\, v_g$

$0.12 = (1 - x) \times 0.001156 + 0.128\, x$

$\therefore$ $x = 0.937$

$\therefore$ steam is wet having D.F = 0.937 **Ans.**

(ii) Corresponding to pressure 9 bar from steam tables. Saturation temperature of dry-saturated steam is 175.4 ^{0}C. But in problem temperature of steam is given as 200 ^{0}C which is greater than temperature of Dry-saturated steam.

$\therefore$ steam is super heated and

degree of super heat = $(t_{sup} - t_{sat})$

$= (200 - 175.4)$

$= 24.6$ ^{0}C

Steam is super heated and degree of super heat = 24.6 ^{0}C **Ans.**

(iii) From steam table corresponding to 27 bar the enthalpy of dry-saturated steam is 2801.9 kJ/kg. But enthalpy of steam given in problem is 2700 kJ/kg which is less than enthalpy of dry-saturated steam. This indicates that steam is wet because enthalpy of wet steam is always less than enthalpy of dry-saturated steam.

At 27 bar from steam table

$h_f = 981.2$ kJ/kg

$h_{fg} = 1820.5$ kJ/kg

$\therefore$ $h = h_f + x\, h_{fg}$

$\therefore$ $2700 = 981.2 + 1820.5 \times x$

$\therefore$ $x = 0.944$

In given problem steam is wet having D.F = 0.944 **Ans.**

14. Find condition of steam and determine enthalpy in following cases.

(i) $P = 40$ bar, $v = 0.062\ m^3/kg$

(ii) $P = 30$ and $t = 260\ ^0C$

(iii) $P = 40$ Bar, Heat added = 2400 kJ/kg

(iv) $P = 30$ bar, $v = 0.666\ m^3/kg$

Take $C_{ps} = 2.0934$ kJ/kg K for super heated steam

Solution: (i) From steam table corresponding to 40 bar

$h_{g_1} = 2800.3$ kJ/kg

$ts_{at} = 250.3\ ^0C + 273 = 523.3$ k

$v_{g_1} = 0.0497\ m^3/kg$

Given volume $v_1 > 0.0497$

$\therefore$ Steam is super heated

$$V_{sup} = v_{g1} \cdot \frac{T_{sup}}{T_{sat}}$$

$$0.062 = 0.0497\ \frac{T_{sup}}{T_{sat}}$$

$\therefore$ $T_{sup} = 652.43$ k **Ans.**

$\therefore$ Enthalpy $= h_1 = h_{g1} + c_{ps}\,(T_{sup} - T_{sat})$

$= 2800.3 + 2.0934\,(652.43 - 523.3)$

$= 3070.62$ KJ **Ans.**

(ii) $P_2 = 30$ and $t = 260\ ^0C$

From steam table corresponding to 30 bar

$h_g = 2802.3$ kJ/kg

$t_{sat} = 233.8\ ^0C$ is less than $260\ ^0C$

∴ Steam is super heated

$T_{sup} = 260^0 + 273 = 533\ ^0k$

$T_{sat} = 233.8 + 273 = 506.8$ k

∴ Enthalpy $= h_g + c_p\ (T_{sup} - T_{sat})$

$= 2802.3 + 2.0934\ (533 - 506.8)$

$= 2857$ KJ/kg

(iii) $P_3 = 40$ bar Heat added = 2400 kJ/kg

Note: Heat is added to water. But feed water temperature is not given in problem. Always assume feed water at 0 °C

At 40 bar from steam table H = 2800.3 kJ/kg

∴ Steam is wet because in problem H = 2400 which is less than Dry saturated enthalpy of steam 2800.3 kJ/kg

$h_{wet} = h_f + xh_f$

At 40 bar

$h_f = 1087.31$ kJ/kg

$h_f = 1714.1$ kJ/kg

$2400 = 1087.31 + 1714.1\ x$

$x = 0.766$ **Ans.**

15. Find Enthalpy internal energy and specific volume.

(i) Mass = 5 kg, P = 10 bar, Quality = 95% dry

(ii) Mass = 2 kg P = 20 bar, Volume = 0.35 m^3/kg

Take $c_{ps} = 2.0934$ kJ/kg K

Solution: (i) At 10 bar from steam table

$h_f = 762.6$ kJ/kg

$h_{fg} = 2013.6$ kJ/kg

$v_g = 0.1943\ m^3/kg$

Specific volume $= 0.95 \times 0.1943 = 0.1846\ m^3/kg$ **Ans.**

Enthalpy $= 5[h_f + x\ h_{fg}]$

$= 5[762.6 + 0.95 \times 2013.6]$

$= 13377.6$ kJ **Ans.**

Internal energy $= h_1 - W.D$

$= 13377.6 - m \times 10^2 . P . x.v_g$

$= 13377.6 - 5 \times 10^2 \times 10 \times 0.95 \times 0.1943$

$= 12454.675$ kJ **Ans.**

(ii) At 20 Bar from steam table

$v_{g_2} = 0.0995 \text{ m}^3/\text{kg}$

$h_{g_2} = 2797.2 \text{ kJ/kg}$

$t_{sat_2} = 212.4 + 273 = 485.4 \text{ K}$

$$v_2 = \text{specific volume of steam} = \frac{0.35}{2} = 0.175 \text{ m}^3/\text{kg}$$

$v_2 > v_{g2}$

(Volume given in problem is 0.35 m^3/kg which is greater than dry-saturated steam $V_g = 0.0995 \text{ m}^3/\text{kg}$)

∴ Steam is super heated

$$v_{sup} = v_g \frac{T_{sup}}{T_{sat}}$$

$$\therefore \quad T_{sup} = \frac{v_{sup} . T_{sat}}{v_g} = \frac{0.175 \times 485.4}{0.0995} = 853.72\text{K}$$

Enthalpy $= 2[h_{g_2} + C_{ps} (T_{sup} - T_{sat})]$

$= 2[2797.2 + 2.0934 (853.72 - 485.4)]$

$= 7136.482$ KJ

I.E $= 7136.482 - 2 \times 10^2 \times 20 \times 0.175 = 6436.482$ kJ **Ans.**

16. A steam turbine obtains steam from a boiler at pressure 15 bar and 0.98 dry. It was observed that steam loses 25 KJ of heat per kg of steam as it flows through pipe line, while pressure remains constant. Calculate Dryness fraction of steam at turbine end of pipe line.

Solution: At 15 bar from steam table

$h_f = 844.6 \text{ kJ/kg}$

$h_{fg} = 1945.3 \text{ kJ/kg}$

$h = 844.6 + 0.98 \times 1945.3 = 2751 \text{ kJ/kg}$

$\therefore$ Total heat/kg of steam at turbine end = 2751 – 25

= 2726 kJ/kg

But $2726 = h_f + x\, h_{fg}$

$= 844.1 + x \times 1945.3$

$$x = \frac{2726 - 844.1}{1945.3} = 0.9671$$

$\therefore$ D.F of steam at turbine end = 0.9671 **Ans.**

17. A vessel of 1.2 m^3 capacity contains steam of 1.0 MP_a (10 bar) and 0.92 Dryness fraction. Steam is blown off until the pressure drops to 0.5 MP_a. The valve is then closed. Determine mass of steam blown off.

Solution: **Step I** Find mass of steam in vessel initially

Step II Because steam is blowing off through valve. It is therefore throttling process

i.e., Total enthalpy before = Total enthalpy after belown off

Step III D.F x_2 of steam in vessel after blown off can be calculated

$$\text{Mass} = \frac{\text{Total volume of vessel}}{\text{Specific volume of steam after blown off}}$$

Step I At 10 bar (1.0 MP_a) $v_g = 0.194$ from steam table

Initial mass m_1 = Mass of steam at 1.0 MP_a

$$= \frac{\text{Volume of vessel}}{\text{Specific volume of steam}}$$

$$= \frac{1.2}{0.92 \times 0.194} = 6.723 \text{ kg}$$

Step II Total heat before blowing/kg (at 1 MPa) i.e., 10 bar and D.F = 0.92

$= h_{f_1} + x_1\, h_{fg_1}$

$= 763 + 0.92 \times 2015$

= 2616.8 kJ/kg

At 10 bar from steam table

$h_{f_1} = 763$ kJ/kg

$h_{fg_1} = 2015$ kJ/kg

At 5 bar (.5 mPa)

$h_{f_2} = 640$ kJ/kg

$h_{fg2} = 2107$ kJ

$v_{g2} = 0.375\ m^3/h$

Total heat after blowing off = $h_{f_2} + x_2 h_{fg2}$

$= 640 + x_2 . 2107$

But total heat before blowing = total heat after blowing

$2616.8 = 640 + x_2 \times 2107$

$\therefore \quad x_2 = 0.9382$

$\therefore$ Specific volume of steam at 5 bar (0.5 MPa)

[From steam table corresponding to 5 bar = 0.375 m^3/kg)

$= 0.9382 \times 0.375$

$= 0.3518\ m^3/kg$

$\therefore\ M_2$ = Mass of steam at 0.5 MP_a = $\dfrac{1.2}{\text{sp.volume}}$

$= \dfrac{1.2}{0.3518}$

$= 3.411$ kg

$\therefore$ Mass of steam blown off $= m_1 - m_2$

$= 6.723 - 3.411$

$= 3.312$ kg **Ans.**

18. A tank of 1 cubic meter capacity is filled with dry-saturated steam at 10 bar pressure, and is then heated to raise the pressure to 12 bar. Find the amount of heat added and degree of super-heat at the end of heating. Take C_{ps} for super heated steam = 2.09 kJ/kg k.

Solution: Closed tank is example of constant volume heating

Step I Find Mass of steam in tank = volume of tank divided by (specific volume at 10 bar).

Step II Specific volume before heating = specific volume after heating from this t_{sup} is can be calculated.

[$\because$ Volume of tank does not change during heating]

Step III Find Enthalpy in KJ at 10 bar also calculate Enthalpy in kJ at 12 bar.

Step I At 10 bar

$v_g = 0.1944$

$h_{f_1} = 763$ kJ/kg

$h_{f_g} = 2015$ kJ/kg

$h_g = 2778$ kJ/kg

$$\therefore \text{ Mass of steam in tank } = \frac{\text{Volume of tank}}{\text{Specific volume of steam}}$$

$$= \frac{1}{0.1944} = 5.144 \text{ kg}$$

Step II Specific volume before heating at 10 bar = specific volume after heating at 12 bar.

$\therefore$ v_g at 10 bar = v_{sup} at 12

[v_g at 10 bar from steam table = 0.19 44

$$0.1944 = v_{g_2} \frac{T_{sup}}{T_{sat}}$$

$v_g = 0.1632$ m^3/kg at 12 bar from steam table

$t_{sat} = 188°$C at12 bar from steam table]

$$0.1944 = 0.1632 \cdot \frac{T_{sup}}{273+188}$$

$\therefore$ $T_{sup} = 549.132$ K

$= 276.132\ ^0$C

$\therefore$ Degree of super heat

$= t_{sup} - t_{sat}$

$= 276.132 - 188$

$= 88.132\ ^0$C **Ans.**

Initial heat at 10 bar = $h_1 = h_f + h_{fg} = h_{g1}$

(As steam is dry-saturated)

$= 763 + 2015$

$= 2778$ kJ/kg

Final heat at 12 and super heated steam = h_2

$\therefore$ $h_2 = h_f + h_{fg} + c_{ps} (t_{sup} - t_{sat})$

$= 798 + 1986 + 2.09 (276.132 - 188)$

$= 2968.2$ kJ/kg

$\therefore$ Heat added = $h_2 - h_1$

= 2968.2 – 2778

= 190.2 kJ/kg

$\therefore$ Total amount of heat added

= m × 190.2

= 5.144 × 190.2 = 978 kJ **Ans.**

19. Calculate the heat required to form 2.5 kg of dry steam at 1.1 MPa from water at 20 ^{0}C. Determine the amount of heat removed at constant pressure to cause the steam to become 0.985 dry. Calculate the specific volume at respective conditions.

Solution: Given

$m_s = 2.5$

P = 1.1 Mp_a = 11 bar

t_w = 20 °C

(i) Dry saturated steam find v_s, v_g corresponding to 11 bar from steam table

(ii) Amount of Heat removed ?

Given x = 0.9

Find v_s

(ii) At 11 bar

h_f = 781 kJ/kg

h_{fg} = 2000 kJ/kg

h_1 = Total heat of Dry saturated steam at 1.1 Mpa (11 bar)

= 2.5 [$h_f + h_{fg}$ – 4.187 × (20 – 0)]

= 2.5 [781 + 2000 – 4.187 × 20]

= 6743.15 kJ **Ans.**

h_2 = Enthalpy of wet steam at 1.1 MPa and D.F = 0.95

= 2.5[$h_f + xh_{fg} - t_w\, c_{pw}$]

= 2.5[781 + 2000 × 0.95 – 4.187 (20–0)]

= 6493.15 kJ

$\therefore$ Heat removed = $h_1 - h_2$ = 6743.15 – 6493.15 = 250 KJ **Ans.**

V_g at 1.1 MPa = 0.1774 m^3/kg

Steam at 1.1 Mpa and 0.95 D.F

V_g = 0.95 × 0.1774 = 0.1685 m^3/kg **Ans.**

20. A 3 kg of water at 40 ^{0}C is heated at constant pressure of 12 Bar to form super heated steam of 190 ^{0}C. Determine

(i) Increase in sensible heat

(ii) Latent heat

(iii) Super heat

(iv) Increase in total heat

(v) Increase in internal energy

Take c_{ps} of super heated steam = 2.1 KJ/kg k

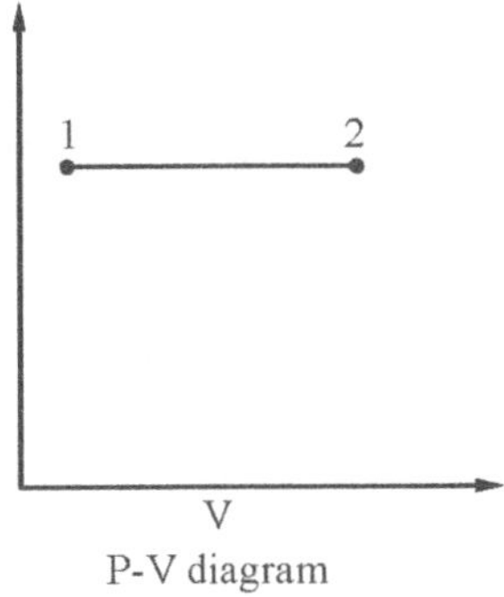

P-V diagram

Solution: (i) Sensible heat of 3 kg of water at 40 ^{0}C

Step I Calculate sensible of water at 40 °C = h_{f1}

Step II Find the sensible heat from steam table corresponding to 12 bar = h_{f2}

Step III Increase in sensible heat = $h_{f2} - h_{f1}$

$$h_{f_1} = C_{pw}(t - 0)$$

$$= 4.187 \times (40 - 0)$$

$$= 167.48 \text{ kJ/kg}$$

$$= 3(167.48)$$

$$= 502.44 \text{ kJ}$$

sensible heat per kg of steam at 12 bar from steam table

$$h_{f_2} = 798.4 \text{ kJ/kg}$$

$$\therefore \; h_{f_2} = 3 \times 798.4 = 2395.2 \text{ kJ}$$

$\therefore$ Increase in sensible heat

$$= h_{f_2} - h_{f_1}$$

$$= 2395.2 - 502.44 = 1892.76 \text{ kJ}$$

(ii) Latent heat of 1 kg of steam at 12 bar from steam table

$h_{fg} = 1986$ kJ/kg

$\therefore$ Total latent heat $= 3 \times 1986 = 5958$ kJ

(iii) At 12 bar $t_{sat} = 188\ ^0$C from steam table

$$\therefore \text{ super heat} = m\, c_{ps}(t_{sup} - t_{sat})$$
$$= 3 \times 2.1\ (190 - 188)$$
$$= 12.6 \text{ kJ}$$

(iv) Increase in total heat (Enthalpy of superheated steam)

$$= h_f + h_{fg} + \text{super heat}$$
$$= 1892.76 + 5958 + 12.6$$
$$= 7863.36 \text{ kJ}$$

(v) Volume of super heated steam

$$= m\, v_g \cdot \frac{T_{sup}}{T_{sat}}$$
$$= 3 \times 0.1633 \times \frac{(273+190)}{273+188}$$
$$= 0.492 \text{ m}^3$$

$\therefore$ External work of evaporation of super heated steam =

$$W = 10^2\, P\, v_{sup} = 10^2 \times 12 \times 0.492 = 590.4 \text{ kJ}$$

Increase in internal energy = Increase in total heat – W.D

$$= 7863.36 - 590.4$$
$$= 7273 \text{ kJ}$$

21. Steam 0.7 dry occupies a volume of 1m^3 at 1.3 MPa. It is heated up to 230 ^{0}C at constant pressure. Find

(i) Mass of steam

(ii) Heat supplied

(iii) Change in volume

Assume specific heat of super heated steam = 2.1 kJ/kg k

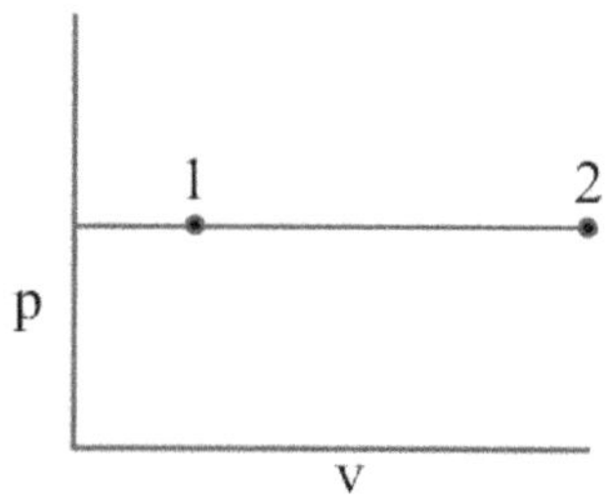

Solution: (i) Mass of steam $= \dfrac{\text{Total volume of wet steam}}{x\, v_g}$

From steam table corresponding to 1.3 mPa $V_g = 0.1512\ m^3/kg$

$$= \frac{1}{0.7 \times 0.1512}$$

$= 9.45$ kg **Ans.**

(ii) heat present before heating $= h_1$

At 13 bar from steam table

$h_f = 814.7$ kJ/kg

$h_{fg} = 1973$ kJ/kg

$t_{sat} = 191.6\ ^0C$

$V_g = 0.1512\ m^3/kg$

$\therefore\ h_1 = h_f + x\, h_{fg}$

$= 814.7 + 0.7 \times 1973$

$= 2195.8$ kJ/kg

Heat of super heated steam per kg $= h_2$

$h_2 = h_f + h_{fg} + cp_s\,(t_{sup} - t_{sat})$

$= 814.7 + 1973 + 2.1\,(230 - 191.6)$

$= 2868.34$ kJ/kg

$\therefore$ Heat supplied per kg of steam

$= h_2 - h_1 = 2868.34 - 2195.8 = 67254$ kJ/kg

$\therefore$ Total heat supplied $= 9.45 \times 672.54 = 6355.5$ kJ **Ans..**

(iii) Specific volume of super heated steam at 13 bar $= v_2$

$$\therefore\ V_{sup} = V_g \times \frac{T_{sup}}{T_{sat}} = 0.1512 \times \frac{(273+230)}{(273+191.6)}$$

$V_g = 0.1637\ m^3/kg$

Increase in volume per kg of steam $= 0.1637 - 0.7 \times 0.1512$

$= 0.0579\ m^3/kg$

$\therefore$ Total change in volume $= m(V_2 - V_1) = 9.45 \times 0.0579$

$= 0.547\ m^3$

22. Find the internal energy of 6 kg of super heated steam at pressure of 8 bar and temperature of 250 ^{0}C. If this steam is expanded to 2 bar and 0.9 dryness fraction, find change in internal energy. Take C_{ps} for super heated steam = 2.09 KJ/kg k.

Solution:

Step I Find Enthalpy of super heated steam at 8 bar for 6 kg of steam

Step II Find W.D of super heated steam at 8 bar for 6 kg of steam

Step II After expansion, find enthalpy of steam at 2 bar and D.F = 0.9 for 6 kg of steam

Step IV Find W.D of wet steam at 2 bar and D.F = 0.9 for 6 kg of steam.

Step I From steam table corresponding to 8 bar

$$h_f = 720.9 \text{ kJ/kg}$$

$$h_{fg} = 2046.5 \text{ kJ/kg}$$

$$h_g = 2767.5 \text{ kJ/kg}$$

$$t_{sat} = 170.4\ ^0\text{C}$$

$$V_g = 0.240 \text{ m}^3\text{/kg}$$

$$\therefore\ h_1 = h_f + h_{fg} + c_p\,(t_{sup} - t_{sat})$$

$$= 720.9 + 2046.5 + 2.09\,(250 - 170.4)$$

$$= 2933.8 \text{ kJ/kg}$$

Step II Work of evaporation/kg of super heated steam

$$w_1 = 10^2\ P_1\ V_{sup}$$

$$= 10^2 \times 8 \times 0.240\ .\ \frac{(273+250)}{(273+170.4)}$$

$$= 226.47 \text{ kJ/kg}$$

Internal energy per kg of super heated steam

$$U_{sup} = h_1 - w_1 = 2933.8 - 226.47$$

$$= 2707.332 \text{ kJ/kg}$$

$\therefore$ Total internal energy = 6 × 2707.294 = 16243.98 kJ

Step III After expansion

At 2 bar from steam table

$$h_f = 504.7 \text{ kJ/kg}$$

$$h_{fg} = 2201.6 \text{ kJ/kg}$$

$$v_g = 0.885$$

$h_2 = h_{f_2} + x\, h_{fg}$

$= 504.7 + 0.9 \times 2201.6$

$= 2486.14$ kJ/kg

w_2 = W.D per kg of wet steam

$= 10^2\, P\, x v_g = 10^2 \times 0.2 \times 0.9 \times 0.885 = 159.3$ kJ/kg

$\therefore$ Internal energy of wet steam at 2 bar

$u_2 = h_2 - w_2 = 2486.14 - 159.3 = 2326.84$

$\therefore$ Change in I.E/kg $= u_2 - u_1$

$= 2326.84 - 2707.332$

$= -380.492$ kJ/kg

$= -2282.95$ kJ **Ans.**

–ve sign indicates the decrease in internal energy

23. A closed vessel of 0.422 m^3 contains steam at a pressure of 10 bar and temperature of 220 ^{0}C. Vessel is cooled to make steam just dry and saturated. Then the vessel is further cooled to a temperature of 165 ^{0}C. Calculate.

(i) Mass of steam

(ii) Pressure of steam at end of first cooling

(iii) Condition and pressure of steam at the end of second cooling

Solution: **Step I** To calculate mass of steam in closed vessel by use of following formula

$$\text{Mass} = \frac{\text{Total volume of steam}}{\text{Specific volume}}$$

(where total volume of steam = volume of given pressure and temperature)

Step II During cooling mass in this case is 2 kg (say) and total volume equal to total volume after cooling.

From this equation v_g can be calculated and hence pressure after cooling can also be found out as condition of steam is dry saturated after first cooling.

Step III (i) Pressure after second cooling can be found from steam table corresponding to 165 ^{0}C

(ii) Total volume of steam (volume of closed vessel) divided by specific volume of steam after cooling

Volume of closed vessel = m.x. V_g

x = condition of steam after second cooling can be calculated

Step I Initial volume of steam/kg

At 10 bar from steam table

$v_g = 0.194 \ m^3/kg$

$t_{sat} = 179.9 \ ^oC$

$$v_{sup} = v_g \cdot \frac{T_{sup}}{T_{sat}}$$

$$= 0.194 \ \frac{(273+220)}{(273+179.9)}$$

$$= 0.211 \ m^3/kg$$

$$\therefore \text{ mass of steam} = m = \frac{\text{volume of steam in closed vessel}}{\text{specific volume of super heated steam}}$$

Mass of steam corresponding to given pressure and temperature in vessel

$$= \frac{0.422}{0.211} = 2 \ kg \qquad \textbf{Ans.}$$

Step II At the end of first cooling steam become dry saturated

Total volume of steam = volume of steam after cooling

$$0.422 = m.v_g$$

$$= 2 \times v_g$$

$$v_g = \frac{0.422}{2} = 0.211 \ m^3/kg$$

Corresponding to 0.211 m^3/kg from steam table,

Pressure = 9.2 Bar **Ans.**

Step III At the end of second cooling saturation temperature t_{sat} = 165 ^{0}C. Corresponding to t_{sat} = 165 ^{0}C from steam table pressure = 7 bar

$\therefore$ Pressure at end of second cooling = 7 bar

At 7 bar specific volume of steam $v_g = 0.273 \ m^3/kg$

But $v_3 = m \cdot x_3 \ v_{g3}$

$0.422 = 2 \cdot x_3 \cdot 0.273$

$\therefore$ Dryness fraction after second cooling is 0.773 **Ans.**

24. A vessel with partition in it contains initially 2 kg of dry-saturated steam at pressure of 7 bar in one compartment and 1 kg of steam with dryness fraction of 0.8 at 3.5 bar in other compartment. After the partition is removed, the pressure of mixture is found to be 5 bar. Neglecting the volume of water, find.

(i) The quality of mixture

(ii) The heat interaction between the vessel and the surrounding.

Solution:

2 kg Dry saturated steam P = 7 bar vg = 0.273	1 kg of steam x = 0.8 P = 3.5 vg = 0.524	→	Mixture of steam at 3.5 bar. Find quality of mixture

Step I Find volume of compartment 1 and 2 before removing partition and Find total volume of vessel

Step II Find mass and specific volume of mixture and find quality of steam.

Step III Find Enthalpy of steam in one compartment, enthalpy of steam in second compartment and enthalpy of mixture after removing partition.

Step IV Enthalpy of compartment one + Enthalpy of second compartment = Enthalpy of mixture + heat exchange from vessel to atmosphere.

Specific volume of dry saturated steam at 7 bar from steam table = 0.273 m^3/kg

(i) Volume of first compartment $= V_1 = 0.273 \times m$

$$= 0.273 \times 2 = 0.546 \text{ m}^3$$

(ii) Volume of second compartment $= V_2$

$$= m \text{ x } v_{g2} \ldots$$

(At 3.5 bar, $V_{g2} = 0.524 \text{ m}^3/\text{kg} = 1 \times .8 \times 0.524 = 0.4192 \text{ m}^3$

(iii) Total volume of vessel $= v_3 = v_1 + v_2 = 0.546 + 0.4192 = 0.9652 \text{ m}^3$

Step II Mass of mixture = 2 + 1 = 3 kg

$$\therefore \text{ Specific volume of mixture} = \frac{\text{volume of mixture}}{3} = \frac{0.9652}{3} = 0.3217 \ \frac{\text{m}^3}{\text{kg}}$$

From steam table corresponding to 5 bar specific volume $v_{g3} = 0.375 \text{ m}^3/\text{kg}$

$$\therefore \quad 0.3217 = x_3 \, vg_3$$

$$x_3 = \text{quality of mixture} = \frac{0.3217}{0.375} = 0.8578 \quad \textbf{Ans.}$$

Step III (i) Enthalpy of Dry saturated steam at 7 bar from steam table $h_1 = h_g = 2762$ kJ/kg

∴ Total enthalpy of steam in first compartment = 2 × 2762

= 5524 kJ

(ii) Enthalpy of wet steam in second compartment = $h_f + x\, h_{fg}$

= 584 + 0.8 × 2147.4

= 2301.92 kJ

(iii) Enthalpy of mixture at 5 bar

= 3[640.1 + 0.8578 ×2107.4]

= 7343.48 kJ

∴ Enthalpy of water in first compartment + second compartment
= Enthalpy of mixture + heat exchange from vessel to surrounding.

$H_1 + H_2 = H_3$ + heat exchange from vessel to surrounding

5524 + 2301.92 = 7343.48 + heat loss to surrounding

∴ heat lost to surrounding = Heat exchange from vessel to surrounding = 482.44 kJ **Ans.**

25. Quantity of steam at pressure of 21 bar and 0.9 dry occupies a volume of 0.427 m^3. It is expanded according to law $pv^{1.25} = C$ to a pressure of 7 bar. Determine.

(i) Mass of steam

(ii) Work transfer

(iii) The change in internal energy

(iv) Heat exchange between steam and surrounding stating the direction of transfer

Solution:

Pressure Bar	Sat temp T sat °C	h_f kJ/kg	h_{fg} kJ/kg	V_g m^3/h	kJ/kg
7	165	697.1	2064.9	0.273	
21	214.9	920	1878.2	0.0949	

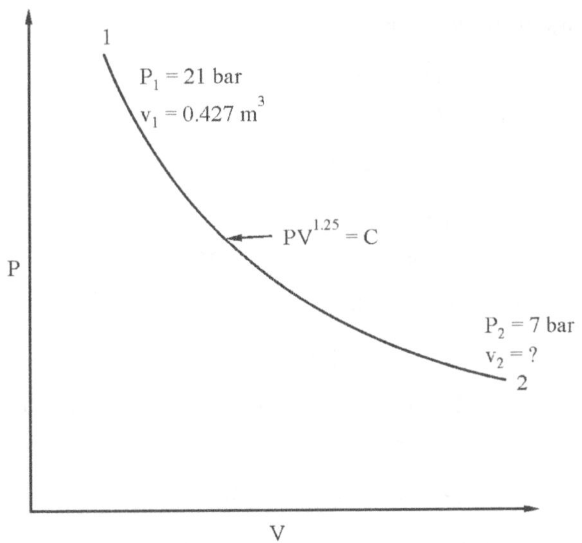

(i) $m = \dfrac{\text{Total volume}}{\text{Specific volume of steam}} = \dfrac{0.427}{0.9 \times 0.0949} = 5 \text{ kg}$ *Ans.*

At pressure 21 bar from steam tube $V = 0.0949 \text{ m}^3/\text{kg}$

(ii) $$P_1 v_1^{1.25} = P_2 v_2^{1.25}$$

$$\therefore \quad \left(\frac{v_2}{v_1}\right)^{1.25} = \frac{P_1}{P_2} = \frac{21}{7} = 3$$

$$\therefore \quad v_2 = (3)^{\frac{1}{1.25}} . v_1 = (3)^{\frac{1}{1.25}} \times 0.427$$

$$\therefore \quad v_2 = 1.0283 \text{ m}^3$$

$$\text{Work transfer} = \frac{10^2 (P_1 v_1 - P_2 v_2)}{n-1} \text{ kJ}$$

(where P_1 and P_2 should be in bar).

$$= \frac{100 \times 21 \times 0.427 - 100 \times 7 \times 1.0283}{1.25 - 1} = 707.56 \text{ kJ}$$ *Ans.*

Work is done by steam because work transfer is positive

(iii) Specific volume $= v_2 = \dfrac{1.0283}{5} = 0.20566 \text{ m}^3/\text{kg}$

$\therefore \quad v_2 = x_2 v_{g2}$ $\quad v_{g2}$ at pressure 7 bar from steam $= 0.273 \text{ m}^3/\text{kg}$

$$x_2 = \frac{v_2}{vg_2} = \frac{0.20566}{0.273} = 0.7533$$

Change in internal energy = $u_2 - u_1$

But $h_1 = m[h_{f1} + x\, h_{fg1})$

$\therefore$ $h_1 = 5[920 + 0.9 \times 1878.2]$

$= 13051.9$ kJ

$h_2 = 5[h_{f2} + x_2\, h_{fg2})$

$= 5[697.1 + 0.7533 \times 2064.9]$

$= 11263.29$ kJ

$\therefore$ $u_2 - u_1 = (h_2 - 10^2\, P_2\, v_2) - (h_1 - 10^2\, P_1\, V_1)$

$= (11263.29 - 10^2 \times 7 \times 1.0283) - (13051.9 - 21 \times 10^2 \times 0.427)$

$= -1611.72$ kJ **Ans.**

(–ve sign indicates that internal energy decreases)

Q = Heat exchange = w + $(u_2 - u_1)$

$= 707.56 - 1611.72$

$= -904.16$ kJ **Ans.**

– ve sign indicates steam loses heat to surrounding

26. Find specific volume, enthalpy, W.D, Internal energy of 2 kg of steam at 25 Bar

(i) When steam is super heated when degree of super heat is 175 ^{0}C

(ii) Wet steam having D.F = 0.8

Take specific heat of super heated steam as 2.1 kJ/kg k

Solution: From steam table corresponding to 25 bar

$h_f = 962$ kJ/kg
$h_{fg} = 1839$ kJ/kg
$h_g = 2801$ kJ/kg
$v_g = 0.0799\ m^3/kg$
$t_{sat} = 223.9$ °C
$t_{sup} = 223.9 + 175 = 398.5$ °C

I (i) When steam is super heated

$$V_{sup} = v_{g.} \frac{T_{sup}}{T_{sat}} = 0.0799 \frac{273 + (223.9 + 175)}{273 + 223.9} = 0.10804\ m^3/kg$$

$\therefore$ Total volume of two kg steam = $2 \times 0.10804 = 0.2161\ m^3$ **Ans.**

(ii) Enthalpy $= h_g + c_{ps} (t_{sup} - t_{sat})$

$= 2801 + 2.1 (175)$

$(\because t_{sup} - t_{sat} = \text{degree of super heat} = 175\ ^\circ C)$

$= 3168.5$ kJ/kg

$\therefore$ Total enthalpy $= 2 \times 3168.5 = 6337$ kJ **Ans.**

(iii) $W.D = 10^2 \text{ m P } v_{sup}$

$= 10^2 \times 2 \times 25 \times 0.10804 = 540.2$ kJ **Ans.**

(iv) Internal energy $= h - 10^2 \text{ m . P.} V_{sup}$

$= 6337 - 10^2 \times 2 \times 25 \times 0.10804$

$= 5796.8$ kJ

II For wet steam DF = 0.8

(i) Volume $= m \text{ x } v_g = 2 \times 0.8 \times 0.0799 = 0.1278 \text{ m}^3$ **Ans.**

Where $v_g = 0.0799$ from steam table at 25 bar

(ii) Enthalpy $= 2[h_f + x\ h_{fg}] = 2\ [962 + 0.8 \times 1839] = 4866.4$ kJ **Ans.**

(iii) $W.D = m.10^2 \text{ P x } v_g$

$= 2 \times 100 \times 25 \times 0.8 \times 0.0799$

$= 319.6$ kg

(iv) Internal energy $= h_{wet} - W.D$

$= 4866.4 - 319.6 = 4546.8$ kJ **Ans.**

27. A quantity of steam at pressure 2 MN/m^2 has a volume of 0.75 m^3 and enthalpy content of 21.5 MJ. Determine (i) D.F of steam (ii) The mass of steam. If some of this steam is throttled to pressure of 1.2 MN/m^2, determine the quality of steam after throttling. Take specific heat of super heated steam = 2.0934 kJ/kg.

Solution: From steam table corresponding to 2 MN/m^2 (20 bar) Pressure

	At 12 bar
$v_g = 0.0995 \text{ m}^3/\text{kg}$	
$h_f = 9086.6$ kJ/kg	$h_f = 798.4$ kJ/kg
$h_{fg} = 1888.6$ kJ/kg	$h_{fg} = 1984.3$ kJ/kg
	$t_{sat} = 188\ ^\circ C$

Step I First find D.F of steam by considering following formula.

(i) $21.5 \times 10^3 \text{ kJ} = m[h_f + x\ h_{fg}]$ .....(i)

(ii) Total volume $= m \times x\ v_g$ .....(ii)

$$\therefore \quad m = \frac{0.75}{v_g x} = \frac{0.75}{0.0995\, x} = \frac{7.538}{x}$$

where x = D.F. of steam

From (i) and (ii)

$$21500 = \frac{7.538}{x}[908.6 + x\ 1888.6] \qquad \because m = \frac{7.538}{x}$$

$$21500 = \frac{6849.027}{x} + 14236.2668$$

$$7263.7332 = \frac{6849.027}{x}$$

$$\therefore \quad x = \frac{6849.27}{7263.773} = 0.9429 \quad \textbf{Ans.}$$

$$\text{But} \quad m = \frac{7.538}{0.9429} = 7.994 \quad \textbf{Ans.}$$

Step II Enthalpy before throttling at 20 bar = Enthalpy of steam after the throttling at 12 bar

$$h_{f_1} + x\ h_{f_{g1}} = h_{f_2} + h_{f_{g2}} + c_{ps}(t_{sup} - t_{sat})$$

$$908.6 + 0.9429 \times 1888.6 = 798.4 + 1984.3 + 2.0934\ (t_{sup} - 188)\ (t_{sup} - t_{sat})$$

$$\therefore \quad t_{sup} = 143.23\ ^0C \quad \textbf{Ans.}$$

28. 10 kg of steam having D.F = 0.8 at 10 bar is heated to temperature of 250 ^{0}C. Calculate heat supplied and change in internal energy. Specific heat of super heated steam = 2.1 kJ/kg k

Solution:

$m_s = 10$ kg

$x = 0.8$

$P_1 = 10$ bar

$C_{ps} = 2.1$ kJ/kg k

At 10 bar from steam table

$t_{sat} = 179.9\ ^0C$

$h_{f1} = 762.6$ kJ/kg

$h_{fg} = 2013.6$ kJ/kg

$h_g = 2776.2$ kJ/kg

$v_g = 0.194\ m^3/kg$

Enthalpy of wet steam = h_1

$$h_1 = h_{f_1} + x\ h_{f_{g1}}$$

$$= 762.6 + 0.8 \times 2013.6$$

$$= 2373.49 \text{ kJ/kg}$$

$$W.D = 10^2\ P\ x\ v_{g1}$$
$$= 10^2 \times 10 \times 0.8 \times 0.194$$
$$= 155.2\ KJ/kg$$

Internal energy $= 2373.49 - 155.2$

$$= 2218.29\ KJ/kg$$

Enthalpy of super heated steam

$$h_2 = h_g + C_{ps}\ (t_{sup} - t_{sat})$$
$$= 2776.2 + 2.1\ (250 - 179.9)$$
$$= 2923.41\ kJ/kg$$

Heat supplied $= 2923.41 - 2373.49 = 549.92$ kJ/kg

$10\ (549.92) = 5499.2$ kJ **Ans.**

$$v_{sup} = v_g \frac{T_{sup}}{T_{sat}}$$
$$= 0.194\ \frac{(250+273)}{(273+179.9)} = \frac{0.194 \times 523}{452.9}\ m^3/kJ$$
$$= 0.2245\ m^3/kg$$

$$W.D = 10^2\ P\ v_{sup}$$
$$= 10^2 \times 10 \times 0.2245 = 224.0\ kJ/kg$$

Internal energy $= h_2 - W.D$

$$u_2 = 2923.41 - 224.0$$
$$u_2 = 2699.41\ kJ/kg$$

$\therefore$ Change in internal energy

$$= m(u_2 - u_1)$$
$$= 10(2699.41 - 2218.29)$$
$$= 4811.2\ kJ$$ **Ans.**

29. Find internal energy of one kg of super-heated steam at pressure of 10 bar and 280 °C. If this steam be expanded to pressure 1.6 bar and 0.8 dry, determine change in internal energy. Assume specific heat of super-heated steam = 2.1 kJ/kg k.

Solution: In this problem law of expansion is not given. It is also not required because condition of steam after expansion is already given.

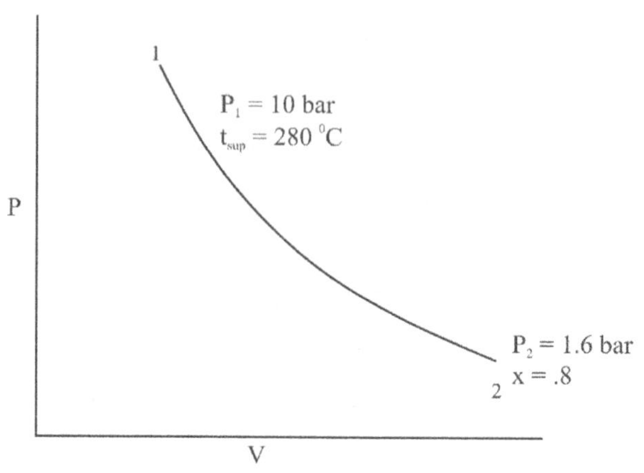

from steam table

At 10 bar	At 1.6 bar
$t_{s1} = 179.9\ ^{o}C$	$t_{s_2} = 111.3\ ^{0}C$
$h_{g_1} = 2776.2$	$h_{f_2} = 475.4$ kJ/kg
$v_{g_1} = 0.194$	$hf_{g_2} = 2220.9$ kJ/kg
	$v_{g_2} = 1.091\ m^3/kg$

h_1 = Enthalpy of super heated steam $= h_{g1} + C_{ps}(t_{sup1} - t_{sat1})$

$= 2776.2 + 2.1\ (280 - 179.9)$

$= 2986.45$ kJ/kg

$$v_{sup} = v_g \cdot \frac{T_{sup}}{T_{sat}}$$

$$= 0.194\ \frac{(273+280)}{(273+179.9)}$$

$$= 0.237\ m^3/kg$$

Internal energy $u_1 = h_1 - W.D$

$= h_1 - 10^2\ P\ v_{sup}$

$= 2986.45 - 10^2 \cdot 10 \times 0.237$

$= 2986.45 - 237 = 2749.2$ kJ/kg

h_2 = Enthalpy of wet steam

$$= h_{f_2} + x\,h_{fg_2}$$

$$= 475.4 + 0.8 \times 2220.9$$

$$= 2252.12 \text{ kJ/kg}$$

u_2 = Internal energy after expansion

$$= h_2 - WD_2 = h_2 - 10^2 \,.\, p_2 \,.\, x_2 \,.\, v_{g_2}$$

$$= h_2 - 10^2 \times 1.6 \times 0.8 \times 1.091$$

$$= 2252.12 - 10^2 \times 1.6 \times 0.8 \times 1.091$$

$$= 2112.472 \text{ kJ/kg}$$

Change in internal energy

$$= u_2 - u_1$$

$$= 2112.472 - 2749.2$$

$$= -\,636.728 \text{ kJ/kg}$$ **Ans.**

–ve sign indicates internal energy decreases

30. A boiler produces 2400 kg/hr of dry saturated steam at 12 bar. Feed water temp. 120 °C. 240 kg of coal having cv of 33500 kJ/kg is fired per hour. It is found that 10% coal is un-burnt. Find efficiency of boiler.

Take C_{p_w} of water as 4.187 kJ/kg.K

Solution:

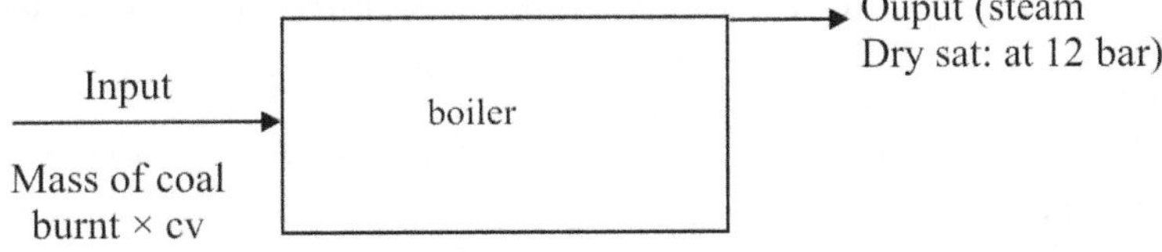

At 12 bar $h_g = 2782.7$

Total heat required/kg at 12 bar $= h_g - 4.187 \times (120 - 0)$

$$= 2782.7 - 4.187 \times 120$$

$$= 2280.26 \text{ kJ/kg}$$

Mass of coal burnt $= 240 \times 0.9 = 216$ kg/hr

$$m_a = \frac{\text{Mass of steam}/\text{hr}}{\text{Mass of coal actually burnt}/\text{hr}}$$

$$= \frac{2400}{216} = 11.11 \text{ kg/kg of coal}$$

$\therefore \quad \eta boiler = m_a(h_g - h_f)$

$$= \frac{11.11(2280.26)}{33500} \times 100$$

$= 75.63\%$ **Ans.**

31. Determine the mass of 0.2 m^3 of wet steam at pressure 4 bar and Dryness fraction 0.9. Also calculate heat of 3 m^3 of steam.

Solution: Given volume of wet steam = 0.2 m^3

P = 4 bar

D.F = 0.9

Note : Calculation are based on 1m^3 of steam

At 4 bar $vg = 0.462 \ m^3/kg$

$h_f = 604.7$ kJ/kg

$h_{fg} = 2133$ kJ/kg

First solution

$$\text{we know that density of wet steam} = \frac{1}{\text{specific volume of wet steam}}$$

$$= \frac{1}{x \, v_g} = \frac{1}{0.9 \times 0.462}$$

$= 2.41 \ m^3/kg$

$\therefore$ Mass of 0.2 m^3 of steam = volume of steam × Density of steam

$= 0.2 \times 2.41$

$= 0.482$ kg **Ans.**

Second solution

$$\text{Mass of } 0.2 \ m^3 \text{ of steam} = \frac{0.2}{0.09 \times 462}$$

$= 0.482$ *Ans.*

Also Mass of 3 m^3 of steam having D.F = 0.9

$= 3 \times 0.9 \times 2.42 = 7.26$ kg

$\therefore$ Total heat of 3 m^3 of steam which has 7.23 kg mass

$= 7.26 \ (h_f + x \ h_{fg})$

$=7.26 \ (604.7 + 0.9 \times 2133)$

$= 18327.14$ kJ **Ans.**

32. Feed water at 65 ^{0}C enters a boiler and steam is generated at pressure of 17 bar and 260 ^{0}C. If 75% of heat of the coal of c.v of 29274 kJ/kg is utilized in producing the steam, determine mass of coal consumed in producing 5000 kg steam per hour. Assume C_{ps} for super heated steam as 2.09 kJ/kg K.

Solution:

$t_w = 65\ ^0C$

$P = 17$

$t_{sup} = 260\ ^oC$

c.v of coal = 29274 kJ/kg

heat of coal utilized only 75% for producing steam

At 17 bar from steam table

$h_f = 871.8$ kJ/kg

$h_{fg} = 1921.5$ kJ/kg

$h_g = 2793.4$ kJ/kg

$t_{sat} = 204.3\ ^oC$

$$h_{sup} = h_g + C_{ps}\,(t_{sup} - t_{sat}) - C_{pw}\,(t - 0)$$

$$= 2793.4 + 2.09\,(260 - 204.3) - 4.187\,(65 - 0)$$

$$= 2909.813 \text{ kJ/kg} - 272.16 = 2637.653 \text{ kJ/kg}$$

$\therefore$ Total heat required to produce 5000 kg/hr = 2909.813 × 5000

= 13188.27 kJ/hr

$$\therefore \qquad \text{Mass of coal/hr} = \frac{13188.27}{29274 \times 0.75}$$

= 600.7 kg of coal/hour **Ans.**

33. A spherical shell of 40 cm radius contains steam at pressure of 88 bar having Dryness fraction 0.95. The shell and its contents cool until the pressure is 9.8 bar. Find the mass of steam the vessel contains and Dryness fraction of steam after cooling.

Solution: Given

Before cooling

$P_1 = 88$ bar

$x_1 = 0.95$

After cooling

$P_2 = 9.8$

Find mass of vessel and D.F of steam

Step I

(i) Find total volume of spherical shell

(ii) Find specific volume of steam at 88 bar and D.F = 0.95

(iii) Mass of steam (contains) = $\frac{\text{Total volume of vessel}}{\text{sp. volume at 88 bar}}$ and D.F = 0.95

(iv) To calculate D.F after cooling

Total vessel volume = m . x v_{g2}

From steam table corresponding to 88 Bar $v_s = 0.0211$

Total volume of spherical shell = $\frac{4}{3}\pi r^3$

$$= \frac{4}{3}\pi(0.4)^3 = 0.268 \text{ m}^3$$

∴ Total volume of steam in vessel = 0.268 m^3

Specific volume of wet steam = 0.95 × 0.0211

(Before cooling at 88 bar $v_{g_1} = 0.0211$)

$$= 0.02 \text{ m}^3/\text{kg}$$

∴ Mass of steam = $\frac{\text{Total volume of vessel}}{\text{Specific volume of steam}}$ having

DF = 95 and pressure 88 bar

$$= \frac{0.268}{0.02} = 13.4 \text{ kg}$$

But, Total volume of vessel = m x_2 v_{s2}

$0.268 = 13.4 \times x_2 \times 0.198$ (At 9.8 bar $v_s = 0.1981$)

$x_2 = 0.1$

34. A closed vessel of 0.5 m^3 capacity contains dry saturated steam at 3.5 bar. The vessel is cooled. The pressure is reduced to 2 bar. Calculate.

(i) The mass of steam in the vessel

(ii) Final Dryness fraction of steam

(iii) Amount of heat transfer during process

Solution:

Before cooling	After cooling
Steam is dry saturated	Pressure is 2 bar
Pressure of steam 3.5 bar	

Step I If volume of closed vessel is given and pressure of steam and its condition is given then

mass of steam in vessel

$$= \frac{\text{Total volume of vessel}}{\text{Specific volume of steam at given pressure and conditon of steam}}$$

Step II By equating volume of steam before cooling and after cooling.

Step III For closed vessel W.D = 0

Heat transfer = $Q = u_2 - u_1$

Step IV $u_1 = h_1 - W.D$ For dry saturated steam $u_1 = h_{g1} - 10^2\, p_1\, v_{sg2}$

For wet steam $u_2 = h_f + x_2\, h_{fg} - 10^2\, p_2\, x_2\, v_{g1}$

(i) From steam table at 3.5 Bar = $v_g = 0.524$

$$\therefore \text{ Mass of steam in vessel} = \frac{\text{Total volume of vessel}}{v_g} = \frac{0.5}{0.524} = 0.9542 \text{ kg}$$

(ii) Volume of steam before cooling (at 3.5 bar) = volume of steam after cooling at 2 bar

$$x_1 v_{g1} = x_2\, v_{g2}$$

$$1 \times 0.524 = x_2 \times 0.885 \qquad (\because x_1 = 1 \text{ for dry-saturated steam})$$

$$\therefore \quad x_2 = \frac{0.524}{0.885} = 0.592$$

(iii) At 3.5 from steam table $v_{g1} = 0.524$, $h_{g1} = 2731.6$

$$u_1 = h_{g1} - 10^2\, P_1\, v_{g1}$$
$$= 2731.6 - 10^2 \times 3.5 \times 0.524$$
$$= 2548.2 \text{ kJ/kg}$$

At 2 bar from steam table

$$v_{g_2} = 0.885 \text{ m}^3\text{/kg}$$
$$h_{f_2} = 504.7 \text{ kJ/kg}$$
$$hf_{g_2} = 2201.6$$

$$\therefore \quad u_2 = h_{f_2} + x_2\, hf_{g_2} - 10^2\, P_2\, x_2\, v_{g2}$$
$$= 505.7 + 0.592 \times 2201.6 - 10^2 \times 2 \times 0.592 \times 0.885$$
$$= 1808.047 - 104.784 = 1703.263 \text{ kJ/kg}$$

$Q = u_2 - u_1$

$= 1703.263 - 2548.2$

$= - 844.9368$ kJ/kg

∴ $Q = - 844.9368$ kJ/kg

–ve sign indicates that heat is rejected

35. In Piston cylinder arrangement, a constant pressure of 8 bar is maintained in cylinder by the weight of Piston. Steam flowing from main pipe at 8 bar and 300 ^{0}C enters into cylinder. Determine final temperature.

Solution: $P_1 = P_2$ ($\because$ constant pressure)

From first law $Q = [(u_2 - u_1) + P_2 (v_2 - v_1)] \times m$ .....(i)

As there is no steam in cylinder in the beginning

∴ $u_1 = 0$, $v_1 = 0$ and $Q = h_1 m$

Putting above values in equation (i), we get

$mh_1 = [(u_2 - 0) + P_2(v_2 - 0)]m$

$h_1 = u_2 + P_2 v_2$

$h_1 = h_2$ $\quad \because h_2 = u_2 + P_2 v_2$

As initial enthalpy is equal to final enthalpy, therefore for same pressure final condition of steam will be same as initial.

∴ $t_1 = t_2 = 300\ ^0$C **Ans.**

36. Determine quality of steam at 15.5 bar if it contains 2500 kJ/kg of heat.

Solution: At 15.5 bar $h_{fg} = 1927.3$ kJ/kg

$h_f = 851.7$ KJ/kg

$h = h_f + x\ h_{fg}$

$2500 = 851.7 + 1927.3\ x$

∴ $x = \dfrac{2500 - 851.7}{1927.3}$

$= 0.855$

∴ Quality of steam = 0.855 Dry (85.5%) **Ans.**

37. In a throttling Calorimeter the pressure before throttling is 9.6 bar. If pressure of steam after throttling is 1 bar, what is the limiting value of the D.F of steam on high side.

Solution :

(i) One of the essential requirements for satisfactory operation of throttling calorimeter is that steam must be super heated after throttling.

(ii) In order to achieve this condition, there is certain minimum value of D.F of steam before throttling, so that super heated steam may be obtained after throttling.

(iii) In this problem after throttling pressure is 1 bar. From steam table corresponding to 1 bar saturation temperature is 99. ^{0}C, which is less than saturation temperature corresponding to pressure 9.6 bar (Before throttling) therefore steam after throttling will be wet instead of superheated.

(iv) In problem it is said that pressure of steam after throttling is 1 Bar i.e., steam after throttling is Dry-saturated.

Enthalpy of steam before throttling = Enthalpy of steam after throttling

Let x = The limiting D.F of steam before throttling.

Corresponding to 9.6 Bar

$h_f = 754.8$ kJ/kg At 1 bar $h_g = 2675.4$ KJ/kg

$h_{fg} = 2019.8$ kJ/kg

$\therefore\ 754.8 + 2019.8\,x = 2675.4$

$x = 0.951$

Therefore D.F of steam at 9.6 bar must not be less than 0.951 in order that throttling calorimeter be used to determine the D.F of steam of main pipe.

38. Steam at 14 bar and D.F = 0.7 is throttled to 1.1 bar. Determine D.F of steam after throttling.

Solution: Note:

- If D.F of steam before throttling is more than 0.9 then after throttling it become super heated. If it is less than 0.9, steam after throttling will not be superheated.

- But in problem D.F before throttling is 0.7 and it is also asked to find D.F of steam after throttling that means the condition of steam after throttling will be wet and not superheated.

P_1 = 14 bar from steam table At P_2 = 1.1

h_f = 830.1 kJ/kg h_f = 428.8 kJ/kg

h_{fg} = 1957.7 kJ/kg f_g = 2250.8

Enthalpy before throttling = Enthalpy after throttling

$830.1+1957.7\, x_1 = 428.8 + 2250.8\, x_2$

$830.1 + 1957.7 \times 0.7 = 428.8 + 2250.8\, x_2$

x_2 = D.F after throttling = 0.787 **Ans.**

39. Following observations were recorded with combine separating-throttling calorimeter.

Water separated 2 kg

Steam Discharged from throttling calorimeter = 20.5 kg

Temperature of steam after throttling = 110 ^{0}C

Initial pressure of steam = 12 bar

Barometer reading = 760 mm of H_g

Final pressure = 5 mm of Hg

Estimate quality of steam supplied.

Take C_{ps} for super heated steam as 2.1 kJ/kg K

Solution:

m_w = 2 kg

m_s = 20.5 kg

P_1 = 12 bar

B.R = 760 m of Hg

P_2 = 5 mm of Hg

T_{sup} after throttling = 110 ^{0}C

C_{ps} = 2.1 kJ/kg K

x_1 = D.F of steam leaving the separating calorimeter

$$x_1 = \frac{m_s}{m_s + m_w} = \frac{20.5}{20.5+2} = 0.911$$

Pressure of steam after = 5 + 760 = 765 mm of H_g absolute

$$\therefore \quad P_2 = \frac{765}{760} = 1.006 \text{ bar}$$

Manometer measure pressure above atmospheric pressure

∴ Actual pressure of steam after throttling = Final pressure + atmospheric pressure

Enthalpy before throttling = Enthalpy after throttling

(Steam is always wet at 12 bar) (After throttling steam is generally Super heated at 1.006 bar say 1 bar)

At 12 bar At one bar

$h_{f1} + x_2\, h_{g1} = h_{g2} + c_{ps}\,(t_{sup} - t_{sat})$

(From steam table at 1 bar t_{sat} = 99.6 °C)

$798.4 + 1984.3\,x = 2675.4 + 2.1\,(110 - 99.6)$

$x_2 = 0.9569$

∴ $x = x_1 \times x_2 = 0.911 \times 0.9569 = 0.872$

40. In a test to find D.F of steam with combined separating and throttling calorimeter, the following observations were recorded

Water collected in separating calorimeter = 4.5 kg

Steam condensed after throttling calorimeter = 45.5 kg

Initial pressure of steam = 11.5 bar (absolute)

Temperature of steam after throttling = 140 ^{0}C

Manometer reading = 75 mm of Hg

Barometer reading = 752 mm of Hg

Estimate the D.F of steam as it enters the throttling calorimeter

Also Determine the D.F of steam before separating calorimeter.

Take c_{ps} for super heated steam = 2 kJ/kg K

Solution: Steam from separating calorimeter flows to throttling calorimeter where it is condensed.

Mass of steam leaving calorimeter = Steam condensed after throttling calorimeter

$$m_1 = 45.5 \text{ kg}$$

Mass of suspended moisture collected in the separating calorimeter m_w = 4.5 kg

$$\therefore \quad x_1 = \frac{\text{Mass of dry steam condensed}}{\text{mass of wet steam}}$$

$$= \frac{\text{Mass of dry steam condensed in throttling calorimeter}}{\text{Mass of steam condensed + mass of suspended water collected in separating calorimeter}}$$

$$= \frac{m_s}{m_w + m_s} = \frac{45.5}{45.5 + 4.5} = 0.91$$

$\therefore \quad x_1 = 0.91$

Absolute pressure of steam after throttling

= (Barometer reading + manometer reading)

= 752 + 75

= 827 mm of Hg

Convert 827 mm of Hg in to bar

We know 1 bar = 750 mm of Hg

$\therefore \quad$ Pressure after throttling $= \frac{827}{750} = 1.1$ bar

At 11.5 from steam table

$h_f = 789.9$ kJ/kg

$h_{fg} = 1991.3$ kJ/kg

$t_{sat} = 186\ ^0C$

At 1.1 bar from steam table

$h_f = 428.8$ kJ/kg

$h_{fg} = 2250.8$ kJ/kg

$h_g = 2679.6$ kJ/kg

$t_{sat} = 102.3\ ^oC$

- Let x_2 be the dryness fraction of steam as it enters the throttle calorimeter.
- It is necessary to know condition of steam after throttling before applying, enthalpy before throttling = Enthalpy after throttling.
- After throttling corresponding to pressure of steam (In this problem 1.1 bar), from steam table saturation temperature is 102.3. But actual temperature of steam after throttling recorded by thermometer is 140 ^0C which is more than saturation temperature of 102.3 ^0C. Therefore steam after throttling is super heated.

$\therefore$ Enthalpy of steam before throttling at 11.5 bar = Enthalpy of steam after throttling at 1.1 bar

$$h_{f_2} + x_2\, h_{fg_2} = h_{g3} + c_p\,(t_{sup} - t_{sat})$$

$$789.9 + 1991.3\ x = 2679.6 + 2(140 - 102.3)$$

$x_2 = 0.987$ **Ans.**

$\therefore \quad x = x_1 \times x_2 = 0.91 \times 0.987 = 0.898$

(where x = D.F of steam before separating calorimeter) **Ans.**

41. In test of wet steam made by the Barrel calorimeter the following observations were made. Calculate Dryness fraction of steam.

Mass of copper calorimeter = 0.98 kg

Mass of calorimeter and water = 3.84 kg

Mass of copper calorimeter, water and steam = 4.04 kg

Specific heat of copper = 0.092 kJ/kg K

Initial temperature of water = 7 °C

Final temperature of water = 44.5 °C

Temperature of steam = 158.8 °C

Solution:

Step I Find water equivalent of calorimeter m_e

Step II Mass of steam condensed and initial mass of water in calorimeter.

Step II Latent heat (h_{fg}) of steam at 158.8 °C (as pressure of steam is not given)

Step I Water equivalent of calorimeter = $m_e = m_c \times c_{pc} = 0.98 \times 0.092 = 0.09$

Step II (i) Mass of steam condensed = (Mass of water + calorimeter + steam) – Mass of calorimeter and water)

$= 4.04 - 3.8$

$= 0.24$ kg

(ii) Initial mass of water in calorimeter = m_w = mass of calorimeter and water – mass of copper calorimeter.

$m_w = 3.84 - 0.98 = 2.86$ kg

Step III Latent heat h_{fg} of steam at 158.8 from steam table = 2085 kJ/kg

Heat lost by steam = Heat gained by water and calorimeter

Latent heat lost by steam = $x\ h_{fg}$ kJ

Sensible heat lost by steam = c_{pw} (t_{sat} – final temperature of water)

Heat lost by steam = Heat gained by water and calorimeter

$\therefore\ m_s[\ xh_{fg} + c_{pw}\ (t_{sat} - t_2)] = c_{pw}\ (m_c + m_e)\ (t_2 - t_1)c_{pw}$

$0.24\ [x \times 2085 + 4.1868(158.8 - 44.5)] = (2.86 + 0.09)$

$(44.5 - 7)\ 4.1868$

$x = 0.696$

42. With a separating calorimeter and throttling calorimeter arranged in series, the following observations were taken.

Water separated = 2 kg

Steam discharged from throttling calorimeter = 20.5 kg

Temperature of steam after throttling = 110 ^{0}C

Initial Pressure = 10.5 bar

Barometer reading = 760 mm of Hg

Final Pressure after throttling = 204 mm of water column

Estimate the quality of the steam sample.

Take specific heat of super heated steam = 2 kJ/kg K

Solution: The quality of steam sample = $x_1\ x_2$

$$\text{But } x_1 = \frac{\text{Steam discharged throttling calorimeter}}{\text{steam discharged from throttling calorimeter + water particles saparated}}$$

$$x_1 = \frac{2.5}{20.5 + 2} = \frac{20.5}{22.5} = 0.911$$

Let us convert final pressure in mm of Hg = $\dfrac{204}{13.6}$ = 15 mm of Hg

$\therefore$ Absolute pressure of steam after throttling = B.R + 15

= 760 + 15 = 775 mm of Hg

1 bar = 750 mm of Hg

$$775 = \frac{775}{750} = 1.033 \text{ bar}$$

Corresponding to pressure 1.033 bar from steam table we can not find value of h_f, h_{fg}, t_{sat} etc directly therefore interpolation is required.

From steam table

Pressure bar	h_g	t_{sat}
1.0	2675.4	99.6
1.1	2679.6	102.3
Deference 0.1	4.2	2.7

$$\text{Enthalpy at 1.033 bar} = \text{Enthalpy at 1 bar} + \frac{h_{g2} - h_{g1}}{1.1 - 1} \times 0.033$$

$$= 2675.4 + \frac{4.2}{0.1} \times 0.033$$

$$= 2675.4 + 1.386$$

$$= 2676.786 \text{ KJ/kg}$$

Saturation temperature at 1.033 bar

$$= \text{saturation temperature} + \frac{2.7}{0.1} \times 0.033$$

$$= 99.6 + 0.891$$

$$= 100.49\ ^0\text{C}$$

Enthalpy before throttling at pressure 10.5 bar = Enthalpy after throttling at pressure 1.033 base

From steam table at 10.5 bar

$h_f = 772.0$ KJ/kg

$h_{fg} = 2005.9$ KJ/kg

$\therefore \quad 772 + 2005.9\, x_2 = 2676.4 + 2\,(110 - 100.49)$

$x_2 = 0.958$

$\therefore \quad x = x_1 \times x_2$

$= 0.91 \times 9.58 = 0.873$

43. Combined separating throttling calorimeter was used to measure the D.F of steam to engine. The following observations were made.

Pressure of steam in mains = 6.8 bar

Temperature of steam after throttling = 110 ^{0}C

Pressure of steam after throttling = 13.6 cm of water above atmosphere

Barometer reading = 75 cm of Hg

Steam supplied to engine = 55 kg

Water collected from separating calorimeter = 1 kg

Steam discharged from throttling calorimeter = 6.5 kg and condensed

The entire steam was passed through separating calorimeter.

Take $C_{ps} = 2.3$ kJ/kg K.

Solution: Pressure of steam after throttling = B.R + $\frac{13.6}{13.6}$

$= 75 + 1 = 76$ cm of Hg

$$75 \text{ cm of Hg} = 1 \text{ bar}$$

$$76 \text{ cm of Hg} = \frac{76}{75} = 1.0133 \text{ bar}$$

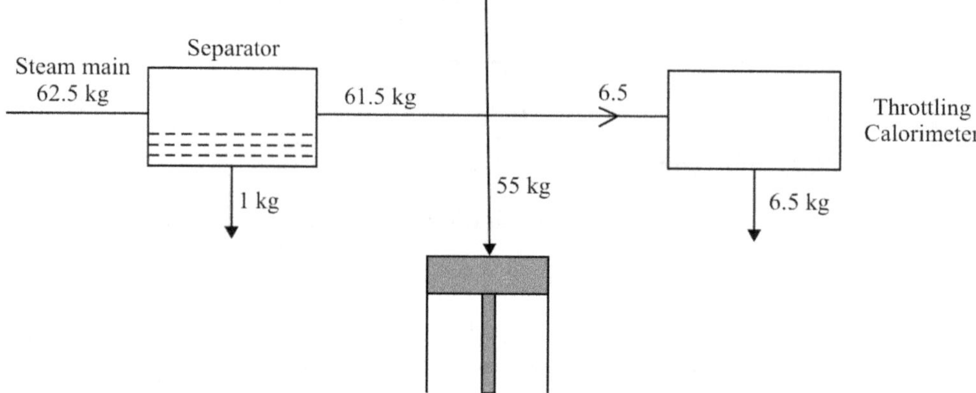

For separating calorimeter

$$x_1 = \frac{61.5}{61.5+1} = 0.985$$

Total heat at 1.0133 = if not given in steam table it can determined by interpolation

Pressure bar	H_g	t_{sat}
1.0	2675.4	99.6
1.1	2679.6	102.3
Difference 0.1	4.2	2.7

$$\therefore \quad h_g \text{ at } 1.0133 = h_g \text{ at 1 bar} + \frac{4.2}{0.1} \times 0.0133$$

$$= 2675.4 + 0.5586$$

$$= 2675.96 \text{ kJ/kg}$$

$$t_{sat} \text{ at } 1.0133 = t_{sat} \text{ at 1 bar} + \frac{2.7}{0.1} \times 0.0133$$

$$= 99.6 + 0.3591$$

$$= 99.9591\ ^0\text{C}$$

Heat before throttling = Heat after throttling

$$h_{f_2} + x_2\, h_{fg_2} = h_g + c_{ps}\,(t_{sup} - ts_{at})$$

(at 10.5 bar) (at 1.0133 bar)

$$772. + 2005.9 \times x_2 = 2675.96 + 2(110 - 99.9591)$$

$$x_2 = 0.959$$

$$\therefore \quad x = x_1 \times x_2$$

$$= 0.911 \times 0.959$$

$$= 0.873 \quad \textbf{Ans.}$$

Questions

1. What is meant by saturation pressure and temperature?
2. Define (a) sensible heat (a) heat of evaporation (a) super heat.
3. What are advantages of super heated steam?
4. Explain specific volume of wet steam, dry steam and super heated steam.
5. Define Dryness fraction and priming
6. Prove that Dryness fraction plus wetness = 1
7. What is meant by internal energy of steam?
8. Describe the process formation of steam from ice at 0 °C
9. Derive expression for external work done during evaporation
   (i) when steam is wet
   (ii) when steam is dry-saturated
   (iii) when steam is super heated
10. Explain with neat sketch the construction and working of Barrel Calorimeter. Why it is not used in practice to find out D.F. of steam from boiler.
11. Draw neat sketch of separating calorimeter and mention its limitations.
12. Can throttling calorimeter measure D.F of steam accurately if not why?
13. Explain with neat diagram the construction and working combined separating throttling calorimeter.
14. For combined separating and throttling calorimeter prove that $x = x_1\, x_2$ with usual notation.
15. What is difference between separating and throttling calorimeter.

Objective Type Questions

1. Locus of saturated liquid line and saturation vapour line meet at

   (a) Boiling Point (b) Critical Point

   (c) Ice point (d) Triple Point

2. What is critical temperature and critical pressure of steam.

   (a) The critical temperature 221.2 ^{0}C and critical pressure 374.15 bar

   (b) The critical temperature 374.15 ^{0}C and critical pressure 221.15 bar

   (c) The critical temperature 374.15 ^{0}C and critical pressure 374.15 Bar

   (d) None of above

3. Latent heat of steam at atmosphere pressure

   (a) 1535 kJ/kg (b) 1880 kJ/kg

   (c) 2257 kJ/kg (d) 2290 kJ/kg

4. D.F of steam is equal to

   (a) $\frac{m_g}{m_g + m_f}$ (b) $\frac{m_f}{m_g + m_f}$

   (c) $\frac{m_g}{m_f}$ (d) $\frac{m_f}{m_g}$

   m_g = mass of dry steam

   m_f = mass of water in suspension

5. Enthalpy of dry saturated steam increase as pressure.

   (a) increases (b) decreases

   (c) remain constant (d) none of above

6. Internal energy of dry saturated steam is given by with usual notations.

   (a) $u = h_g + 100\ Pv_g$ (b) $u = \frac{h_g}{100Pv_g}$

   (c) $u = h_g - 100\ PV_g$ (d) $\frac{100pv_g}{hg}$

Answers

1. (b)	2. (b)	3. (c)
4. (a)	5. (a)	6. (c)

Fill in the Blanks

1. Internal energy is a .... function and heat is a .... function.
2. The change in internal energy in isothermal process is ....
3. The change in work done in constant volume process is ....
4. A reversible adiabatic process takes place with either .... nor ....
5. All heat transferred or supplied during constant volume process will be stored in steam ....
6. The state of substance whose evaporation from its liquid state is completed is known as ....
7. .... measures more accurately Dryness fraction of steam.
8. If pressure of steam increases than enthalpy will ....
9. Enthalpy is equal to .... + ...
10. Saturation temperature of steam .... with increase in pressure.
11. Latent heat of steam .... with increase in pressure.
12. Value of D.F for dry saturated steam is ....
13. One kg of wet steam having D.F = 0.85 will have .... water and .... Kg of dry steam.
14. At critical pressure latent heat will become ....
15. Enthalpy remain constant during .... process
16. Entropy remain constant during .... process.
17. The difference between temperature of super heated steam and the saturated steam for given pressure is known as ....
18. One kg of water having its temperature 50^0 will have approximately .... kJ of heat.

Answers

1. Point, path
2. zero
3. zero
4. change in entropy, heat transfer
5. as Internal energy
6. vapour
7. separating throttling calorimeter
8. increase
9. internal energy + PV
10. increases
11. decrease
12. one

13. 0.15 kg of water and 0.85 kg of dry steam
14. zero
15. throttling process
16. isentropic
17. degree of super heat
18. 209.3 kJ

Exercises

1. Find D.F of steam if 3 kg of water is suspended in 95 kg of dry saturated steam.

   [D.F = 0.97]

2. Determine the Dryness fraction of steam having mass 5.2 kg. If total heat of steam is 12000 kJ, super heated steam pressure is 5 bar.

   [D.F = 0.791]

3. A pressure cooker contains 1.3 kg of steam at 5.2 bar pressure and 0.88 dry. Find the quantity of heat that must be rejected so as the quality of steam becomes 0.475 dry.

   [Heat rejected = 1090.4 kJ]

4. Steam at pressure of 1.2 Mpa and with dryness fraction of 0.8 occupies a volume of 1.2 m^3. It is heated till its temperature reaches 250 ^{0}C while pressure remains constant. Find.

   (i) Mass of steam

   (ii) Quantity of heat supplied

   (iii) Change in volume

   Take specific heat of super-heated steam = 2.09 KJ/kg k

   [(i) 9.23 kg (ii)4868 KJ (iii) 0.5089 m^3]

5. 4000 m^3/kg of wet steam having Dryness fraction = 0.95 at 10 bar pressure is to be supplied by boiler for processing work. Calculate.

   (a) Mass of steam supplied per hour

   (b) Quantity of coal of c.v = 16000 KJ/kg, to be burnt in boiler if overall efficiency of boiler is 29%

   [(a) 21703.744 kg/hr (b) 12514.83 kg of coal/hr]

6. 2 kg of water at 47.8 ^{0}C is heated under constant pressure of 14 bar until it is converted into steam with 200 ^{0}C of super heat. Determine.

   (i) Increase in sensible heat

   (ii) Latent heat

(iii) Super heat

(iv) Total heat supplied

(v) Internal energy change

Take c_{ps} of super heated steam = 2.09 KJ/kg k

[(i) 1260 KJ (ii) 3920 KJ (iii) 836 KJ (iv) 6016 KJ (v) 5456 KJ]

7. Find the internal energy of 2 kg of super heated steam at pressure of 10 bar and temperature of 280 ^{0}C. If this steam is expanded to 1.6 Bar pressure and dryness 0.80, find change in internal energy.

[5522 KJ, –1318 KJ (decreases)]

8. Steam at 19 bar is throttled to 1 bar and temperature after throttling found to be 150 ^{0}C. Calculate initial dryness fraction of steam

[0.989]

9. 4 kg of 0.5 dry steam at 6 bar pressure is heated so that it becomes

(a) 0.95 dry at 6 bar pressure

(b) dry and saturated at 6 bar pressure

(c) Super heated to 300 ^{0}C at 6 bar pressure

(d) Super heated to 250 ^{0}C 8 bar pressure. Determine in each case quantity of heat to be supplied.

[(a) 3753 KJ (b) 4170 KJ (c) 5397.6 KJ (d) 4950 KJ]

10. Two boilers generates steam, delivers equal amount of steam into common main. One boiler is fitted with super heater and other without super heater. Both the boilers operate at pressure of 15 bar which is also the pressure in main. The temperature of steam in the super heater boiler is 300 ^{0}C. It is desired to maintain a temperature of 225 ^{0}C in main. What will be quality of steam being raised in the boiler without super heater? Assume specific heat of super heated steam = 2.1 KJ/kg k.

[0.944]

11. Two boilers of unequal capacities generate steam at same pressure of 14 bar absolute. One boiler produces steam with 45 ^{0}C super heat and the other boiler having evaporative capacity half that of first boiler. If the main is just dry saturated, determine quality of steam of the second boiler. Take C_p for super heated steam as 2.2 KJ/kg k.

[0.90]

12. Steam at 10 bar and 0.925 dry is contained inside vessel having capacity $1m^3$. The delivery valve is opened and steam is blown off. The period of blown off is so regulated that pressure drops to 5 bar. Delivery valve is then closed and vessel is then cooled until pressure becomes 4 bar. Calculate (1) Mass of steam blown off (2) Dryness fraction of steam in the vessel after cooling.

[mass = 2.745 kg, D.F = 0.7654]

13. Following observation were recorded in an experiment on separating throttling calorimeter.

Initial pressure = 12.7875 bar gauge

Pressure after throttling = 0.00675 Bar gauge

Barometer reading = 760 mm of Hg

Temperature after throttling = 118 ^{0}C

Water collected in separating calorimeter = 0.45kg

Steam condensed after throttling = 6.75kg

Determine the D.F. of steam sample

Take specific heat of super heated steam as 2 KJ/kg k

[x = 0.904]

14. Steam at pressure of 7 bar and Dryness fraction 0.5 is throttled to pressure of 4.5 Bar. Find final condition of steam.

[x = 0.52]

15. Steam from boiler enters into barrel calorimeter and after few minutes, observations were taken of mass and temperature of water. From collected data given below, calculate the D.F of the steam.

Mass of empty barrel = 25 kg

Mass of barrel and cold water = 135 kg

Mass of barrel × cold water × steam = 140 kg

Initial temperature of water = 15 ^{0}C

Final temperature of water = 42 ^{0}C

Steam pressure = 8 bar

Specific heat of barrel material = 0.4 KJ/kg k

[0.96]

16. 5 kg of steam at 6.8 bar absolute were sampled and passed into a barrel calorimeter whose water equivalent at 20 ^{0}C was 10 kg. The barrel contained 140 kg of water at 20 ^{0}C. The temperature of water when steam condensed into it was recorded at 40 ^{0}C. Determine Dryness fraction of steam.

[x = 0.9669]

CHAPTER 5

Heat Engines

5.1 Thermal Prime Movers

A device used to convert natural source of energy into mechanical work is known as a prime mover. A prime mover which converts thermal (heat) energy into mechanical work is known as thermal prime mover. It is also known as heat engine. Thermal prime mover is the most important of all the prime movers and is widely used for various applications. They have made much contribution in the field of transportation and communication on land, on sea and in the space and in power generation. Steam engine, petrol engine, diesel engine, steam turbines, gas turbines and rockets are some examples of thermal prime movers.

5.2 Elementary Heat Engines

As we have seen thermal prime mover is commonly known as heat engine where heat energy is converted into mechanical work. The general arrangement of an elementary heat engine is shown in Fig. 5.1. The main elements of a heat engine are 1. source of heat; 2. medium to carry heat energy; 3. converting machine; 4. sink of heat.

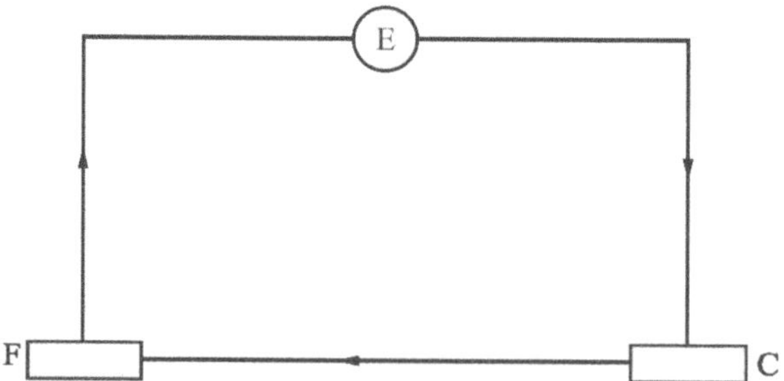

Fig. 5.1 Elementary heat engine.

A source of heat is a furnace or combustion chamber or a reservoir of heat from which heat is continuously supplied to the medium. F indicates a source of heat. Medium is also known as working substance or working fluid. It may be in the form of a gas or vapour that takes heat energy from the source and flows to the converting machine or engine E where some heat energy is converted to mechanical work. Then, the working fluid is discharged into the sink of heat C. Sink or cold body is a low temperature reservoir where heat is rejected by the working fluid. Then, working fluid may go back to the source of heat again. Thus the cycle is completed and these operations are repeated over and over again.

5.3 Source of Heat

The source of heat is a heat reservoir where heat is produced and supplied to working fluid. There are various sources of heat. e.g., 1. burning of fuel in furnace of boiler or in combustion chamber of I.C. engine; 2. fission of atom; 3. fusion of atom etc.

Heat energy can be produced by burning of fuel like coal, petrol, diesel, crude oil, natural gas, wood etc. The nuclear energy produced by fission of atom and thermo-nuclear energy by fusion of atom like hydrogen provides huge quantity of heat energy.

From practical point of view, the source of heat should be so selected that heat is continuously available at low cost and by simple means. From this consideration, burning or combustion of fuel is the chief source of heat. At present, nuclear energy is also employed as a powerful source of heat.

5.4 Working Substance

Working substance or working fluid is a medium which receives heat from heat source, transmits it to the heat engine for doing work and rejects some heat to the sink. It should be non-poisonous, cheap, easily available and of greater thermal capacity. Steam and hot gases are the most common working fluids. Mercury vapour is also used as a working fluid.

Steam is the most important vapour used as a working fluid in steam engines and steam turbines. Water is converted into steam by application of heat in a closed vessel known as a boiler. It is then supplied to steam engine or steam turbine by a pipe.

Hot gases are produced by burning of fuel and are used in the engine.

5.5 Converting Machines

A machine which converts heat energy into mechanical work is called converting machine. It may be reciprocating, rotary or jet type.

The reciprocating type consists of a hollow cylinder into which a piston reciprocates due to expansion and compression of working fluid. The rotary type machine consists of a wheel having blades or vanes and mounted on a shaft. Wheel rotates due to action of working fluid on the blades. In jet machines, working fluid is discharged from the machine in form of jet and motion is caused due to reaction of jet.

5.6 Sink of Heat

Sink is a reservoir of heat at a temperature lower than that of the source. Heat is rejected by the working fluid in the sink.

5.7 Classification of Heat Engines

Heat engines can be classified in many ways:

(a) According to the place of combustion

1. External combustion engines
2. Internal combustion engines (I. C. engines)

In external combustion engine, fuel is burnt outside the engine cylinder and the working fluid is heated outside. Examples of external combustion engine are steam engine, steam turbine, gas turbine etc.

In steam engines and steam turbines, combustion of fuel takes place in boiler where high pressure steam is produced. This steam is expanded in steam engine or steam turbine which produces mechanical work.

In I.C. engines, combustion of fuel takes place inside the engine cylinder with the help of air. Combustion product itself is the working fluid of the engine. Petrol engine and diesel engine are the examples of I.C. engines.

(b) According to the type of converting machines

1. Reciprocating engines
2. Rotary engines or turbines
3. Jet engines

In reciprocating engine, piston reciprocates in the cylinder. Steam engine, petrol engine and diesel engine are reciprocating engines.

In turbines, working fluid is issued on the blades. Steam turbine and gas turbine are rotary engines.

In jet engine, working fluid is discharged in the form of jet. Rocket is an example of jet engine.

5.8 Heat Engine Cycles

A cycle may be defined as a series of processes performed in a definite order or sequence such that the original conditions are restored at the end of processes. There are various cycles which are performed in different heat engines.

Cycles which provide the basic concept of the performance of a heat engine in an ideal condition is called ideal cycles. Cycles which explain the working of heat engine in actual condition is called actual cycles.

These cycles are easily studied by diagrams such as pressure-volume (p-V) diagram, temperature-entropy (T-s) diagram etc.

First, we will consider an ideal heat engine working on ideal cycle.

5.9 Carnot Cycle

Carnot conceived an ideal cycle in which all processes are reversible. This cycle consists of four processes – two isothermals and two adiabatics. The Carnot engine and pressure-volume (p-V) diagram of Carnot cycle are shown in Fig. 5.2.

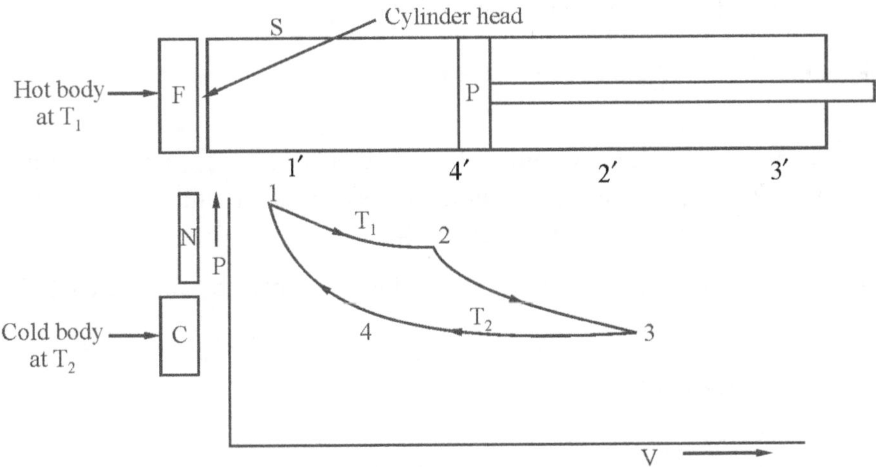

Fig. 5.2 Carnot engine and p-v diagram of carnot cycle.

The carnot engine consists of 1. a cylinder S and a piston P made of perfectly non-conducting material but the head of the cylinder is made of perfectly conducting material, 2. a movable cover N, a perfect non-conductor of heat, 3. a movable hot body F, a source of heat kept at constant absolute temperature T_1, 4. a movable cold body C, a sink kept at constant absolute temperature T_2 ($T_2 < T_1$).

The unit mass of working fluid, assumed to be a perfect gas, is taken inside the engine cylinder.

Processes of a carnot cycle are as follows:

Isothermal Process 1-2: The source of heat F is brought near to the cylinder head. Heat is added to the working fluid at constant temperature T_1 from the source and working fluid is expanded at constant temperature T_1. Piston moves from 1' to 2'. This is denoted by curve 1-2 on p-V diagram. Heat received which is equal to work done during isothermal expansion is

$$Q_1 = P_1V_1 \, ln\frac{V_2}{V_1} = RT_1 \, ln\frac{V_2}{V_1} \qquad .....(5.1)$$

Adiabatic Process 2-3: The hot source is removed and the cover N is applied to the head of the cylinder. Adiabatic expansion of the working fluid occurs with decrease of temperature from T_1 to T_2. Piston moves from 2' to 3'. This is denoted by curve 2-3 on p-v diagram.

Isothermal Process 3-4: Sink C is brought near the cylinder head. Isothermal compression of working fluid occurs at temperature T_2 and heat is rejected to sink. Piston moves from 3' to 4'. This is denoted by curve 3-4.

Heat rejected $$Q_2 = P_3V_3 \, ln\frac{V_3}{V_4} = RT_2 \, ln\frac{V_3}{V_4} \qquad .....(5.2)$$

Adiabatic Process 4-1: Sink C is removed and the cover N is applied to the head of the cylinder. Adiabatic compression occurs with increase in temperature from T_2 to T_1. Piston moves from 4' to 1'. This is denoted by curve 4-1.

Thus the cycle is completed. The net work done is shown by the area 1-2-3-4 in p-V diagram.

Efficiency of Carnot cycle:

For adiabatic processes 2-3 and 4-1

$$\frac{T_1}{T_2} = \left(\frac{V_3}{V_2}\right)^{\gamma-1} \text{ and } \frac{T_1}{T_2} = \left(\frac{V_4}{V_1}\right)^{\gamma-1} \qquad .....(5.3)$$

$$\therefore \quad \frac{V_3}{V_2} = \frac{V_4}{V_1}$$

$$\therefore \quad \frac{V_3}{V_4} = \frac{V_2}{V_1} = \text{Say } \gamma$$

Now efficiency $$\eta = \frac{\text{heat converted to mechanical work}}{\text{heat added}}$$

$$= \frac{\text{heat added} - \text{heat rejected}}{\text{heat added}}$$

$$= \frac{Q_1 - Q_2}{Q_1}$$

$$= \frac{RT_1 \ln\frac{V_2}{V_1} - RT_2 \ln\frac{V_3}{V_4}}{RT_1 \ln\frac{V_2}{V_1}}$$

$$= \frac{T_1 - T_2}{T_1} \qquad .....(5.4)$$

$$= 1 - \frac{T_2}{T_1} \qquad .....(5.5)$$

This shows that efficiency of Carnot engine depends on temperatures only. Carnot cycle has the highest possible efficiency. Carnot cycle is practically impossible.

5.10 Carnot Cycle with Vapour

In this cycle, steam or any other vapour is used as a working fluid. This cycle is appropriate for steam power plant. Components of this plant are shown in Fig. 5.3. p-V diagram of this cycle is shown in Fig. 5.4.

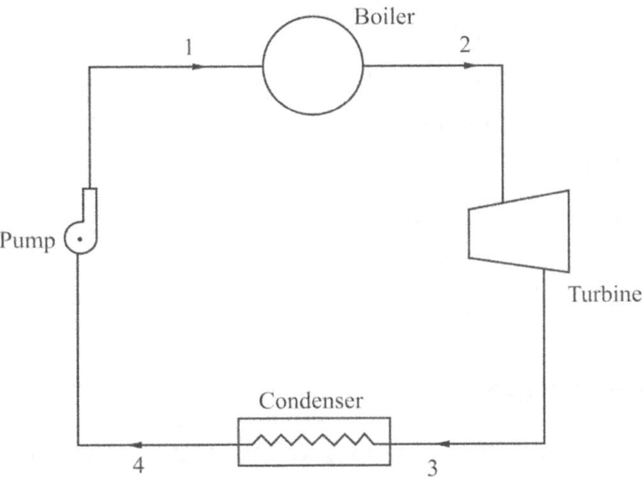

Fig. 5.3 Components of Carnot cycle with vapour.

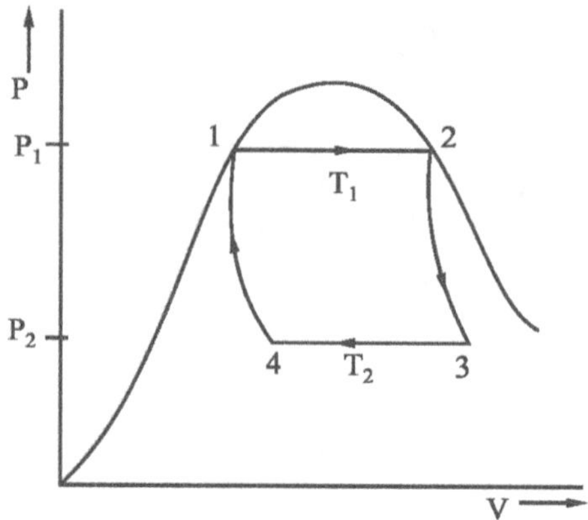

Fig. 5.4 p-v diagram for Carnot cycle with vapour.

Unit mass of liquid at its saturation temperature T_1 is taken inside the boiler at point 1. During process 1-2, heat is added to the liquid in boiler at constant pressure. Temperature remains constant during this process. Thus process 1-2 is isothermal process in wet region. At point 2, it is dry steam. During process 2-3, steam expands adiabatically in the turbine with decrease in temperature from T_1 to T_2 and work is done. During process 3-4 heat is rejected to cooling medium in the condenser at constant pressure and constant temperature T_2. Thus process 3-4 is isothermal. Process 4-1 is adiabatic compression of wet steam from condenser pressure to boiler pressure with increase in temperature from T_2 to T_1. Thus, the cycle is completed.

Here, when the vapour is expanded (process 1-2) or compressed (process 3-4) at constant temperature, the pressure remains constant. So, the isothermal lines are horizontal in p-V diagram.

Here also, the efficiency of this Carnot cycle is

$$\eta = \frac{T_1 - T_2}{T_1} \qquad \text{.....(5.6)}$$

5.11 Rankine Cycle

Rankine cycle also is used in steam turbines. Components of the plant are already shown in Fig. 5.3.

P-V diagram of this cycle is shown in Fig. 5.5.

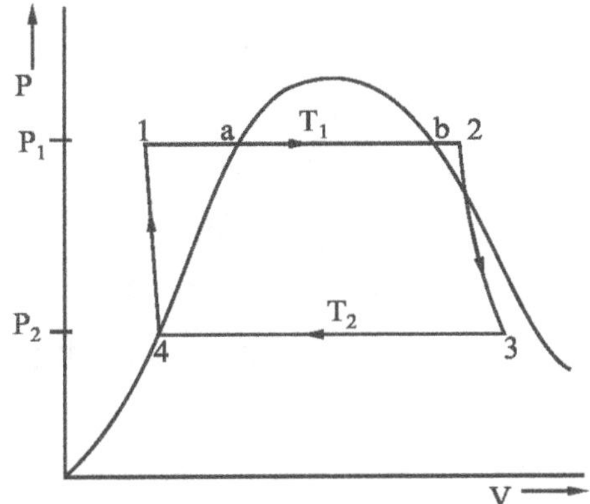

Fig. 5.5 P-V diagram of Rankine cycle.

Unit mass of water is taken inside the boiler at point 1. Rankine cycle consists of four processes.

Process 1-2: Heat is supplied to water at constant pressure in boiler. During process 1-a, sensible heat is added to water and its temperature is increased to saturation temperature corresponding to boiler pressure. During process a-b, latent heat is added to water and saturated water is converted into saturated vapour. During process b-2, steam is superheated.

Heat supplied in boiler

$$Q_1 = (h_{f1} + h_{fg1} + C_{ps} \text{ superheat}) - h_{f4}$$
$$= h_2 - h_{f4}$$
$$= h_2 - h_{f3} \quad .....(5.7)$$

where C_{ps} = specific heat of superheated steam

h_{f_1} = enthalpy of water (sensible heat) at 1

h_{fg_1} = latent heat at 1

$h_{f_4} = h_{f_3}$ = sensible heat of water at 4

h_2 = enthalpy of steam at 2.

Process 2-3: Superheated, dry saturated or wet steam is expanded isentropically into steam turbine from boiler pressure P_1 to condenser pressure P_2.

Work done by the turbine $W_t = h_2 - h_3$ .....(5.8)

Where h_3 = enthalpy of steam at point 3.

Process 3-4: Exhaust steam coming out from turbine is condensed in condenser and leaves as saturated liquid. Heat from the steam is rejected to cooling water.

Heat rejected $Q_2 = h_3 - h_4$ .....(5.9)

Where h_4 = enthalpy of water at point 4.

Process 4.1: Saturated water at condenser pressure P_2 is pumped to boiler pressure P_1

Work required for pumping $W_p = h_1 - h_4$ .....(5.10)

It is seen that in Carnot cycle, it is difficult to pump mixture of vapour and liquid and deliver it as saturated liquid. This difficulty is eliminated in Rankine cycle by complete condensation of vapour to point 4 and then pumping it isentropically to P_1.

We see that Rankine cycle is made up of two constant pressure processes and two isentropic processes.

Efficiency of Rankine cycle

$$\eta = \frac{\text{Net work output}}{\text{Heat supplied}}$$

Net work output $W = W_t - W_p$

$$= (h_2 - h_3) - (h_1 - h_4)$$

Also, net work output W = heat supplied – heat rejected

$$= Q_1 - Q_2$$

$$= (h_2 - h_1) - (h_3 - h_4)$$

$$= (h_2 - h_3) - (h_1 - h_4) \quad .....(5.11)$$

$$\therefore \quad \eta = \frac{(h_2 - h_3) - (h_1 - h_4)}{h_2 - h_{f3}} \quad .....(5.12)$$

Generally pump work is very small as compared to work done in turbine and hence it is neglected.

$$\therefore \quad \eta = \frac{h_2 - h_3}{h_2 - h_{f3}} \quad .....(5.13)$$

5.12 Otto Cycle

Otto cycle, also known as constant volume cycle is used in petrol or gas engine. The working fluid is a mixture of air and fuel. The ratio of air to fuel in a mixture may vary from 10:1 to 24:1. Hence properties of working fluid are similar to properties of air. So such cycle may be analysed by ideal cycle known as air standard cycle where air is used as a working fluid.

Otto cycle consists of four processes, two isentropic and two constant volume processes. p-v diagram of air standard Otto cycle is shown in Fig. 5.6.

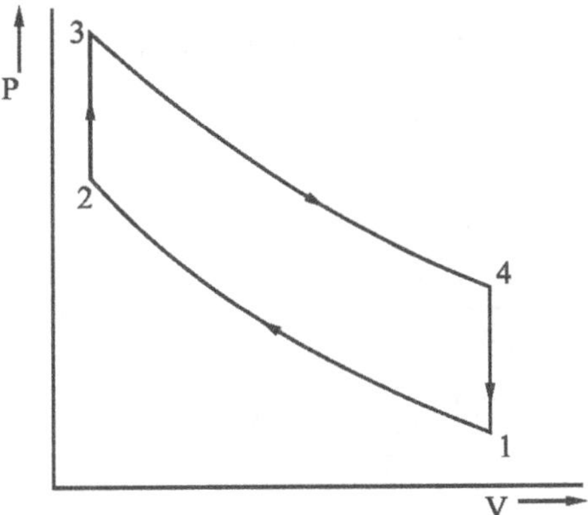

Fig. 5.6 Otto cycle.

Isentropic process 1-2: Piston in the cylinder of an engine compresses the air from 1 to 2 isentropically. Pressure and temperature of air will increase.

Constant Volume Process 2-3: Heat is added to air at constant volume from 2 to 3 by burning of fuel. Further rise in pressure and temperature takes place.

Isentropic Process 3-4: High pressure hot gases expand in the cylinder isentropically from 3 to 4. So work is done during this process. Pressure and temperature of gases decrease.

Constant volume process 4-1: Heat is rejected at constant volume from 4 to 1 as the gases are exhausted from the engine. Pressure and temperature restore to the original values at 1.

Thus, the cycle is completed

Efficiency of Otto cycle:

Heat added at constant volume during process 2-3

$$Q_1 = c_v (T_3 - T_2) \qquad .....(5.14)$$

where c_v = specific heat of air at constant volume

Heat rejected at constant volume during process 4-1

$$Q_2 = c_v (T_4 - T_1) \qquad .....(5.15)$$

Work done in cycle W = heat added – heat rejected

$$= Q_1 - Q_2$$

$$= c_v (T_3 - T_2) - c_v (T_4 - T_1) \qquad .....(5.16)$$

Efficiency

$$\eta = \frac{\text{work done}}{\text{heat added}}$$

$$= \frac{c_v (T_3 - T_2) - c_v (T_4 - T_1)}{c_v (T_3 - T_2)}$$

$$= 1 - \frac{T_4 - T_1}{T_3 - T_2} \qquad .....(5.17)$$

Express T_2 interms of T_1 and T_3 interms of T_4.

For isentropic process 1-2

$$\frac{T_2}{T_1} = \left(\frac{V_1}{V_2}\right)^{\gamma-1}$$

$$= r^{\gamma - 1}$$

where $r = \frac{V_1}{V_2}$ = compression ratio

$$\therefore \quad T_2 = T_1 \, r^{\gamma - 1} \qquad .....(5.18)$$

For isentropic process 3-4

$$\frac{T_3}{T_4} = \left(\frac{V_4}{V_3}\right)^{\gamma-1}$$

$$V_4 = V_1 \text{ and } V_3 = V_2$$

$$\therefore \quad \frac{T_3}{T_4} = \left(\frac{V_1}{V_2}\right)^{\gamma-1}$$

$$= r^{\gamma-1}$$

$$\therefore \quad T_3 = T_4 . \, r^{\gamma-1} \qquad .....(5.19)$$

Substituting T_2 and T_3 in eq. 5.17,

$$\eta = 1 - \frac{T_4 - T_1}{T_4 r^{\gamma-1} - T_1 r^{\gamma-1}}$$

$$= 1 - \frac{T_4 - T_1}{(T_4 - T_1) r^{\gamma-1}}$$

$$= 1 - \frac{1}{r^{\gamma-1}} \quad \text{.....(5.20)}$$

Following points should be remembered:

1. Volume $V_2 = V_3$ is the clearance volume denoted by V_c.

   i.e. $V_c = V_2 = V_3$ .....(5.21)

2. Volume swept by piston in one stroke is known as a swept volume.

   It is denoted by V_s.

   $$V_s = V_1 - V_2 = V_1 - V_c \quad \text{.....(5.22)}$$

   Also $V_s = \frac{\pi}{4} D^2 L$ .....(5.23)

   where D = bore (diameter) of the cylinder

   L = stroke length

3. Work done is also equal to area enclosed in p-V diagram.
4. Maximum pressure and temperature occur at point 3 of the cycle. Minimum pressure and temperature occur at point 1.
5. Mean pressure of the cycle is called mean effective pressure (mep). It can be obtained from

   $$\text{mep} = \frac{\text{work done}}{\text{swept volume}} \quad \text{.....(5.24)}$$

Thus, efficiency of Otto cycle depends on compression ratio only. Efficiency increases with increase in compression ratio. In actual engine working on Otto cycle, compression ratio varies from 5 to 8. The efficiency is also known as thermal efficiency.

5.13 Diesel Cycle

This cycle is used for Diesel engine. It is also known as constant pressure cycle.

Diesel cycle consists of four processes – two isentropic, one constant pressure and one constant volume process. p-V diagram of air standard Diesel cycle is shown in Fig. 5.7.

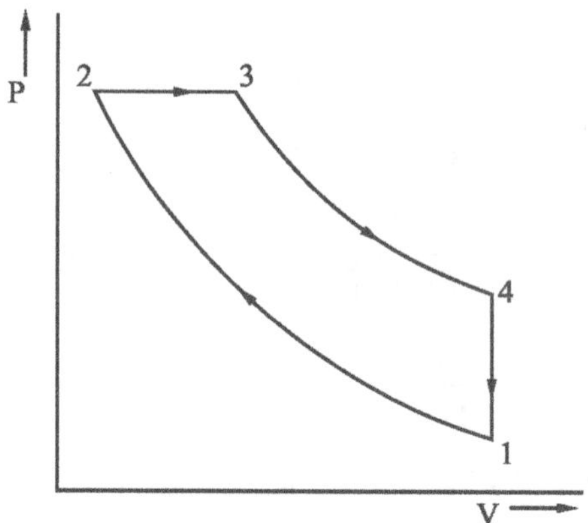

Fig. 5.7 Diesel cycle.

Isentropic process 1-2: Piston in the cylinder of an engine compresses the air from 1 to 2 isentropically. Pressure and temperature of air will increase.

Constant pressure process 2-3: Heat is added to air at constant pressure from 2 to 3 by combustion of fuel (diesel). During this process, pressure of air remains constant and volume and temperature will increase. Fuel is cut off at paint 3. Volume ratio $\frac{V_3}{V_2}$ is called cut-off ratio, generally denoted by ρ.

Isentropic process 3-4: Hot gases produced due to combustion of fuel expand in the cylinder isentropically from 3 to 4. So work is done during this process. Pressure and temperature of gases will decrease.

Constant volume process 4-1: Heat is rejected at constant volume from 4 to 1 as the gases are exhausted from the engine. Pressure and temperature restore the original values at 1.

Thus, the cycle is completed.

Efficiency of Diesel Cycle:

Heat added at constant pressure during process 2-3

$$Q_1 = c_p (T_3 - T_2) \quad .....(5.25)$$

where c_p = specific heat of air at constant pressure.

Heat rejected at constant volume during process 4-1

$$Q_2 = c_v (T_4 - T_1) \quad .....(5.26)$$

Work done in cycle $W = Q_1 - Q_2$

$$= c_p (T_3 - T_2) - c_v (T_4 - T_1) \qquad .....(5.27)$$

Efficiency $\eta = \dfrac{\text{work done}}{\text{heat added}}$

$$= \frac{c_p(T_3 - T_2) - c_v(T_4 - T_1)}{c_p (T_3 - T_2)}$$

$$= 1 - \frac{c_v(T_4 - T_1)}{c_p (T_3 - T_2)}$$

$$= \frac{c_p}{c_v} = \gamma$$

$$\therefore \qquad \eta = 1 - \frac{T_4 - T_1}{\gamma(T_3 - T_2)} \qquad .....(5.28)$$

We will substitute T_2, T_3 and T_4 in terms of T_1. For isentropic process 1-2

$$\frac{T_2}{T_1} = \left(\frac{V_1}{V_2}\right)^{\gamma-1}$$

$$= r^{\gamma-1}$$

$$\therefore \qquad T_2 = T_1 . r^{\gamma-1} \qquad .....(5.29)$$

For constant pressure process 2-3

$$\frac{T_3}{T_2} = \frac{V_3}{V_2} = \rho$$

$$\therefore \qquad T_3 = T_2 . \rho$$

$$= T_1 . r^{\gamma-1} . \rho \qquad .....(5.30)$$

For isentropic process 3-4

$$\frac{T_4}{T_3} = \left(\frac{V_3}{V_4}\right)^{\gamma-1}$$

$$\therefore \qquad T_4 = T_3.\left(\frac{V_3}{V_2}.\frac{V_2}{V_4}\right)^{\gamma-1}$$

$$= T_3.\left(\frac{V_3}{V_2}.\frac{V_2}{V_1}\right)^{\gamma-1} \quad \text{as } V_4 = V_1$$

$$= T_3.\left(\rho.\frac{1}{r}\right)^{\gamma-1}$$

$$= T_1.\, r^{\gamma-1}.\, \rho\left(\frac{\rho}{r}\right)^{\gamma-1}$$

$$= T_1.\, \rho^{\gamma} \qquad .....(5.31)$$

Substituting for T_2, T_3 and T_4 in eq. 5.28,

$$\eta = 1 - \frac{T_1.\rho^{\gamma} - T_1}{\gamma\left(T_1.r^{\gamma-1}\rho - T_1 r^{\gamma-1}\right)}$$

$$= 1 - \frac{1}{r^{\gamma-1}}\left[\frac{\rho^{\gamma-1}}{\gamma(\rho-1)}\right] \qquad .....(5.32)$$

Since $\gamma = 1.4$ for air and ρ is always greater than 1, the quantity in the bracket is always greater than 1. So, for same compression ratio, efficiency of Otto cycle is greater than that of Diesel cycle.

Solved Problems

1. A hot air engine working on Carnot cycle works between the temperature limit of 400 °C and 20 °C. Find the efficiency of the cycle.

Solution:

$$T_1 = 400 + 273 = 673 \text{ K}$$

$$T_2 = 20 + 273 = 293 \text{ K}$$

$$\eta = \frac{T_1 - T_2}{T_1}$$

$$= \frac{673 - 293}{673}$$

$$= \frac{380}{673}$$

$$= 0.565$$

$$= \mathbf{56.5\ \%}$$

2. A power plant is supplied with dry saturated steam at a pressure of 16 bar. Steam is exhausted at 0.2 bar. Using steam table, find the efficiency of Carnot cycle.

Solution: $P_1 = 16$ bar, $P_2 = 0.2$ bar

From steam table

Corresponding to a pressure of 16 bar,

Saturation temperature $T_1 = 201.4°C = 201.4 + 273 = 474.4$ K

Corresponding to a pressure of 0.2 bar,

Saturation temperature $T_2 = 60.1°C = 60.1 + 273 = 333.1$ K

Efficiency of Carnot cycle

$$\eta = \frac{T_1 - T_2}{T_1}$$

$$= \frac{474.4 - 333.1}{474.4}$$

$$= 0.298$$

$$\mathbf{= 29.8\%}$$

3. A steam turbine working on Rankine cycle is supplied with dry saturated steam at a pressure of 20 bar. The steam is exhausted at a pressure of 0.2 bar with a dryness fraction of 0.72. Find the efficiency of the cycle.

Solution: From steam table

At a pressure of 20 bar, $h_2 = 2799.5$ kJ/kg

At a pressure of 0.2 bar, $h_3 = h_{f3} + xh_{fg3}$

$$= 251.4 + 0.72 \times 2358.3$$

$$= 1949.4 \text{ kJ/kg}$$

$$h_{f3} = 251.4$$

$$\eta \text{ of Rankine cycle} = \frac{h_2 - h_3}{h_2 - h_{f3}}$$

$$= \frac{2799.5 - 1949.4}{2799.5 - 251.4}$$

$$= \frac{850.1}{2548.1}$$

$$= 0.3336$$

$$\mathbf{= 33.36\%}$$

4. In an Otto cycle, the temperatures at the beginning and end of isentropic compression are 43 °C and 250 °C. Determine compression ratio and air standard efficiency. Take $\gamma = 1.4$

Solution:

$$T_1 = 43°C = 43 + 273 = 316 \text{ K}$$

$$T_2 = 250°C = 250 + 273 = 523 \text{ K}$$

$$\frac{T_2}{T_1} = \left(\frac{V_1}{V_2}\right)^{\gamma-1}$$

$$\therefore \quad \frac{523}{316} = r^{1.4-1}$$

$$\therefore \quad r^{0.4} = 1.655$$

$$\therefore \quad r = (1.655)^{\frac{1}{0.4}}$$

$$= (1.655)^{2.5}$$

$$= \mathbf{3.5}$$

$$\eta = 1 - \frac{1}{r^{\gamma-1}}$$

$$= 1 - \frac{1}{3.5^{0.4}}$$

$$= 1 - 0.604$$

$$= 0.396$$

$$= \mathbf{39.6\%}$$

5. Clearance volume of Otto cycle is 0.3 cm³ and stroke volume is 1.8 cm³. Calculate efficiency of the cycle.

$$V_c = V_2 = 0.3 \text{ cm}^3$$

$$V_s = 1.8 \text{ cm}^3$$

$$V_1 = V_2 + V_s$$

$$= 0.3 + 1.8 = 2.1 \text{ cm}^3$$

$$r = \frac{V_1}{V_2} = \frac{2.1}{0.3} = 7$$

$$\eta = 1 - \frac{1}{r^{\gamma-1}}$$

$$= 1 - \frac{1}{7^{1.4-1}}$$

$$= 0.54$$

$$= \mathbf{54\%}$$

6. An engine working on air standard Otto cycle has bore of 140 mm and stroke length of 160 mm. Clearance volume for the engine is 370 cm^3. Calculate air standard efficiency of the cycle.

Solution:

$$V_c = V_2 = 370 \text{ cm}^3$$

$$V_s = \frac{\pi}{4} D^2 L = \frac{\pi}{4} \times (14)^2 \times 16 = 2462 \text{ cm}^3$$

$$V_1 = V_2 + V_s = 370 + 2462 = 2832 \text{ cm}^3$$

$$r = \frac{V_1}{V_2} = \frac{2832}{370} = 7.65$$

$$\eta = 1 - \frac{1}{r^{\gamma-1}} = 1 - \frac{1}{7.65^{1.4-1}}$$

$$= 1 - 0.443$$

$$= \mathbf{0.557 = 55.7\%}$$

7. An engine working on Otto cycle has temperature and pressure at the beginning of compression are 25 °C and 1.5 bar respectively. Air standard efficiency is 48%. Calculate compression ratio, pressure and temperature at the end of compression.

Solution:

$$\eta = 1 - \frac{1}{r^{\gamma-1}}$$ (Assume $\gamma = 1.4$ for air)

$$\therefore \quad 0.48 = 1 - \frac{1}{r^{1.4-1}}$$

$$\therefore \quad \frac{1}{r^{0.4}} = 1 - 0.48 = 0.52$$

$$\therefore \quad r^{0.4} = \frac{1}{0.52} = 1.923$$

$$\therefore \quad r = (1.923)^{\frac{1}{4}}$$

$$= \mathbf{5.13}$$

Now, for isentropic compression 1-2

$$\frac{T_2}{T_1} = \left(\frac{V_1}{V_2}\right)^{\gamma-1} = r^{\gamma-1}$$

$$\therefore \quad T_2 = T_1 \, . \, r^{\gamma-1}$$

$$= 298\,(5.13)^{0.4}$$

$$= 573 \text{ K}$$

$$= \mathbf{300\ ^\circ C}$$

$$\frac{P_2}{P_1} = \left(\frac{V_1}{V_2}\right)^{\gamma}$$

$\therefore$ $P_2 = P_1 \times r^{\gamma}$

$= 1.5 \times (5.13)^{1.4}$

$= \mathbf{14.8\ bar}$

8. An Otto cycle has compression ratio of 7. Temperature at the beginning of compression is 20°C and heat supplied by combustion is 2000 kJ/kg. Find temperature at remaining three points, efficiency and work output.

Take $c_v = 0.718$ kJ/kg. K for air.

Solution: $T_2 = T_1 . r^{\gamma-1}$

$= 293 \times 7^{1.4-1}$

$= \mathbf{638\ K}$

Heat supplied $Q_1 = c_v (T_3 - T_2)$

$\therefore$ $2000 = 0.718 (T_3 - 638)$

$\therefore$ $T_3 = 3423$ K

$= \mathbf{3150\ ^{o}C}$

$$\frac{T_3}{T_4} = r^{\gamma-1}$$

$\therefore$ $T_4 = \dfrac{T_3}{r^{\gamma-1}} = \dfrac{3423}{7^{0.4}} = 1572$ K

$= \mathbf{1299\ ^{o}C}$

$$\eta = 1 - \frac{1}{r^{\gamma-1}} = 1 - \frac{1}{7^{0.4}}$$

$= 1 - 0.46$

$= 5.4$

$= \mathbf{54\%}$

Work output $= \eta \times$ heat supplied Q_1

$= 0.54 \times 2000$

$= \mathbf{1080\ kJ/kg}$

9. An Otto cycle having compression ratio 8 has pressure and temperature at the beginning of compression are 1 bar and 27 °C respectively. If heat transfer per cycle is 1900 kJ/kg, find pressure and temperature at the end of each process. Take $c_v = 0.718$ kJ/kg. K

Solution:

$$\frac{T_2}{T_1} = \left(\frac{V_1}{V_2}\right)^{\gamma-1} = 8^{0.4} = 2.3$$

$\therefore$ $T_2 = 300 \times 2.3 = 690\text{ K} = \mathbf{417\ ^\circ C}$

$$\frac{P_2}{P_1} = \left(\frac{V_1}{V_2}\right)^{\gamma} = 8^{1.4} = 18.4$$

$\therefore$ $P_2 = P_1 \times 18.4 = 1 \times 18.4 = \mathbf{18.4\ bar}$

$Q_1 = C_v (T_3 - T_2)$

$\therefore$ $1900 = 0.718 (T_3 - 690)$

$\therefore$ $T_3 = 3336\text{ K} = \mathbf{3063\ ^\circ C}$

Applying characteristic equation between point 2-3

$$\frac{P_3V_3}{T_3} = \frac{P_2V_2}{T_2}$$

$$\therefore \quad \frac{P_3}{P_2} = \frac{T_3}{T_2}$$

$$\therefore \quad P_3 = P_2 \times \frac{T_3}{T_2} = 18.4 \times \frac{3336}{690}$$

$$= \mathbf{89\ bar}$$

$$\frac{T_3}{T_4} = \left(\frac{V_4}{V_3}\right)^{\gamma-1} = r^{\gamma-1}$$

$$\therefore \quad T_4 = \frac{T_3}{r^{v-1}} = \frac{3336}{8^{0.4}}$$

$$= 1450\text{ K}$$

$$= \mathbf{1177^\circ C}$$

$$\frac{P_4}{P_3} = \left(\frac{V_3}{V_4}\right)^{\gamma} = \left(\frac{V_2}{V_1}\right)^{\gamma} = \frac{1}{r^{\gamma}}$$

$$= \frac{1}{8^{1.4}} = 0.054$$

$$\therefore \quad P_4 = P_3 \times 0.054$$

$$= 89 \times 0.054$$

$$= \mathbf{4.8\ bar}$$

10. A petrol engine has swept volume of 550 cm^3 and clearance volume of 60 cm^3. Pressure and temperature at suction point are 1 bar and 27°C respectively. Maximum temperature in the cycle is 1400 °C. Calculate air standard efficiency and mean effective pressure of the cycle. Take c_v = 0.71 kJ/kg. K and R = 287 J/kg. K for air.

Solution:

$$V_1 = V_s + V_c = 550 + 60 = 610 \text{ cm}^3$$

$$V_2 = V_c = 60 \text{ cm}^3$$

$$r = \frac{V_1}{V_2} = \frac{610}{60} = 10.2$$

$$\eta = 1 - \frac{1}{r^{\gamma-1}} = 1 - \frac{1}{10.2^{0.4}}$$

$$= 1 - 0.4 = 0.6 = \mathbf{60\%}$$

$$\frac{T_2}{T_1} = r^{\gamma-1}$$

$$\therefore \quad T_2 = 300 \times 10.2^{0.4} = 760 \text{ K}$$

Maximum temperature $T_3 = 1400 + 273 = 1673$ K

Heat supplied $Q_1 = c_v (T_3 - T_2)$

$$\therefore \quad Q_1 = 0.71\,(1673 - 760) = 648 \text{ kJ/kg}$$

Work $W = \eta \times Q_1 = 0.6 \times 648 = 389$ kJ/kg

$$\text{Mean effective pressure } P_m = \frac{\text{Work output}}{\text{Swept volume}} \; \frac{\text{kJ / cycle}}{\text{m}^3\text{ / cycle}}$$

w is in kJ/kg. We need to convert it into kJ/cycle

$$P_1V_1 = mRT_1$$

$$\therefore \quad m = \frac{P_1V_1}{RT_1}$$

$$= \frac{1 \times 100 \text{ kN/m}^2 \times 610 \times 10^{-6} \text{ m}^3\text{/cycle}}{0.287 \text{ kJ/kg K} \times 300 \text{ K}} = 0.0007 \quad \text{kg/cycle}$$

Work $W = 389$ kJ/ kg $\times$ 0.007 kg/cycle = 0.2723 kJ/cycle

So,
$$P_m = \frac{0.2723}{550 \times 10^{-6}} \text{ kPa}$$

$$= 495 \text{ kPa} = \mathbf{4.95 \text{ bar}}$$

11. In air standard Otto cycle, the maximum and minimum temperatures are 1400 °C and 30°C. Heat supplied per kg of air is 900 kJ. Find compression ratio and ratio of maximum pressure to minimum pressure.

Solution:

$$T_3 = 1400 + 273 = 1673 \text{ K},$$

$$T_1 = 30 + 273 = 303 \text{ K}$$

Heat added $Q_1 = c_v (T_3 - T_1)$

$$\therefore \quad 900 = 0.71\,(1673 - T_2)$$

$$\therefore \quad T_2 = 1673 - \frac{900}{0.71}$$

$$= 1673 - 1268 = 405 \text{ K}$$

$$(r) = \frac{V_1}{V_2} = \left(\frac{T_2}{T_1}\right)^{\frac{1}{\gamma-1}}$$

$$= \left(\frac{405}{303}\right)^{\frac{1}{0.4}}$$

$$= (1.337)^{2.5} = \mathbf{2.07}$$

Ratio of maximum pressure to minimum pressure $= \dfrac{P_3}{P_1}$

Now $$\frac{P_2}{P_1} = \left(\frac{V_1}{V_2}\right)^{\gamma} = 2.07^{1.4}$$

$$= 2.77$$

$$\frac{P_3}{P_2} = \frac{T_3}{T_2} = \frac{1673}{405} = 4.13$$

$$\frac{P_3}{P_1} = \frac{P_3}{P_2} \times \frac{P_2}{P_1}$$

$$= 4.13 \times 2.77$$

$$= \mathbf{11.44}$$

12. An oil engine working on Diesel cycle has ratio of adiabatic compression 16. The ratio of adiabatic expansion is 9. Find thermal efficiency of the cycle.

Solution:

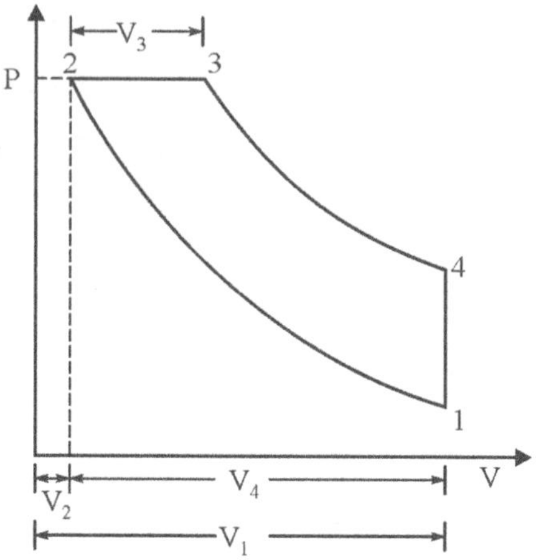

$$r = \frac{V_1}{V_2} = 16$$

$$\frac{V_4}{V_3} = 9 = \frac{V_1}{V_3}$$

$$\text{Cut off ratio } \rho = \frac{V_3}{V_2}$$

$$= \frac{V_3}{V_1} \times \frac{V_1}{V_2}$$

$$= \frac{1}{9} \times 16$$

$$= 1.78$$

$$\eta = 1 - \frac{1}{r^{\gamma-1}}\left[\frac{\rho^{\gamma}-1}{\gamma(\rho-1)}\right]$$

$$= 1 - \frac{1}{16^{0.4}}\left[\frac{1.78^{1.4}-1}{1.4\,(1.78-1)}\right]$$

$$= 1 - \frac{1}{3.03}\left[\frac{2.24-1}{1.4\times 0.78}\right]$$

$$= 1 - \frac{1}{3.03} \times \frac{1.24}{1.4\times 0.78}$$

$$= 1 - 0.375$$

$$= 0.625$$

$$\mathbf{= 62.5\ \%}$$

13. In an ideal Diesel cycle, temperatures at the beginning and end of compression is 50 °C and 600 °C respectively. The temperatures at the beginning and end of expansion are 1980 °C and 880 °C respectively. Determine air standard efficiency of the cycle.

Solution:

$T_1 = 50 + 273 = 323$ K,

$T_2 = 600 + 273 = 873$ K,

$T_3 = 1980 + 273 = 2253$ K,

$T_4 = 880 + 273 = 1153$ K

$$\eta = 1 - \frac{T_4 - T_1}{\gamma (T_3 - T_2)}$$

$$= 1 - \frac{1153 - 323}{1.4\,(2253 - 873)}$$

$$= 1 - \frac{830}{1.4 \times 1380}$$

$$= 1 - 0.43 = 0.57$$

$$= \mathbf{57\%}$$

14. In a Diesel cycle 1 kg of air at a pressure of 1 bar and temperature of 15 °C is compressed through a ratio of 15. Heat added is 1850 kJ. Find air standard efficiency. Take $c_p = 1.005$ kJ/kg K

Solution:

$P_1 = 1$ bar, $\quad T_1 = 288$ K

$$\frac{T_2}{T_1} = \left(\frac{V_1}{V_2}\right)^{\gamma-1} = r^{\gamma-1} \qquad \therefore\ T_2 = T_1 \,.\, r^{\gamma-1}$$

$\therefore$ $\quad T_2 = 288 \times 15^{0.4}$ (γ is not given in problem assume $\gamma = 1.4$)

$= 851$ K

Heat added $\quad Q_1 = c_p\,(T_3 - T_2)$

$\therefore$ $\quad 1850 = 1.005\,(T_3 - 851)$

$\therefore$ $\quad T_3 = 1840.8 + 851 = 2691.8$ K

Now $\quad \rho = \dfrac{V_3}{V_2}$

For process 2-3, $\quad \dfrac{V_3}{V_2} = \dfrac{T_3}{T_2}$ $\quad$ (For process 2.3 $\dfrac{P_3V_3}{T_3} = \dfrac{P_2V_2}{T_2}$)

$$\therefore \quad \rho = \frac{V_3}{V_2} = \frac{T_3}{T_2}$$

$$= \frac{2691.8}{851}$$

$$= 3.16$$

Air standard efficiency

$$\eta = 1 - \frac{1}{r^{\gamma-1}}\left[\frac{\rho^{\gamma}-1}{\gamma(\rho-1)}\right]$$

$$= 1 - \frac{1}{15^{0.4}}\left[\frac{3.16^{1.4}-1}{1.4(3.16-1)}\right]$$

$$= 1 - 0.3385\left[\frac{4.0068}{1.4\times 2.16}\right]$$

$$= 1 - 0.45$$

$$= 0.55$$

$$= \mathbf{55\%}$$

15. An engine operating on air standard Diesel cycle has pressure and temperature at the beginning of compression are 1 bar and 25 °C. Maximum pressure and temperature of the cycle are 40 bar and 1200 °C. Find air standard efficiency of the cycle

Solution:

$P_1 = 1$ bar, $T_1 = 298$ K, $P_3 = P_2 = 40$ bar, $T_3 = 1473$ K

$$\frac{V_1}{V_2} = \left(\frac{P_2}{P_1}\right)^{\frac{1}{\gamma}}$$

$$= \left(\frac{40}{1}\right)^{\frac{1}{1.4}}$$

$$= 40^{0.714}$$

$$= 13.93$$

i.e. $r = 13.93$

$$\frac{T_2}{T_1} = \left(\frac{V_1}{V_2}\right)^{\gamma-1} = r^{\gamma-1}$$

$$= 13.93^{0.4}$$

$$= 2.87$$

$\therefore$ $T_2 = 298 \times 2.87 = 855$ K

$$\rho = \frac{V_3}{V_2} = \frac{T_3}{T_2} = \frac{1473}{855}$$

$$= 1.72$$

$$\eta = 1 - \frac{1}{r^{\gamma-1}}\left[\frac{\rho^{\gamma}-1}{\gamma(\rho-1)}\right]$$

$$= 1 - \frac{1}{13.93^{0.4}}\left[\frac{1.72^{1.4}-1}{1.4\,(1.72-1)}\right]$$

$$= 1 - 0.3487\left[\frac{1.1367}{1.4\times 0.72}\right]$$

$$= 1 - 0.39 = 0.61$$

$$= \mathbf{61\ \%}$$

16. In an ideal Diesel cycle, compression ratio is 15 and expansion ratio is 7.5. Pressure and temperature at the beginning of compression are 1 bar and 47 °C. The pressure at the end of expansion is 2.6 bar. Determine cut-off ratio, maximum temperature and thermal efficiency of the cycle.

Solution:

$$\gamma = 15 = \frac{V_1}{V_2} = \frac{V_4}{V_3} = 7.5,\ P_1 = 1\text{ bar},\ T_1 = 320\text{ K},P_4 = 2.6\text{ bar}$$

$$\rho = \frac{V_3}{V_2} = \frac{V_3}{V_1}\times\frac{V_1}{V_2} = \frac{V_3}{V_4}\times\frac{V_1}{V_2}$$

$$= \frac{\mathbf{15}}{\mathbf{7.5}} = \mathbf{2}$$

$$\frac{T_2}{T_1} = r^{\gamma-1}$$

$$\therefore\quad T_2 = 320\times 15^{0.4}$$

$$= 945\text{ K}$$

For process 2-3

$$\frac{T_3}{T_2} = \frac{V_3}{V_2} = 2 \qquad \because \frac{V_3}{V_2} = \rho$$

$$\therefore\quad T_3 = 945\times 2 = \mathbf{1890\ K}$$

This is maximum temperature

$$\eta = 1 - \frac{1}{r^{\gamma-1}}\left[\frac{\rho^{\gamma}-1}{\gamma(\rho-1)}\right]$$

$$= 1 - \frac{1}{15^{0.4}}\left[\frac{2^{1.4}-1}{1.4\,(2-1)}\right]$$

$$= 1 - 0.3385\left[\frac{1.639}{1.4 \times 1}\right]$$

$$= 1 - 0.396$$

$$= 0.604$$

$$= \mathbf{60.4\ \%}$$

17. An ideal Diesel engine has a diameter 150 mm and stroke 200 mm. Clearance volume is 10% of the swept volume. Cut-off takes place at 6% of the stroke. Determine compression ratio and air standard efficiency.

Solution: D = 150 mm, L = 200 mm

$$V_s = \frac{\pi}{4} D^2 L = \frac{\pi}{4} \times (15)^2 \times 20 = 3530 \text{ cm}^3$$

$$V_c = 0.1 \times V_s = 0.1 \times 3530 = 353 \text{ cm}^3 = V_2$$

$$V_1 = V_s + V_c = 3530 + 353 = 3883 \text{ cm}^3$$

$$\therefore \quad r = \frac{V_1}{V_2} = \frac{3883}{353} = \mathbf{11}$$

$$\frac{V_3 - V_2}{V_1 - V_2} = 0.06$$

$$\therefore \quad V_3 = 0.06\,(3530) + 353$$

$$= 211.8 + 353$$

$$= 564.8 \text{ cm}^3$$

For process 2-3

$$\rho = \frac{V_3}{V_2} = \frac{564.8}{353} = 1.6 \quad \text{where } \rho = \text{cut off}$$

$$\eta = 1 - \frac{1}{r^{\gamma-1}}\left[\frac{\rho^{\gamma}-1}{\gamma(\rho-1)}\right]$$

$$= 1 - \frac{1}{11^{0.4}}\left[\frac{1.6^{1.4} - 1}{1.4\,(1.6 - 1)}\right]$$

$$= 1 - 0.3832\left[\frac{0.931}{1.4 \times 0.6}\right]$$

$$= 1 - 0.42$$

$$= 0.58$$

$$= \mathbf{58\ \%}$$

18. A compression ignition engine working on Diesel cycle has bore of 150 mm and stroke 250 mm. Clearance volume is 400 cm^3. Fuel injection takes place at constant pressure for 5% of the stroke. Find air standard efficiently. If fuel cut off is delayed from 5% to 8% of the stroke, find loss in efficiency if compression ratio is same.

Solution:

$$D = 150 \text{ mm}, \quad L = 250 \text{ mm}, V_c = V_2 = 400 \text{ cm}^3$$

$$V_s = \frac{\pi}{4} D^2 L = \frac{\pi}{4} \times (15)^2 \times 25 = 4416 \text{ cm}^3$$

$$V_1 = V_s + V_c = 4416 + 400 = 4816 \text{ cm}^3$$

$$r = \frac{V_1}{V_2} = \frac{4816}{400} = 12.04$$

cut-off = 5% of the stroke

$$\therefore \quad \frac{V_C}{V_S} = \frac{V_3 - V_2}{V_1 - V_2} = 0.05$$

$$\therefore \quad V_3 = 0.05\,(4816 - 400) + 400 = 620.8 \text{ cm}^3$$

$$\rho = \frac{V_3}{V_2} = \frac{620.8}{400} = 1.552$$

$$\eta = 1 - \frac{1}{r^{\gamma-1}}\left[\frac{\rho^{\gamma} - 1}{\gamma(\rho - 1)}\right]$$

$$= 1 - \frac{1}{12.04^{0.4}}\left[\frac{1.552^{1.4} - 1}{1.4\,(1.552 - 1)}\right]$$

$$= 1 - 0.37\left[\frac{0.85}{1.4 \times 0.552}\right]$$

$$= 1 - 0.41$$

$= 0.59$

$= 59\ \%$

Now consider cut-off = 8% of the stroke

$$\therefore \quad \frac{V_3 - V_2}{V_1 - V_2} = 0.08$$

$$\therefore \quad V_3 = 0.08\ (4816 - 400) + 400 = 753.3\ \text{cm}^3$$

$$\rho = \frac{V_3}{V_2} = \frac{753.3}{400} = 1.883$$

$$\eta = 1 - \frac{1}{12.04^{.4}}\left[\frac{1.883^{1.4} - 1}{1.4\,(1.883 - 1)}\right]$$

$$= 1 - 0.37\left[\frac{1.425}{1.4 \times 0.883}\right]$$

$= 1 - 0.427$

$= 0.573$

= 57.3 %

$\therefore$ Loss in efficiency = 59 – 57.3

= 1.7%

19. In a Diesel cycle, pressure and temperature at the beginning of compression are 1 bar and 27 °C respectively. Compression ratio is 16 and heat added per kg of air is 2500 kJ. Find (i) maximum pressure and temperature (ii) air standard efficiency (iii) power output for 0.25 kg/s of air flow.

Take $c_p = 1.005$ kJ/kg K.

Solution: $P_1 = 1$ bar, $T_1 = 300$ K, $r = 16$, $Q_1 = 2500$ kJ/kg

$$\frac{T_2}{T_1} = r^{\gamma-1} \qquad \therefore\ T_2 = T_1 \cdot r^{\gamma-1}$$

$$\therefore \quad T_2 = 300 \times (16)^{0.4} = 909\ \text{K}$$

$$Q_1 = c_p\,(T_3 - T_2)$$

$$\therefore \quad 2500 = 1.005\,(T_3 - 909)$$

$$\therefore \quad T_3 = 2487.6 + 909 = \mathbf{3396.6\ K.}$$

This is maximum temperature

Now $$\frac{P_2}{P_1} = \left(\frac{V_1}{V_2}\right)^{\gamma} = 16^{1.4}$$

$\therefore \quad P_2 = 1 \times 48.5 = 48.5$ bar

$\therefore \quad P_3 = P_2 =$ **48.5 bar**, which is maximum pressure

$$\rho = \frac{V_3}{V_2} = \frac{T_3}{T_2} = \frac{3396.6}{909} = 3.74$$

$$\eta = 1 - \frac{1}{r^{\gamma-1}}\left[\frac{\rho^{\gamma}-1}{\gamma(\rho-1)}\right]$$

$$= 1 - \frac{1}{16^{0.4}}\left[\frac{3.74^{1.4}-1}{1.4\,(3.74-1)}\right]$$

$$= 1 - 0.33\left[\frac{5.34}{1.4\times 2.74}\right]$$

$$= 1 - 0.46 = 0.54 = \mathbf{54\%}$$

$$W = \eta \times Q_1 = 0.54 \times 2500$$

$$= 1350 \text{ kJ.kg}$$

$$\text{Power output} = 0.25\frac{\text{kg}}{\text{s}} \times 1350\frac{\text{kJ}}{\text{kg}} = \mathbf{337.5\ kW}$$

Objective Type Questions

1. Carnot cycle consists of
   (a) two constant volume and two isentropic processes
   (b) two isothermal and two isentropic processes
   (c) two constant pressure and two isentropic processes
   (d) one constant pressure, one constant volume and two isentropic processes
2. The efficiency of Carnot cycle may be increased by
   (a) increasing the highest temperature
   (b) decreasing the highest temperature
   (c) increasing the lowest temperature
   (d) decreasing the highest and increasing the lowest temperatures

3. Rankine cycle is used in
   (a) petrol engine (b) diesel engine
   (c) steam turbine (d) gas engine
4. Petrol engine is
   (a) a rotary engine (b) reciprocating engine
   (c) jet engine (d) none of the above
5. Otto cycle is also known as
   (a) constant pressure cycle (b) constant temperature cycle
   (c) constant enthalpy cycle (d) constant volume cycle
6. Air standard efficiency of an Otto cycle is given by
   (a) $1+\frac{1}{r^{\gamma-1}}$ (b) $1-r^{\gamma-1}$
   (c) $1-\frac{1}{r^{\gamma-1}}$ (d) $1-\frac{1}{r^{\gamma-1}}$
7. In diesel cycle, heat is added
   (a) at constant pressure (b) at constant volume
   (c) at constant enthalpy (d) at constant of temperature
8. For the same compression ratio, efficiency of Diesel cycle is ______ Otto cycle.
   (a) greater than (b) less than
   (c) equal to (d) greater than or equal to
9. Maximum pressure in Otto cycle occurs
   (a) at the beginning of compression (b) at the end of compression
   (c) at the end of heat addition (d) at the end of expansion

Answers

1. (b) 2. (a) 3. (c) 4. (b) 5. (d) 6. (c) 7. (a) 8. (b) 9. (c)

Short Questions

1. Define

   (a) thermal prime mover; (b) heat engine; (c) working fluid; (d) external combustion engine; (e) internal combustion engine; (f) heat engine cycle; (g) air standard cycle; (h) compression ratio; (i) cut-off ratio; (j) clearance volume; (k) swept volume.

2. Write names of main components of a heat engine.
3. Which processes are practically impossible?
4. Why Carnot cycle is practically impossible?
5. Which processes are included in Otto cycle?
6. Why Otto cycle is known as constant volume cycle?
7. What is the range of air-fuel ratio for Otto cycle?
8. What is the range of compression ratio for Otto cycle?
9. Why diesel cycle is known as constant pressure cycle?
10. Why Otto and Diesel cycles are considered as air standard cycles?
11. Show that efficiency of Otto cycle is greater than that of Diesel cycle for same compression ratio.

Long Questions

1. Write short notes on

   (a) elementary heat engine; (b) source of heat; (c) working substance;

   (d) converting machines

2. Classify heat engines
3. Explain working of

   (a) Carnot cycle; (b) Carnot cycle with vapour; (c) Rankine cycle; (d) Otto cycle; (e) Diesel cycle

4. Derive expression for efficiency of Carnot cycle
5. Derive expression for efficiency of Rankine cycle
6. Derive expression for efficiency of Otto cycle
7. Derive expression for efficiency of Diesel cycle
8. Show that efficiency of Otto cycle is a function of compression ratio only

Exercises

1. A hot air engine works on Carnot cycle with thermal efficiency of 60%. If final temperature of air is 20 °C, find the initial temperature. [460 °C]

2. A simple Rankine cycle steam power plant operates between the temperatures of 260 °C and 95 °C. The steam is supplied to the turbine at a dry saturated condition. It expands isentropically and is exhausted with a dryness fraction of 0.77. Determine the efficiency of the Rankine cycle and of the Carnot cycle operating between the same temperature limit.

   [27%, 30.96%]

3. For the engine working on ideal Otto cycle, temperatures at the beginning and at the end of compression are 50 °C and 400 °C respectively. Find compression ratio and air standard efficiency.

   [6.27, 52%]

4. In an ideal constant volume cycle, pressure and temperature at the beginning of the compression are 97 kPa and 50 °C respectively. Compression ratio is 5. Heat supplied during the cycle is 930 kJ/kg. Determine
   (i) Maximum temperature attained in the cycle
   (ii) Thermal efficiency of the cycle
   (iii) Work done during the cycle

   [1637 °C, 47.5%, 442 kJ/kg]

5. Air at a pressure of 1 bar and temperature 70 °C is compressed isentropically up to the pressure of 7 bar in an Otto cycle engine. 460 kJ of heat per kg of air is added at constant volume. Determine (i) compression ratio (ii) temperature at the end of compression (iii) temperature at the end of heat addition.

   Take $c_v = 0.707$ kJ/kg K.

   [3.97, 331 °C, 982 °C]

6. In an engine working on Otto cycle, air has a pressure of 1 bar and temperature at 27°C at suction.

   Air is compressed with a compression ratio of 7. Heat is added at constant volume until the temperature rises to 2000 K. Find (i) air standard efficiency (ii) pressure and temperature at the end of compression (iii) heat supplied

   (iv) mean effective pressure. Take $c_v = 0.718$ kJ/kg K, R = 287 J/kg K for air.

   [54.1%, 15.24 bar. 653.4 K, 967 kJ/kg, 7.1 bar]

7. For an air standard Otto cycle, maximum and minimum temperature are 1350 °C and 30 °C. Heat supplied is 750 kJ/kg of air. Calculate (i) compression ratio (ii) Air Standard efficiency (iii) work done (iv) Ratio of maximum to minimum pressure.

   [5.03, 47.6%, 3.57 kJ/kg, 26.9]

8. An engine operating on Otto cycle has bore of 100 mm and stroke of 120 mm. Compression ratio is 6.5. Pressure and temperature at the beginning of compression are 1 bar and 19 °C. If heat supplied is 1900 KJ/kg, find (i) Pressure, volume and temperature at all points (ii) Mean effective pressure

[$V_1 = V_4 = 1114$ cm^3, $V_2 = V_3 = 171$ cm^3, $P_2 = 13.7$ bar, $T_2 = 617.4$ K, $T_3 = 3263.6$ K, $P_3 = 72.4$ bar, $P_4 = 5.3$ bar, $T_4 = 1547.6$ K, $P_m = 14.13$ bar]

9. In an Otto cycle, compression ratio is 6. The intake conditions are $P_1 = 1$ bar and $T_1 = 27$ °C. Heat added at constant volume is 1046 kJ/kg. Find thermal efficiency and mean effective pressure.

[51.16%, 7.46 bar]

10. A Diesel engine has compression ratio 19 and expansion ratio 9.1. Calculate air standard efficiency.

[63.55%]

11. In an ideal Diesel cycle, temperatures at the beginning and end of compression are 57 °C and 603 °C respectively. The temperatures at beginning and end of expansion are 1950 °C and 870 °C respectively. Determine efficiency of the cycle. If pressure at the beginning is 1 bar, find maximum pressure in the cycle.

[70%, 30.5 bar]

12. In a Diesel cycle, temperature at the beginning of compression is 87 °C. If $r = 14$, find temperature at the end of compression. If temperatures at the beginning and end of expansion are 1795 °C and 677 °C respectively, find thermal efficiency of the cycle.

[761.6 °C, 59.2%]

13. An air standard Diesel cycle has bore 160 mm and stroke 240 mm. The clearance volume is 340 cm^3. Find loss of efficiency if fuel cut-off is delayed from 6% to 10% with same compression ratio.

[35.15%]

14. A Diesel engine has compression ratio of 19. Pressure and temperature at the beginning of compression are 1 bar and 17°C. Heat added at constant pressure is 730 kJ/kg. Find (i) Pressure, volume and temperature at all points. (ii) Work done in kJ/kg (iii) Air standard efficiency (iv) Mean effective pressure.

[$V_1 = 0.8323$ m^3/kg, $V_2 = 0.0438$ m^3/kg, $P_2 = P_3 = 61.7$ bar, $T_2 = 941.7$ K, $T_3 = 1958.4$ K, $V_3 = 0.0911$ m^3/kg $= V_4$, $T_4 = 808.3$ K, $P_4 = 2.79$ bar, $W = 649.6$ kJ/kg, $\eta = 63.58\%$, $P_m = 8.24$ bar]

15. An air standard Diesel cycle has compression ratio of 16. Pressure and temperature at the beginning of compression are 1 bar and 20 °C. The maximum temperature of the cycle is 1431 °C. Determine thermal efficiency and mep for this cycle.

[61.75%, 6.42 bar]

CHAPTER 6

Steam Boiler

6.1 Introduction

- A steam boiler is closed pressure vessel in which steam is generated from water at required pressure and temperature.
- We can also define boiler according to the Indian Boiler act 1923, a boiler is closed pressure vessel with capacity exceeding 22.75 litres used for generation of steam under pressure. It includes all mountings fitted to such vessels which remain whole or partly under pressure.
- Let us see difference between boiler and generator. Boiler is old name used for unit which generates saturated steam i.e., it is not provided with superheater, Economizer and air-pre-heater. But boiler with these accessories is called generator. Generator always produces superheated steam which boiler can not produce. So now-a-days boiler is called generator.

Let us clearly understand the following terms generally used in connection with various types of boilers.

Boiler Shell (steam drum) : It consists of one or more steel plates bent into cylindrical form and riveted or welded together. The ends of shell are closed by means of end plates or heads which are made flat or concave. The shell with closing plates is called drum.

Grate : It is platform in furnace upon which fuel such as coal or wood is fired. Grate consists of iron bars, which are placed apart so that air for combustion can pass through them. Also provides passage for ash to fall into ash pit constructed below Grate.

Furnace : Furnace is called fire box in which fuel is burnt.

Volume of shell : It is the volume occupied by water (known as water space) and steam space. Steam space is equal to the entire shell volume less that occupied by water and tubes.

Water level : It is level of water in boiler shell, remaining space above water is called steam space. Water level in boiler is indicated by water level indicator.

Heating surface : It is part of boiler surface that is exposed to the fire and to hot gases. As they (hot gasses) pass from furnace to chimney in order to transfer heat to working fluid to generate steam. A following heating surface, boiler may have:

(i) Radiant heat surface – water wall receiving direct radiant flame heat.

(ii) Convective heating surface – located in convective zone of furnace and receiving heat entirely by convection.

(iii) Radiant and convective heating surface–heated partly by radiant and partly by convection heat.

For, same specifications water tube boiler have more heating surface area than fire tube boiler.

Mountings : They are fittings primarily used for safety of boiler and control of steam generation process completely according to India Boiler Act 1923. They are

1. Pressure gauge
2. Safety valve (2)
3. Water level indicator (2)
4. Feed check valve
5. Steam stop valve
6. Fusible plug
7. Blow off cock
8. Man holes
9. Mud holes

Special provision is always made on boiler to mount them. Boiler cannot function safely without above mentioned mountings.

Accessories : Accessories form integral part of the boiler but they are not mounted on the boiler, they are super-heater, Economizer, air-pre-heater, feed pump etc. They increase the efficiency of boiler.

Blow off

(i) It is act of removing the floating impurities that floats on the surface of water and is termed as surface blow off. This is done by means of some kind of surface blow of appliance.

(ii) Removal of mud and other impurities of water from lowest part of boiler where they usually settle is termed as blowing off. They are removed with help of Blow off cock.

Foaming : It is formation of stable foam on the evaporating surface in boiler drum. Foaming is chiefly due to contamination of boiler water with oil, soap, organic matter, suspended matter or other foreign particles.

It is continuous formation of bubble possessing sufficiently high surface tension to prevent them from *coalescing* and breaking, with the result they retain bubble form, even after their disengagement from water surface.

Scale : It is general name ascribed to sufficiently dense deposits appearing on heating or cooling surface and reduces the normal heat transfer in a boiler unit or heat exchanger.

Scale are formed because of dissolved impurities in water, which has lowest solubility, product under particular condition, crystallize first, and deposit on heating surface and form scaling.

Lagging : Blocks of asbestos or magnesia insulation wrapped on out side of boiler shell and steam piping.

Refractory : A heat insulation material such as fire brick or plastic fire clay used for lining combustion chamber. It protects the walls from the products of combustion.

6.2 Classification of Boiler

6.2.1 Relative Positions of Fluid

Boiler are classified on basis of relative position of water and the hot furnace gases.

Fire tube boiler : Fire tube boiler, in which hot gases pass through the tubes and water surrounds the tubes.

The examples of fire tube boilers are–1. Cochran 2. Lancashire 3. Carnish 4. Locomotives.

Water Tube boiler : In water tube boiler, water flows inside the tubes and hot gases out side the tubes. Examples of water tube boilers are–(i) Bebcock and Wilcox boiler which usually have inclined straight tubes (ii) Stilling boiler which has bent water tubes.

6.2.2 Methods of Firing

On the basis of methods of firing boiler are classified as:

(a) Internally fired boiler in which furnace is placed inside boiler shell and it is completely surrounded by water cooled surface. Boilers of this type are Lancashire, Locomotive, Cochran and scotch boilers.

(b) Externally fired boiler

In this type of boiler the furnace is placed out side the boiler shell. The furnace region (space) is simple and easy to construct. Examples of this type boiler are Bebcock and Velox.

6.2.3 Method of Circulation of Water

On the basis of method of circulation boilers are classified as:

(a) ***Natural circulation boiler :*** Natural circulation is closed circuit in which working fluid circulates by virtue of its density difference. As shown in Fig. 6.1(a) the baffle separates out the riser from unheated down comer and therefore creates temperature difference between two tube systems. The saturated water flows down the unheated down comer and receives heat in riser where part of it gets converted into steam. Difference in densities of saturated water in down comer and steam water mixture in the riser bring about natural circulation.

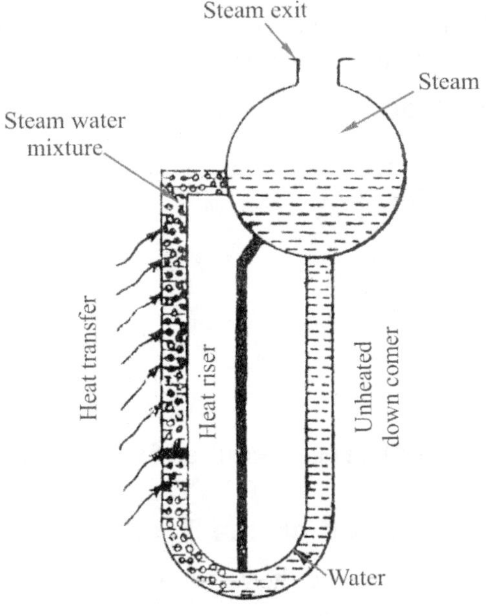

Fig. 6.1(a)

(b) ***Forced circulation :*** In this type, circulation of water is done by pump, which increases the pressure and circulating water in the circuit. Examples are Lamont, Velox boiler. i.e., only high pressure boilers are force circulation type.

6.2.4 Nature of Service

Boilers are also classified according to service as below:

(a) ***Stationary boilers :*** They are land based boiler. As its name indicates, it is not require to be transported from one place to other place. They are (i) industrial boiler (ii) boiler used in power generation (iii) residential boiler (hot water).

(b) ***Portable boilers :*** Boilers which can be readily dismantled and transported from one place to other are called portable boilers.

(c) ***Mobile boilers :*** They are mounted on marine vessels and steam locomotive boilers.

6.2.5 Pressure of steam

Boilers may be classified according to pressure:

(a) ***Low pressure boiler :*** Boiler which generate steam at a pressure below 80 bars is called as low pressure boiler. Example of low pressure boilers are Cochran, Carnish, Lancashire, and Locomotive boiler.

(b) ***High pressure boiler :*** Boilers which generate steam at a pressure more than 80 bars are known as high pressure boilers. Example Lamont, Velox, Benson etc.

6.2.6 Nature of Fuel Used

(a) Coal fired

(i) Pulverized coal fired

(ii) Stoker fired

(iii) Hand fired

(b) Gas fired

(c) Oil fired

(d) Wood fired

(e) Baggasse fired (fluidized bed baggasse fired furnace have recently been developed)

6.2.7 Boiler can be Classified on the Bases of Heat Source

(a) ***Fuel fired boiler :*** They derive their heat energy by combustion of fuel, may be solid, liquid or gaseous.

(b) ***Waste heat boiler :*** These boilers recover heat from hot waste gases of other chemical reactions.

(c) ***Nuclear powered boiler :*** Utilize the energy of controlled thermo-nuclear fission reactions to generate steam.

6.2.8 Classification Based on Manufacturer's Trade Name

(i) Benson boiler

(ii) Sulzer boiler

(iii) Lamont boiler

(iv) Volex boiler

(v) Lo-effrer boiler

(vi) Bebcock boiler

(vii) Wilcox boiler

(viii) Yarrow boiler

(ix) Manning boiler

6.2.9 According to Axis of the Shell

The steam boilers according to the axis of shell may classified as:

(i) ***Vertical steam boilers:*** In this type of boilers the axis of shell is vertical. For example Cochran boiler and simple vertical boiler.

(ii) ***Horizontal steam boiler:*** In this type axis of shell is horizontal. For example Lancashire boiler.

6.2.10 Nature of Draught Boilers

According to this, boilers may be classified as:

(a) Natural draught boilers, in this the draught is produced by natural circulation of air by providing chimney.

(b) Force draught boiler, here the drought is produced by mechanical means such as fan.

6.2.11 Single tube and multi Tube Boilers

(a) Single tube boilers: The single tube boiler generally have one fire tube or water tube e.g.,

 (i) Simple vertical boiler

 (ii) Carnish boiler

(b) In multi tube steam boilers there are two or more fire tubes or water tubes example: locomotive boiler, Cochran boiler, Bebcock and Velox etc.

6.2.12 Fluid Used

On the basis of fluid used, boilers are classified as:

(a) ***Steam boiler :*** In this water is the working fluid

(b) ***Mercury boilers :*** In this mercury is working fluid.

6.2.13 Boiler Shell Material

On the basis of material used boiler are classified as:

(a) Case iron boiler – suitable for low pressure

(b) Steel boiler – suitable for high pressure

(c) Copper and stainless steel boiler – they have high heat and scale resistance at elevated temperature.

6.3 What Basic Factors will you consider in Selecting a Boiler?

1. Power required to be generated and the working pressure.

2. Rate at which steam is to be generated and condition of steam (wet, dry, superheated).
3. Type of fuel to be used
4. Location of power house
5. Water availability and its quality
6. Cost of installation and erection
7. Cost of operation and maintenance
8. Probable load factor
9. Availability of labour and technical staff
10. Comparative initial cost
11. Availability of floor space

6.4 What are the Essential Qualities of Good Boiler

1. It should be capable of quick start
2. It should occupy less space
3. It should meet large load fluctuations

6.5 Comparison

(Comparison between water tube boiler and fire tube boilers OR advantage and disadvantage of water tube boilers and fire tube boilers).

Fire Tube Boiler		Water Tube Boiler	
1.	Hot gases flow inside the tubes while water surrounding the tubes.	1.	Water flows inside the tubes while the hot gases outside the tubes.
2.	It has large ratio of water content to rate of steam generation (i.e., steam capacity) Therefore (i) slow rate of evaporation (ii) failure in water supply is less effective and may not over heat the boiler due to large capacity of water.	2.	It has comparatively small ratio of water content to steam generated and therefore (i) quick evaporation (ii) failure in water supply system may cause, explosion of boiler due to over heating because of low capacity of water.
3.	It generates steam at pressure in range of 17.5 bar to 24.5 bar thus it is suitable for low pressure.	3.	It generates steam at a high pressure up to 200 bar.
4.	Rate of generation of steam is low i.e., 9 tones per hour due to small heating surface area, resulting in slow rate of steam generation.	4.	The rate of generation of steam is high i.e., 450 tones per hour. Due to large heating surface area resulting, in high rate of steam generation.
5.	It is not suitable for steam power plant.	5.	It is suitable for large power plant.

Contd...

Fire Tube Boiler		Water Tube Boiler	
6.	Its overall efficiency is less i.e., up to 75%.	6.	Overall efficiency is high i.e., up to 90%.
7.	Floor area require is more i.e., 8 square meter per tone per hour of steam generation.	7.	Floor area required is less i.e., about 5 square meter per tones per hour of steam generation.
8.	Risks of damage due to explosion is more because its large water capacity.	8.	Because of small drum size the water is uniformly spread in large number of tubes- Thus the failure of one tube does not cause any disastrous explosion.
9.	All parts are not easily accessible for cleaning, inspection or repair.	9.	All parts are easily accessible for cleaning, inspection or repair.
10.	Water treatment is not very necessary as minor scaling would not go for enough to cause overheating and tube bursting.	10.	Water treatment is required as scaling will lead to tube bursting.
11.	Operating cost is less.	11.	Operating cost is high.
12.	The bursting produces greater risk to the damage of the property.	12.	Bursting does not produces any destruction to the whole boiler.
13.	The water does not circulates in definite direction.	13.	The direction of water circulation is well defined.
14.	The construction is simple, rigid, and compact hence initial cost is less.	14.	Design is complex hence initial cost is more.
15.	Skill required for efficient operation is less	15.	Skill required for efficient operation is more.
16.	It can also cope reasonably with sudden increase in load but for a shorter period.	16.	It is referred for widely fluctuation loads.
17.	Transportation from one place to other place is inconvenient.	17.	It can be transported and erected easily as its various parts can be separated.
18.	The construction of furnace and combustion chamber does not permit alterations to suit various classes of fuel and system of firing.	18.	The construction of furnace and combustion chamber permits alteration to suit various classes of fuel and system of firing.

6.6 Essential Qualities of Good Boiler

Following are the important qualities of good boiler.

1. It should produce maximum mass of steam with minimum fuel combustion.
2. It should be economical to install.
3. It should be quick starting.
4. Should meet large load fluctuations.
5. It should be light in weight.
6. It should occupy less floor space.

7. Should afford easy maintenance and inspection.
8. Tubes should be sufficiently strong to resist wear and corrosion.
9. Mud and other deposits should not collect on heated plates.
10. Velocity of water and that of flue gases should be minimum.
11. It should comply with safety regulations as laid down in Boiler Act.
12. Joints should be minimum and accessible for inspection and should away from direct flame.
13. It should be efficient.
14. Friction loss in various parts of boiler should be minimum when water and steam passes through them.

6.7 Simple Vertical Boiler

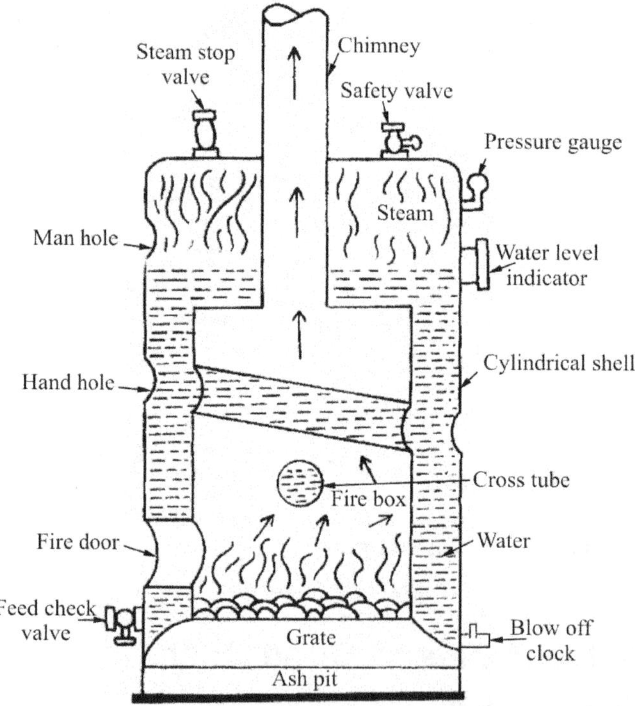

Fig. 6.1(c) Simple vertical boiler.

Introduction

It is the simplest form of internal furnace vertical fire tube. The working pressure ranges from 7.5 to 10 bar and steam production is about 2500 kg/hr. These boilers have less water capacity so they required less time to raise steam from water. The simplest form of vertical boiler is shown in Fig. 6.1(c).

Construction

The main components of the boiler are (i) cylindrical shell (ii) fire box (iii) cross tubes (iv) chimney (v) the grate (vi) ash pit (vii) man hole (viii) hand holes. Cylindrical shell is provided with various mountings such as pressure gauge, water level indicator, steam stop valve, safety valve, feed check valve, fusible plug, and blow off cock.

Generally no accessories are fitted with this type of boiler.

(i) ***Cylindrical shell :*** It is vertical shell surrounding fire box which is also cylindrical in shape. Greater portion of cylindrical shell is full of water and remaining is steam space.

(ii) ***Fire box :*** It is provided with one or two cross tubes, this increases the heating surface and circulation of water. This fire box is fitted with grate and ash-pit. Combustion takes place inside fire box.

(iii) ***Cross tubes :*** These tubes are for increasing heating surface and improving water circulation. The cross tubes may be one or more than one. They are flanged or riveted to box. They are inclined to horizontal axis of fire box. (about 15°)

(iv) ***Chimney :*** It is also known as stack. Through chimney flue-gases passes and discharged to atmosphere. It is fitted on top of fire box.

(v) ***Grate :*** It is platform on which coal is placed and fired. It is made up of steel or C.I. bars. They are placed apart from each other so that atmospheric air may pass through grate. Also ash may fall down through this space in to ash pot.

(vi) ***Ash pit :*** Ash is collected in ash pit, so that ash collected can be removed from time to time.

(vii) ***Manhole :*** It is for cleaning and maintenance of the interior of boiler shell and exterior of combustion chamber and chimney.

(viii) ***Handhole :*** For cleaning the cross tubes.

Simple vertical boiler is provided with following mountings as stated below.

(a) *Pressure gauge :* It is mounted in the region (space) of steam. Its function is to indicate the pressure of steam in the shell.

(b) *Water level indicator :* It is fitted in the front of shell in such a way that upper tube lies in steam space and lower tube in water space. It indicates the level of water in the shell.

(c) *Steam stop valve :* It is fitted on the top of shell. It regulates the amount of steam taken out of boiler for further utilization.

(d) *Safety valve :* It is also fitted on top of shell. It maintains constant pressure inside the boiler. When pressure of steam increases than preset value, the valve is lifted from its seat and excess steam is discharged in to atmosphere and the boiler is saved from bursting.

(e) *Feed check valve :* Required quantity of water fed in to boiler is regulated by this valve.

(f) *Fusible plug :* When level of water falls inside the boiler, the shell and tubes of boiler may be surrounded by hot gases (instead of water) and get over heated resulting in an explosion.

(g) *Blow off cock :* It is provided at bottom of shell by a bent pipe, to blow off mud and dirty water at regular interval from boiler.

Working

Water is fed into boiler through feed check valve as per requirement. Coal is added to grate from fire hole. Coal is fired, hot gases so produced travels straight up from furnace to the chimney upward moving hot gases come in contact with inclined cross tube walls of fire box and transfer heat to water inside tubes and water surrounding fired box, resulting in formation of steam. This steam gets collected at the top of boiler shell.

Since the cross tubes are inclined, water moves by natural circulation from the lower end of the cross tubes and come out from higher end.

Flue gases then discharged in to atmosphere through chimney.

Uses : These boilers are largely used on cranes, excavators and general contract-work.

Advantages

- Low cost of construction.
- Solid fuel such as coal, wood, agricultural waste etc., can be used.
- It requires small floor area for their installation.
- It is portable therefore can be transported easily from one place to other.
- They are known as low pressure and Economical Boiler.
- They are internal fired boiler.
- It is vertical fire tube boiler.

Disadvantages (Laminations)

- It requires high head room.

- As no air-preheater, Economiser and superheater is provided, flue gases carry away much of the heat
- Efficiency of boiler is less compare to other boilers as gases move directly upward towards the chimney.

6.8 Cochran Boiler

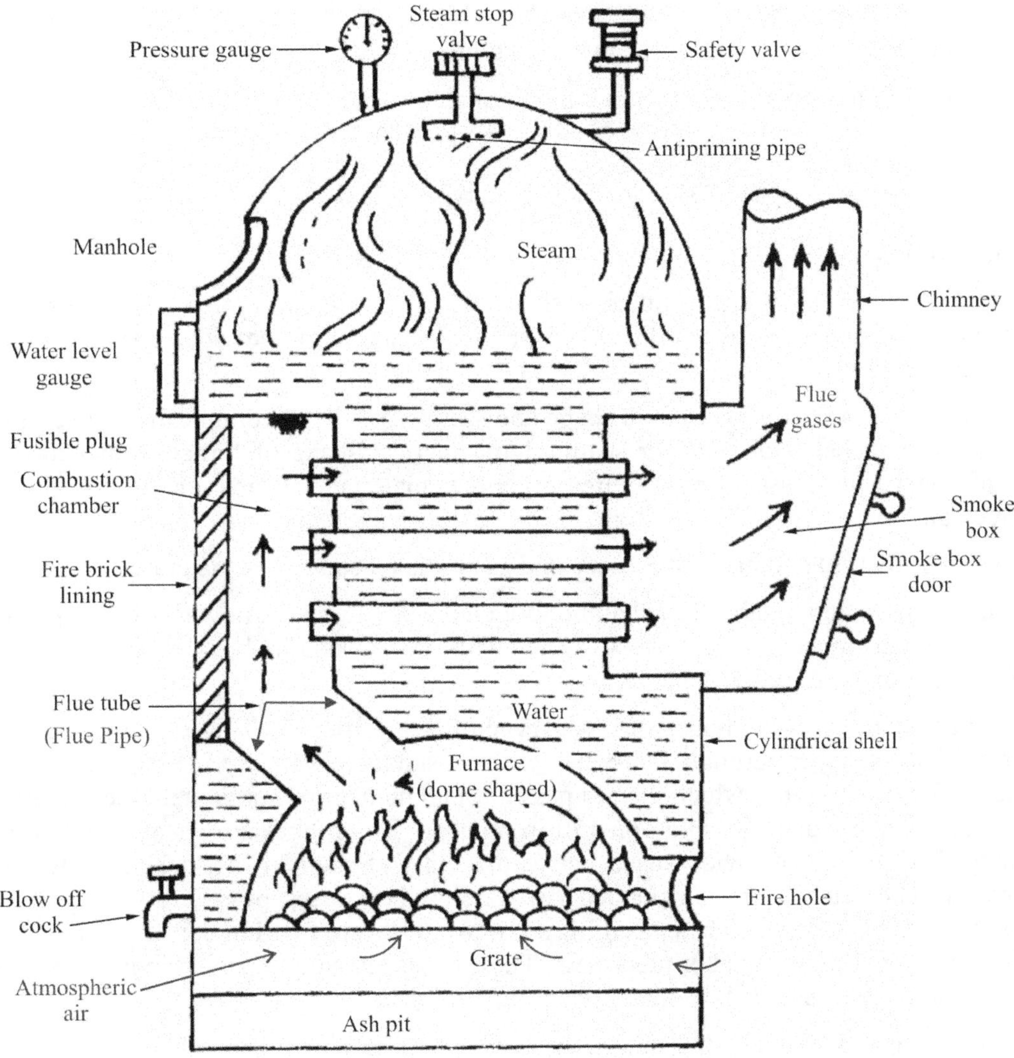

Fig. 6.2 Cochran boiler.

It is one of best types of vertical multi tubular boiler and has a number of horizontal fire tubes and natural circulation. These boilers are manufactured by firm Cochran and Co, Annam Ltd, Annam Scotland. The shell diameter ranges from 0.9 m to 2.75 m.

The specification of 2.75 m shell diameter are as follows:

Shell height = 5.79 m

Working pressure = 6.7 bar (maximum pressure 15 bar)

Steam capacity = 3500 kg/hr to 4000 kg/hr.

Economical rating = ¾ of maximum rating.

Heating surface = 120 m^2.

Efficiency = 70% to 75% (depends upon type of fuel used)

Construction

Cochran boiler consists of a cylindrical shell with hemispherical (dome shaped) top-shell is made up of mild steel plates. The hemispherical shape of top provides maximum value of space for steam.

Furnace is also hemispherical in shape and seamless as it is hydraulically pressed from one plate without single weld or seam. Hemispherical shape of furnace can withstand intense heat and pressure and requires minimum material with maximum strength and space-value.

Grate is located at bottom of furnace and ash pit is placed below the grate.

The furnace and combustion chamber are connected by pipe known as flue pipe. The combustion chamber and smoke box is lined with fire bricks. Fire brick lining is provided on one side of combustion chamber.

When coal is fired, hot gases are produced in furnace. These gases enter the combustion chamber through flue-pipe with considerable velocity. These gases strike against fire-brick linings which then to pass through number of horizontal tubes generally 62.5 mm external diameter and 165-170 in number. There tubes are equal in length and arranged in group with wide space between them and shell, so as to help convection currents. There tubes are surrounded by water, when gases pass through the tubes, transfer heat to the water. The ends of these smoke tubes are fitted in the smoke-box tube-plate and combustion chamber tube-plate. These tubes are simply passed in holes of tube-plates and expanded at ends to make steam tight joints.

Smoke box is built of steel plates and fitted with hinged door which gives an easy access to smoke tubes for cleaning and inspecting.

Flue gases from the fire tubes after giving heat to water enter in to smoke box for discharging into atmosphere through chimney provided on the top of smoke box.

The furnace is surrounded by water on all sides except at the opening of fire door and combustion chamber smoke box tubes are also completely surrounded by water.

Coal or oil can be used as fuel in this boiler. If oil is used as fuel no grate is provided but the bottom of the furnace is lined with fire bricks, oil burners are fitted at a suitable location below the fire door.

Manhole near the top of shell is provided for cleaning. In addition to this number of hand-holes are also provided around the outer shall for cleaning purposes. Smoke box is provided with doors for cleaning of the interiors of fire tubes.

The airflow through the grate is caused by means of the draught produced by chimney. The damper is placed inside the chimney to control the discharge of hot gases from chimney thereby the supply of air to the grate is controlled. Chimney may be provided with steam nozzle to discharge the flue gases faster through the chimney and steam to the nozzle is supplied by the boiler.

Working

Hot gases produced by burning fuel on grate, rise up through flue pipe and enter the combustion chamber. Hot gases pass through fire tubes, heat is transferred to water and steam is generated which is collected at dome-shaped top of shell. The steam formed passes through the steam stop valve for utilization.

The flue gases are collected in smoke box and finally discharged to atmosphere through the chimney.

Advantages

1. Low cost of manufacturing.
2. Minimum floor area is required as it is vertical and compact.
3. Any type of fuel can be used with this boiler.
4. It gives about 70% thermal efficiency with coal firing and 75% thermal efficiency with oil firing.
5. It can be transported from one place to other and can be easily installed at site.
6. Spherical top of the boiler shell and spherical shape of fire box are special features of this boiler. These shapes give higher area to volume ratio and required least material.

Disadvantages

1. Due to vertical position of boiler, the interior is not convenient to inspect, clean and repair.
2. It is not suited for high capacity requirement.

6.9 Lancashire Boiler

It is stationary fire tube, internally fired, horizontal and natural circulation boiler.

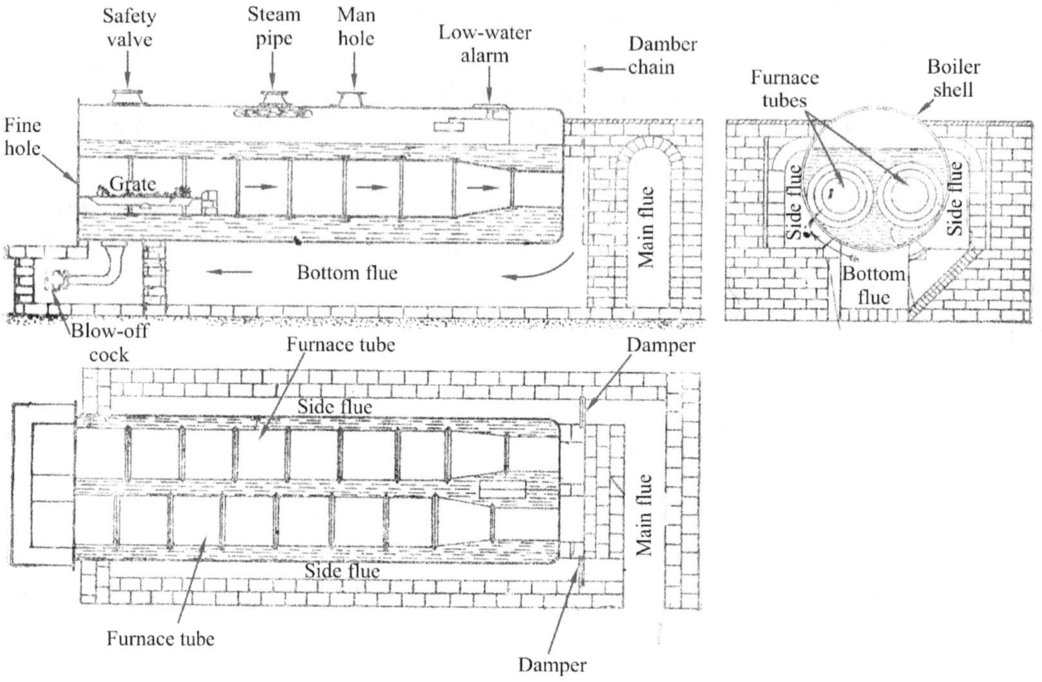

Fig. 6.3 Lancashire boiler.

Specification

Shell diameter	=	2 m to 3 m
Length of shell	=	8 m to 10 m
Maximum working pressure	=	16 bar
Steam generating capacity	=	8000 to 9000 kg/hr
Efficiency	=	50% to 70%

Construction

Lancashire boiler consists of a cylindrical shell length of shell is 8 m to 10 m and diameter varies from 2 to 3.5 m. Boiler shell is made of steel plates, the plates are of 10 mm to 12.5 mm thickness. It has two large internal flue tubes having diameter between

80 cm to 1000 cm in which grate is situated. A grate is of sufficient area and minimum length. The flue tubes are tapered being larger in diameter at front and smaller at back. The diameter is made larger at front end to accommodate grate. The smaller diameter at end will increase velocity of hot gases coming out of flue pipes.

The cylindrical shell is placed over brick work. The brick work forms one bottom flue and two side flues. (Flue is nothing but channel for flow of gases). Both the flue tubes which carry hot gases lay-down below the water level as shown in Fig. 6.3.

The Dampers in form of sliding doors placed at the end of the side flues to control the flow of gases by restricting the passage of the flow. This controls the generation of steam and combustion rate. These dampers are operated by chain passing over pulley at front of boiler.

This boiler is fitted with usual mountings. The pressure gauge and water level indicator are provided at front of boiler whereas steam stop valve, safety valve, low water level and high steam safety valve, and man hole are provided at top of the shell.

The blow off cock is situated beneath (bottom) the front portion of boiler shell for removal of sediments and mud. It is also used to empty the water in boiler whenever required for inspection.

The fusible plugs are mounted on the top of the flue-pipes, just over the grates, to prevent the over heating of boiler by extinguishing the fire, when the water level falls below a particular level. A low water level alarm is usually mounted in the boiler to give a warning in case the water level goes below preset value.

Feed check valve with feed pipe is fitted on the front of end plate. The feed pipe projecting in to the boiler is perforated so that water is uniformly distributed in to shell.

Anti-priming device provided below steam stop valve inside the shell of boiler. It is for separating out suspended moisture and allowing, as for as possible the dry saturated steam through stop valve.

Superheater is placed at end of main flue tubes. The hot gases before entering to bottom channel are passed over superheater tubes and steam from steam stop valve enter into superheator, steam absorbs heat from hot gases and become superheated.

Economiser is placed at the end of side flue before exhausting the gases to chimney. The water before being fed in to boiler is passed through Economiser. Temperature of feed water is increased as it absorbs the heat from exhaust gases, thus leading to better boiler efficiently. Generally chimney is used for providing the draught and discharging exhaust gases in atmosphere at some height.

Working

The fuel is burnt on the grate as a result hot gases are produced. Flue gases (hot gases) move along the furnace tubes (known as main flue) and are deflected up by the fire bridge. As soon as the flue gases reach the back of main flue they deflected downwards and travel through bottom flue. The bottom flue is situated below water shell and heats the lower (Bottom) portion of shell. They move from front of boiler where they are divided in to two streams and pass to side flues. Now they travel from front to back in side flues and thus heat the side of the water shell. After this these two streams of the flue gases meet again. The main flue passing through damper, from where they are discharged to atmosphere through the chimney.

Uses : (i) It is commonly used in sugar-mills and textile industries where both power and steam for the process work is needed.

Advantages

1. Less operating and maintenance costs.
2. This boiler is reliable, has simplicity of design.
3. It is easy to inspect and clean.
4. Its heating surface area per unit volume of the boiler is considerably large.
5. Superheater and economiser can be easily reduce gap in the system. Therefore over all efficiency of boiler is increased, up to 80%–85%.
6. It is suitable where large reserve of hot water is needed.
7. Load fluctuation can easily be met by this boiler due to large reserve capacity.

Limitations (disadvantages)

1. Grate area is restricted because grate is situated inside flue tube.
2. It takes more time to deliver steam at required rate and pressure because of large water capacity of boiler.
3. The maximum pressure is limited to 16 bar because of huge size of shell and brick work setting.
4. Occupies large floor area due to brick work and large length of shell.
5. Due to temperature difference bottom brick setting facing hot gases on one side and atmospheric temperature on other side. The cracks may develop on walls and leakage of gases takes place.
6. Brick work setting is expensive in initial cost and troublesome in maintenance.

6.10 Comparison of Cornish and Lancashire Boilers

	Cornish		Lancashire
1.	Diameter of shell is 1.25 to 2 m.	1.	Diameter of shell is 2 m – 4 m.
2.	Length of shell 5 m to 8 m.	2.	Length of shell 8 m to 10 m.
3.	Steam pressure 10-14 bar.	3.	Pressure of stem 15-20 bar.
4.	Number of internal flue tube is one.	4.	Number of internal flue tubes are two
5.	Steam capacity ranges from 2000 to 4000 kg per hour.	5.	Steam capacity ranges from 8000 kg per hour to 10000 kg per hour.
6.	Initial cost is less.	6.	Initial case is more
7.	This boiler is less popular.	7.	This boiler is more popular.
8.	Path of flue gases is from front to back through one internal flue tube, back to front through two side flues and front to back through one bottom flue. (channel)	8.	Path of flue gases is from front to back through two internal flue tubes, back to front through one bottom flue and front to back through two side flues.

6.11 Locomotive Boiler

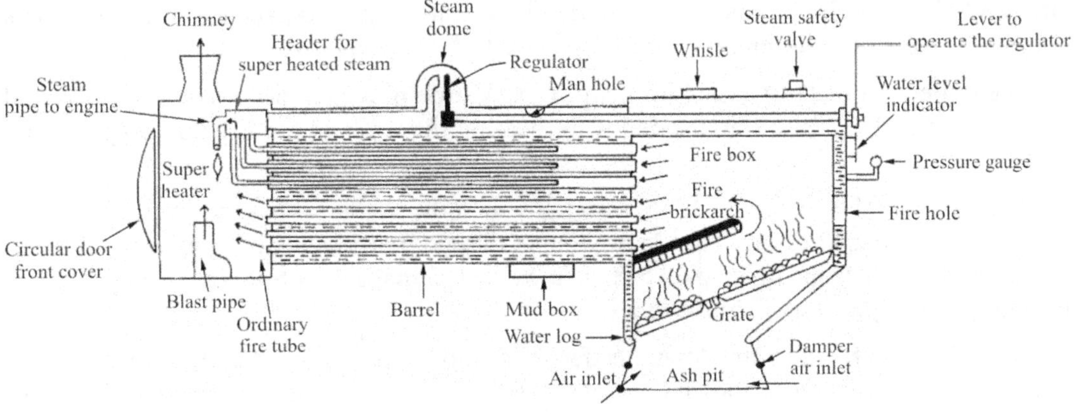

Fig. 6.4 Locomotive boiler.

Introduction

Locomotive boiler is the horizontal, multi-tubular, natural circulation, artificial draught, internally fired, portable, fire tube boiler.

Dimensions and specification of Locomotive boiler (made at Chitranjan works in India) are as below.

Barrel diameter	=	2.095 m
Length of barrel	=	5.206 m
Size to tube	=	14 cm (superheater)
Number of ordinary tube	=	38 (superheater)
Size of ordinary tube	=	5.7 cm
Number of ordinary tubes	=	116
Working pressure	=	14 bar (maximum = 25 bar)
Grate area	=	4.27 m^2
Coal burnt per hour	=	1600 kg
Heating surface	=	271 m^2
Efficiency	=	70%
Steam production rate	=	55 to 70 kg/hr/m^2 of heating surface area

Construction

It consists of cylindrical barrel with a rectangular fire box at one end and smoke box at other end. Fire box forms a combined grate and combustion chamber. The firebox is surrounded by water space on all three sides except the bottom. The ash pit which is situated below the fire box is fitted with dampers at its front and back ends. The dampers control the flow of air to grate.

The deflection of hot gases with help of brick-arch prevents the flow of ash and coal particles with the gases and it also helps for heating the walls of the fire box properly and uniformly.

There are number of smoke tubes through which the hot gases pass from furnace (fire) box to smoke box. They are fixed between end plate of barrel and are immersed in water. A larger door in front of smoke box is provided. This door provides an opening for cleaning and inspection of smoke tubes.

A mud box is fitted to the down header. The impurities and mud particles from water are collected in mud box and they blown off from time to time by means of blow off valve (cook).

Height of chimney must be low to facilitate the locomotive to pass through tunnels and bridges. Because of short chimney, artificial draught must be provided to drive out hot gases. The draught is created with help of exhaust steam when locomotive is moving and with help of live steam when locomotive is stationary.

Pressure gauge and water level indicator are located in driver's cabin at front end of fire box.

The spring loaded safety valve and fusible plug are located as shown Fig. 6.5. Blow-off cock is provided at bottom of water wall to remove the sediments and muddy water.

Dome is fitted to boiler shell at a reasonable distance above water level from which steam is taken nearly free of water. The function of steam dome is to increase release capacity. Steam dome is located at centre so that water level at the dome does not change due to inclinations of rail line or otherwise.

Working

Through feed check valve, water is fed in to boiler shell to required level, manually coal is added to grate. Coal is burnt, hot gases are produced due to combustion of the coal first heat the water surrounding the fire box. The hot gases then pass from fire box to smoke box through series of fire tubes and then they are discharged in to atmosphere through short chimney with help of a strong draught which is created by steam jet from blast pipe. Heat of hot gases is transmitted to water in shell and steam generated in shell is collected over water surface and domed shape chamber which is fitted on top of shell, away from water surface. From dome steam is taken in to steam header which is divided into two parts one part of the steam header is known as saturated steam header. The saturated wet steam through pipe enters into saturated steam header and other part is known as superheated steam header. The saturated wet steam through pipe enters the into saturated steam header then it is passed through superheated steam coming out of superheated tubes is collected into superheated header and then fed to steam engine. A steam stop valve also serving as regulator for steam flow is provided inside the dome. This is operated by driver through regulator shaft (Rod) passing from front of the boiler.

Supply of air to the grate is obtained by creating artificial draught. The draught is produced by exhaust steam which is discharged through the blast pipe which is placed below chimney. The air flow caused by this method is known as induced draught. When locomotive is not moving and no exhausts steam from engine is available, live steam from boiler is used for creating draught.

Use : This type of boiler is used in locomotive engines

Advantages

1. It is very compact and economical.
2. It is free from brick work, special foundations and long chimney.
3. Large rate of steam generation per square meter of heating surface.
4. High steam capacity and it is capable to meet fluctuating demand of power and speed.
5. Due to portability it can easily be moved and installed.

Disadvantages

1. It is difficult to clean some water spaces due to long shell of boiler.
2. Leakage may takes place where tubes join the plate.

3. Large flat surfaces need bracing (supports).
4. Due to accumulation of sediments and mud particles in water legs, there are changes of corrosion and scale formations.
5. More wastage of heat from shell
6. Water circulation is not so proper.

Note: Locomotive boilers manufactured at Chittaranjan locomotive works. Produce 18700 lb/hr (8550 kg/hr) steam at 14.5 bar gauge and super heated to 370 °C at a rate of 1590 kg/hour.

6.12 Babcock and Wilcox Boiler

Introduction

Babcock and Wilcox is water tube boiler with water flowing inside tubes and hot gases over the tubes. It has longitudinal drum. It is externally fired and natural circulation type stationery boiler. It is suitable for small size thermal power plant and other industrial work. It is suitable for all types of fuel and for hand stoker firing. Evaporative capacity of such boilers ranges form 20 tons to 40 tons per hour and operating pressure form 11.5 to 17.5 bar are quite common. But operating pressure may be as high as 42 bar.

Specification : Of this boiler are given below.

Diameter of drum	=	1.22 m to 1.83 m
Length of the drum	=	6.096 m to 9.144 m
Size of water tubes	=	7.62 m to 10.16 m
Size of superheated tubes	=	3.84 cm to 5.71 cm
Working pressure	=	40 bar (maximum)
Steaming capacity	=	40000 kg/hr (maximum)
Efficiency	=	60% to 80%

Construction

The boiler shell known as water and steam drum is made up of high quality steel. The heads of drum are forged by hydraulic press and are dished to radius equal to diameter of the drum. From each end of drum connection are made by short tube with upper header or riser and longer tubes to the down take header. Large number water tubes connect the uptake and down take headers and are kept inclined at an angle of 5 degree to 15 degree to horizontal. Inclination of tubes promote water circulation. Water tubes are 10cm in diameter and made up of steel. They are expanded into bored holes of headers to ensure proper fixing. Water tubes are arranged in staggered manner and this exposes complete surface to the gases. The headers are provided with hand-hole in front of the tubes and covered with the cap. The arrangements help in cleaning of the tubes.

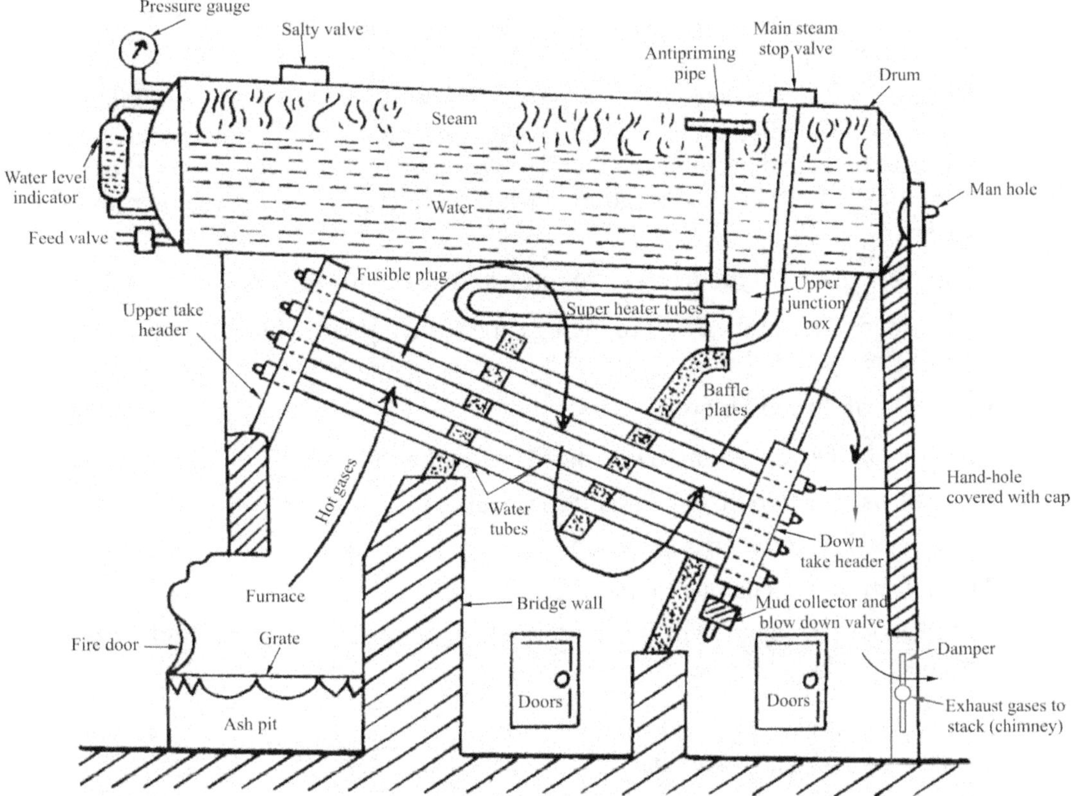

Fig. 6.5 Babcock and Wilcox boiler.

The heating surface of this boiler is the outer surface of the water tubes and half of the cylindrical surface of the water drum which is exposed to hot gases. A mud box is attached to the bottom of the down take header.

The impurities and mud particles from water are collected in the mud box and they are blown off from time to time by means of cock as shown in the Fig. 6.7 The headers are provided with hand holes in front of the tubes and covered with caps. This arrangement helps in cleaning of tubes from inside.

To access the interior of the boiler it is provided with the doors. This is necessary for cleaning the outer surface of water tubes and to remove soot from their surfaces. Some part of shell also can be cleaned and soot from its surface can be removed.

Whole of assembly of water tubes is hung along with drum from steel girder frame by steel straps (called slings) rests on masonry work. Masonry work is lined with fire bricks. The sling passes around each of drum and thus entirely independent of brick work setting.

Therefore boiler can expand or contract freely without causing strain to the brick work setting.

The furnace (grate) is arranged below uptake header. Coal is fed to chain grate stoker through fire door. Two fire brick baffles are arranged in such a manner that hot gases from grate are compelled to move upward. Two metallic baffle plates are provided as shown in the Fig. 6.5, which provides three passes of hot gases. Chimney is provided to exit the gases.

The damper is placed at the inlet of chimney to regulate the draught.

There are superheated tubes for producing superheated steam. Superheater is placed between drum and water tubes.

The **advantages** of this boiler are :

- The draught loss is minimum compared with other boilers.
- It can be used with both solid and liquid fuels.
- The circulation of water is natural.
- The defective tubes can be replaced easily.
- The evaporation capacity of this boiler is high compare with other boiler (20000 to 40000 kg per hour).

Uses

- Suitable for thermal power stations.
- Heating installation.
- Industrial processes.
- Rolling mills.
- Wood working factories.

Working

Through the feed check valve water is admitted into the boiler shell. Before starting the boiler water is filled up, in all water tubes, headers and shell. Shell is generally half filled, rest space is kept for steam.

The coal is kept on the chain grate stoker manually or mechanically through the fire door. The coal is fired. Hot gases are produced and necessary air for combustion of coal and circulation of gases, dampers are provided. Front portion of water tubes comes in contact with the hot gases which are at higher temperature. The water from this portion rises in upper direction due to decrease in density and passes into the drum through uptake header. Here water and steam is separated and steam being lighter is collected on the surface of the water in the shell. From back portion of the drum water enter into the down-take header and passes into the inclined water tubes and heated by the hot gases.

Thus a continuous circulation of water is maintained by convective currents and is known as natural circulation.

The bridge wall deflector which deflects the hot gases upward, baffles are also arranged, cross water tubes to act as deflector for flue gases and to provide necessary gas path. Here two baffles are placed which provides three passes of the gases. A chimney is provided for discharging these gases to atmosphere.

With provision of baffles the gases are deflected downward and complete the second pass. Again they rise upward due to baffle and complete the third pass. And finally come out of the chimney. The sine wave moment of hot gases enables maximum contact with surface of the water tube and superheater tube and maximum amount of heat is transferred to water. Flow path of the combustion gases is shown in the Fig. 6.7. During their travel, they give heat to water and steam is formed.

To get superheated steam, the steam is accumulated in boiler shell, allowed to enter in superheater. During the first turn (path) the hot gases, are passed over superheater tubes and heat is absorbed by steam from hot gases and steam leaves the superheater in superheated condition. The steam coming out from superheater is supplied through steam pipe and steam stop valve to turbine or for different uses.

When the steam is being raised from cold water in boiler, the superheater is filled with water to prevent over heating of superheater tubes. The superheater remain flooded with water until the steam reaches the working pressure. Once the rated pressure of steam is reached in the boiler, then water from the superheater is drained and steam is fed to it for superheating.

6.12.1 Advantages of High Pressure Boilers

Following are the advantages of the high pressure boilers:

1. Due to uniform heating of water inside tubes, the danger of over heating and thermal stress is reduced.
2. Due to high velocity of water inside the tube reduces scale formation.
3. Reduction in number of drums or no drums at all.
4. Lighter for given output as there are no drums.
5. Heat of combustion is utilized, more efficiently by using small diameter tubes in large numbers and multiple circuits.
6. Greater freedom for disposing of heating surfaces and hence greater evaporation for given size.
7. Space and weight of high pressure is minimized which reduces the erection cost and time.
8. Use of high pressure and temperature steam is economical.
9. Efficiency of plant is increased up to 40 to 42% by using high pressure and temperature steam.

10. Rapid changing of load can be met without the use of complicated control devices.
11. Because of uniform expansion due to uniform temperature reduces the possibility of gas and air leakages.
12. Pressured combustion is used which increases rate of firing of fuel thus increases the rate of heat release.
13. Components can be arranged horizontally, as high head required for natural circulation is eliminated by using forced circulation. There is greater flexibility in components arrangement.

Examples of high pressure boilers are:

1. Lamont Boiler; 2. Benson Boiler; 3. Loeffler Boiler; 4. Velox Boiler; 5. Schmidt – Hartman Boiler; 6. Critical and Supercritical Boilers.

6.13 Scotch Boiler (Scotch Marine Fire Tube Boiler)

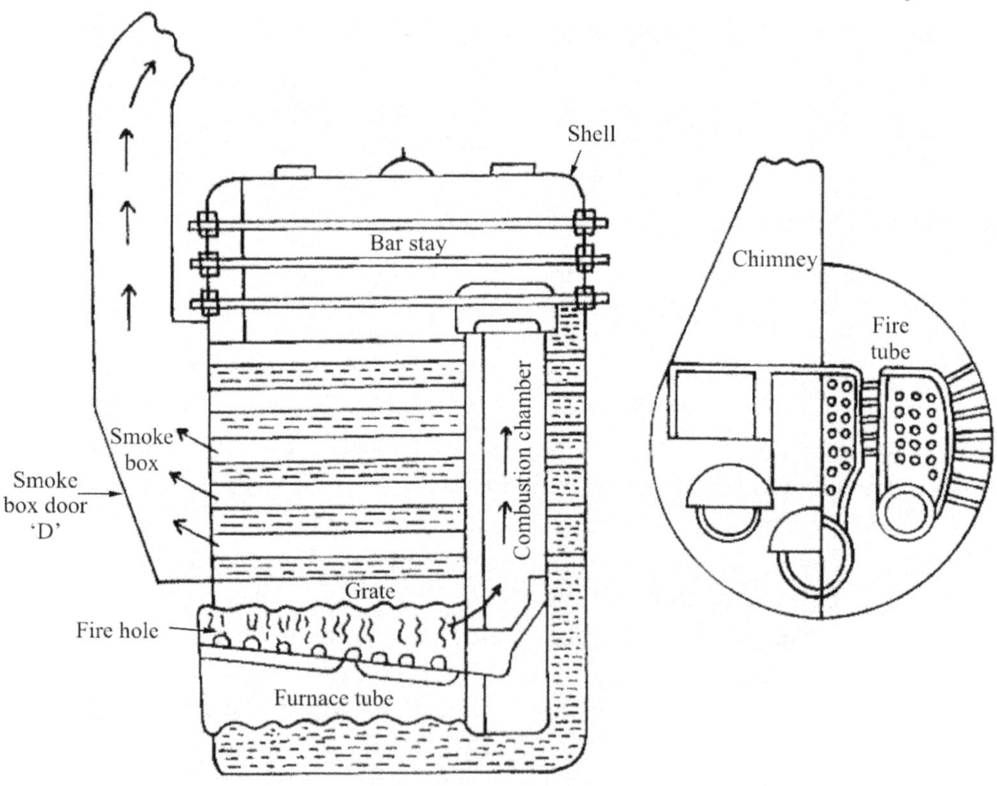

Fig. 6.1(b) Scotch boiler.

Introduction

It is marine fire tube boiler. It is horizontal multi-tubular, internally fired, natural circulation, fire tube boiler. It provides a large area of heating surface for space occupied.

On ships, fire tube or water tube boilers are used. In heavy ships high speed is not required therefore we can use fire tube boiler. In passenger ships and naval ships, fast steaming and high pressure steam is required, the water tube boiler is used.

Common types of marine boilers are single and double ended scotch boilers, Yarrow marine boilers and Babcock and Wilcox marine boilers.

Construction

Fig. 6.6 shows a single ended marine boiler known as scotch type. The cylindrical shell of this boiler contains one to four corrugated steel furnace. The furnaces are internally fired and surrounded by water. The combustion chamber is located at the back end of furnace, is also surrounded by water and therefore does not require a fire brick lining. Each furnace usually has its own combustion chamber but in some cases two or more furnaces open into common combustion chamber. A large number of fire tube run from the front tube plate to the back tube plate. The flat ends of the shell are stayed by longitudinal stays. Some of the smoke tubes are screwed in to the tube plates and work as stays for the flat surfaces of the tube plates. The Flat Plate of the combustion chamber are stayed by the screw stays, while their flat tops are strengthened by Girinder stays as shown in the Fig. 6.1(b). Door "D" is provided in the front end for cleaning the smoke tubes.

There is a provision for uptake passage from the boiler front to the stack (chimney).

A thin ash plate rests below the grate, at the bottom of ash pit to facilitate the removal of ashes at regular intervals. Access to the boiler is through three main holes placed on front end, one near the tip and the one between the center and each side of furnace.

Various mounting and accessories are provided on boiler.

Working

The hot gases produced by combustion of fuels on the grate are deflected upward by fire brick bridge and pass to the combustion chamber due to draught. From the combustion chamber the hot gases enter into the fire tubes to smoke box situated at front end of boiler from where they move through uptake to the chimney and discharged into atmosphere.

During their travel from grate to chimney the hot gases give up heat to water and steam is produced. Steam is collected at top of shell and through steam stop valve it is taken away from boiler for utilization purpose.

Double-ended Marine Boiler

They have furnace at each end. They look like single ended boiler placed back to back. The furnace tubes at each end open into centrally placed combustion chamber, from which the hot gases pass through fire tubes to the smoke boxes one at each end of the boiler. After they enter the uptake and are discharged through stack (chimney) into the atmosphere.

A double ended boiler for same evaporation capacity is cheaper, lighter and occupying less space as compared with single-ended boiler.

Advantages

- It is portable boiler.
- It is of compact size and occupies less floor space.
- As combustion chamber and furnace are surrounded by water, minimum amount of brick work is required.
- As height of the boiler is not so high therefore it requires low head room.
- It provides a large heating surface area for the space occupied.

Disadvantages

- As there are large number of tubes in each furnace and there are more than one furnace the cleaning and inspection are difficult.
- The construction of boiler is such that fuel change requirements are difficult to meet.
- There is practical limit to capacity (1000 kg/hr) and pressure (17 bar).
- During starting, circulation difficulties are observed due to large diameter of shell (2.5 to 3.5 m).

Application (use) : It is used in marine applications, especially heavy ships.

6.14 Lamont Boiler

Water Circuit

The hot feed water from hot well is supplied by feed pump to economiser where temperature of feed water is increased by absorbing heat from flue gases.

From Economiser feed water is supplied to storage and steam separator drum. A centrifugal pump is employed to circulate water from storage and steam separating drum to radiant (Primary) evaporator and afterward it flows to convective (Secondary) evaporator. The heated water is returned to storage and separating drum. Here steam being lighter than water is separated and collected above water.

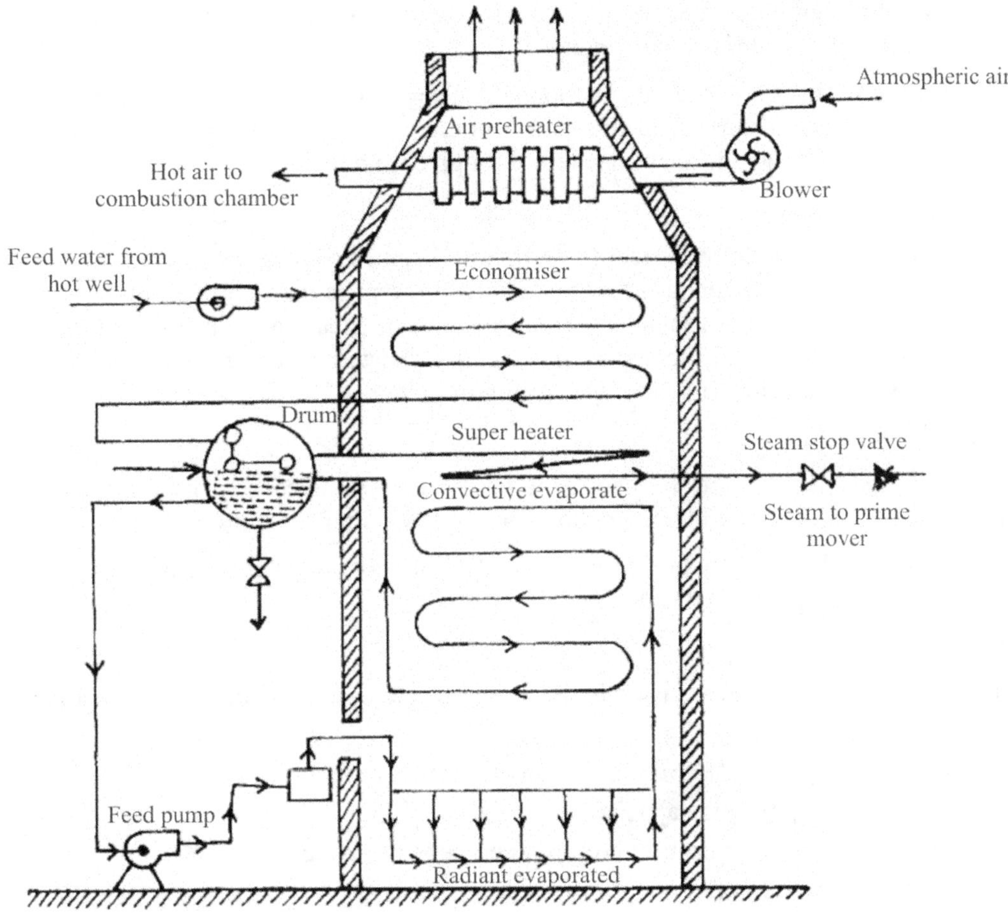

Fig. 6.6 Lamont boiler.

The quantity of water circulated is eight to ten times the weight of steam separated in the drum. This reduces the over heating of evaporator tubes. The pressure delivered by circulating pump is 2 to 5 bar above the drum pressure. The distributor header distributes the water through nozzles in to evaporator.

Steam collected in drum is farther supplied to superheater. Steam picks up heat from hot gases and superheated to required temperature and through steam stop valve, steam is supplied to steam turbine.

Air and Flue Gas Circuit

The Air is necessary far combustion of fuel (coal or oil). The air from atmosphere enters the air preheated through blower. Air preheater is located in between chimney and economiser. Air is heated in air preheater by flue gases. Heated air is then supplied to boiler furnace or combustion chamber located at bottom of boiler for combustion of coal, hot air increases the combustion efficiency.

When coal burns hot gases are produced. They travel upward, supplying heat to primary evaporator, secondary evaporator, superheater, economizer and air-preheater, before they are discharged into atmosphere through chimney.

Steam Circuit

Partially evaporator water (wet steam) enters into drum, and steam is separated and gets collected on the top of water in the drum. The wet steam is then passed into superheater coil which is located above secondary evaporator where steam absorbs the heat from hot gases and steam becomes superheated. This superheated steam is supplied to steam turbine through steam stop valve for generation of power.

Cool and Ash Circuit

Coal enters the stocker and burned. Hot gases are produced. By burning fuel after combustion of coal, ash is collected at bottom of stocker in the ash pit .

These boilers have been built to generate 45 to 50 tones of steam at pressure of 130 bar and at a temperature of 500 °C.

Advantages

1. Small diameter tubes are used in this boiler to increase evaporator surface area.
2. It is light in weight.
3. There is less possibility of scale formation inside the tubes.
4. It has high evaporative capacity.
5. It has small diameter drum therefore occupies less space.
6. It is capable to meet fluctuation of load.
7. It can be rapidity started from cold condition.

Limitations

1. Power consumption increases due to pumping equipments.
2. Maintenance cost increases.
3. It operates below critical pressure.
4. Bubbles evolving inside water tubes due to density difference between water and steam, bubbles get collected on inner surface of tubes and reduces heat transfer to water. (Because of high thermal resistance than water film adjacent to the wall of tube.

6.15 Benson Boiler

We know the bubbles forming inside water tubes of Lamont boiler, offers resistance to heat flow having high thermal resistance than water film adjacent to the wall of the tube.

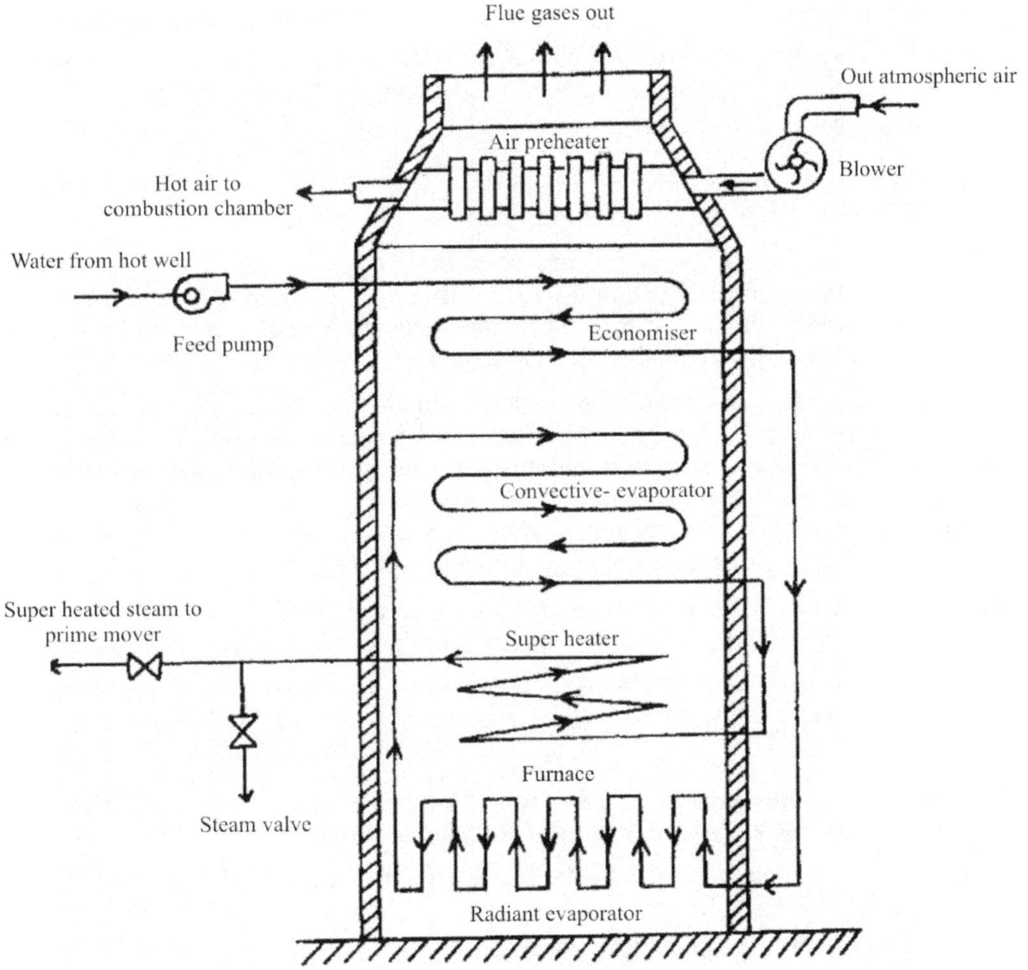

Fig. 6.7 Benson boiler.

This problem created, can be eliminated if steam generation is carried out at critical pressure so that densities of water and steam bubbles become same and bubbles will not be separated from water and cannot get collected on walls of tube.

In 1927 AD first Benson boiler was developed and constructed in Germany.

Construction

It consists of following components.

1. Feed pump
2. Economiser
3. Radiant evaporator
4. Superheater
5. Convective evaporator
6. Air preheater
7. Air blower
8. Furnace
9. Ash pit

Working

Water-Steam Circuit: Feed water supplied by feed pump is passed into economiser situated in between air preheated and superheater, water absorbs heat from flue gases (exhaust gases) as a result temperature of water is increased. From economiser water enters into the Radiant evaporator where majority of water is converted into steam, rest of water is evaporated in convective evaporator by hot gases. Saturated steam generated at 225 bar is then passed into superheater and finally supplied to turbine.

The salt present in soluble state, are separated in the convective evaporator and deposited there, (wall of tube) creating major difficulty i.e., it reduces heat transfer and block up the diameter of evaporator. It becomes necessary to flush out the boiler after every 4000 working hours to remove the salt deposited.

Air fuel circuit : Through blower air enters into the air-preheater, located between economiser and chimney. Air is heated by flue gases before discharging through chimney to atmosphere. Hot air is used to burn oil, due to burning of oil, hot gases are produced in combustion chamber. They move upward, and supply heat to primary (Radiant) evaporator, superheater, secondary (convective) evaporator, economiser and air preheater and through stack flue gases are discharged into atmosphere.

This boiler can supply steam at temperatures as high as 650 °C. The maximum working pressure obtained so for in this boiler is 500 bar. It produces 150 tons per hour of steam.

Advantages

1. This is drumless boiler which reduces the total weight by 20 percent than other boilers.
2. Forced circulations of water permits welded joints in boiler, saving the cost of expansion joints, necessary in natural circulation boilers.
3. Boiler components are easy to transfer at site for (assembly), as it is drum less boiler.
4. Compact therefore requires less space.
5. The water wall constructed by using closely pitched smaller diameter tubes, can protect furnace walls more efficiently.
6. It does not require special arrangements for superheater as it is an integral part of forced circulation system.
7. It can be rapidly started due to welded joints.
8. During over load and partial loads, the Benson boiler can be operated most economically by varying the pressure and temperature. It is also easy to maintain temperature constant at any pressure.
9. The blow off losses are hardly 4 percent that of natural circulation boilers of same capacity.
10. It has less hazards, as it consists of only tubes of small diameter with very little storage capacity compare to drum type boiler.
11. The efficiency of boiler is about 90%.

Limitations

1. It requires salt free pure water to avoid scale deposition in evaporator tubes.
2. It is difficult to operate it below critical pressure (221 bar).
3. Operation and maintenance cost of forced circulation system is moderately high due to more power required to overcome friction resistance of water circulation through smaller diameter and longer tubes.

6.16 Loffler Boiler

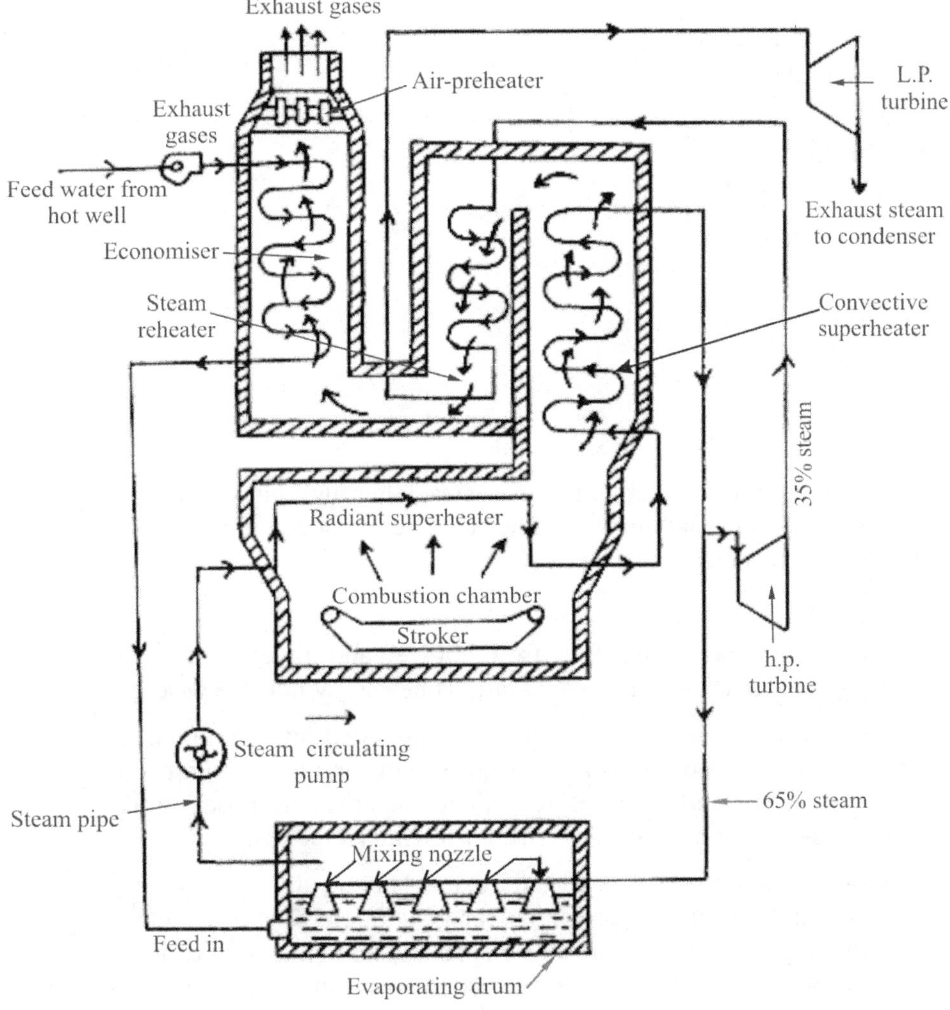

Fig. 6.8 Loffler boiler.

The problem created due to the salt deposition in evaporator, increases thermal resistance, causing danger of overheating of tubes in Lamont boiler and to some extend in Benson Boiler. This difficulty was overcome by preventing water to flow in boiler tubes i.e., water is not converted into steam in radiant evaporator and connective evaporator but it is heated by steam in evaporative drum or chamber.

Construction

It consists of following components:

1. Economiser
2. Economiser feed pump
3. Evaporating drum with drain
4. Steam circulating pump
5. Furnace
6. Radiant superheater
7. Convective superheater
8. Steam reheaters
9. Air-preheater
10. Air blower
11. Casing of boiler

Working

Air Circuit

Cold air from atmosphere, enters into air preheater through blower. It is heater up in air preheater. The resulting hot air is directed to combustion chamber, to be used as combustion air for combustion of the fuel.

Water Steam Circuit

The boiler feed pump draws feed water from hot well and forces it into evaporator drum through economiser. In economiser feed water is heated by hot flue gases.

The heated feed water in this drum is evaporated by mixing about 65 percent superheated steam into the evaporator drum, steam circulating pump draws the steam from evaporator drum and supplies to radiant superheater. From radiant superheater steam is then passed to convective superheater for final superheating.

Remaining 35 percent of superheated steam, leaving the superheater is supplied to H.P. (high pressure) steam turbine for power generation. Steam leaving H.P. turbine is reheated to the required temperature in reheater by flue gases before it enters in to L.P. stage of the turbine. Exhausted steam after condensing in condenser is discharged in to hot well.

The specially designed steam nozzles provided in the evaporator drum mix superheated steam with water without noise.

Drain is provided to clean the drum. The superheated tubes carry dry saturated steam raised in drum therefore no salt deposition on superheater tubes in this boiler.

Advantages

1. No Problem of salt deposition in the boiler tubes as only steam is passing through tubes.
2. No danger of overheating of tubes as heat transfer is efficient.
3. This boiler can use water with dissolved impurities without any problem. Thus saves cost of water treatment plant.
4. It can be utilized to generate power on land or sea. Loeffler boiler are generally constructed of capacity 100 ton per hour at operating pressure 140 bar.
5. Co-efficient of heat transfer of super heated steam to water is more than hot gases.

Limitations

1. Requires large quantity of steam as 35 percent of total quantity is utilized for power generation.
2. The steam circulating pump increases power consumption, cost and maintenance.
3. It can not be operated at critical pressure as density of steam and water inside evaporator drums becomes equal. Therefore steam can not be separated in drum.
4. It is low capacity boiler.
5. It requires superheated steam from outside sources during starting or evaporator drum must be provided electric heater to heat the water in drum and generate the wet or dry steam.

6.17 Performance of Steam Boilers

6.17.1 Introduction

Performance of the boiler is measured in terms of evaporative capacity. The evaporative capacity of boiler may be expressed in anyone of the form as mentioned below:

1. kg of steam/hr
2. kg of steam/hr/m^2 of heating surface
3. kg of steam/kg of coal fired

The boiler cannot be compared only by knowing their evaporative capacity, hence to compare the performance of different boilers working under different condition, some common base must be provided, that is, water must be evaporated under standard conditions, they are:

(i) Feed water temperature at 100 °C

(ii) Standard atmosphere pressure 1.01325 bar

(iii) Water at 100 °C should be converted in to dry saturated steam at same temperature (100 °C).

6.17.2 (i) Equivalent Evaporation

It is quantity of water evaporated from and at 100 °C to produce dry saturated steam at 100 °C and at atmospheric pressure. As water is already at boiling temperature, it requires only latent heat at 100 °C (1.01325 bar) to convert it into steam at 100 °C.

The latent heat at 100 °C = 2257 kJ/kg (From steam table)

$$M_e = \frac{M_a (h - h_f)}{2257} \text{ or } \frac{M_a (h - h_w)}{2257}$$

where M_e = equivalent evaporation.

M_a = Mass of steam generated per unit mass of fuel

$$= \frac{\text{kg of steam}}{\text{kg of fuel}} \text{ OR } \frac{\text{kg}}{\text{kg of fuel}}$$

$h_f = h_w$ = sensible heat of water

h = total specific Enthalpy of steam $kJ/_{kg}$ under operating conditions

$= h_f + x\, h_{fg}$ for wet steam

$= h_f + h_{fg}$ for dry steam

$= h_f + h_{fg} + Cp_s (T_{sup} - T_{sat})$ OR

$= h_f + h_{fg} + Cp_s (t_{sup} - t_{sat})$

where T_{sup} = Superheated temperature of steam in degree kelvin

T_{sat} = saturated temperature of steam corresponding to actual pressure of steam given in problem in degree kelvin.

$h_f = h_w$ = sensible heat of water at feed water temperature

= 4.18 × (Feed water temperature in (t°C – 0 °C)) per kg

= 1 × 4.187 × t °C

= 4.187t kJ/kg of steam

Cp_s = specific heat of superheated steam, may be taken as 2.1 kJ/kgK

6.17.3 Evaporation Rate

Amount of steam produced in boiler per hour

$$\text{Unit of evaporation rate} = \frac{\text{kg of steam}}{\text{hr}} = \text{kg/hr} \quad \text{..... (i)}$$

$$= \text{kg/hr/m}^2 \text{ of heating surface} \quad \text{..... (ii)}$$

$$= \text{kg/hr/kg of coal} \quad \text{...... (iii)}$$

6.17.4 Actual Evaporation

Amount of steam produced per kg of fuel burnt. It is denoted by M_a

$$M_a = \frac{\text{Total evaporation in kg per hour}}{\text{Fuel used/hr}} = \frac{\text{Total evaporation in kg}}{\text{Fuel used in kg}}$$

6.17.5 Factor of Evaporation

It is the factor to be multiplied with quantity of steam generated under working condition to get equivalent evaporation. It is denoted by f

$$M_e = \text{Actual evaporation} \times f$$

$$= M_a \times f$$

But we known that $M_e = \dfrac{M_a(h - h_w)}{2257}$

$$\therefore \quad M_e = \frac{M_a(h - h_w)}{2257} = M_a \times f$$

$$\therefore \quad f = \frac{(h - h_w)}{2257}$$

Variable Factors affecting boiler efficiency

They are as follows:

1. Fuel condition as it is fired
2. Change in draught from the rated, due to atmospheric conditions
3. Excess air fluctuations
4. Actual firing rate
5. Humidity and temperature of combustion air
6. Incomplete combustion and combustibles in the refuse
7. The condition of heat absorbing surface.

6.17.6 Heat Loses in Boiler

Following are the heat loses in a boiler

1. ***Heat loss in dry flue gases per kg of fuel*** $= m_g \, Cp_g \, (t_g - t_b)$ .....(1)

where m_g = mass of dry flue gasses per kg of fuel

Cp_g = specific heat of dry flue gases

t_g = temperature of flue gases leaving the chimney

t_b = temperature of boiler room

This loss is maximum in the boiler.

2. *Heat lost in moisture present in the fuel*

It is assumed that the moisture is converted in to Superheated steam at atmospheric pressure of 1.013 bar.

Heat lost in moisture present in fuel = $M_m (h_{sup} - h_b)$

$= M_m (h_g + Cp_s (t_g - t) - h_b)$

$= M_m [2676 + Cp_s (t_g - 100) - h_b]$ .....(2)

where

h_g	=	enthalpy of 1 kg of dry saturated steam
h_g	=	2676 kJ/kg from steam table corresponding to p = 1.013 bar
t	=	100 °C from steam table corresponding to p = 1.0 13 bar.
Cp_s	=	specific heat of superheated steam in flue gases
t_g	=	temperature of flue gasses leaving the chimney
t_b	=	temperature of boiler room
h_b	=	sensible heat of water at boiler room temperature
M_m	=	mass of moisture per kg of fuel

3. *Heat lost of steam formed by combustion of hydrogen per kg of fuel*

H_2 = mass of hydrogen per kg of fuel

∴ Mass of steam formed = $9H_2$

Then heat loss of steam per kg of fuel = $9H_2 [2676 + Cp_s (t_g - 100) - h_b]$ .....(3)

Combining 2 and 3 i.e., heat lost by steam formed by combustion of H_2 and moisture per kg of fuel

$$= (9H_2 + M_m) [2676 + Cp_s (t_g - 100 - h_b]$$

4. *Heat lost due to unburnt carbon per kg of fuel*

$$= m_1 \times c_1$$

where m_1 = Mass of carbon in ash pit per kg of fuel

C_1 = Calorific value of carbon

One kg of carbon releases 10120 kJ of heat, if it burns to CO.

5. ***Heat lost due to incomplete combustion of carbon to carbon monoxide (CO)***

$$= M_2 C_2$$

M_2 = Mass of carbon monoxide in flue gas per kg of fuel

C_2 = C.V. of CO.

If carbon burns to CO instead of CO_2 it releases 10120 kJ

But if carbon burns to CO_2 it will release 33800 kJ/kg

∴ Heat lost due to incomplete combustion of 1 kg of carbon

$$= 33800 - 10120 = 23680 \text{ kJ}$$

∴ Heat lost due to incomplete combustion of carbon per kg of fuel CO_2 and CO are expressed.

$$= \frac{CO \times C}{CO_2 + CO} \times 23680 \text{ kJ/kg of fuel}$$

CO_2 and CO = Percentage combustion by volume in flue gases.

6. ***Heat lost due to Radiation***

There is no direct method for finding heat lost due to radiation. This loss is calculated by subtracting the heat utilized in raising steam plus all heat losses. From the heat supplied.

C = Fraction of carbon in one kg of fuel

We know that heat utilized in raising steam/kg of fuel = $m_a (h - h_w)$

The ***heat balance sheet for a boiler trail per kg of fuel*** is drawn below.

Heat supplied	**kJ**		**Heat consumed**	**kJ**
Heat supplied by =	x	1.	Heat utilized in raising steam	a
1 kg of dry fuel		2.	Heat lost by dry flue gases	b
		3.	Heat lost in moisture in fuel	C
		4.	Heat lost to steam by combustion of hydrogen	D
		5.	Heat lost due to un-burnt carbon in fuel	E
		6.	Heat lost due to incomplete combustion	F
		7.	Heat lost due to radiation by difference	x – (a + b + c + D + E + F)
Total	**X**	**Total**		**X**

6.17.7 Boiler Efficiency

$$\text{Boiler efficiency} = \frac{\text{Heat utilized by steam}}{\text{Heat generated by burning fuel}}$$

$$= \frac{m_s(h - h_w)}{m_f \times \text{C.V of coal}} = \frac{m_a(h - h_w)}{Cv}$$

m_s = steam produced in kg/hr

h_w = Enthalpy of feed water

h = Enthalpy of steam at working condition of boiler (pressure and temperature)

m_f = Mass of fuel used/hr

C.V. = Calorific valve of coal (Gross)

Note

1. If m_a is taken in kg of steam per hour then m_f must be taken in kg of fuel burnt per hour.
2. If m_a is taken kg of steam/kg of fuel then.

   M_f = 1 kg is to taken in above formula.

Factors influencing Boiler efficiency are as follows.

1. **Fixed factors**
   (i) Boiler design which includes the arrangement and effectiveness of heating surfaces, shape and volume of furnace, the arrangement of flues, the arrangement of steam and water circulation.
   (ii) Heat recovery equipments which include the economiser, superheater, air preheaters and feed water heaters.
   (iii) Built in losses which include heat transfer properties of the setting and construction material, flue gas and ash heat losses.

6.17.8 Economiser Efficiency

It is defined as the ratio of heat absorbed by B.F.W. in the economiser to heat supplied by the flue gasses in economiser.

$$\eta_{econ} = \frac{m_s\,(T_2 - T_1)Cp_w}{m_{f_1}\,Cp_{f_1}\,(T_f - T_{air})}$$

where

m_s = Actual evaporation in kg.

T_1 = Feed water inlet temperature entering to economiser

T_2 = BFW leaving the economiser

M_{fl} = Mass of flue gases in kg

T_f = Temperature of flue gases

T_{air} = Boiler room temperature

Cp_{fl} = specific heat of flue gases

Cp_w = specific heat of water.

6.17.9 Boiler Power

$$\text{Boiler power} = \frac{\text{Equivalent evaporation/hour}}{21.196} \text{ KW}$$

Note : generally we are expressing equivalent evaporation in kg of steam/kg of fuel but in 'Boiler Power' equivalent evaporation is expressed in equivalent evaporation/hr.

6.18 Examples

1. Boiler is supplied with feed water at 13 °C and produces 2400 kg of steam/hr at pressure of 1.2 MPa it consumes 190 kg of oil per hour and dryness fraction = 0.95 having C.V. = 44886 kJ/kg.

Calculate = (i) equivalent evaporation (ii) thermal efficiency of boiler take $Cp_w = 4.1868$ kJ/kg K.

Solution:

Given

Feed water temp = $t_w = 13\ ^\circ C$

$$m_s = 2400\ \frac{kg}{hr}$$

$$P = 1.2\ MPa = 1.2 \times 10^6\ Pa = 1.2 \times 10^6\ N/m^2 = 12\ bar$$

$$x = 0.95$$

Mass of oil/hr = m_f = 190 kg/hr

C.V. = 44386 kJ/kg

From steam table corresponding to 12 bar

h_f = 798.4 kJ/kg

h_{fg} = 1984.3 kJ/kg

$\therefore$ h = Enthalpy of steam (from °C) = $h_f + x\, h_{fg}$

$= 798.4 + 0.95 \times 1984.3$

$= 2683.485$ kJ/kg

$h_w = 4.1868 \times (13 - 0) \times 1 = 54.428$ kJ/kg

Let us find M_a in $\dfrac{\text{kg of steam}}{\text{kg of oil}}$

$$M_a = \frac{2400 \text{ kg of steam}/_{hr}}{190 \text{ kg of oil}/_{hr}}$$

$$= \frac{2400}{190} \frac{\text{kg}}{\text{kg of oil}}$$

$$= 12.623 \frac{\text{kg}}{\text{kg of oil}}$$

$$Me = \frac{Ma\,(h - h_w)}{2257} = \frac{12.632\,(2683.485 - 54.428)}{2257}$$

= 14.714 kg of steam/kg of oil **Ans.**

Thermal efficiency

$$\frac{m_a\,(h - h_w)}{m_f \times C.V}$$

$$= \frac{12.632\,(2683.485 - 54.428)}{1 \times 44386}$$

$= 0.7482$

$= 74.82\%$ **Ans.**

2. Find efficiency of boiler from following data.

Steam supplied	=	30,000 kg/hr
Coal used	=	3000 kg/hr
C.V. of coal	=	29400 kJ/kg
Pressure of dry saturated steam	=	12 bar
Temperature of water supplied	=	37 °C
Specific heat of water	=	4.2 kJ/kg K. [G.U. 1995]

Solution:

Given

m_s = mass of steam = 30000 kg/hr

m_f = mass of coal = 3000 kg/hr

C.V. of coal = 29400 kJ/kh
P = 12 bar

Feed water bat temperature = t_w = 37 °C

$$Cp_w = 4.2 \text{ kJ/kg k.}$$

We known boiler efficiency $= \dfrac{m_s(h - h_w)}{m_f \times C.V}$

$$= \frac{m_a(h - h_w)}{CV}$$

$$\therefore \quad m_a = \frac{m_s}{m_f} = \frac{\text{kg of steam / hr}}{\text{kg of fuel/hr}}$$

$$= \frac{\text{kg of steam}}{\text{kg of coal}}$$

$$m_a = \frac{30000}{3000} = 10 \frac{\text{kg of steam}}{\text{kg of coal}} \quad \text{or 10 kg/kg of coal}$$

h for dry saturated steam = hg corresponding to 12 bar from steam table

= 2782.7 kJ/kg

$h_w = Cp_w$ (t. – 0) 1

$h_w = 4.2 \times 37 \times 1 = 155.4$ kJ/kg

$$\therefore \quad \eta \text{ boiler} = \frac{10(2782.7 - 155.4)}{29400}$$

= 0.8936 i.e., 89.36% **Ans.**

3. Find heating surface and grate area required for a boiler if it generates 10,000 kg of dry saturated steam/hr at 7 bar. Feed water temperature 40 °C heating surface permits heat at a rate of 2720 kJ/m^2/min.

The ratio of heating surface to grate area is 25

Take Cp_w = 4.1868 kJ/kg K

Solution:

Given

$$M_s = \text{ mass of steam/min} = 10000 \frac{\text{kg}}{\text{hr}} = \frac{10000}{60} = 166.667 \frac{\text{kg}}{\text{min}}$$

t_w = 40 °C

P = 7 bar and steam is dry-saturated

$h_g = 2762$ kJ/kg (from steam table)

$h_w = 4.1868 \times (40 - 0) = 167.427) \dfrac{kJ}{kg}$

Heat required to produce 1 kg of steam from feed water at 40°

$$= (h_g - h_w) = 2762 - 167.472 = 2591.528 \frac{kJ}{kg}$$

Total heat required $= 2594.528 \times 166.667 = 432422.2 \dfrac{kJ}{min}$

Total heating surface and grate are =

$$\frac{\text{Total heat required}}{\text{Heating surface permit heat per m}^2\text{per min}}$$

$$= \frac{432422.2}{2720} = 158.978 \text{ m}^2$$

$$\frac{\text{Heating surface area}}{\text{Grate area}} = 25$$

∴ heating surface area = 25 grate area

But heating surface area + grate area = 158.978

∴ 25 grate area + grate area = 158.978

∴ 26 grate area = 158.978

∴ grate area $= \dfrac{158.978}{26} = 6.115 \text{ m}^2$

∴ **surface heating area = 25 × 6.115 = 152.863 m²** **Ans.**

4. A boiler generates 8 kg of steam per kg of coal burnt at a pressure 10 bar and 45 °C super heat. If C.V. of coal is 30,000 kJ/kg, feed water 35 °C calculate.

   (i) Equivalent evaporation from and at 100 °C

   (ii) Factor of equivalent evaporation

   (iii) The boiler efficiency.

   Take specific heat of superheated steam as 2.1 kJ/kg K and specific heat of water as 4.187 kJ/kg K.

Solution:

Given $m_a = 8 = \dfrac{m_s}{m_f} = \dfrac{8}{1}$

$P = 10$ bar

Degree of super heat = 45 °C = $(t_{sup} - t_{sat})$ = 45 °C

C.V. of coal = 30, 000 kJ/kg

$t_w = 35$ °C

At 10 bar $h_g = 2776.2$ kJ/kg (From steam tables)

∴ enthalpy of 1 kg of superheated steam at 10 bar and 45 °C superheat

$$h = h_g + Cp_s\,(t_{sup} - t_{sat})$$
$$= h_g + Cp_s\,(\text{degree of super heat})$$
$$= 2776.2 + 2.1 \times 45$$
$$= 2870.7 \text{ kJ/kg}$$

$h_w = Cp_w\,(t - 0)$ as m = 1 kg

$= 4.187 \times 35 = 146.545$ kJ/kg

∴ M_e = equivalent evaporation = $\dfrac{m_a(h - h_w)}{2257}$

$$= \frac{8(2870.7 - 146.545)}{2257}$$

= 9.656 kg/kg of coal

Factor of equivalent evaporation = $\dfrac{h - h_w}{2257} = \dfrac{(2870.7 - 146.545)}{2257}$

= 1.207

% Boiler efficiency = $\dfrac{m_a(h - h_w)}{C.V} \times 100 = \dfrac{8(2870.7 - 146.545)}{30000}$

= 72.64% Ans.

5. Steam is generated at 10 bar and 270 °C in a boiler from feed water at 30 °C. The C.V of coal is 33707 kJ/kgK. Find (a) The factor of evaporation. (b) equivalent evaporation. (c) Boiler power if 260 kg of coal/hr is burnt Also calculate over all efficiency of plant if this boiler supplies steam to an engine in which the heat equivalent of brake work is 3010 kJ/kg of steam supplied. Steam generated per kg of coal is 9 kg.

Solution:

Given

P = 10 bar

$t_{sup} = 270$ °C

$t_w = 30$ °C

C.V. of coat = 33707 kJ/kg K

$m_a = 9 =$ units of actual evaporation $\frac{\text{kg of steam}}{\text{kg of coal}}$

$h = h_g + Cp_s\,(t_{sup} - t_{sat})$

$h_g = 2776.2$ kJ/kg at 10 bar from steam table

$t_{sat} = 179.9\ ^{o}C$ at 10 bar from steam table

$= 2776.2 + 2.1\,(270 - 179.9) = 2965.4$ kJ/kg K

(In problem Cp_s is not given assume 2.1 kJ/kg K)

$h_w = 125.8$ from steam corresponding to $30\ ^{o}C = h_f = 125.8$

OR

$h_w = 2.1 \times 30 = 125.6$ [by formula $Cp_s\,(t - 0) \times 1$]

$$\text{Equivalent evaporation} = \frac{m_a(h - h_w)}{2257}$$

$$= 9\frac{(2866.3 - 125.6)}{2257}$$

$= 10.929$ kg of steam/kg of coal

$$\text{But factor of evaporation} = \frac{h - h_w}{2257} = \frac{(2866.3 - 125.6)}{2257} = 1.214$$

Note : Factor of evaporation can also be calculated by formula i.e., equivalent evaporation $= m_a.f.$

$m_e = m_a f$

$$\therefore \quad f = \frac{10.929}{9} = 1.214 \quad \textbf{Ans.}$$

$$\text{Boiler power} = \frac{\text{Equivalent evaporation/hr}}{21.296}$$

= equivalent evaporation in kg of steam kg of coal × 260

$$= \frac{10.929 \times 260}{21.296} \times \frac{\text{kg of steam}}{\text{hr}}$$

$= 133.431$ kwh

$$\text{Overall efficiency} = \frac{\text{Heat equivalent of brake power}}{\text{Heat supplied in coal}}$$

$$= \frac{3010 \times (\text{kg of steam produced /kg of coal})}{\text{mf} \times \text{Heat supplied/ kg of coal}}$$

$$= \frac{3010 \times 9}{1 \times CV}$$

$$= \frac{3010 \times 9}{33707}$$

= 0.80369 i.e., 80.369% **Ans.**

6. Steam used by turbine is 6 kg/kWh at a pressure of 55 bar and temperature of 340 °C. The efficiency of boiler is 82% with feed water temperature 140 °C.

Calculate

(i) How many kg of coal are required per kWh

(ii) If the coal per ton is Rs. 5000 what is fuel cost/kWh

Take C.V. of coal = 28100 kJ/kg of coal

Solution: Given mass of steam used = m_g = 6 kg/kWh

P = 55 bar

t_{sup} = 340 °C

η boiler = 82%

Feed water temps = t_w = 140 °C

(i) $$\eta \text{ boiler} = \frac{m_s (h - h_w)}{m_f \times CV}$$

Note

(i) In above equation of m_s is taken in kg/kWh, we will get m_f also in kg/kWh.

(ii) Some time we are taking m_s in $\frac{kg}{hr}$ then we have to take m_f also in $\frac{kg}{hr}$.

From steam table at 55 bar, h_g = 2789.9 kJ/kg

t_{sat} = 269.9 °C

Therefore $h = h_g + Cp_s (t_{sup} - t_{sat})$

For superheated steam = h; = 2789.9 + 2.1 (340 – 269.9)

assuming Cp_s = 2.1 kJ/kg K

= 2937.11 kJ/kg

$$h_w = 4.1868 \times (140 - 0) = 586.152 \text{ kJ}$$

$\therefore$ $$0.82 = \frac{6(2937.11 - 586.152)}{m_f \times 28100}$$ [because ms is in kg/kWh

m_f = 0.612 kg/kWh we get m_f also in kg/kWh **Ans.**

(ii) Cost of fuel (coal)/kWh

$$= \frac{0.612}{1000} \times 5000 \quad [\because \text{ one tone} = 1000 \text{ kg}]$$

$$= 0.306 \text{ Rs.}$$

= **30.6 paise Ans.**

7. A boiler generates 7 kg of steam/kg of coal at a pressure of 15 bar from feed water temperature of 75 °C. The efficiency of boiler is 80% and factor of evaporation is 1.16. Specific heat of steam is 2.3 kJ/kg K and Cp_w = 4.18 kJ/kg.

Calculate

(i) Degree of super heat and temperature of steam generated.

(ii) C.V of coal in kJ/kg

(iii) Equivalent evaporation in kg of steam per kg of coal.

Solution: Given

$$m_a = 7 \text{ kg of steam/kg of coal}$$

$$P = 15 \text{ bar}$$

Temperature of feed water = t_w = 75 °C

Boiler efficiency = 80%

(i) Factor of evaporation = 1.16

$$Cp_s = 2.3 \text{ kJ/kg K}$$

At 15 bar from steam table h_g = 2789.9

Note : Whenever degree of super heater is asked and factor of evaporation is given in problem, apply the formula as given below and get the required answer.

$$\text{Factor of evaporation} = \frac{h - h_w}{2257}$$

$$= [h_f + h_{fg} + Cp_s (t_{sup} - t_{sat}) - h_w]$$

$$1 \ 1.16 = \frac{h_g + Cp_s (t_{sp} - t_{sat}) - h_w}{2257}$$

$$1.16 = \frac{2789.9 + 2.3(t_{sup} - t_{sat}) - 4.18 \times 75}{2257}$$

$$\text{Degree of super heat} = (t_{sup} - t_{sat}) = \frac{1.16 \times 2257 - 2789.9 + 4.18 \times 75}{2.3}$$

$$= \mathbf{259.92\ ^{\circ}C} \qquad \textbf{Ans.}$$

$$\text{(ii)} \quad \eta \text{ boiler} = \frac{m_a (h - h_w)}{C.V} = \frac{7[(h_g + Cp_s(t_{sup} - t_{sat}) - 4.18 \times 75)]}{C.V}$$

$$C.V. = \frac{7[2789.9 + 2.3(61.62) - 313.5]}{0.80}$$

$$= \mathbf{22908.6\ kJ/kg\ of\ coal} \quad \textbf{Ans.}$$

8. Boiler produces 2000 kg of dry saturated steam per hour at a pressure of 19 bar and feed water is heated by an economiser to temperature of 110 °C. 240 kg of coal of calorific value of 33500 kJ/kg are fired per hour. And it is found that 10% of coal is unburnt. Find thermal efficiency of boiler and also of boiler and grate combined.

Take specific heat of water as 4.187 kJ/kg k.

Solution: Given

$m_s = 2000$ kg/hr of dry saturated steam is produced

$P = 19$ bar

$t_w = 110$ °C

$m_f = 240$ kg/hr

C.V. of coal = 33500 kJ/kg

Find (i) boiler efficiency

(ii) boiler and grate combined

$$\text{Boiler efficiency} = \frac{m_s(h - h_w)}{m_f \times C.V} = \frac{m_a(h - h_w)}{1 \times C.V} \qquad [m_f = 1]$$

If m_s is kg/hr then m_f must be taken in kg/hr actually burnt.

From steam table corresponding to 19 bar and dry-saturated steam,

$h_g = 2796.1$ kJ/kg (from steam table corresponding to 19 bar)

$$h_w = 4.187 \times (110 - 0) = 460.57$$

m_f = coal actually burnt

$$= \text{mass of coal fire} - \frac{10}{100} \times \text{mass of unburnt}$$

$$= 240 - \frac{10 \times 240}{100} = 240 - 24 = 216 \text{ kg/hr}$$

$$\eta \text{ boiler} = \frac{2000 \times (2796.1 - 460.57)}{216 \times 33500}$$

= 0.645 i.e., 64.5% Ans.

Thermal efficiency of the boiler and grate combined

$$= \frac{m_a (h - h_w)}{C.V}$$

$$= \frac{2000 \times (2796.1 - 460.57)}{240 \times 33500}$$

= 0.5809 i.e., 58.09% Ans.

9. A boiler generates 1600 kg of steam per hour at a pressure of 11 bar and with 60 °C superheat and burns 200 kg of coal per hour. If C.V. of coal is 30000 kJ/kg, feed water temperature 35 °C,

Calculate

(i) The factor of equivalent evaporation

(ii) Equivalent evaporation per kg of coal

(iii) The boiler efficiency

Take specific heat of steam = 2.1 kJ/kg K

Specific heat of water as 4.187 kg/kg K

Solution: Given $m_s = 1600$ kg/hr

$P = 11$ bar

$m_f = 200$ kg/hr

CV of coal = 30000 kJ/kg K

$Cp_s = 2.1$ kJ/kg K

$Cp_s = 4.187$ kJ/kg K

$T_w = 35$ °C

From steam table at 11 bar $h_g = 2779.2$ kJ/kg

$h_w = 4.187 \times (35 - 0) \times 1$

$= 146.545$ kJ/kg

Factor of equivalent evaporation = $(h - h_w) / 2257$

$$= \frac{h_g + Cp_s(t_{sup} + t_{sat}) - h_w}{2257}$$

$$= \frac{2779.7 + 2.1 \times 60 - 146.545}{2257}$$

= 1.224 Ans.

$$M_e = \frac{m_a(h - h_w)}{2257} \text{ OR}$$

$$M_e = m_a.f = \frac{\text{Mass of steam / hr}}{\text{Mass of coal / hr}} \times f$$

$$m_e = \frac{1600}{200}.f$$

$= 8f$

$= 8 \times 1.224$

= 9.794 Ans.

$$\eta \text{ boiler} = \frac{M_a(h - h_w)}{1 \times C.V}$$

$$= \frac{8 \times \lfloor h_g + Cp_s(t_{sap} - t_{sat}) - h_w \rfloor}{1 \times C.V}$$

$$= \frac{8[2779.7 + 2.1 \times 60 - 146.545]}{30000}$$

$= 0.7358$

= 73.58% Ans.

10. Calculate the power of boiler which produced 1600 kg of steam per hour. The feed water temperature 110 °C. The dry saturated steam produced at 9 bar.

Sol: Given $m_s = 1600$/kg of coal per hour

$t_w = 110$ °C

Pressure = 9 bar

From steam table the enthalpy of dry saturated steam at 9 bar,

$$h_g = 2772.1 \text{ kJ/kg}$$

$$h_w = 1 \times 4.1867 \times (110 - 0)$$

$$= 460.57 \text{ kJ/kg}$$

$$\text{Power} = \frac{m_a (h - h_w)}{2257 \times 21.296}$$

$$= \frac{1600 (2772.1 - 460.57)}{21.296 \times 2257}$$

$$= 76.95 \text{ kw} \qquad \textbf{Ans.}$$

11. The following data are provided for two boilers:

	Steam pressure	Quality of steam	Evaporation rate
Boiler 1	8 bar	Dryness fraction 0.9	8.5 kg/kg of coal
Boiler 2	19.5 bar	Superheated to 295 °C	7 kg/kg of coal

Both the boilers are supplied with feed water temperature at 40 °C. Determine which boiler has the comparatively higher heat utilization rate per kg of coal fired and with same thermal efficiency.

Take $Cp_w = 4.1867$ kJ/kg K and $Cp_s = 2.1$ kJ/kg K

Solution: For boiler 1

$$P = 8 \text{ bar}$$

$$D.F = x = 0.9$$

$$m_a = 8.5 \, \frac{\text{kg of steam}}{\text{kg of coal}}$$

$$t_w = 40 \text{ °C}$$

At 8 bar from steam table

$$h = h_f + xh_{fg}$$

$$h_f = 720.9 \text{ kJ/kg}, \; h_{fg} = 2046.5 \text{ kJ/kg}$$

$$= 720.9 + 0.9 \times 2046.5$$

$$= 2562.75 \text{ kJ/kg}$$

$$h_w = 1 \times 4.187 \times 40 = 167.48 \text{ kJ/kg}$$

Equivalent evaporation for boiler 1

$$= \frac{m_a (h - h_w)}{2257}$$

$$= \frac{8.5(2562.75 - 167.48)}{2257} = 9.02$$

For boiler 2 $P = 19.5$ bar

$t_{sup} = 295\ ^\circ C$

$m_a = 7$ kg/kg of coal

$t_w = 40$

From steam table corresponding to 19.5 bar

$h_g = 2796.7$ kJ/kg

$t_{sat} = 211.1\ ^\circ C$

$h_w = 167.468$ kJ/kg

Equivalent evaporation for boiler 2

$$= \frac{m_a \left[h_g + Cp_s (t_{sup} - t_{sat}) - h_w\right]}{2257}$$

$$= \frac{7\left[2796.7 + 2.1(295 - 211.1) - 167.46811\right]}{2257}$$

$$= 8.7 \frac{\text{kg of steam}}{\text{kg of coal}}$$

Boiler 1 has comparatively higher utilization rate and thermal efficiency because equivalent evaporation of boiler 1 is greater than boiler 2.

12. The following were the observations made on boiler plant during one hour test.

Steam pressure 18 bar, the steam temperature 270°C, steam generated 37500 kg. Temperature of water leaving the economiser 90°C, fuel used 4400 kg, energy of combustion of fuel 30000 kJ/kg. If inlet temperature of water to economiser is 15 °C,

calculate

(i) Equivalent evaporation per kg of fuel

(ii) Thermal efficiency of plant

(iii) Factor of equivalent evaporation

(iv) Percentage heat energy of the fuel energy utilized by economiser

Solution:

$$P = 18 \text{ bar}$$

$$t_{sup} = 270\ ^\circ C$$

$$m_s = 37500\ \frac{kg}{hr} \qquad (\because \text{ observations were made for one hour})$$

Economiser inlet temperature = 15 °C

Economiser outlet temperature = 90 °C

$$m_f = 4400 \text{ kg/hr} \quad (\because \text{ duration of observation is one hour})$$

CV of fuel = 30000 kJ/kg

$$m_a = \frac{m_s}{m_f} = \frac{37500}{4400} = 8.52 \text{ kg of steam/kg of coal}$$

Note: In this problem Cp_w is not given we have to assume as 4.1867 kJ/kg or corresponding to given temperature of water sensible heat h_f can be seen from steam table. Similarly, Cp_s is also not given in problem, we have to remember it, Cp_s = 2.1 kJ/kg K or from steam table corresponding to 18 bar and 210 °C enthalpy can be seen directly.

From steam table, corresponding to 15 °C feed water temperature.

$$h_{w1} = h_{f1} = 63 \text{ kJ/kg}$$

OR by formula assuming Cp_w = 4.1867 kJ/kg K,

$$h_{w1} = C_{pw}(15 - 0) \times 1 = 4.1867\,(15 - 0) \times 1 = 62.8 \text{ kJ/kg}$$

(Value seen from steam table gives little higher value than calculated by formula as seen here)

$$h_{f2} \text{ or } h_{w2} = 376.9\ \frac{kJ}{kg} \quad \text{[from steam table corresponding to 90°C]}$$

By formula h_{f_2} or h_{w_2} = 4.1867 (90 – 0) × 1 = 376.803 KJ/kg

h_g at 18 bar from steam table = 2794.8 kJ/kg and t_{sat} = 207.1 °C

Therefore enthalpy of 1 kg of superheated steam

$$h = 2794.8 + Cps\,(t_{sup} - t_{sat})$$

$$= 2794.8 + 2.1\,(270 - 207.1)$$

(where t_{sat} = 207.1 °C from steam table at 18 bar]

$$= 2926.89 \text{ kJ/kg}$$

(i) Equivalent evaporation $= \dfrac{m_a(h - h_w)}{2257}$

$$= \frac{8.52(2926.89 - 62.8)}{2257} = 10.8 \text{ kg/kg of coal}$$

Note: h_w should be corresponding feed water temperature at 15 °C and not 90 °C when economiser is given in problem.

(ii) Thermal efficiency $= \dfrac{m_a(h - h_w)}{C.V} \times 100$

$$= \frac{8.52(2926.89 - 62.8)}{30000} \times 100$$

= 8134% **Ans.**

(iii) Me = ma.f where f = factor of evaporation

$\therefore$ $\mathbf{f = \dfrac{10.8}{8.52} = 1.268}$ **Ans.**

(iv) Heat energy utilized by economiser per kg of fuel

$= m_a (h_{f2} - h_{f1})$ OR $m_a(h_{w2} - h_{w1})$

$= 8.52\ (376.9 - 62.8)$

$= 2675$ kJ

Percentage heat utilized by economizer per kg of fuel $= \dfrac{2675}{30000}$

= 0.089 or 8.9% **Ans.**

13. A boiler generates 1000 kg/hr of dry saturated steam at 9 bar. Feed water is heated by economiser before it is supplied to boiler. If feed water temperature is 34°C and leaves at 94°C, find the percentage saving in heat by the use of economiser.

If 2500 kg per hour of flue gas pass over the economiser at 300°C and leaves the economiser at 160 °C, calculate effectiveness (efficiency) of economiser. Take specific heat of flue gases as 1.005 kJ/kg K and specific heat of water as 4.187 kJ/kg K.

Solution: Given $m_s = 1000$ kg /hr

Dry saturated steam is produced at P = 9 bar

Feed water temp. inlet to economiser = 34 °C

Feed water temp. leaving the economiser = 94 °C

Find % saving in heat by economiser

Note: In order to find % saving in heat by economiser:

(i) First calculate heat required to produce 1 kg of dry saturated steam without economiser, i.e. Feed water at 34 °C

(ii) Find heat required to produce 1 kg of dry saturated steam with the help of economiser, when temperature of water entering in boiler is at 94 °C.

(iii) Saving in heat due to economiser

= Enthalpy required to produce 1 kg of steam when water temperature is 34 °C – enthalpy of steam when feed water temperature is 94 °C

(iv) % saving in heat by use of economiser

$$= \frac{\text{Saving in heat due to economiser}}{\text{Total heat of steam without economiser}} \times 100$$

Without Economiser

From steam table corresponding to 9 bar, $h_g = 2772.1$ kJ/kg.

h_f = Enthalpy of entering water without economiser

$= C_{pw}(t - 0) \times 1$

$= 4.187\ (34)$

$= 142.36$ kJ/kg

Net Enthalpy required and to produce I kg of Dry saturated steam at 9 bar when feed water temperature is 34 °C $= h_g - h_f$

$= 2772.1 - 142.36$

$= 2629.74$ kJ/kg

Also net enthalpy required to produce 1 kg of steam at 9 bar when feed water temperature is 94 °C.

$= 2772.1 - 4.187 \times 94$

$= 2378.52$ kJ/kg

% saving in heat by use of economiser

$$= \frac{\text{saving in heat due to economiser}}{\text{enthalpy of steam without economiser}}$$

$$= \frac{2629.74 - 2378.52}{2629.74}$$

$= 0.0955 =$ **9.55%** **Ans.**

(i) Heat received by feed water in Economiser from flue gases

$$= 1000 \times 4.187\ (94 - 34) = 251220 \text{ kJ/hr.}$$

(ii) Heat rejected by flue gases in economiser

$$= 2500 \times 1.005\ (300 - 160)$$

$$= 351750 \text{ kJ/hr}$$

Effectiveness (efficiency) or economiser

$$= \frac{\text{Actual heat utilized by economiser}}{\text{Heat siupplied by flue gas in economiser}}$$

$$= \frac{251220}{351750} \times 100$$

= 71.42 % Ans.

14. A boiler evaporates 9 kg of water per kg of coal fired from feed water 45°C working at 9.6 bar absolute. Determine equivalent evaporation from and at 100°C per kg of coal fired.

1. The steam produced is 0.95 dry
2. If the steam produced is dry saturated
3. If the steam produced is superheated to 230 °C

Assume (Cp_s) specific heat of steam = 2.1 kJ/kg K

Solution:

Given $m_a = 9$ kg/kg of coal

$t_w = 45°C$

$P = 9.6$ bar

$h_w = 4.1867\ (45 - 0) \times 1 = 188.4$ KJ/kg

At 9.6 bar from steam table

$h_f = 754.8$ kJ/kg K

$h_{fg} = 2019.8$ kJ/kg

$h_g = 2774.6$ kJ/kg

$t_{sat} = 178.1°C$

1. $$M_e = \frac{m_a (h - h_w)}{2257}$$

$$= \frac{m_a \left[h_f + xh_{fg} - 4.1867 \times 45\right]}{2257}$$

$$= \frac{9(754.8 + 0.95 \times 2019.8 - 4.1867 \times 45)}{2257}$$

= 9.9 kg/kg of coal **Ans.**

2. $$M_e = \frac{M_a (h - h_w)}{2257}$$

$$= \frac{9(2774.6 - 188.4)}{2257}$$

= 10.31 kg/kg of coal **Ans.**

3. $$M_e = \frac{M_a \left[h_g + Cp_s (t_{sup} + t_{sat}) - h_f\right]}{2257}$$

$$= \frac{9[2774.6 + 2.1(230 - 178.1) - 188.4]}{2257}$$

= 10.75 kg/kg of coal **Ans.**

6.19 Mountings

They are fittings primarily used for safety of boiler and control of steam generation process completely according to Indian Boiler Act 1923, they are as follows:

1. Safety valves
2. Water level indicator (known as water gauge)
3. Pressure gauge
4. Steam stop valve (known as junction) valves
5. Fusible plug
6. Feed check valve
7. Blow off cock.

1. **Safety Valves**

   Function: The function of a safety is to blow off the steam from boiler to atmosphere when pressure of steam exceeds the rated pressure (working pressure). The safety valve automatically opens and excess steam rushes out from boiler to atmosphere till the pressure in boiler drops down to working pressure or rated pressure.

   Generally valve is pressed against seat through spring, screw or strut by external weight.

The blow off steam through the safety valve creates a noise which serves as warning to the operator. As per boiler regulation every boiler must be fitted with two safety valves. There are following types of safety valves.

1. Dead weight safety valve.
2. Lever safety valve
3. Spring loaded safety valve
4. High steam and low water safety valve.

Dead Weight Safety Valve

Function : The function of a Dead weight safety valve is to release excess steam when the pressure of steam inside boiler exceeds working pressure.

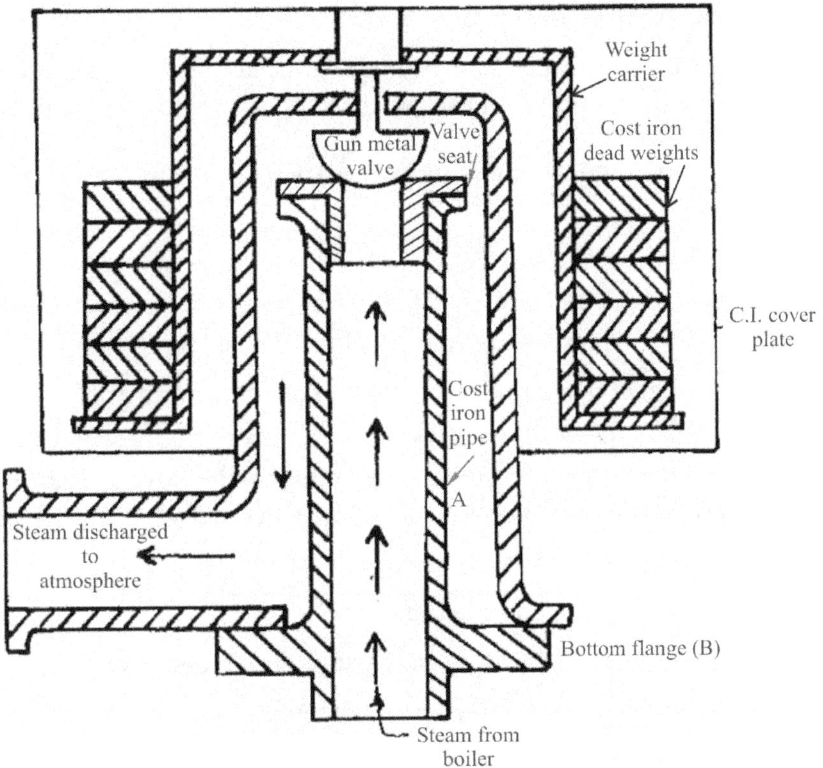

Fig. 6.9 Dead weight safety valve.

Location

It is mounted to the mounting Block on the top of the boiler shell communicating to the steam space.

Construction

"A" is vertical cast iron pipe through which steam pressure acts. "B" is bottom flange of vertical steam pipe is attached at top of the boiler shell. At other end of steel pipe a gun metal valve seat is screwed to it and gun metal valve rest on it. The valve is secured by means of nut to a large cast iron weight carrier which hangs, freely over vertical pipe like cover, which acts as weight carrier. The dead weight consists of cast iron rings are placed on weight carrier to keep the valve on its seat against the working steam pressure inside boiler.

Working

At normal working of boiler, the upward force exerted by steam is balanced by the downward force plus weight of valve equivalent to weight put on valve. Total load on valve is equal to weight of carrier, plus weight of cover. The steam will not blow off from boiler.

When the steam pressure in boiler exceeds the working pressure, it lifts the valve with its weight and allows steam to scape from boiler to the atmosphere.

For equilibrium condition $W = \frac{\pi}{4} d^2 P$

where W = Total weight in Newton

d = diameter of valve in meter

P = pressure of steam in N/m^2

Lift of valve is controlled by studs. The moment of valve is guided by feather (not shown Fig. 6.9).

The center of gravity of dead weight of safety valve is considerably below the valve, which ensures that the load hangs vertically.

Advantages of dead weight safety valve

1. Very simple in construction
2. Give quite satisfactory performance during operation
3. From the pressure adjustment point of view, it can not be easily tempered.

Disadvantage

1. Unsuitable for use on any boiler where extensive vibrations and movement are experienced (Locomotive and marine).
2. It is not suitable for high pressure boiler because large amount of weights are required to balance the steam pressure.

Uses

It is mainly used for low pressure, low capacity and stationary boilers. For example Cornish and Lansachire boilers.

Spring Loaded Safety Valve (Rams Bottom Spring Loaded Safety Valve)

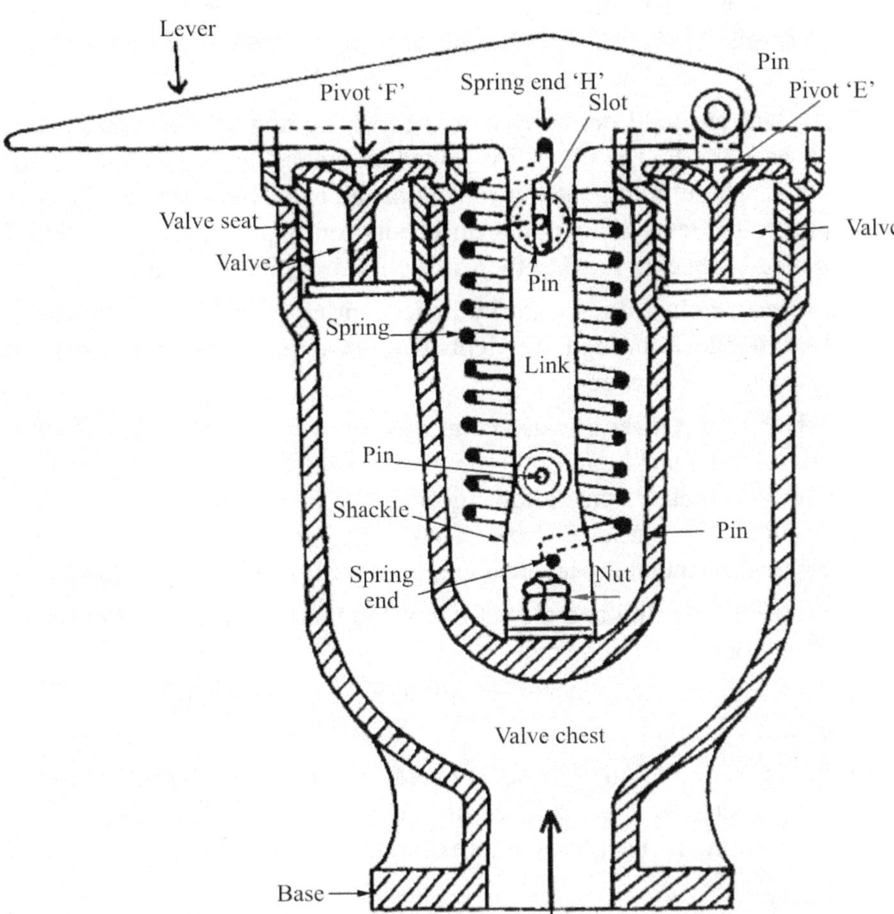

Fig. 6.10 Spring loaded safety valve.

Function

Its function is to maintain constant steam pressure inside the boiler. Apart from this function, the discharge capacity should be equal to evaporative capacity so that frequent built up of pressure does not occur which is placed in boiler.

The flange of body of valve is bolted to a mounting block of boiler shell at top of shell.

Construction

It consists of a cast iron body connected to top of boiler. Cast iron body has two valve chests for flow of steam, it has two separate valves of same size and they have separate seats in upper ends of hollow valve chest.

The valves are hold down by means of spring and a lever. The lever has two pivots at E and F. The pivot E is joined by pin to the lever while F is forged on the lever. These pivots rest on centers of the valves. The upper end of spring is hooked to the lever at its mid span H between two pivots. While the lower end is hooked to shackle fixed to the valve chest by studs and nuts.

The lever is pulled downward by helical spring on half of spring force acts on each of the valve through the pivots because upper end of the spring is fixed to lever to midway between valves.

The shackle and lever are also connected by two links, one end of which is pin jointed to the shackle and other end has a slot cut in to it, to allow for pin to slide in it vertically, thus allowing the lever to be lifted and retaining the link connection. The purpose of these links is to act as safety measure in case of spring breaks so that the lever should not fly off its seat.

The tension of spring and the setting of the blow off pressure can be altered by loosening or screwing the nut.

The checking of valve is done by lifting the lever manually and blowing off the steam to ensure and test that seats are not sticking.

Spring is made of round or square steel rod in helical form, spring may be in tension or compression as pressure acts along the axis of spring. In actual practice, spring is placed in compression.

Working

When the pressure of steam exceeds the safe limits, the upward thrust of steam raises the valve from its seat against the spring force. This allows the steam to escape till the pressure of steam, falls back to normal value. The valve then returns back to its original position.

The lever has an extension which projects in to driver cabin. The driver can release the pressure if required just by raising the lever.

Uses

- This type of safety valve is commonly used now a days for stationary as well as mobile boilers.
- The safety valve is unaffected by jerks and vibration therefore this type of valve is used for locomotive and marine services.

Advantages

1. This valve is much light and compact compare with other safety valves.
2. Elimination of heavy weight.
3. Easy maintenance and examination.
4. It is not affected by jerk and vibration.

High Steam low water safety valve

(Happkinson's high steam and low water safety valve)

Function

1. Its function is to maintain constant steam pressure inside the boiler. If the boiler pressure increases the working pressure, it automatically discharges the excess steam in the atmosphere.
2. It also automatically maintains constant water level inside the boiler. If the water level in the boiler goes below the certain required level, then steam automatically discharged into atmosphere. This raiser alarm so that corrective action can be taken by pumping in water.

Location

It is fitted on mounting block on top of the boiler shell.

Construction and Working

It is actually combination of two valves. One of which is lever safety valve which blow off steam when working pressure of steam exceeds than normal value. Second valve operates by blowing off steam when water level in the boiler falls below the predetermined level. This raises an alarm, so that corrective action is taken by pumping in water.

It consists of two valves "V" and "U", of which "V" is resting on valve seat VS and "U" loaded with "W" rests on the valve "V". The valve "V" acts as seat for valve "U".

To maintain the constant pressure of steam in the boiler it acts like simple lever safety valve. Lever L_1 is hinged at one end and weight W_1 is hung from the other end as shown in the Fig. 6.11. F_1 is fulcrum of lever L_1. Pivot P. is placed on valve "V" and when pressure exceeds than normal limit valve "V" is lifted along with Lever L_1 and the valve U. So steam escapes through passage between the valve seat and the valve V till the pressure becomes normal.

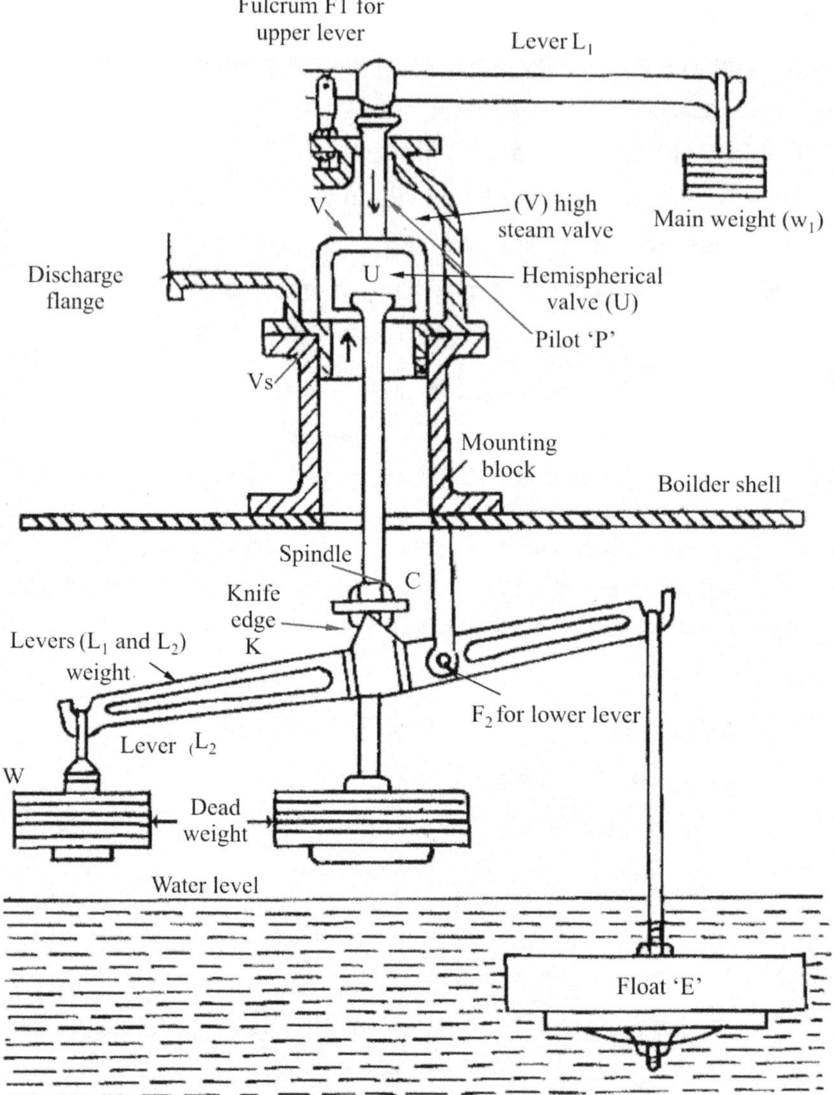

Fig. 6.11 High steam low water safety valve.

Inside boiler shell a lever L_2 is hinged at the point F_2 as shown in the Fig. 6.15, Weight W is connected to one end of the lever and on the other end float E is attached.

The knife edge K touches a collar C of the spindle of the valve "U". Under normal level the Float E is submerged in water, the level L_2 is balanced about fulcrum F_2. When level of water falls and Float E is uncovered therefore the weight of float increases i.e., (weight of Float in air is more than when it is submerged in water. So buoyancy for it is decreased as it is partly uncovered.) Lever L_2 becomes unbalanced. To balance the lever L_2 Float moves downward direction. This causes the knife edge K lift the collar C with valve "U" in upward direction. So, steam escapes through the passage between valves V & valve U. The escaping steam causes noise. Noise warns the operator about the low level of water in the boiler, so that corrective action may be initiated.

Uses

Generally used in Lancashire boiler.

Lever Safety Valve

Function

The function of lever safety valve is to blow off the steam from boiler to atmosphere when pressure of steam exceeds the working pressure. It maintains constant safe pressure inside the steam boiler.

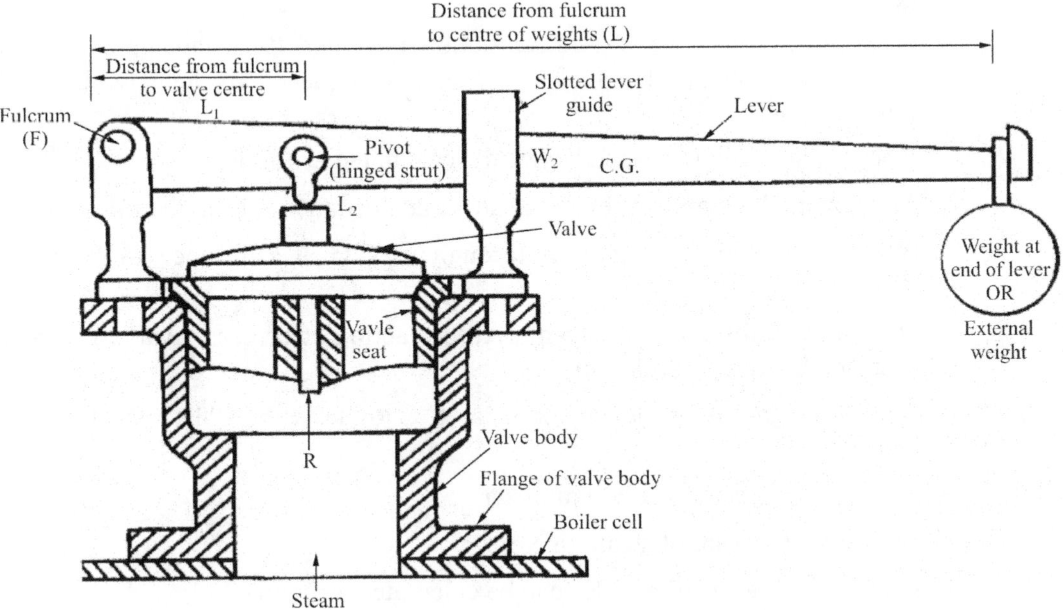

Fig. 6.12 Lever safety valve.

Location

It is mounted on top of boiler shell communicating to steam space.

Construction

It consists of valve body with flange fixed on steam boiler. Bronze steam valve-seat is screwed to the body. The valve also made of bronze is placed over valve seat. It may be noted that by using the valve and seat of same material, rusting is considerably reduced.

Valve is held by mild steel or wrought iron lever fulcrum at one end and loaded at the other end by an external weight. The thrust is applied to the valve through strut, placed near to fulcrum and hinged to the lever. The Guide keeps the lever in vertical plane. The load is properly adjusted at the other end of the lever according to steam pressure below the value.

The weight is firmly secured to lever by pin and locked, so that the unauthorized person should not displace it.

Working

When the pressure of steam exceeds the safe limit than the upward thrust of steam lifts the valve from its seat and allows the steam to escape till the pressure falls back to the normal value. The valve returns back to its original closed position.

Uses

This type of safety valve is used for stationery boiler. Its advantage over dead weight safety valve is that its large weight is replaced by lever with small weight at its end.

The external weight suspended can be calculated as follows

W = external weight suspended on lever at a distance L from the fulcrum

W_1 = weight of valve passing through its C.G at a distance L_1 from the fulcrum.

R = upward force due to steam pressure acting at center of valve i.e., at a distance L_1 from the fulcrum.

W_2 = Weight of the lever acting at C.G of lever at a distance L_2 from fulcrum.

where d = diameter of valve in meter

P = Pressure of steam in N/m^2

Magnitude of external force "R" can be calculated as below:

Taking moment about fulcrum "F" we have

Clockwise moment due to weights = Anticlockwise moment due to steam pressure

$$\downarrow WL + \downarrow W_1L_1 + \downarrow W_2L_2 = RL_1 \uparrow$$

2. Water Level Indicator (Also Known as Water Gauge)

Function

The function of water level indicator is to indicate level of water in the boiler constantly.

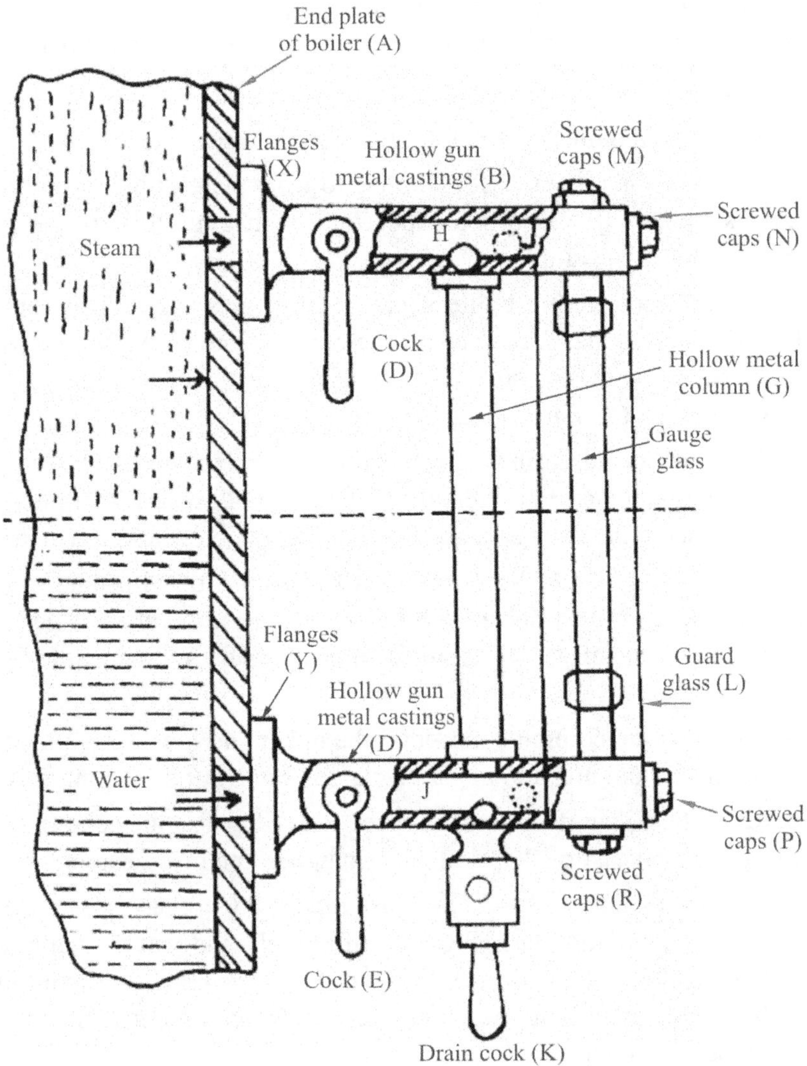

Fig. 6.13 Water level indicator.

Location

It is mounted in front of boiler in such a way that it is visible to the operator.

Generally two water level indicators are used in boiler having capacity 135 kg/hr of steam or more. Smaller boilers having steam capacity less than 135 kg/hr are provided with one water level indicator because there is no sufficient space to mount two water indicators on the boiler.

If boiler drum is situated at considerable height from floor, the water level indicator is fitted in inclined position to make the water level visible from any position.

Construction

A – end plate of boiler to which level indictor is attached.

B/C – Hollow Gun metal casting bolted to front plate of boiler by two flanges. One flange is bolted to plate portion exposed to steam and other flange is bolted to plate portion exposed to water.

D/E – Cocks connecting boiler shell space to glass tube. The vertical downward position indicates that D & E cocks are open. Steam cock "D" keeps the glass tube in connection with the steam space and water cock E puts the glass tube in connection with the water space in boiler. .

K – This cock is closed in vertical position and opens in horizontal position. Drain cock to blow out water at intervals so as not to allow any sediment to accumulate. This also ascertains the steam and water cocks are clear and in working condition.

F – Gauge glass should be very tough and strong to withstand at the pressure of steam in boiler. When cocks D & E are open, the level of water is indicated by gauge glass. Normally there is mark on gauge glass showing safe level of water in boiler.

G – Hollow metal column connecting the upper and lower gun metal casting. Level of water in this column and in gauge glass is same but in column it is not visible.

H and J – The rectangular passage at the ends of glass tube contain two balls, one in lower gun metal casting "C" and other in upper gun metal casting "B".

L – It indicates guard glass. The glass tube (gauge glass) is protected with guard glass. It is very tough glass. Property of this glass is that it crumbles and does not throw out splinters on braking. Thus when gauge glass breaks and this guard glass which normally will hold the flying pieces, also give away, thus pieces will not fly out and hurt the operator or attendant.

M, N, P and R – They are screwed caps for internal cleaning of passages after dismantling the water level indicator.

X – Y: are flanges for bolting the indicator on plate A. Front plate is provided with holes with inside threads for screwing castings for the observation of water level in the boiler. The steam and water cocks are opened and drain cock is closed. In this case all three handles are put in vertical position as shown in Fig. 6.12. The steam enters from the upper end of the glass tube and water enters the lower end of glass tube. So, water level inside the boiler will be same as seen in the glass tube.

A red mark is put on glass which indicate safe water level in the boiler.

Working: There is an arrangement for Automatic Shutoff steam and water supply to the glass tube in event of damage to glass tube. For this gun metal tube is connected between upper and lower gun metal castings.

The rectangular passage at the end of the glass tube contains two balls. In case glass tube is broken, the water rushes from the bottom casting and steam comes out from the top casting.

The flow of water carries bottom ball into the dotted position and shut off the water. Similarly the flow of water rises in the hollow gun metal column G. This water and steam rushing through the upper passage will carry the ball "H" from its seat and placed in new position shown by dotted circle and shutoff steam pressure passing in to the glass tube. Operator can now close the cocks B & C for repair and replacement of glass tube.

When the steam boiler is not working the water level indicator can be taken out from boiler for cleaning purpose. The water tube is kept free from leaking by means of conical ring and the gland nut.

Following are the sequences of operations performed, during line up of gauge glass.

1. Drain cock is opened.
2. Water cock is crack opened.
3. Steam cock crack opened.
4. Drain cock is fully closed
5. Steam cock and water cock are fully opened.

Why boiler is provided with two dependable gauge glasses (or water level indicator) : If one fails to accidental breakage the other one will continue to be in line and show the water level.

Pressure Gauge

Function

- The function of pressure gauge is to indicate the steam pressure of the boiler in kg/cm^2 or bar gauge.
- It reads pressure of steam above atmosphere
- The actual pressure of steam = gauge pressure + Barometer reading
- The common type of pressure gauge used is known as Bourdon pressure gauge.

- Graduation on dial is made twice the pressure it is normally to work. The graduation are so made that the normal pressure is marked with read colour at centre of graduation.

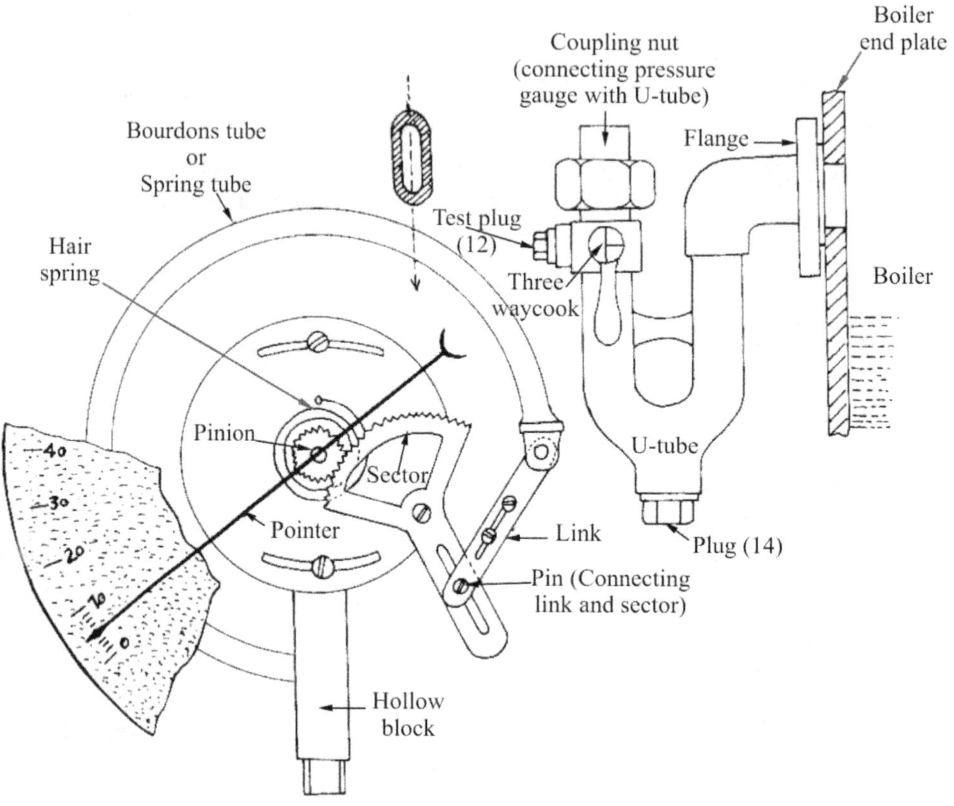

Fig. 6.14 Pressure gauge.

Location in Boiler

It is fixed in front of steam boiler and in steam space in drum.

Construction

It consists of various components a as below

1. Spring tube (Known as Bourdon tube)
2. Hollow Block
3. Link
4. Sector

5. Pinion
6. Pin (connecting link and sector)
7. Pointer
8. Hair spring
9. U-tube
10. Flange
11. Coupling nut (for connecting pressure gauge with U-tube)
12. Test plug (for connecting inspector's gauge with U-tube)
13. Three way cock
14. Plug

Spring tube is also known as elliptical elastic tube or Bourdon's tube. It is made up of high quality phosphor bronze. It is elliptical in cross section, one open end of spring tube is connected to hollow block and other end is closed and is connected with sector by mean of link and pin. The toothed sector meshes with pinion and pinion is fixed on spindle. The pointer as shown in Fig. 6.13 is also attached to the spindle to read the pressure on the dial.

There are two types of pressure gauges (i) Bourdon's tube pressure gauge (ii) Diaphragm type pressure gauge.

A pressure gauge is known as compound pressure gauge, as it is designed in such fashion so as to measure pressure above and below the atmosphere on the same dial.

U-tube siphon which connects the pressure gauge to the boiler. U-tube siphon is connected to steam space of boiler and contains condensed steam which enters in gauge tube. In short whole siphon and gauge tube is full of condensed water.

Condensed water in the U-tube siphon prevents the steam from coming in contact with spring tube otherwise it could over heat the spring-tube and induce thermal stresses in it. This may ultimately spoil accuracy of the pressure gauge and reading obtained from pressure gauge will be false. Test plug "12" is used for connecting the inspector's standard pressure gauge and testing accuracy of boiler pressure gauge while in service. Plug "14" is employed for cleaning the siphon.

Working

The pressure of steam acting through water on the inside of spring tube tries to make the tube circular in cross-section, since one end is fixed other end moves. The moment is proportional to the difference between internal and external pressure on the spring tube. (in side of tube subjected to steam pressure and outside of tube is subject to atmospheric pressure).

The movement of closed end of tube is transmitted and magnified by link and sector. This causes pinion to move but pinion is attached to spindle so spindle rotates and pointer is fixed on spindle, therefore pointer moves. The magnitude of the movement is indicated by pointer on a calibrated scale which directly gives the gauge pressure of steam. To get actual (known absolute) pressure of steam, the atmospheric pressure (1.013 bar) is added in the reading.

4. Steam Stop Valve (Junction Valve)

The function of a steam stop valve is:

1. To control the flow of steam from boiler to the main steam pipe.
2. To shut off the steam completely when required.

Located in Boiler

Steam stop valve is usually fitted to highest part of boiler shell i.e., Top of shell.

Construction

Steam stop valve and junction valve are essentially the same. The smaller sizes are called steam stop valve and the larger sizes are called junction valves. When valve is mounted on top most portion of boiler shell, it is called junction, valve when it is connected to steam pipe, leading to steam turbine or steam engine is called steam stop valve.

Commonly used steam stop valve is shown in Fig. 6.13. Main body is made up of cast steel. The valve, seat and the nut, through which the valve spindle works, is made up of brass or gun metal for smooth working.

The valve in form of disc rests over valve seat. The valve disc is screwed to the spindle by nut. The spindle is rotated by means of a hand wheel. The upper portion of the spindle is screwed and made to pass through nut in a cross head, carried by two pillars. The pillars are screwed in cover of the body as shown in Fig. 6.13.

Working

Under normal working condition, the valve is opened with help of a hand wheel, and the steam flows from boiler to the steam pipe. But when the steam is not required, the valve is closed by operating the handle.

Due to the rotation of hand wheel, valve may move up and down; it may open the passage fully or partially for the flow of steam.

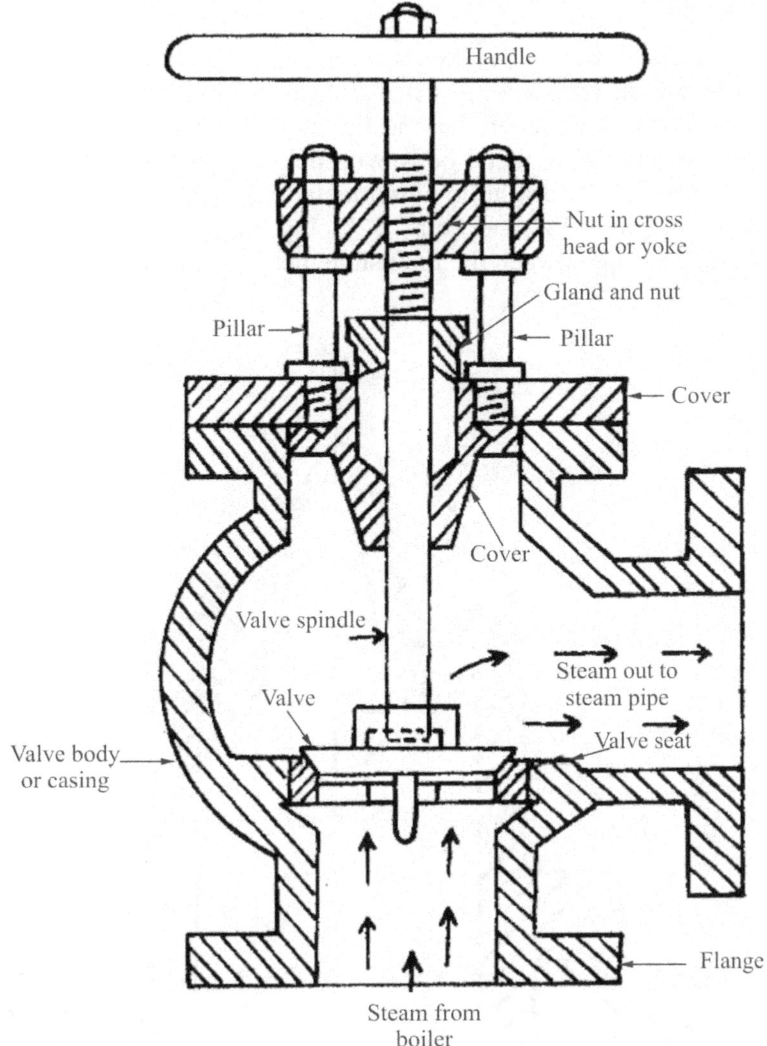

Fig. 6.15 Steam stop valve (junction valve).

5. **Fusible plug**

   Function

   The function of fusible plug is to put-off the fire in furnace of the boiler, when water level in the boiler falls below an unsafe level and thus avoids the explosion which may takes place due to over heating of tubes and shell.

   Location in Boiler

   It is fitted near crown of furnace or over combustion chamber at its appropriate place.

Construction

Fig. 6.16 shows a fusible plug. It consists of hollow gum-metal body "P_1", screwed into furnace crown (fire box crown plate). This body has flange for fixing it in position with help of spanner. Second hollow Gun-metal plug "P_2" screwed in to the upper portion of Gun-metal body with help of a hexagonal flange provided in plug P_2. There is also third hollow gun metal plug "P_3" with conical top which is kept in hole of plug "P_2" and both "P2" plug and "P3" plug are held firmly by putting fusible metal between them. Fusible metal may be tin or lead which has low fusing point.

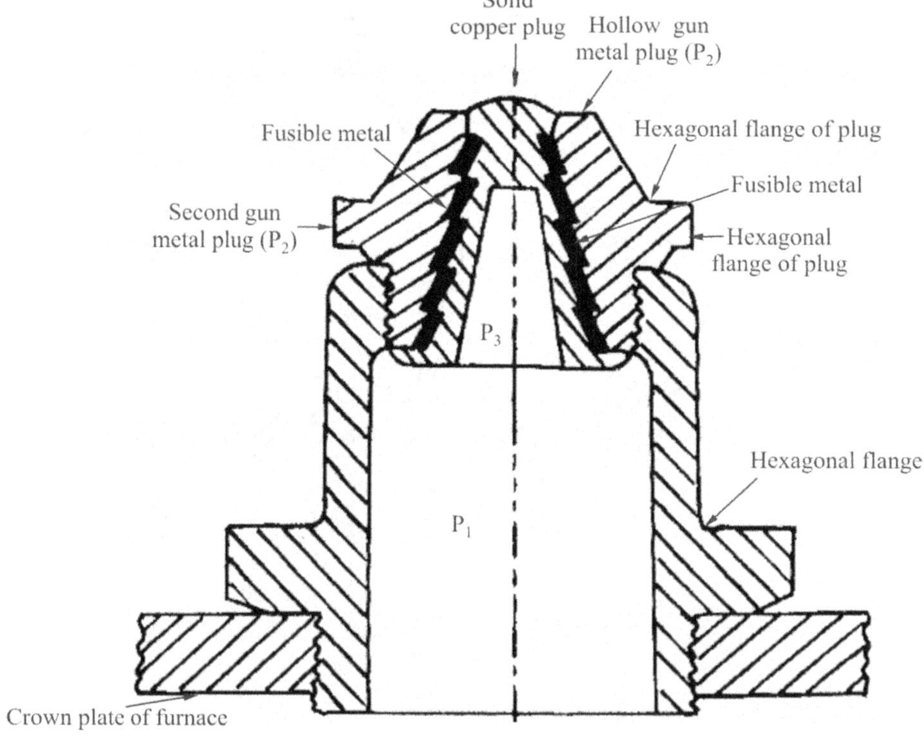

Fig. 6.16 Fusible plug.

Working

Under normal working condition the fusible plug is submerged in water. In this case, the heat from the fusible plug is being conducted to water, which keeps the fusible metal at almost constant temperatures and below its melting point.

When the water level falls below the fusible plug, it gets uncovered from water and exposed to steam. Heat conducted by fusible plug to steam is very little compare to water. Hence fusible plug becomes over heated and it melts at 560 K (287 °C) and gun metal "P_3" is released and rests with gun metal body by ribs. This opens way between steam space and furnace, thus water and steam rush to the fire box and extinguish the fire.

Before starting the boiler again, the operator fixes new fuse.

The fusible plug should renew after period of about two year as they are liable to become defective over long period of use.

6. Feed check valve

Function

Its function is to allow and control flow of boiler feed water from feed pump to boiler at a higher pressure than boiler pressure continuously, secondly to prevent back flow of water from boiler to the pump as pressure is less than boiler pressure or when it is not working.

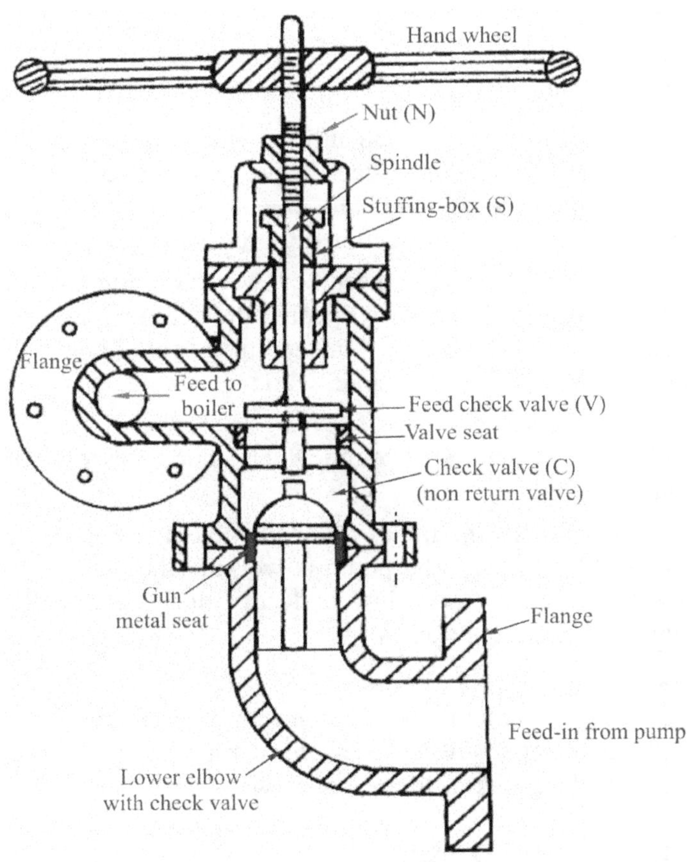

Fig. 6.17 Feed check valve.

Location in Boiler

It is placed down stream of B.F.W. pump and is fitted to the shell slightly below normal water level of boiler. Inlet end of flange is connected to delivery end of feed pump while the outlet end of flange is connected to the pipe which leads to boiler.

Construction

Feed check valve consist of various component as mentioned below.

1. Check valve
2. Feed valve
3. Flange
4. Stuffing box
5. Nut
6. Cross bar
7. Pillar
8. Spindle
9. Hand wheel

Check valve (C) moves automatically up and down under the pressure of the water, on its gun metal seat. Check valve is also known as non return valve. Check valve operates automatically due to pressure difference of water. It allows water to flow in one direction i.e., from feed pump to boiler side.

Feed valve (V) which can be raised or lowered on gun metal seat with help of spindle and hand wheel. Thereby opening the delivery passage, also control the lift of check valve. It can be kept shut and the lower elbow with check valve can be removed for cleaning or repair while the boiler can be retained at high pressure.

Stuffing box with gland (S), through which spindle passes, gland ensures steam seal against leakage.

Nut (N) which adjusts the lift of the feed valve and consequently check valve. In normal practice when feed pump is working the feed valve is kept open to desired extend depend upon flow of feed water required. Actually the lift of check value regulates the flow.

Working

At normal working conditions, the non-return valve (check valve) is lifted due to pressure difference between pressure of water from pump and feed water in boiler. The water enters in to the boiler via check valve and then through feed valve.

But when pump pressure falls below boiler pressure or if the pump stops, the non return valve closes automatically due to pressure of water from boiler and prevents escape of water from boiler.

7. Blow off cock

Function

The principal functions of blow off cock are:

(i) To empty the boiler when necessary for cleaning, repair and inspection.

(ii) To discharge the mud and sediments carried with feed water and accumulated at bottom of the boiler.

Blow off cock should be operated only when boiler is working and if sediments are to be removed. This is because the sediments are forced out quickly due to the high steam pressure in the boiler.

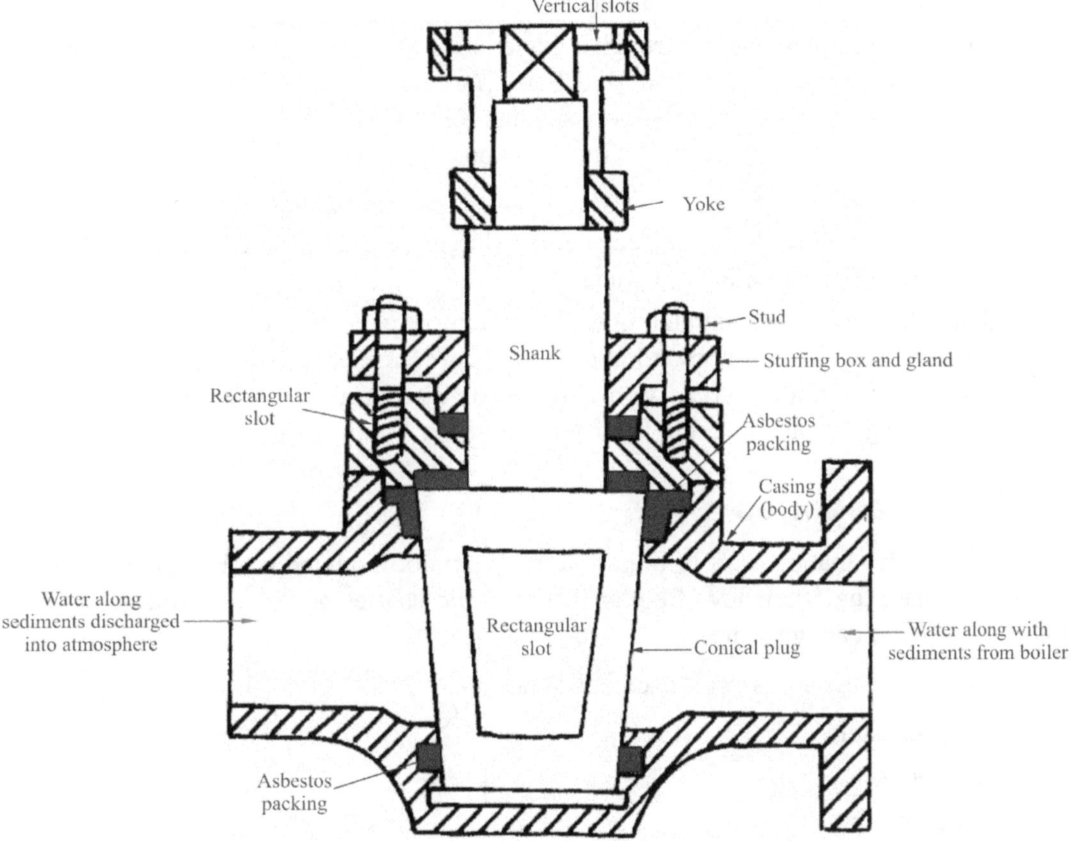

Fig. 6.18 Blow off cock.

Location in Boiler

Device is fitted to bottom of boiler drum directly or to the blow down pipe (small pipe) connected with bottom of boiler.

But when more than one boiler is working and they drain in the same waste pipe line, an isolating valve is necessary to prevent the discharge of one boiler, from entering in to the other.

Construction

It consists of conical plug fitted to casing. Plug has rectangular opening. The casing is packed with asbestos packing in groves around the top and bottom of the plug. The asbestos packing is made tight and plug bears on the packing. It may be noticed that the cock placed in this way keep the grip between under high pressure and easily operated than unpacked.

The shank of plug passes through a gland and stuffing box in the cover. The plug is held down by yoke and two stud bolts (not shown in Fig. 6.16). there are two vertical slots for fixing the box spanner on the top of yoke. The plug shank (spindle) is generally rotated by means of box spanner.

Working

As shown in Fig. 6.16 the rectangular slot in plug is perpendicular to paper or perpendiculars to the flow of water.

The plug has a rectangular opening which may be brought in line with flow passage of the casing by rotating the plug. This causes the water along with sediments to be discharged from boiler. The discharging of water may be stopped by rotating the plug again to its original position.

6.20 Accessories

Accessories form integral part of boiler but they are not mounted on the boiler. They are used to increase the efficiency of boiler. Without accessories boiler can work but without mountings boiler can not work.

Following are the accessories used in boiler.

1. Air preheaters
2. Economiser
3. Superheater
4. Steam injector
5. Steam separator
6. Steam traps

(a) Air Preheater

Introduction

There are two kinds of air-preheaters.

1. Recuperative air heaters (Tubular type and plate type)
2. Regenerative air heaters (Storage type)

1. **Recuperative air heaters :** These are mostly tubular heat exchanger in which heat from combustion products or some other heat source is transferred continuously to air as the air flows through shell-side and combustion products through tube side.

2. **Regenerative Air Preheaters**

   It is a kind of heat exchanger in which the heat transfer surface is alternatively heated and cooled by changing it alternately with combustion products and air.

   (i) Direct heat transfer from flue gases. (As in a tubular air heater)

   (ii) Indirect heating through an intermediate heat transfer agent (Solid packing as in regenerative air heater)

   (iii) Combine heating in which air is heated up by low temperature waste steam or hot water as well as by combustion of gases.

Flow scheme of air in the airheater of boiler of moderate capacity is generally single flow scheme and flow scheme of air in airheater of high capacity boilers is split flow scheme in which it is distributed in heater, two or more separate flows.

In single stage circuit air is preheated maximum up to 250 °C to 350 °C. But air can be heated to temperature 350 °C to 450 °C by two stage arrangements in which economiser is placed between the two stages of the air preheater.

Hot section of economiser is placed before the second stage of the airheater because the tubes of air preheaters are made of carbon steel that limits the allowable surface temperature to 500 °C. Therefore this means that for air preheating to 400 °C - 425 °C the temperature of gases should not exceed 570 °C - 580 °C. But the temperature of combustion products down stream of superheater is higher 600 °C - 650 °C. In order to protect the metal of air-heater tubes, the combustion products must be cooled to 580 °C, before allowing it to pass through the air heater. Therefore hot section of economizer is mounted before the second stage of the air preheater.

By incorporating economiser and air preheater in boiler it is possible to bring down flue gas temperature at the stack outlet as low as 150 °C and there by the boiler efficiency 88-90% is attained.

Heating surface area required by air preheater is generally range from 30 m^2 to 300000 m^2 such a large heating area is required because of very low value of heat transfer co-efficient between flue gases and air.

1. Tubular Type Air Preheater

It is cuperative type of heat exchanger. There are three types of Air preheaters (i) tubular type (ii) plate type (iii) storage or regenerating.

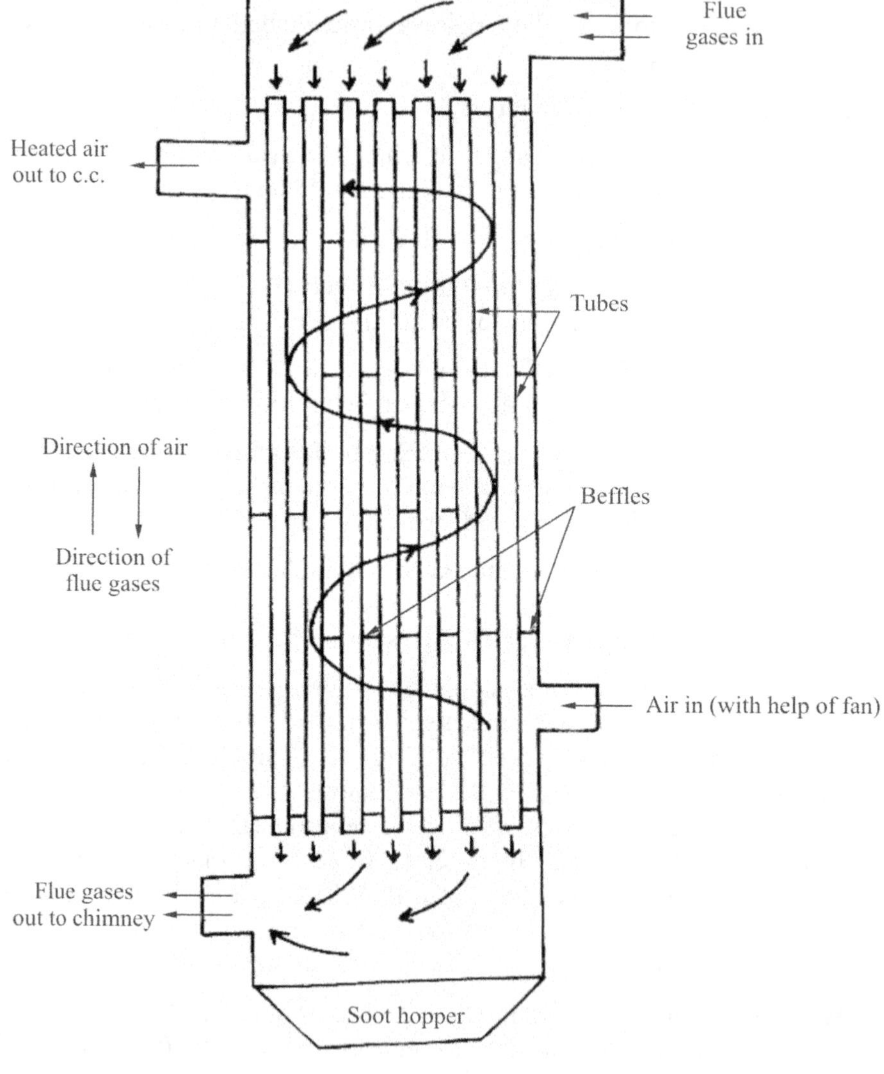

Fig. 6.19 Tubular type air preheater.

Function

The function of air preheater is to increase the temperature of air before it enters in to furnace.

Location

It is placed down stream (after) of economiser and at base of chimney so that flue gases pass through the economiser and then to air preheater. The flue gases then pass to chimney and is discharged in to atmosphere.

Construction

It consists of following parts

1. Flue gas inlet pipe.
2. Flue gas outlet pipe.
3. Air inlet pipe.
4. Air outlet pipe.
5. Baffle plates.
6. Tubes attached to plates on both end side of tubes.
7. Casing.
8. The soot hooper provided at bottom of casing.

It consists of number of tubes. The bank of tubes is enclosed in insulated casing provided with inlet and outlet connection for flue gases as well as for air, in such a way that gases and air moves in opposite direction as shown in Fig. 6.23. Baffles are provided inside air preheater in such away that air has to travel several passes before leaving the air preheater. Due to this there is sufficient time available for heat transfer between flue gases and air and reduce its friction resistance as flow of air is in organised manner.

Tube is 30-40 mm and thickness 1.25-1.5 mm, upper plate is 20-25 mm thick while lower plate is 15-20 mm thick.

Working

After leaving the economiser the flue gases enters into tubes of air preheater from top and leaves at bottom end.

The air at room temperature is admitted to heating surface at lower end of air preheater with help of fan. Air travels upward in opposite direction to that of flue gases around the tubes and heated air leaves the airheater at upper end of the tubes of banks. Thus air gets initially heated before being supplied to furnace and flue gases are discharged in to atmosphere through chimney. The soot hopper is fitted to bottom of airheater casing to collect soot. Soot is caused due to cleaning of tube banks.

Uses

It supplies hot air to the furnace as a result efficient combustion of fuel take place and overall efficiency of boiler increases by 2 to 9%.

Advantages Air Preheater

1. Improved combustion and efficient use of fuel.
2. Stabilized fuel ignition which improves low head combustion.
3. Increase steam generation capacity and higher boiler efficiency. Which increases 1 % for every 20 °C rise of preheated air temperature.
4. Higher thermal efficiency as most of the heat is from combustion products is recovered.
5. Better utilization of low grade, high ash fuel.
6. Greater load flexibility.
7. Reduction in boiler heating surface due to increase in overall efficiency because of improved combustion.
8. Greater the preheat, lower the SO_3 in flue gas and therefore stack outlet temperature can be lowered, resulting in greater heat utilization.
9. Air preheating ensures complete burning of fuel resulting in less slogging and clear flue gas, this reduces boiler outage for cleaning.

Disadvantages of Air Preheater

1. Increases maintenance cost of stocker and furnace refractory.
2. Entails the risk of serious damage due to decomposition and ignition of combustibles.
3. Requires an F.D. fan and ducts for air passage and needs careful design calculation for draught loss.
4. Any air leakage necessitates higher electricity consumption by F.D. Fan.
5. Leakage is difficult to detect until corrosion is in advance stage, requiring a major repairing or total replacement and greater outage time.

2. Plate Type Air Preheater

Location

It is placed between economiser and chimney.

Function

To supply hot air to the furnace for efficient combustion of fuel.

Construction and Working

In plate type air preheater, alternate gas and air passages are formed between closely spaced (1.5 to 2.5 cm part), parallel vertical plates. They are arranged such a way that the gas passes vertically upward through narrow passage and air enters at side passes downward in one or two passes.

This type of air preheater is not used in modern installation as its initial cost and maintenance cost compare to tubular air preheater is more.

Advantage

1. Boiler efficiency is increased by 2 to 10%
2. Low grade fuel can be burned.
3. Better combustion with less soot, smoke and ash.

Disadvantages

1. Higher running and maintenance cost due to addition ID and FD fan.
2. Increase in capital cost of the preheater, induced draught fan for removing flue gases and forced fan to force cold air in preheater.

3. Regenerative Type Air Preheater

It is a kind of heat exchanger in which the heat transfer surface is alternatively heated and cooled by changing it alternatively with flue gases and air.

Function

To supply hot air to the furnace for efficient combustion and increase overall efficiency of boiler.

Location

It is located between economiser and stack (chimney) of boiler.

Construction and Working

It consist of following parts.

1. Rotor
2. Packing
3. Air inlet pipe
4. Air outlet pipe
5. Flue gas inlet pipe
6. Flue gas outlet pipe
7. Driven gear and driving gear
8. Electric motor
9. Casing
10. Induced Fan and force fan.

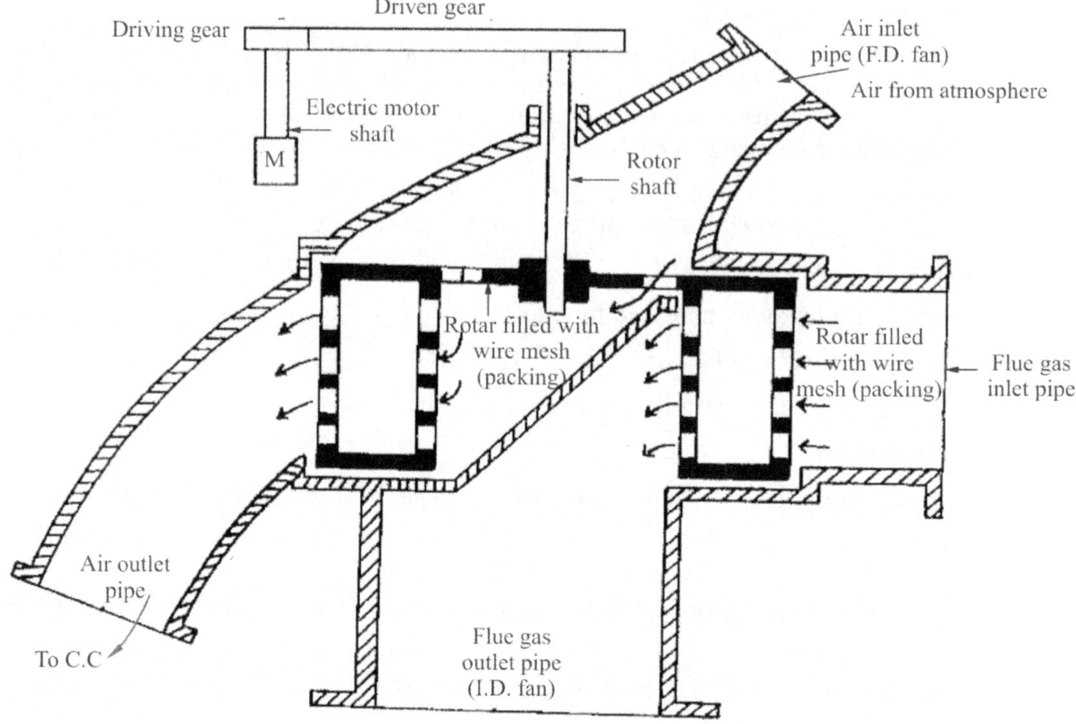

Fig. 6.20 Regenerative type air preheater.

This consists of concentric rings of alternate flat corrugated (sonidoal curve) plates arranged in such a manner that they from vertical passages for upward flow of flue gas through one half and downward flow of air through other opposite half.

The diameter of drum is 10 m but in some cases rotor diameter is as large as 15 to 18 m in high capacity power plants. Packing used in rotor are made up of steel sheets or ceramic or glass.

Thickness of packing sheet is 0.6 to 1.25 mm. The contact time of air/gas with packing is usually less then half a minute.

The drum is rotated about vertical shaft at less than 3 RPM (0.5 + 0.3 RPM) by means of small motor and gear arrangement. So that they are alternately heated by the flue gases and cooled by combustion air which in turn is heated. A force fan is used to force air into the rotor from its top and gases enter bottom of rotor and moves down up, during this heat is picked by air from packing agent which is heated previously by hot gases.

The heated air is delivered to the furnace through suitable duct arrangement and gases are removed through chimney to atmosphere.

Types of Economisers

There of two types of Economisers.

1. Boiling type 2. Non boiling type

1. **Boiling type :** In this type of economiser the part of (about 25%) boiling feed water is converted into steam.
2. **Non-boiling :** In this type of economiser the boiler feed water (BFW) is heated below saturation temperature.

Membrance Type Economiser

It is non boiling type economiser. They are made up of plain tubes with steel membrances (steel sheet 2.3 mm thick) welded in between the straight portion of tubes.

Following are the advantages of Membrance type economiser are:

1. High efficiency
2. More reliable operation
3. Requires less metal for the same amount of heat absorption.

2. Economiser (Green's Economiser)

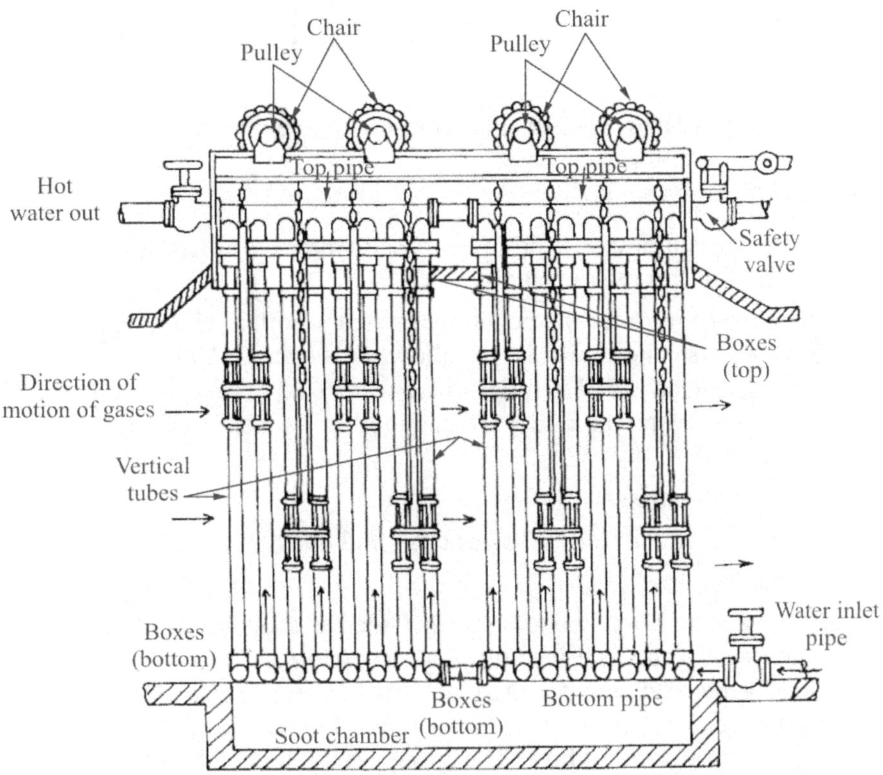

Fig. 6.21(a) Economiser

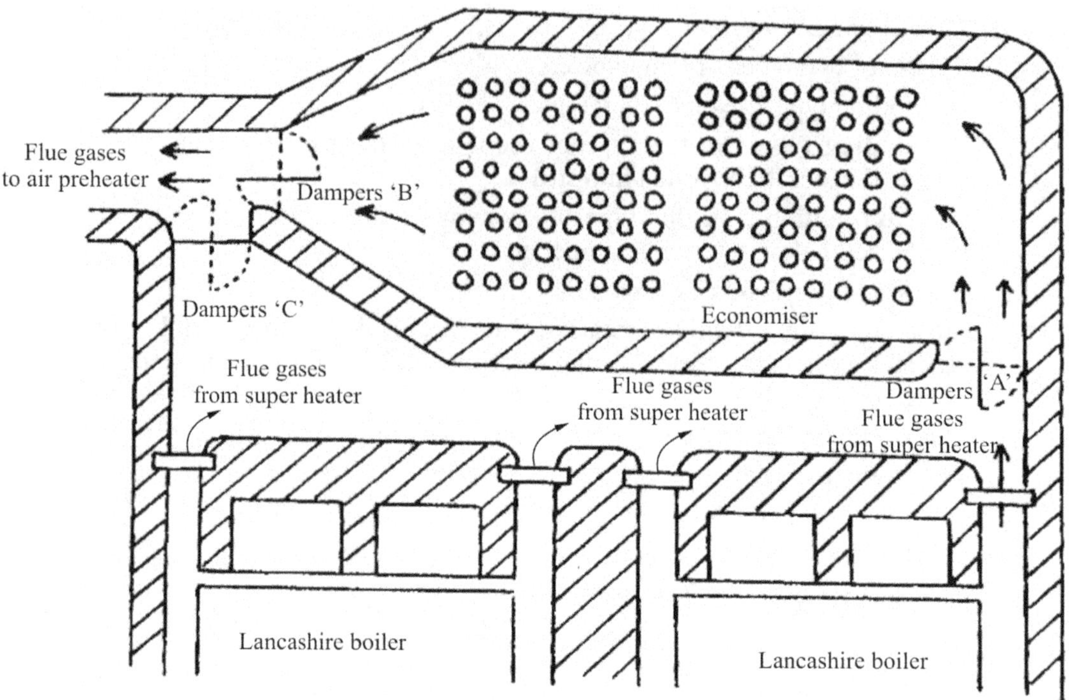

Fig. 6.21(b) By-pass arrangement of flues for economizer.

Function

An economiser is a device used to heat feed water by utilizing the heat in exhaust flue gases before leaving the chimney.

The economiser improves the economy of steam boiler. The temperatures of water leaving economiser normally is not allowed to exceed 105 °C, however in some high pressure plant BFW (Boiler feed water) is heated to 200 °C. Generally temperatures of water at outlet of economiser is kept 10° less then saturation temperature.

Location of Economiser

It is placed downstream (after) superheater and upstream (before) air preheater.

Construction

A well known type of Economiser is Green's economiser. It consists of large number of vertical tubes. These tubes are available 2.75 m - 4 m in length and are hydraulically pressed into top and bottom headers. Tube I.D: 20 - 30 mm and thickens 2.5 - 3.5 mm. Small diameter tubes are chosen to maximize rate of heat transfer and minimize fouling. The group of vertical tubes is generally built in section 4, 6, 8, 10, and 12 tubes wide.

Tubes are joined to horizontal pipes or boxes or header at bottom and top respectively.

It is essential that vertical pipes (tubes) may be kept free from deposit of soot as the deposition of soot reduces the heat flow rate from gases to water and efficiency of economiser is reduced, each tube is provided with scraper.

1. Vertical tubes
2. Boxes for top of tubes
3. Boxes for bottom of tubes
4. Top pipe
5. Bottom pipe
6. Water inlet pipe
7. Hot water outlet pipe
8. Safety valve
9. Pulley
10. Chain
11. Soot chamber

The soot deposited over tubes is removed by scrapers which are kept slowly moving up and down to clean the surface. A pair of scrapers for adjacent tubes is coupled together by chain which passes over pulley. Mechanical type drive moves the pulley i.e., each pulley is connected to warm wheel and driven by warm mounted on electric motor shaft. The scraper automatically reverse when they reach the top or bottom end of tubes. They are kept in motion continuously when economiser is in use. Speed of scraper is slow about 46 m/hr. The soot thus removed is collected in soot chamber from where is it disposed off. Every economiser is provided with safety valve, blow off cock (Drain valve) and air Release valve. According to Indian Boiler Regulation, the economiser should be fitted with explosion doors. To measure the temperature of inlet and outlet water thermometers are provided.

A safety valve is fitted on top header for safety of pipes against any high pressure water that may be developed. The blow off cock is fitted (not shown in Fig. 6.19(a)) at lowest point on the economiser to discharge the sediments collected from feed water to atmosphere.

It may be noticed that temperature of feed water should not be less than 35 °C otherwise there is danger of corrosion due to moisture and SO_2 contained in the flue gases deposited in cold tubes.

Working

Condensate and make up water enters into bottom header and moves up toward top header and during this heat is absorbed from flue gases as gases are moved across the tubes. The temperature of water is raised. Hot water gets collected at top header and enters the boiler through discharge valve provided on the top header.

Flue gases passing heat to economiser move to air preheater and from air preheater flue gases enter in to chimney and discharged to atmosphere. When economiser is in use, the dampers A and B are open and dampers C is closed. When economiser is to be cleaned or repaired, the damper A and B must be closed and C must be opened. Flue gases directly pass in to chimney without heating the tubes of economiser. Fig. 6.19(b) of by pass arrangement for economiser.

Advantages of Economiser

1. Increase in thermal efficiency of boiler plant. For every 6 °C rise in feed water temperature in economiser, the efficiency increases maximum 1%. Fuel cost savings upto maximum 20% can be achieved.
2. Temperature range between various parts of boiler is reduced which results in reduction of thermal stresses and therefore longer life of boiler.
3. Increase in evaporating capacity or for same evaporative capacity there is saving in fuel.
4. Cuts down load on the furnace and burners, more effectively.
5. Endows quicker steam generation capacity when operated with on-off cycle.
6. Lower requirement of auxiliary power.

Disadvantages

1. BFW delivered to economiser is at 10% to 20% over the boiler pressure because there is pressure drop of BFW as it passes through the economiser tubes, piping and valves.
2. Entails the risk of serious damage due to deposition and ignition of combustibles.
3. Greater the preheat, lower the SO_3 in flue gas therefore the stack outlet temperature will be lower.
4. There is loss of draught for flue gases therefore natural draught will not be sufficient for circulation of air and flue gases. The plant must be provided with artificial draught.

Use

Green's economiser commonly used in Lancashine boiler for heating feed water.

Examples

S = Heat recovered per kg of water/Heat required to convert unit mass of water into steam

$$S = \frac{C_p(T_2 - T_1)}{(h - h_w)} \times 100$$

where S = Percentage saving in fuel consumption

T_2 = Feed water temperature leaving the economiser

T_1 = Feed water temperature entering the economiser

h = Enthalpy of steam in kJ/kg

h_w = Sensible heat or enthalpy of water initially (kJ/kg) of water

C_p = Specific heat of water in kJ/kg k

1. Calculate the percentage of fuel saving in economiser by installing an economiser with Lancashire boiler, if steam leaves dry saturated at 14 bar. Feed water enters the economiser at 30 °C and leaves at 115 °C. Assume C_p of water as 4.18 kJ/kg K.

Solution: Given

$t_1 = 30\ ^\circ C$

$t_2 = 115\ ^\circ C$

Pressure of steam = 14 bar

Condition of steam = Dry saturated

First Method

(i) Without Economiser

Enthalpy of 1 kg of Feed water at $30^0 = 4.187 \times 30 = 125.61$ kJ/kg

Enthalpy of Dry-saturated steam at 14 bar from steam table = 2790 kJ/kg

∴ Net heat required to produce 1 kg of Dry-saturated steam = 2790 – 125.61 = 2664.39 kJ/kg

(ii) With Economiser

Enthalpy of 1 kg of Feed water at 115 ^{0}C $= m\,c_{p_w}\ (115 - 0)$

$= 1 \times 4.187 \times 115$

$= 481.5$ kJ/kg

Enthalpy of 1 kg of Dry-saturated steam at 14 bar from steam table

= 2790 kJ/kg

∴ Net heat required to produce 1 kg of Dry saturated steam at 14 bar from Feed water at 115 °C = 2790 – 481.5 = 2308.5 kJ/kg

Saving of heat by using feed water at 115 °C instead of 30 °C

= Net enthalpy without Economiser – Net Enthalpy with Economiser

= 2664.39 – 2308.5

= 355.89 kJ/kg

(iii) Percentage saving in heat = Saving of heat by using feed water at 115 °C/ Net heat required to produce 1 kg of Dry-saturated steam at feed water temperature to inlet to economiser.

$$= \frac{355.89}{2664.39}$$

$$= 0.1336$$

$$= 13.36\%$$

Second method by using formula

$$\text{i.e. \% solving in heat} = \frac{c_{p_w}(t_2 - t_1)}{(h - h_w)}$$

where h_w is feed water temperature at inlet i.e., 30 °C

$$\therefore \quad \text{\% solving in heat} = \frac{4.187(115-30)}{(h-h_w)} \times 100$$

$$= \frac{4.187(85) \times 100}{2790 - c_{p_w}(t_1 - 0)}$$

$$= \frac{4.187 \times 85}{2790 - 4.187 \times 30} \times 100$$

$$= \frac{355.895}{2790 - 125.61} \times 100$$

$$= \frac{355.895}{2664.39} \times 100$$

$$= 13.36\ \%$$

Superheaters

Function

Superheaters are used in boilers to increase the temperature of steam above its saturation temperature to superheated temperature. The difference between superheated temperature and saturation temperature is known as degree of super heat. Superheated steam is produced by passing wet or dry saturated steam through small groups of tubes and hot gas passes over them.

Location of Superheater

They are located in the path of furnace gases so that heat is absorbed by steam in superheater from hot gases.

(a) Convective superheaters are placed in the convective gas space which absorb heat mainly by convection.

(b) *Radiant superheaters :* They are arranged on furnace walls. They absorb heat by direct radiation from flame (fire) and re-radiation from refectory walls behind then.

(c) Platen superheaters (combination superheaters) are mounted on the top of furnace and partially in flue gas duct between convective and radiant heating surface.

They absorb both radiant and convective heat, steam, first enters into radiant superheater and the convective superheater.

Classification of Superheaters

1. Convective superheater
2. Radiant superheater
3. Combination (platen) superheater.

They are also classified according to arrangement of superheater tubes.

1. *Over deck :* It is placed over (above) the water tube as in case of Babcock and Wilcox boiler.
2. *Inter deck :* It is placed between the water tubes which are located near the furnace.
3. *Inter tubes :* They are placed between bank or row of water tubes.

   Material of superheated tubes : They are made up of high-alloyed Cr-Ni. Austenitic are

Table showing Material for Superheater for different Temperature of Superheated steam

Sr No.	Cr	Ni	W	Mo	Ti	Nb	B	Fe	Working temperature
1.	13 – 15	13 – 15	2 – 3	0.45.0 – 0.6	–	–	–	Rest	925 °k
2.	17 – 19	11 – 13	–	–	0.65	–	–	Rest	925 °k
3.	15 – 17	15 – 17	2 – 3	0.5 – 1	–	0.005	0.005	Rest	980 °K
4.	15 – 17	12 – 15	2 – 3	–	–	1 – 1.3	0.005	Rest	980 °K

- They are made from austenitic-class because this steels have high heat and corrosion resistance.

- It have corrosion resistance chromium metal, produces hard layer of Cr_2O_3 on tube surface exposed to hot gases. This prevents further penetration of oxygen or corrosive gases to base metal.
- Molybdenum and tungsten are added for high heat resistance.
- Titanium and niobium prevent austenitic steel from inter granular corrosion.
  What is characteristic of superheater or relationship between Boiler load and superheated steam temperature?
- For radiant superheaters as load increases superheated temperature falls but in case of convective superheaters as load increases the superheated temperature of steam increases.
- In combined superheaters as load increases or decreases the superheated constant because it is combination of Radiant and convective superheater.

As Boiler load is increased the quantity of fuel burning increases. However this increased rate of fuel burning influences very little on those parameters which govern the quantity of heat absorbed by radiant heating surfaces. Therefore the heat absorption in radiant superheater does not increase in same scale as the flow rate of steam through it. As result, the heat absorption per unit steam flow rate decreases. The parameters which govern the absorption of radiant superheater are combustion temperature, flame emissivity and thermal efficiency of water wall.

1. Sudgen's Hair Pin Superheater

This type of super heater is used in low pressure boilers such as Lancashire, Cornish and Babcock and Wilcox.

The position of sudgen's superheater in different boilers are different as given blow.

Boiler	**Position of superheater**
Lancashire and carnish	In down take
Locomotive	In internal fire tube
Babcock and Wilcox	In over deck section
Scotch marine	In internal fire tube

Location

It is located in path of furnace gases where temperatures of gases not less than 550°C. The superheater is located downward of boiler just before the gases enters the bottom flue.

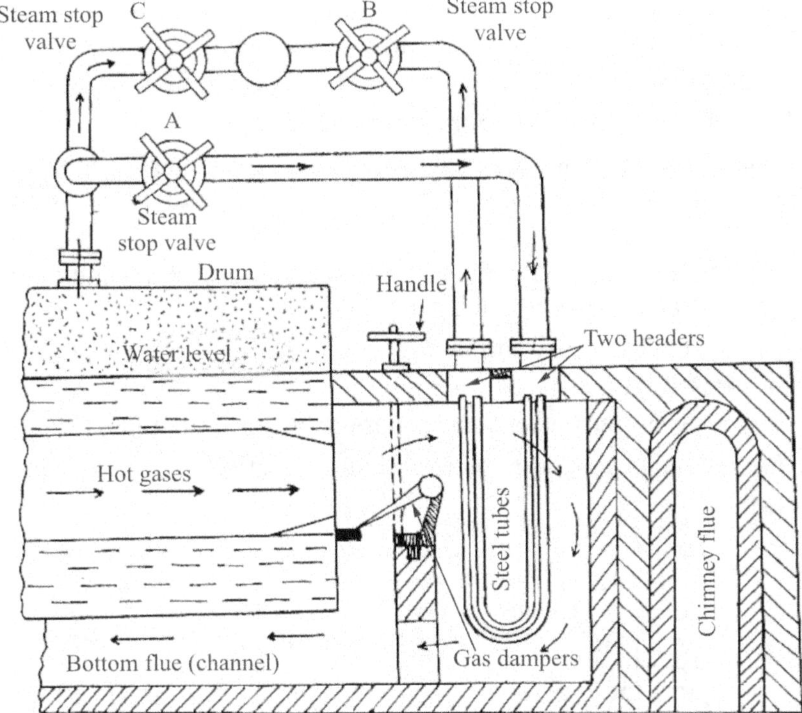

Fig. 6.22 Sudgen's hair pin super heater.

Construction and Working

The superheater consists of two headers, and set of superheat tubes. Tubes are in the form of "U".

The amount of hot gases (at 550°C) passed over the superheater tubes should be proportion to amount of superheated steam passing through the tubes otherwise, the tubes would be over heated.

The superheater is put out of action by turning the damper upward to vertical position. In this position of the damper, hot gases coming out of fire tube pass directly in to bottom flue (channel) without passing over to the superheated tubes.

For getting superheated steam the valves A and B are opened and valve "C" is closed the damper is opened as per the quantity of heat required for superheating the steam in superheater pipes. Steam through valve "A" enters in to right hand header from this header steam enters in to U tubes where it is superheated by absorbing heat from hot gases. The Superheated steam is collected in left hand header and superheated steam pass to turbine through valve B.

If we want wet steam then superheater is to be cut off from line, that is damper is closed and valves A and B are also closed valve "C" is opened. Steam from boiler passes through valve C to turbine or engine or for process work.

Advantages of Superheater (or Superheated Steam)

1. Due to superheating of steam temperature and heat contained per kg of steam is increased. This increases the power output of turbine or steam engine.
2. The condensation is reduced in turbines or steam engine. This reduces corrosion and erosion problems in turbines.
3. Efficiency of steam turbine is increased also efficiency of steam plant increases.
4. It reduces the steam consumption for same power output.
5. Losses due to condensation in cylinder and steam pipes are reduced.

4. Steam Injector

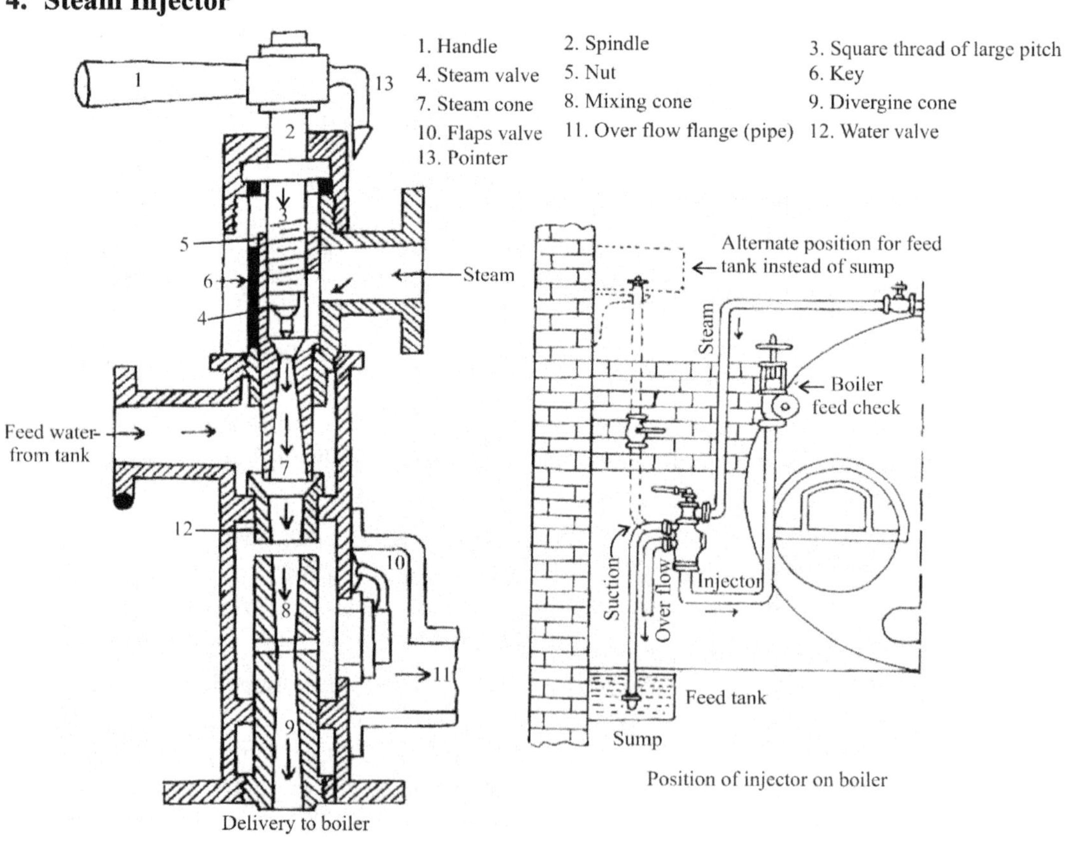

Steam injector

Position of injector on boiler

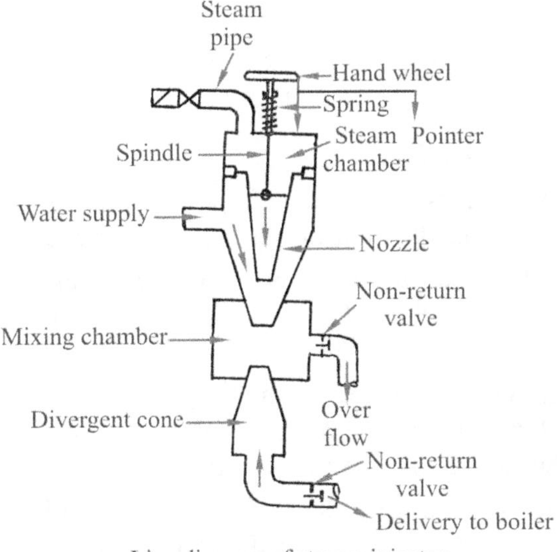

Line diagram of stream injector

Fig. 6.23 Steam injector.

There are three types of pumps generally used for feeding water in boilers

1. Reciprocating pump (duplex direct acting steam pump)
2. Rotary pump (centrifugal type)
3. Injector

Function

Function of injector is same as that of pump which delivers feed water in to boiler under pressure.

Location

Location of injector in boiler plant is shown in Fig. 6.23.

Construction

1. Handle
2. Spindle
3. Square thread of larger pitch
4. Steam valve
5. Nut
6. Key
7. Steam cone
8. Mixing cone
9. Diverging cone

10. Flap valve
11. Over flow Flange pipe
12. Water valve
13. Pointer (not shown figure)

The spindle upper end is provided with handle and lower end is provided with valve. The pointer on handle indicates the shut and open position of valve.

Key is provided to check the rotation of nozzle (steam cone). A collar is fitted which prevents axial motion of valve. When handle is rotated the spindle will rotate only.

Lower part of spindle has a screw which works in nut which is integral part of steam cone. With rotation of the handle, steam cone moves up and down as result valve controls the steam flow through steam cone.

Lower end of steam cone form the water valve, seat of which is entrance to the combining cone. It will be seen from figure that water valve remains open when steam is in shut position.

Working

Steam coming through the steam pipe operated by steam valve steam enters in nozzle, the out let of which is in a converging nozzle. In the nozzle section, velocity of steam increases forming jet of steam while pressure decreases below atmosphere. Due to decrease of pressure of steam at the exit of converging nozzle than the feed water from tank, enters in to mixing chamber. When steam and water is mixed and condensation of steam takes place, Due to this strong vacuum is created in mixing tube and the mixing chamber. Due to strong vacuum more and more water rushes from the feed tank into mixing chamber. Velocity of water in mixing tube is greatest and mixing chamber itself, there is high velocity jet.

Jet of water then enters into the divergent tube where the K.E. of water is converted into pressure energy. Pressure of water at exit of divergent tube is sufficient high to force the water into boiler through feed check valve.

When steam valve is just opened for starting the injector, the water and uncondensed steam escape to atmosphere through the non return valve. But when steam has heated water, the mixture gains velocity to enter into mixed-up chamber or cone and this result in fall of pressure in over flow gap below atmospheric pressure and overflow flap valve is closed.

To ensure working of the injector, a definite relation between the quantity of steam and water entering the injector for particular value of steam pressure is established.

So long as the handle is kept in open position, this type of injector will automatically restart even after water supply or steam is interrupted temporary due to some reason.

Used

It is normally used on vertical boilers and locomotive boilers. It is not used for large capacity high pressure boilers.

Advantages

1. It occupies minimum space.
2. It is cheap as compared to other feed pump devices.
3. Maintenance cost is negligible.
4. Easy in operation.
5. Less moving parts, less friction losses and whole of heat of steam is returned back to boiler. Thermal efficiency is very high about (100%).
6. Simple design.
7. Thermal stresses in boiler are reduced because injector supplies hot water to boiler (compare to economiser).
8. There is no necessity of lubrication.
9. It's life is long.
10. It is reliable. It does not go out of work easily.

Disadvantage

1. When load is fluctuating to great extend it is difficult to maintain sufficient flow of water to boiler.
2. Low pumping efficiency.
3. It can not be used for larger capacity high pressure boilers.
4. Irregularity in operation under extreme variation in steam pressure.

5. Steam Separator (Steam Driers)

According to the principle of separation the steam separator are classified as under.

1. Impact or Baffle type.
2. Reverse Current type.
3. Centrifugal type.

Baffle Type Separator

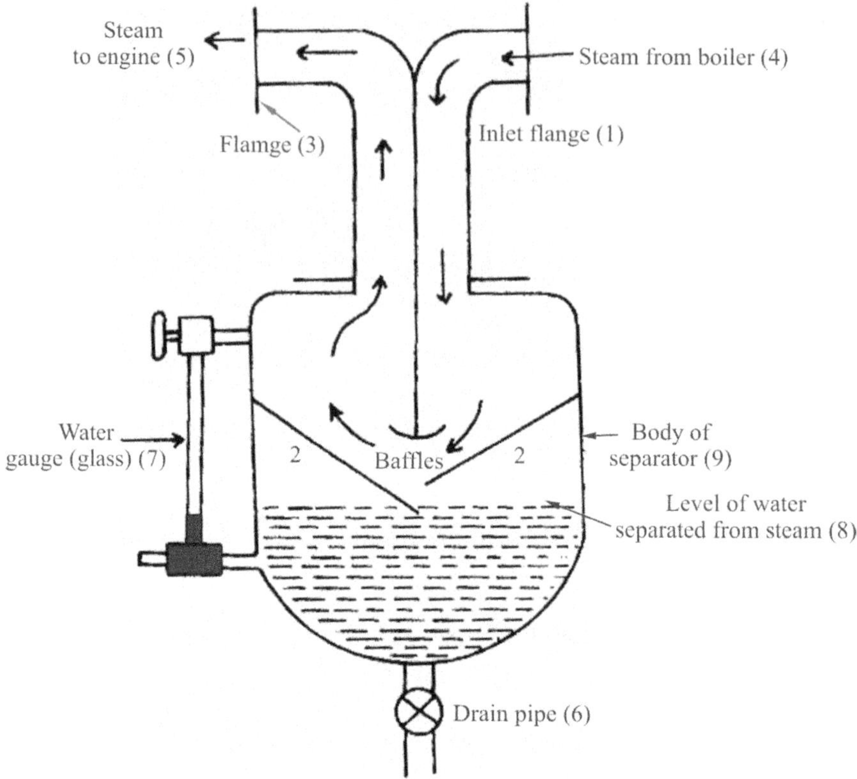

Fig. 6.24 Baffle type steam separator.

The function of steam separator is to separate suspended water particles carried by steam, on it's way from boiler to steam turbine or steam engine.

Location

It is installed near to steam turbine or steam engine in the main steam pipe.

Construction **:** Various components of baffle type separator are as under.

1 = Connection for inlet pipe from boiler.

2 = Baffles

3 = Discharge flange

4 = Steam from boiler

5 = Steam to engine or turbine

6 = Drain pipe

7 = Water level indicator (water gauge)

8 = Water separated from steam

9 = Body of steam separator.

Working

The steam from boiler enters the steam separator through Flange 1 and flow down. In its passage downward, it strikes the Baffles 2 and is deflected upward, but water particles having greater density and greater inertia fall to the bottom of separator. The nearly dry steam deflected up and comes out through the Flange 3. The level of water collected in separator can be seen on water gauge. The water collected in separator is removed at intervals through drain pipe 6.

Reverse current type

In this type of separator the steam enters from the boiler through inlet flange. The direction of steam is changed generally by 180°. The water having greater density and greater inertia is collected to the bottom of separator. The dry steam is separated out in reverse direction and discharge to turbine or steam engine.

Centrifugal Steam Separator

The steam available from boiler may be, either, wet, dry or superheated. But in many cases there will be loss of heat from it, when it passes through steam pipe from boiler to steam turbine or steam engine tend to become wet. The use of wet steam in steam engine or turbine is not advisable as it cause erosion and corrosion inside turbine blades or engine parts.

Function

To remove the entrained water particles from the steam send to steam turbine or engine.

Location

It is placed close to steam turbine and on the main steam pipe.

Construction

It consists of

1. Inlet flange connected to pipe from boiler
2. Steam outlet pipe to turbine
3. Spiral plate
4. Steam pipe
5. Body of separator.
6. Steam inside pipe
7. Water drain pipe

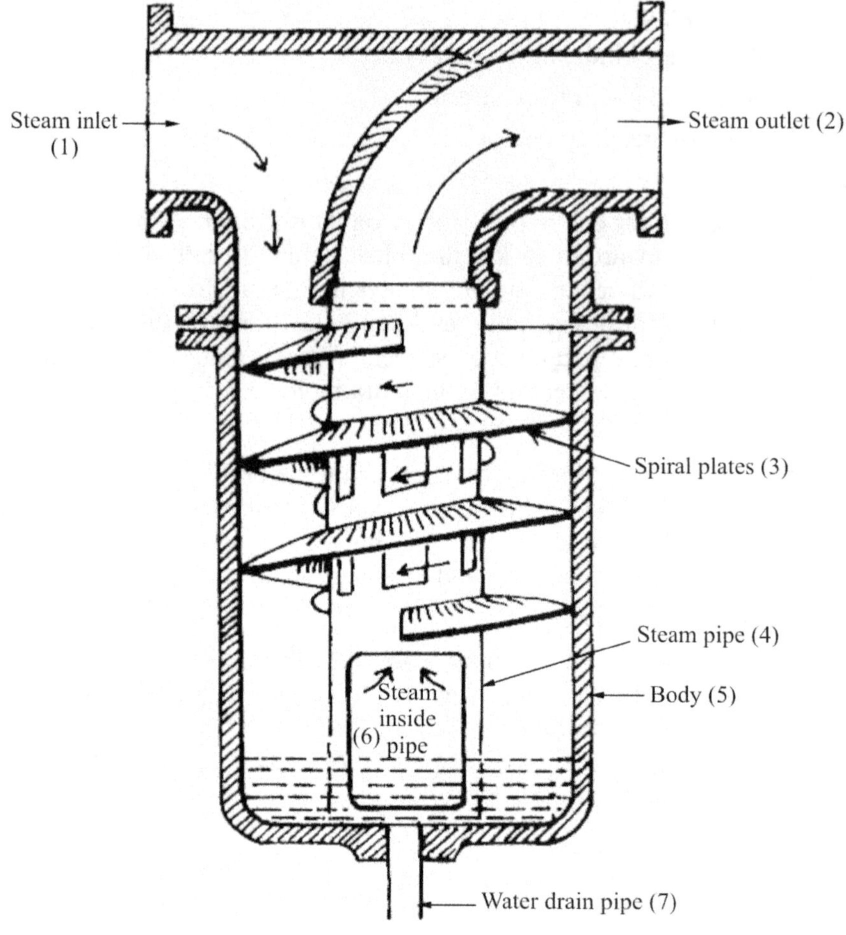

Fig. 6.25 Centrifugal steam separator.

Working

Fig. 6.25 shows centrifugal steam separator when steam enters the separator steam strikes over spiral plates and steam is imparted whirling velocity due to shape of plates.

Due to centrifugal force water particles which are suspended with steam are thrown out from the steam. They are collected at the bottom, steam move up and discharged to pipe line going to steam turbine or steam engine. Water collected at bottom of separator is removed at intervals through drain pipe to atmosphere.

Note: Water from all separators is discharged through steam traps into atmosphere.

6. Steam Traps

The function of steam trap is to drain off water resulting from partial condensation of steam from steam pipe and steam jackets automatically without allowing steam to escape through it.

It is arranged in steam pipe after steam separator and near to steam turbine or steam engine.

There are three types of steam traps generally used in practice.

1. Bucket type
2. Thermal expansion type
3. Mechanical float type.

Bucket type steam traps

Function

The function of steam trap is to automatically drain away the condenced steam from steam pipe, steam-jackets and steam separators without permitting any steam to escape.

Location

It is placed in steam pipe near steam turbine or steam engine.

Construction

1. Inlet passage – "A"
2. Water-casing – B
3. Floating bucket – C
4. Spindle – D
5. Conical shaped valve – F
6. Outlet pipe – G (exit pipe or discharged pipe).
7. Seat E for valve – F
8. Vertical tube – J
9. Clearing door – L
10. Knob – K.

Working

Bucket type steam trap is so arranged (placed) that water resulting from condensation of steam flows by gravity to it. Thus water enters through inlet pipe "A" and gets collected in casing "B".

Casing always contains sufficient water. Float Bucket "C" is provided in water casing which Floats in water. Spindle is attached to the float bucket at its bottom and carries conical shaped valve E at the top. F is seat fixed to outlet pipe G, on which valve E presses (resting). The bottom of the spindle carries spiral valves which can rotate or move up and down freely along with bucket in vertical tube J. vertical tube J is attached to out let pipe G.

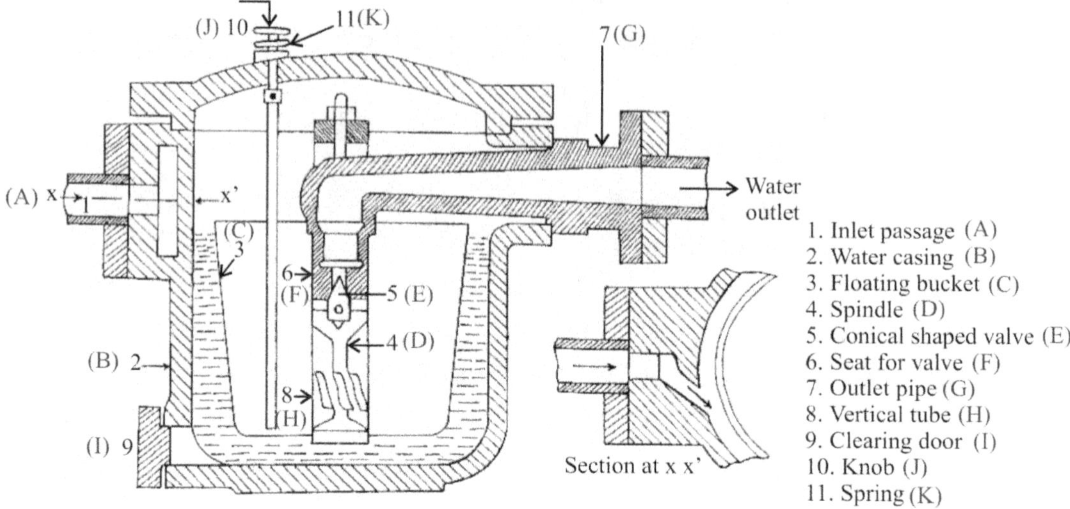

Fig. 6.26 Bucket type steam trap.

The condensed water enters the casing and level of water in casing, rises which rises the float bucket and ensures proper closing of valve.

When water level rises in chamber, there is chance of water entering the bucket to fill it. This causes bucket to come down valve opens, steam pressure acts on the water in bucket as a result water is forced up the guide tube through vanes and discharged through exit pipe.

The flow of water through spiral vanes rotates the bucket. As soon as practically all water from bucket is forced out, it rises up and floats. This closes the valve.

At the time of closing the valve, the rotating bucket serves to grind the valve and valve seat.

Clearing door is provided for cleaning and removing the sediments deposited in casing.

For testing the working of bucket with spindle and valve, the knob is pressed by hand to push the float bucket down and water and steam blow down. It indicates that valve is not stick to the valve seat.

Fig. 6.24 shows section x-x hot water and steam is entering tangentially which assist in rotation of the bucket at time of blowing off.

Expansion Steam Trap

Function

The function of steam trap is to drain away automatically the condensed steam from steam pipe, steam jackets and steam generators without permitting any steam to escape.

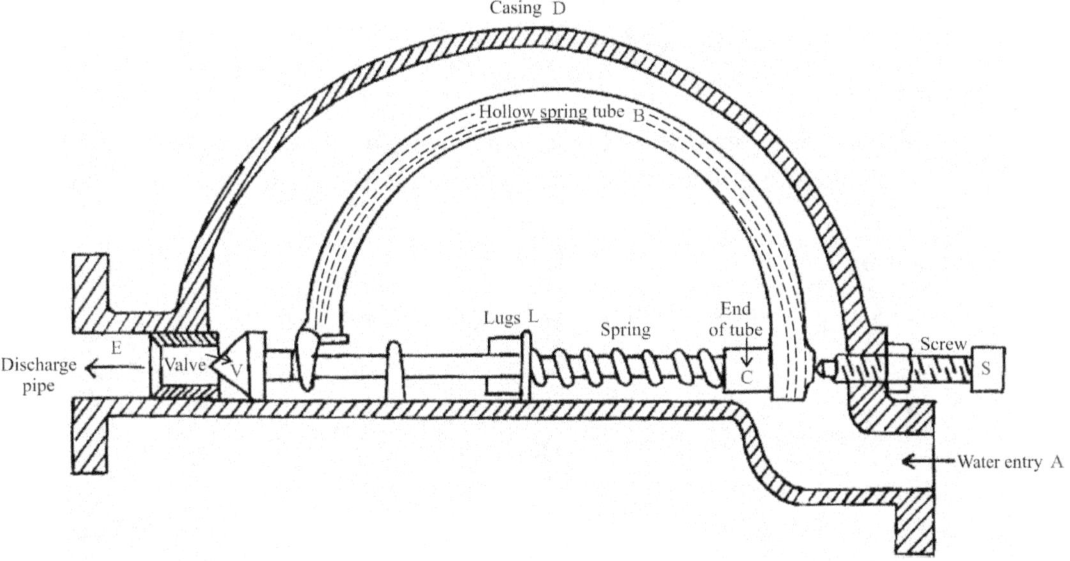

Fig. 6.27 Expansion steam trap (Sirius type).

Location

It is located in steam pipe near to steam turbine or steam engine.

Construction

1. Water entry – A
2. Hollow spring tube – B
3. End of spring tube – C
4. Screw – S
5. Valve – V
6. Lugs – L
7. Casing – D
8. Outlet pipe – E

Working

Fig. 6.27 shows expansion steam trap (Sirius type). A hollow spring tube made up of nickel steel contains liquid which becomes gas at temperature to that of steam is likely to have.

Initially there is steam inside the casing and spring tube is surrounded by steam. Liquid in the spring tube, due to increased temperature, the liquid gets converted into gas inside spring tube. The curved tube tends to straighten itself and consequently valve remains in closed position.

But when water enters at "A" being lower in temperature than steam, it causes tube 'B' to contract and valve is opened. When all condensed water is discharged the steam enters the trap the increased temperature converts liquid in tube B in to gas, the curved tube "B" tend to straighten and consequently valve is closed. The end C of tube is held against adjusting screw by spring, one end of which passes against lugs fixed to casing.

Mechanical Float Steam Trap

Function

Its function is to remove condensed water resulting from partial condensation of steam from steam pipe, jackets and separator without allowing steam to escape through it.

Location

It is located in steam pipe near steam turbine or steam engine.

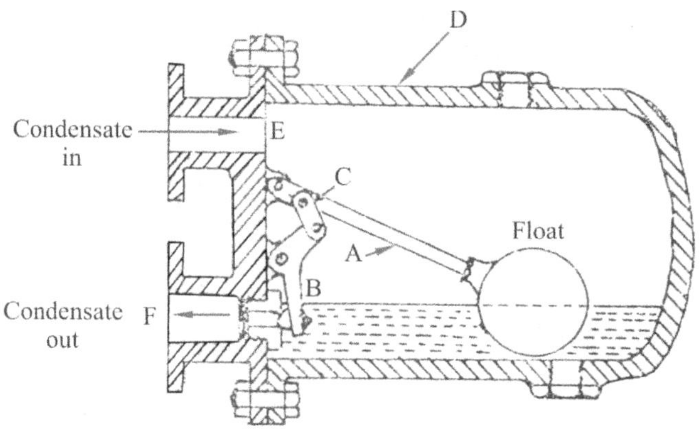

Fig. 6.28(a) Mechanical float steam trap.

Construction

1. Casing – D
2. Inlet pipe connection – E
3. Out let pipe connection – F
4. Lever A with Float –
5. Valve arm - B
6. Toggle link - C.

Working

It is working on the principle that Float is floating in water but immersed in the steam. When level of condensate in casing increases, the float moves up and valve will open.

The steam pressure is acting on the condensate, due to this condensate is discharged through valve. The level of water in casing, falls and float moves downward and valve is closed. In this way balance is maintained between condensate collected in casing and condensate going out from casing through valve.

7. Steam collecting pipe or anti priming pipe or dry pipe

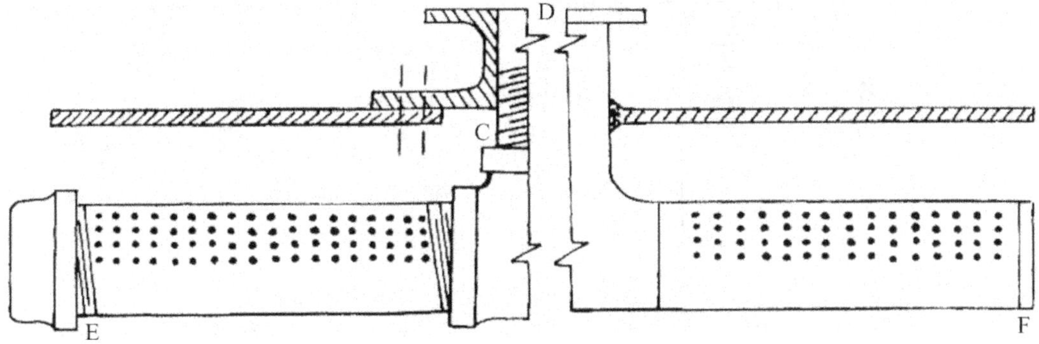

(C) Portion of collecting pipe

(D) Stand pipe for steam stop valve

(E), F Collecting pipe

Fig. 6.28(b) Anti priming pipe.

Function

As steam leaves the boiler, it carries with it certain amount of water (in suspension). The water has much greatest specific gravity than the steam passes through. Anti priming pipe having holes on its surface and both ends of it are closed. Due this construction the water particles being heavier are separated of steam and steam become nearly dry.

Location

It is located below steam stop valve, inside the boiler in steam space.

Construction and Working

EF is collecting pipe, made up of cast-iron. Portion "C" of EF fits into stand pipe D. On stand pipe the valve is fitted. The anti priming pipe is fixed to the boiler. On the upper surface of E F there are large numbers of holes. Steam enters the anti-Priming pipe through these holes. Due to this suspended water particles are separated from steam. This separated water returns back to boiler through drains holes. Nearly dry steam is taken out through steam stop valve.

8. Feed Pump

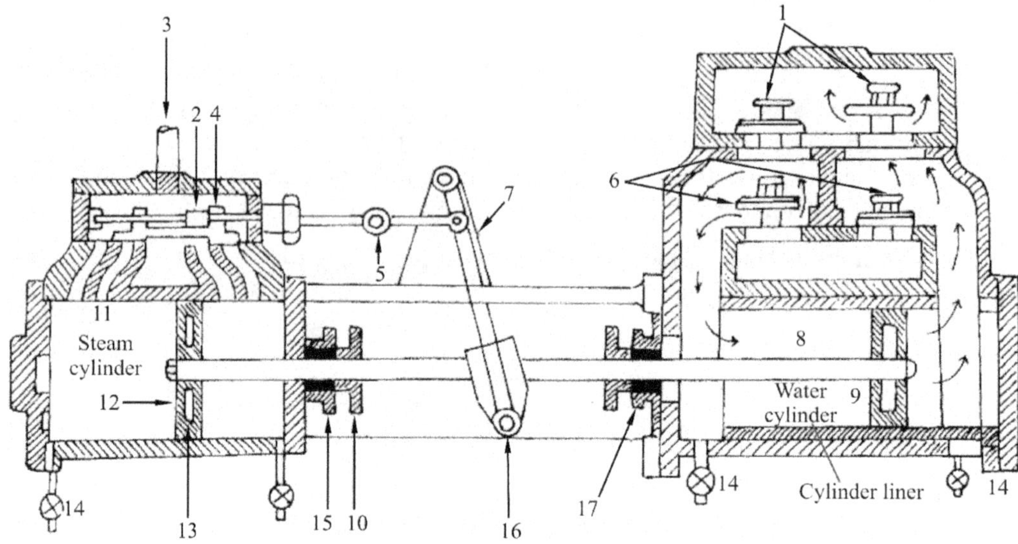

1. Delivery valves
2. Nut for adjusting stroke
3. Steam
4. D-slide valve
5. Valve rod
6. Suction valves
7. Lever
8. Water cylinder
9. Piston
10. Gland
11. Steam cylinder
12. Piston
13. Piston rings
14. Drain cocks
15. Stuffing box
16. Cross head
17. Stuffing box
18. Steam chest

Fig. 6.29(a) Feed pump.

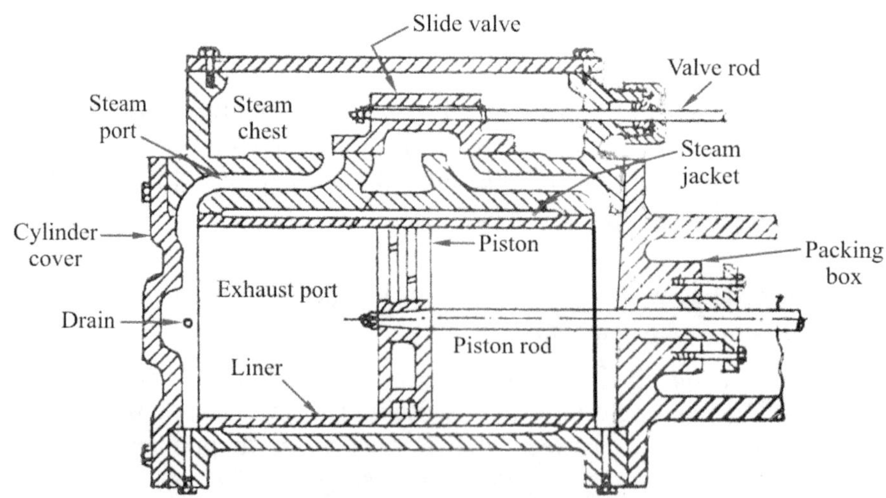

Fig. 6.29(b) Duplex direct-acting feed water pump.

Introduction

There are three types of feed pumps commonly used. They are 1. Reciprocating Pumps 2. Injectors. 3. Rotary Pumps. Rotary Reciprocating pumps are farther classifieds as (i) single acting pump (ii) double acting pump. In single acting pump the water is discharged by one side of piston only so water is discharged in alternate strokes.

In double acting reciprocating pump water is discharged continuously. Single acting reciprocating pumps are suitable for small size and low pressure while Duplex and Triplex Pumps are useful for medium and high pressure boilers respectively. Pumps are continuously run by steam turbine or by electric motor. In large steam power plant, the feed pump used is multi-stage centrifugal pump having four to eight stages. Injectors are used for very small stationary power plants and portable Locomotives. Rotary feed pumps are of centrifugal type and are commonly run either by small steam turbine or by an electric motor.

Rotary pump consists of a casing and rotating element known as impeller which fitted over a shaft. It utilises the centrifugal force of rotating impeller for increasing the pressure of water.

Duplex Feed Pump Function

Feed pump is to deliver pressurised feed water to the boiler. The quantity of water supplied should be at least equal to that evaporation (steam) required.

Location

It is kept near to feed check valve and on the side of boiler.

Construction and working

It is a double acting reciprocating type Feed Pump. It consists of two pumps mounted side by side one is water pump and other is steam pump. Water pump has a water cylinder and steam pump has steam cylinder. Both cylinders are connected to their own piston rod and connected to a common cross head so that the steam pump also serves the driver of water pump. In steam cylinders there are four ports two for steam inlet called steam ports and two for exhaust steam called the exhaust port.

A D-slide valve is used to control the admission and exhaust of steam, D-slide valve is connected by valve rod and valve rod is connected to cross head through lever and rocker arm (lever).

One valve rod moves in the same direction as the piston while other valve moves in opposite direction.

Feed pump is generally double acting i.e., it delivers water on each forward and backward stroke. There is, one suction and one discharge valve on each side of pump plunger. Two pumps work alternatively and continuous flow of water is maintained.

Packing glands and stuffing box are provided for piston rod and valve rod where they enter the cylinder head and valve chest respectively. The glands prevent water and steam from water cylinder and steam cylinder.

There are two steam ports for each end of the cylinder. The steam enters the cylinder through outer port and discharged to atmosphere or condenser through inner port, after it has done work on the piston. D-slide valve is operated by lever connection. Motion to valve is given by rocker arm. The rocker arm gets motion from piston rod. A nut is provided in steam valve chest. This helps to adjust the stroke length. The water cylinder is fitted with drain plugs. The suction and deliver valves are placed above the water cylinder.

The piston in water cylinder moves to and fro by piston of steam cylinder through piston rod and crosshead when piston moves towards right hand side inside water cylinder. The suction valve on right hand side is closed and delivery valve is opened and water is delivered to feed check valve. At the same time suction valve on left hand side of piston is opened and water is admitted in to cylinder. That is on one side of piston suction valve is closed and delivery opens and on other side of piston suction valve is opened and delivery valve is closed.

When piston moves to left hand side, the suction valve is closed and delivery valve is open on that side (L.H.S.). Water is delivered to feed check valve. At the same time suction valve is open on right hand side of piston and delivery valve remain closed.

Advantages of Feed Pump

1. It uses much less steam
2. It can handle hot water
3. Exhaust steam from feed pump can be utilized to heat feed water.
4. Because of simple construction, there are less chances of failure.

Uses

Duplex feed pump is commonly used for medium size steam boiler.

Exercises

1. Define steam boiler.
2. How the boilers are classifieds?
3. What are requirements of boiler?
4. Give the comparison between fire tube boilers and water tube boilers.

5. Explain principle of fire tube and water tube boilers.
6. Describe with neat sketch the working of Cochran boiler, show the different position of mounting.
7. Give an out line sketch showing the arrangement of water tubes and furnace of Babcock and Wilcox boiler. Indicate on it the path of flue gases and water circulation, show the position of the Blow off cock and superheater.
8. Draw line sketch showing a flue passage for Lancashire boiler show also mounting fitted on a Lancashire boiler.
9. Describe with neat sketch the working of Lancashire boiler, show the position of different mountings and accessories.
10. Describe construction and working of locomotive boiler.
11. Why steam jet draught is used in locomotive boiler?
12. What are advantages of high pressure boiler?
13. Describe the construction and working principle of Lamont, Benson and Loeffler high pressure boilers.
14. What are limitations of Lamont boiler?
15. Name only the various mountings used on Boilers.
16. Name various accessories used on boilers.
17. Describe with neat sketch, water level indicator for boiler.
18. Why boiler is provided with two water level indicators?
19. Explain how the flow of steam and water is automatically stopped when glass tube brakes.
20. Why safety valves are provided on boilers. Name various types of safety valves generally used?
21. Mention the function and place of location only of feed check valve, pressure gauge, blow-off cock, fusible plug and steam stop valve.
22. Describe function, location, construction and working of high steam and low water safety valve.
23. Name the boiler on which high steam and low water safety valve is used.
24. Explain the difference between lever safety valve and dead weight safety valve.
25. Sketch and describe a Rams bottom spring loaded safety valve.
26. Discuss the construction and working of pressure gauge. Which pressure is indicated by pressure gauge.
27. Why "U" tube is used for connection pressure gauge with boiler and mention its location on boiler.
28. Compare Cornish boiler with Lancashire boiler.

29. Explain with neat sketch the working of scotch marine boiler.
30. What is purpose of steam stop valve? Explain with neat diagram its working.
31. What is function of superheater? Describe with line sketch Sudgen's superheater.
32. What is economiser? Locate its position on boiler and what are its advantages and disadvantage.
33. Draw a neat diagram of Greens Economiser and explain its working.
34. Explain why Air preheaters are used on high pressure boilers.
35. What are various type of Air preheaters? Give example of Recuperative and Regenerative air preheaters.
36. Draw neat diagram of regenerative air preheater, and explain its working.
37. Explain with neat sketch in brief,
    (i) Steam separator
    (ii) Recuperative airheater (a) Tube type (b) plate type.
38. What is difference between steams trap and steam separator?
39. Discuss the function, construction and working of Duplex feed pump.
40. Explain with line diagram the construction and working of injector.
41. What is difference between feed pump and injector?
42. Explain with neat sketch the function, construction of bucket type steam trap.
43. Discuss function, construction and working of expansion type steam trap, draw necessary line diagram.
44. What is difference between bucket type steam trap and expansion type steam trap?
45. What is function of water heater what are various types of water heaters?
46. Explain construction and working of
    (i) Closed type water heater (ii) Open type water heater.
47. What is function of a steam separator? Describe working of steam separator.
48. What do you understand by the evaporative capacity of boiler?
49. Define the term "factor of evaporation".
50. Explain clearly the equivalent evaporation from and at 100 °C of a boiler.
51. Define boiler efficiency.
52. Enlist the various losses in boiler, which loss is the biggest one?
53. Draw heat balance sheet of a boiler.
54. Explain boiler Power.

Objective Type Questions

1. A device attached to steam chest for preventing explosions due to excessive internal pressure of steam is called.
   (a) Safety valve (b) Steam stop valve
   (c) Fusible plug (d) Blow off cock.
2. A device used to heat feed water by exhaust flue gases before leaving chimney is known as.
   (a) Superheater (b) Ecnomiser
   (c) Expansion type steam trap (d) All of the above
3. Function of blow of cock in boiler while it is working
   (a) To blow off steam to maintain constant pressure in boiler
   (b) To maintain water level in boiler by draining excess water
   (c) To remove sludge and sediments collected in drum
   (d) To empty water drum of boiler.
4. Air Preheater
   (a) Increases evaporative capacity of the boiler
   (b) Increases efficiency of the boiler
   (c) Enables low grade fuel to be burnt
   (d) All of the above
5. Which of the following are boiler accessories
   (a) Safety valve (b) Blow off cock
   (c) Economiser (d) Superheater,
   (e) Both (a) and (b) (f) both (c) and (d)
6. Device used to increase the temperature of saturated steam without increasing the pressure
   (a) Economiser (b) Fusible-plug
   (c) Superheater (d) Air preheater
7. Safety valve mainly used with locomotive and marine boiler
   (a) Lever safety valve (b) Dead weight safety valve
   (c) High pressure and low safety valve (d) Spring loaded safety valve
8. Pressure of feed water to boiler is increased by device known as
   (a) feed check valve (b) steam stop valve
   (c) injector (d) feed pump

9. Which of the following are mountings of boiler
   (a) Economiser (b) Feed pump
   (c) Injector (d) Safety valves

10. A device which is used to put off fire in furnace of boiler when level of water in the boiler falls to an unsafe limit is called
   (a) Fusible plug (b) Blow offcock
   (c) Feed pump (d) Economiser

11. Which device controls the steam discharged from boiler to main pipe or steam is completely stopped from boiler to main pipe.
   (a) Blow off cock (b) Feed check valve
   (c) Steam stop valve (d) Fusible plug

12. Equivalent evaporation as referred to boiler, means
   (a) Ratio of heat actually used in producing steam to heat produced in furnace
   (b) Quantity of steam generated per hour at rated pressure
   (c) Quantity of steam generated from and at STP
   (d) Amount of water evaporated from and at 100 °C in to dry saturated steam at pressure 1.0132 bar.

13. Ratio of heat actually used in producing the steam to heat liberated in furnace due to burning fuel is called
   (a) Evaporative capacity (b) Boiler power
   (c) Equivalent evaporation (d) Boiler efficiency

14. When enthalpy of steam is h kJ/kg and sensible heat of water is h_w and mass of steam produced per kg of coal is m_a. Then equivalent evaporation is given by
   (a) $\frac{m_a(h-h_w)}{2257}$ (b) $\frac{m_a(h+h_w)}{2257}$
   (c) $\frac{(h \times h_w)m_a}{2257}$ (d) $\frac{h}{h_f \times 2257}$

15. In boiler various losses takes place, the biggest loss is due to
   (a) Moisture in fuel (b) Steam formation
   (c) Dry flue gases (d) Unburnt carbon

16. Which boiler has highest efficiency
    (a) Equivalent evaporation = 9.5 kg/kg of coal and rated pressure of boiler = 20 bar
    (b) Equivalent evaporation = 9 kg/kg of coal and rated pressure of boiler = 30 bar
    (c) Equivalent evaporation = 8 kg/kg of coal and rated pressure of boiler = 9 bar
    (d) Equivalent evaporation = 6 kg/kg of coal and rated pressure of boiler = 60 bar
17. Factor of evaporation of steam boiler is given by
    (a) $\dfrac{\text{Acutal evaporation in kg per hour}}{\text{Equivalent evaporation in kg per hour}}$
    (b) $\dfrac{\text{Equivalent evaporation in kg per hour}}{\text{Actual evaporation in kg per hour}}$
    (c) $\dfrac{\text{Actual evaporation in kg per sec}}{\text{Boiler power}}$
    (d) None of above
18. Evaporation factor for steam boiler may be approximately
    (a) 0.8 (b) 0.7 (c) 0.56 (d) 1.2
19. Thermal efficiency of good boiler in power station may in range of
    (a) 80% – 90% (b) 60 – 70% (c) 40 – 55 (d) 25 – 55 %
20. The best suited steam generator for fluctuating demand is
    (a) Lancashire (b) Cochran
    (c) Babcock-Wilcox boiler (d) Locomotive
21. The temperature of flue gases entering the air Preheater of steam generator installation is recommended as
    (a) Less than wet bulb temperature (b) More than wet bulb temperature
    (c) Less than dew point temperature (d) More than dew point temperature
22. Overall efficiency of thermal plant is
    (a) Combine boiler, turbine and alternator efficiency
    (b) Reheating cycle efficiency
    (c) Carnot cycle efficiency
    (d) Regenerative cycle efficiency

23. Thermal efficiency of well maintained boiler is
    (a) 40% (b) 50% (c) 70% (d) 90%
24. Boiler specification are expressed in terms of
    (a) Equivalent evaporation. (b) Thermal efficiency of boiler
    (c) Pressure and temperature of steam (d) All of the above
25. For same diameter and thickness of tube a water tube boiler compare to fire tube boiler has
    (a) More heating surface (b) Less heating surface
    (c) Equal heating surface (d) None of the above
26. Relative percent of heat absorbed through the heat transfer of
    (a) Furnace wall (b) Boiler bank
    (c) Superheater (d) Economiser
    (e) Air preheater, of typical boiler would be of the order of
    A : 40 : 20 : 15 : 8 : 10
    B : 10 : 8 : 15 : 20 : 47
    C : 20 : 47 : 8 : 15 : 10
    D : 0 : 8 : 15 : 20 : 10 : 47

Answers

1.	(a)	2.	(b)	3.	(c)	4.	(d)	5.	(f)	6.	(c)
7.	(d)	8.	(d)	9.	(d)	10.	(a)	11.	(c)	12.	(d)
13.	(d)	14.	(a)	15.	(c)	16.	(a)	17.	(b)	18.	(d)
19.	(a)	20.	(d)	21.	(d)	22.	(a)	23.	(c)	24.	(d)
25.	(a)	26.	(a)								

Exercises

1. The following readings were obtained during a boiler trial of 6 hours duration.

   Steam pressure 12 bar, mass of steam generated = 40000 kg;

   Mean dryness fraction = 0.85, Feed water temperature = 30 °C,

   Coal used = 4000 kg,

   C.V of coal = 33400 kJ/ kg. Take specific heat of water = 4.18 kJ/kg K.

   Calculate,

   1. Factor of equivalent evaporation

2. Equivalent evaporation from and at 100 °C
3. Efficiency of the boiler.

Ans : [fe = 1.045, M_e = 10.45 kg of steam/kg of coal, efficiency of boiler = 70.65%]

2. In trial on a boiler the observations recorded are

Feed water temperature = 50 °C, Boiler pressure = 10 bar,

Quality of steam = 95%,

Coal consumption = 500 kg/hr,

C.V of coal = 30500 kJ/kg,

Feed water supplied = 4000 kg/hr

Determine the equivalent evaporation from and at 100 °C in kg per kg of coal fired and evaporation factor. Take specific heat of water = 4.1868 kJ/kg K.

Ans : [M_e = 8.7475 kg/kg of coal, evaporation factor f = 1.0935]

3. A boiler generates steam at the rate of 9 kg per kg of coal at a pressure of 10 bar with dryness fraction 0.95. The boiler receives its feed water at 110 °C from an economiser which has received water at 30 °C. The steam then passes through a superheater where temperature is raised to 310 °C at constant pressure.

The coal has C.V of 33900 kJ/kg

Determine

(i) The equivalent evaporation at standard conditions in kg per kg of coal

(ii) The efficiency of combined plant

(iii) The percentage of available heat in one kg of coal utilized by boiler, economiser and superheater separately.

Take specific heat of Superheated .steam at 2.1 kJ/kg K and specific heat of water as 4.187 kJ/kgK

Ans : [(i) 11.66 kg/kg of coal (ii) 77.63% (iii) heat utilized by boiler 59%, heat utilize by economiser = 8.89%, heat utilized by superheater = 9.93%]

4. Steam boiler evaporates 5625 kg of water per hour at a pressure of 11 bar and at a temperature of 250 °C. It consumes 750 kg of coal per hour having calorific value of 32000 kJ/kg. Feed water temperature 30 °C.

Find

(i) Actual evaporation per kg of coal

(ii) Factor of equivalent evaporation

(iii) The equivalent evaporation

(iv) Thermal efficiency of boiler

Take specific heat of Superheated steam as 2.3 kJ/kg K of coal and specific heat of water as 4.187 kJ/kg K.

Ans : [(i) 7.5 kg/kg of coal (ii) 1.247 (iii) 9.328 kg/kg of coal (iv) 65.83 %]

5. A boiler raises 3.7 kg of water per kg of coal from feed water temperature at 45 °C. The steam pressure is 10.5 bar if dryness fraction is 0.98. Determine equivalent evaporation from and at 100 °C.

Ans : [4.52 kg/kg of coal]

6. The following observation were made in the case of boiler fitted with economiser.

Rate of steam generated 5 ton/ton of coal.

Equivalent evaporation from and at 100 °C = 5tons/ton of coal

Boiler feed water temperature inlet to economiser = 100 °C

Temperature of feed water inlet to boiler = 180 °C

Temperature of flue gases entering the economiser = 400 °C

Weight of flue gases produced per ton of dry coal = 15 t

Mean specific heat of flue gases = 0.837 kJ/kg K

Calorific value of coal = 22604 kJ/kg.

Determine

(i) Boiler efficiency

(ii) The economiser efficiency

(iii) The combined efficiency of the whole plant.

Ans : [(i) 54.92% (ii) 36.045% (iii) 62.33%]

7. During test on boiler the following data were collected

Steam pressure = 0.5 Bar

Steam temperature = 260°C

Feed water temperature = 38°C

Evaporation = 10 kg/kg of coal

C.V of coal = 33.7 10^3 kJ/kg K

Find equivalent evaporation from and at 100 °C and thermal efficiency of boiler. Take specific heat of superheated steam = 2.1 kJ/kg K and

Specific heat of water = 4.1868 kg/ kg K.

Ans : [12.68 kg/kg of coal, 85%]

8. The following results were obtained in boiler trial

   Feed water per hour = 700 kg

   Temperature of feed water = 27 °C

   Steam pressure = 8.2 bar (gauge)

   Barometer pressure = 1 bar

   Condition of steam = 0.95

   Coal used per hour = 90 kg

   C.V of coal = 27405 kJ/kg

   Ash and unburnt coal collected in ash pit/hr = 7 kg

   C.V of ash unburnt coal collected = 2761 kJ/kg

   Quantity of flue gas = 17.3 kg per kg of coal

   Discharge gas temperature = 325 °C

   Boiler house temperature = 25 °C.

   Mean specific heat of flue gas = 1.025 kJ/kg K

   Determine

   (i) The boiler efficiency

   (ii) The percentage heat loss in flue gases

   (iii) The percentage heat loss to ashes

   (iv) The percentage heat loss unaccounted for

Ans : [(i) 72.52%, (ii) 19.38%, (iii) 7.318%, (iv) 0.782%]

9. During trial of boiler following data was obtained

   Steam pressure = 12.0 bar

   Degree of superheat = 75 °C

   Temperature of feed water = 84°C

   Water evaporated = 3000 kg/hr

   Coal fired = 410 kg/hr

   Percentage combustible in ash = 9%

Moisture in coal = 4%

C.V of dry coal/kg = 30500 kJ/kg

Calorific value of carbon = 33860 kJ/kg of carbon

Specific heat of superheated steam = 2.1 kJ/kg K

Calculate (i) Efficiency of boiler plant with superheater

(ii) Efficiency of boiler and furnace combined.

Ans : [(i) 64.7%, (ii) 65.4%]

10. A boiler produces 1000 kg of steam per hour at pressure of 9 bar from feed water temperature 48°C. If 10% of coal is unburnt at the grate and if efficiency of boiler is 75%, find efficiency of the boiler and grate combined. Take C.V of coal as 31000 kJ/kg.

Ans : [68.2%]

11. Boiler produces 220 tons of dry saturated steam per hour at a pressure of 59 bar absolute from feed water 120°C, ton/day coal combustion 1200 ton/day C.V of coal 17573 kJ/kg, I % of coal escapes unburnt.

Determine

(i) The equivalent evaporation per ton of coal fired

(ii) Efficiency of boiler

(iii) Overall efficiency

Take CP_w = 4.186 kJ/kg K

Ans : [(i) 4.452 $\frac{\text{Ton of steam}}{\text{Ton of coal}}$, (ii) 57.76%, (iii) 57.18%]

12. A boiler consumes 224 tons of coal to produce 1864 tons of steam per day. Steam is dry saturated at 90 atm. (91.193 bar) absolute. Calculate the boiler thermal efficiency and equivalent evaporation per ton if C.V. of coal is 22609 kJ/kg and specific enthalpy of feed water being 425.036 kJ/kg of water.

Ans : $\left[87.7\%,\ 8.55\ \frac{\text{ton of steam}}{\text{ton of coal}}\right]$

CHAPTER 7

Internal Combustion Engines

7.1 Introduction

Internal combustion engine (I.C. engine) is an important prime mover which has wide applications in various fields. It is used in scooters, bikes, cars, trucks, buses, ships, power generation and many industrial applications. It is a heat engine which converts heat energy of fuel into mechanical work. Heat engines are classified into two categories

1. External combustion engine and
2. Internal combustion engine.

In external combustion engine, combustion of fuel takes place outside the engine. Steam engine and steam turbine fall under this category. In internal combustion engine, combustion of fuel takes place inside the engine cylinder. Petrol engine, Diesel engine, gas engine etc. are internal combustion engines.

7.2 Classification of I.C. Engines

I.C. engines are classified as follows:

1. According to the type of the fuel used

   (a) Petrol engine (b) Diesel engine or oil engine (c) Gas engine

2. According to number of strokes required to complete the cycle

   (a) Four stroke cycle engine (b) Two stroke cycle engine

   In four stroke cycle engine, the cycle is completed in four strokes of the piston. Cycle is completed in two revolutions of crank shaft. In two stroke cycle engine, cycle is completed in two strokes of the piston. It is completed in one revolution of crank shaft. Petrol, Diesel or gas engine may work on four stroke cycle or two stroke cycle.

3. According to the cycle of operation

   (a) Otto cycle or constant volume cycle (b) Diesel cycle or constant pressure cycle

   (c) Duel cycle or semi diesel cycle

In Otto cycle, combustion of fuel takes place at constant volume, so heat is added at constant volume. Petrol and gas engines mostly operate on Otto cycle.

In Diesel cycle, combustion of fuel takes place at constant pressure, so heat is added at constant pressure. Diesel engines mostly operate on diesel cycle.

4. According to the method of ignition of fuel

   (a) Spark ignition engines (S.I engines) (b) Compression ignition engines (C.I. engines).

   The engines in which mixture of fuel and air is ignited by an electric spark, is called spark ignition engine. Petrol and gas engines are S.I engines.

   In C.I engines, fuel is ignited as it comes in contact of hot compressed air in the cylinder. Diesel engine is C.I engine.

5. According to the method of cooling the cylinder

   (a) Air cooled engines (b) Water cooled engines

   Small capacity engines are air cooled e.g., scooter engine. Large capacity engines are water cooled e.g., car, bus, truck engines.

6. According to the speed of the engine

   Engines may be classified as low, medium and high speed engines. Petrol engines are generally high speed engines while diesel engines are low or medium speed engines.

7. According to number of cylinders

   I.C. engines may be classified as single cylinder or multicylinder engine

8. According to arrangement of cylinders

   (a) Horizontal engine (b) Vertical engine (c) V-type (d) Radial engine (e) Inline engine

   Scooters and bikes have vertical engines. Earlier aeroplane engines were of radial type.

9. According to application

   (a) Stationary (b) Automobile (c) Air craft (d) Marine (e) Locomotive

7.3 Engine Details

Fig. 7.1 shows sectional view of I.C. engine. Main components of the engine are cylinder, piston, cylinder head, connecting rod, crank shaft, valves, cam shaft, fly wheel, carburettor, spark plug, fuel injector, fuel pump, frame etc.

1. ***Cylinder :*** Cylinder of an engine is cut in a cylinder block. It is open at both ends. The block contains passage for cooling water known as water jackets. Cylinder liner is provided in the cylinder. A separate cylinder head is fitted over the cylinder block with studs and nuts. Cylinder block is made from cast iron.

2. ***Piston :*** It is a gas tight movable cylindrical part which reciprocates in the cylinder under the action of hot gases. Thus, the heat energy is converted into mechanical work. Piston rings are provided around the piston to prevent leakage of hot high pressure gases and to transmit heat from the piston to cylinder wall. Piston is generally made from aluminium alloys.

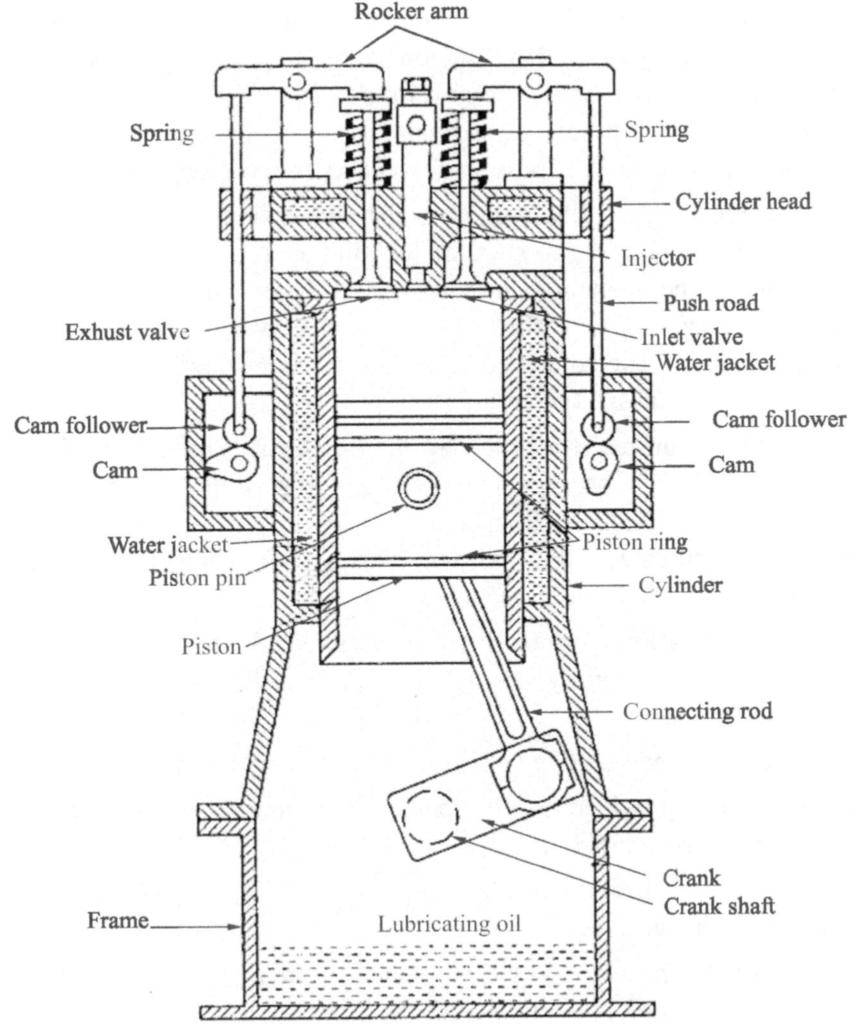

Fig. 7.1 I.C. engine.

3. ***Cylinder head :*** The top of the cylinder is closed by means of a cylinder head. Cylinder head contains inlet and exhaust valves. These valves are kept closed by springs and kept opened by means of rocker arm, push rod and cam.

4. ***Connecting rod :*** It connects piston and crank. Small end of the connecting rod is fitted with piston by piston pin or gudgeon pin or wrist pin. The big end of connecting rod is connected to crank by crank pin. Reciprocating motion of piston is converted to rotary motion of crank shaft by connecting rod and crank. Connecting rod is made from forged steel.
5. ***Crank shaft :*** One end of crank is fitted on crank shaft. Crank shaft is the main rotating part of the engine. It is supported in main bearings. Mechanical work is available at the crank shaft. It drives various accessories like cam shaft, lubricating oil pump, cooling water pump, fuel pump, dynamo, spark distributor etc. Fly wheel is mounted on crank shaft. Crank shaft is made from forged steel.
6. ***Valves :*** There are generally two valves for each cylinder – inlet valve and exhaust valve. The charge i.e., the mixture of fuel and air or only air is admitted into the cylinder due to the opening of inlet valve and the exhaust valve. These valves are operated by means of cam shaft, cam, cam follower, push rod, rocker arm and spring mechanism. The combination of these parts is known as valve gear. Valves are made from nickel steel or chrome steel.
7. ***Cam shaft :*** It is driven by crank shaft through gear. It rotates at half of the crank shaft speed. This is because each valve needs to be operated once in two revolutions of crank shaft. It operates the valve gear mechanism. One cam is provided for each valve. Cam shaft also carries a cam which operates fuel pump or distributor. It is made from forged steel.
8. ***Fly wheel :*** It is a heavy wheel fitted on crank shaft. It minimizes fluctuations of cyclic variation in speed by storing the energy during power stroke and using it during other strokes.
9. ***Carburettor :*** It is provided in petrol engine. It properly mixes the petrol and air.
10. ***Spark plug :*** It is provided in petrol engine. It is fitted in cylinder head. It provides spark to ignite air-fuel mixture.
11. ***Fuel injector :*** It is provided in diesel engine. It is fitted in cylinder head. It injects diesel in the cylinder in the form of fine atomised spray under high pressure.
12. ***Fuel pump :*** It is provided in diesel engine. It pumps the diesel fuel from storage tank to injector at high pressure.

    In petrol engine fuel pump is used to pump the petrol from storage tank to carburettor if storage tank is below the carburettor level.
13. ***Frame :*** It supports cylinder block of the engine. It consists of bed plate and crank case. Bed plate is rigidly fixed with foundation. Lower part of the crank case contains lubricating oil.

7.4 Important Terms Used in I.C. Engines

1. ***Dead centres :*** They are the positions occupied by the piston at the end of stroke where crank and centre line of connecting rod are in a straight line.
2. ***Top dead centre (T.D.C) :*** It is top most positions of piston to cylinder head side (cover end side) in vertical engine.
3. ***Bottom dead centre (B.D.C) :*** It is the bottom most position of piston to crank end side in vertical engine.
4. ***Inner dead centre (I.D.C) :*** It is the inner most position of piston to cylinder head side in horizontal engine.
5. ***Outer dead centre (O.D.C) :*** It is the outer most position of piston to crank end side in horizontal engine.
6. ***Bore :*** Diameter of the cylinder is known as bore. It is denoted by D.
7. ***Piston area :*** It is the area of piston on which products of combustion act. It is denoted by A

$$A = \frac{\pi}{4} D^2 \qquad .....(7.1)$$

8. ***Stroke :*** Travel of piston from one dead centre to other dead centre in an engine is known as stroke.
9. ***Stroke length :*** It is the distance traveled by piston from one dead centre to other dead centre. It is equal to twice the crank radius or crank throw. It is denoted by L.

$$\text{Stroke length } L = 2 \times \text{crank radius} \qquad .....(7.2)$$

10. ***Swept volume :*** Volume swept away by piston in one stroke is known as swept volume. It is denoted by V_s. Its unit is m^3 or litre

$$V_s = \frac{\pi}{4} D^2 L \qquad .....(7.3)$$

$$= AL$$

11. ***Clearance volume :*** Volume entrapped between piston and cylinder head when piston is at T.D.C or I.D.C is known as clearance volume. It is denoted by V_c.
12. ***Compression ratio :*** It is the ratio of volume of cylinder at the beginning of compression stroke to the volume at the end of the stroke. It is denoted by CR or r. Volume at the beginning of compression stroke is $V_s + V_c$. Volume at the end is V_c. So, the compression ratio is

$$r = \frac{V_s + V_c}{V_c} \qquad .....(7.4)$$

For petrol engine, its value varies from 6 to 10, while for diesel engine, it varies from 16 to 20.

13. ***Engine speed :*** Rotational speed of the crank shaft is termed as engine speed. It is measured in revolutions per minute (rpm) or revolutions per second (rps). It is denoted by N.

14. ***Piston speed :*** It is the average piston speed measured in m/s. During one stroke, piston travels a distance equal to stroke length and piston completes two strokes in one revolution of crank shaft, so piston speed will be

    Piston speed S = 2LN m/s, if N is in rps. .....(7.5)

7.5 Otto Four Stroke Cycle Engine

Petrol engine works on Otto cycle. Petrol fuel is vaporized in the carburettor and a gaseous mixture of petrol and air is prepared here.

Spark plug is provided in petrol engine to produce spark for ignition of air fuel mixture.

The cycle is completed in four strokes of the piston. These are 1. Suction stroke 2. Compression stroke 3. Expansion or power or working stroke 4. Exhaust stroke.

Fig. 7.2 shows working of four stroke petrol engine.

Fig. 7.3 shows p-V diagram for this engine.

1. ***Suction stroke :*** As shown in Fig. 7.2(a), inlet valve remains open and exhaust valve remains closed during this stroke. Piston moves from TDC to BDC. Air fuel mixture (charge) from the carburettor is sucked into the cylinder at nearly atmospheric pressure as piston moves down.

   During this stroke, pressure remains constant with increase in volume. This is denoted by line 0-1 on p-V diagram (Fig. 7.3). Crank shaft makes half revolution during the stroke. Inlet valve is closed when piston reaches BDC.

2. ***Compression stroke :*** As shown in Fig. 7.2(b), both valves remain closed during this stroke. Piston moves from BDC to TDC and crank shaft makes other half revolution. Since both valves are closed during upward motion of piston, air fuel mixture is compressed. Its pressure and temperature are increased and volume is decreased. This is shown by line 1-2 on p-V diagram. The compression is adiabatic. At or near the end of stroke, the charge is ignited by means of electric spark produced from spark plug. This combustion is almost instantaneous. So, there is sudden rise in pressure at constant volume. Temperature also rapidly increases. This heat addition is represented by lines 2-3on p-V diagram.

3. ***Expansion stroke :*** A shown in Fig. 7.2(c), both inlet and exhaust valves remain closed during this stroke. Piston moves from TDC to BDC. High pressure combustion product acts on piston and piston moves downward. Thus, work is done during this stroke, so this is a working or power stroke. The expansion of gas is adiabatic. Pressure decreases and volume increases. This is represented by line 3-4 on p-V diagram. At the end of the stroke, exhaust valve opens and the hot gases are discharged into the atmosphere at constant volume with decrease of pressure. This is represented by line 4-1 on p-V diagram.

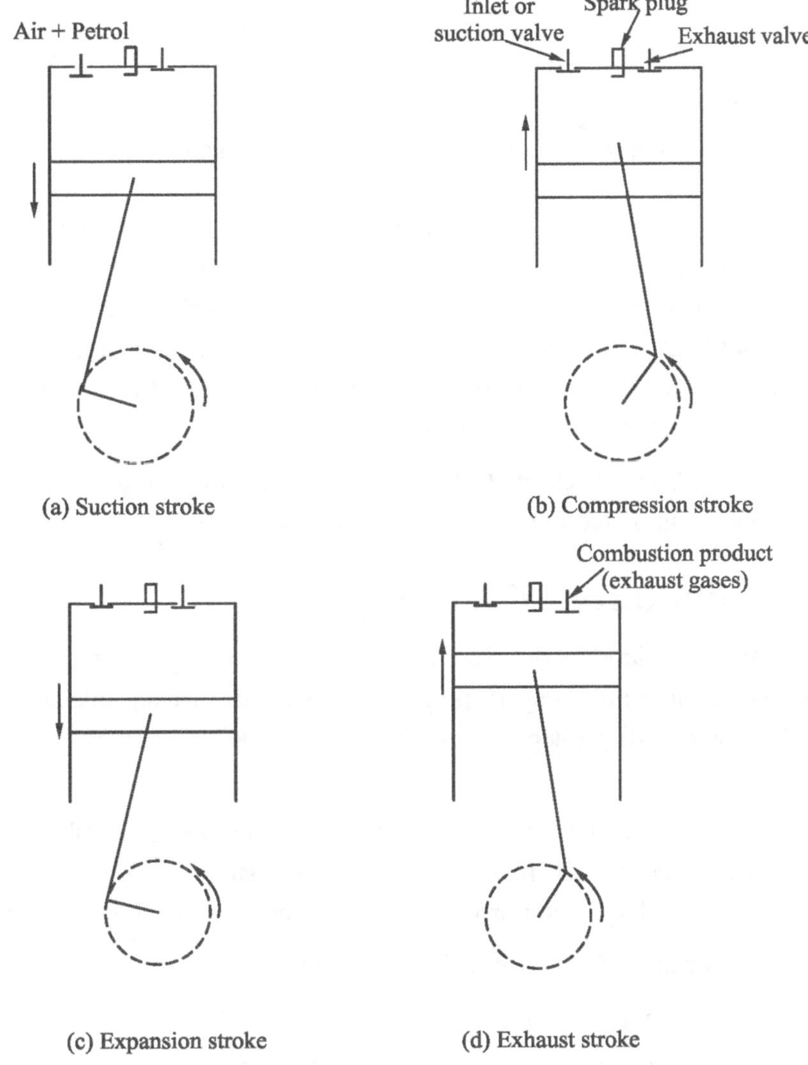

Fig. 7.2 Working of four stroke petrol engine.

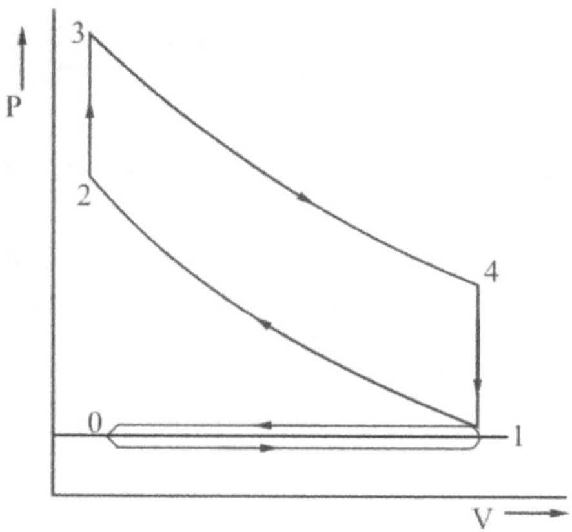

Fig. 7.3 p-v diagram of 4-stroke petrol engine.

4. ***Exhaust stroke :*** As shown in Fig. 7.2(d), inlet valve remains closed and exhaust valve is open during this stroke. Piston moves from BDC to TDC. During upward movement of piston, it pushes the remaining exhaust gases out of the cylinder at nearly atmospheric pressure. Pressure remains constant with decrease in volume. This is represented by line 1-0 on p-V diagram. At the end of the stroke, exhaust valve is closed and inlet valve is opened.

   There will be again a suction stroke and the cycle will be repeated. One cycle of four strokes is completed in two revolutions of crank shaft.

7.6 Diesel Four Stroke Cycle Engine

Diesel engine works on diesel cycle. Here ignition of fuel takes place in presence of high pressure high temperature air, so spark plug is not required. Instead, injector is provided to inject the fuel with high pressure. Fuel pump is provided to increase the pressure of fuel.

The cycle is completed in four strokes and two revolutions of crank shaft. In each stroke, crank shaft makes half revolution. The four strokes are 1. Suction stroke 2. Compression stroke 3. Expansion or working or power stroke 4. Exhaust stroke.

Fig. 7.4 shows working of four stroke diesel engine.

Fig. 7.5 shows p-V diagram for this engine.

1. ***Suction stroke :*** As shown in Fig. 7.4(a), inlet valve remains open and exhaust valve remains closed during this stroke. Piston moves from TDC to BDC. Due to

downward motion of piston, pressure inside the cylinder decreases slightly below atmospheric, so air is sucked into the cylinder from the atmosphere through inlet valve. Pressure remains constant with increase in volume. This is represented by line 0-1 on p-V diagram. Inlet valve is closed when piston reaches BDC.

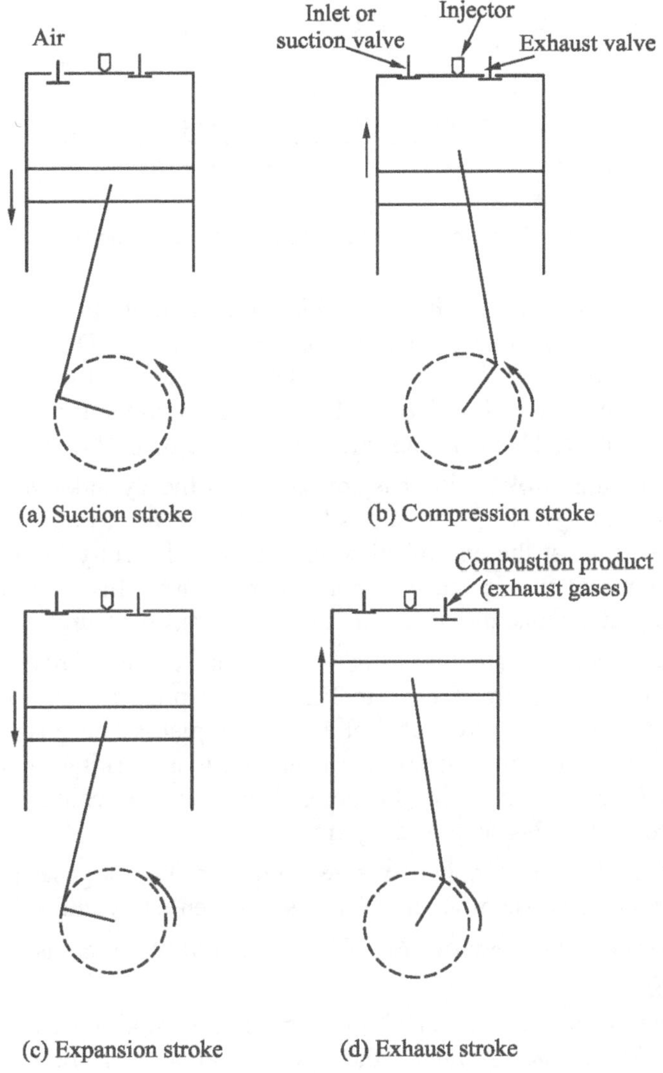

Fig. 7.4 Working of four stroke diesel engine.

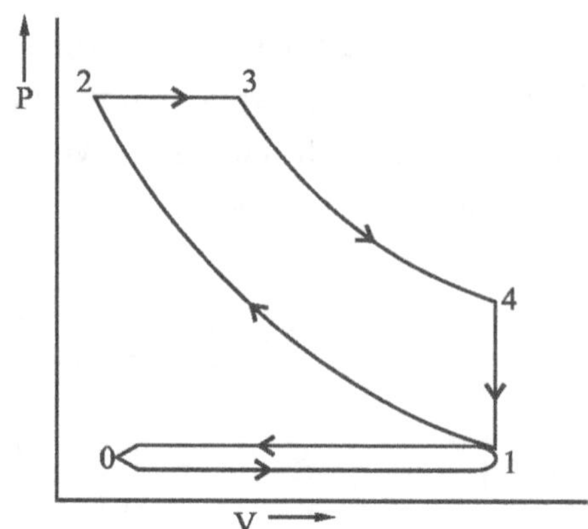

Fig. 7.5 p-v diagram for four stroke diesel engine.

2. ***Compression stroke :*** As shown in Fig. 7.4(b), both inlet and exhaust valves remain closed during this stroke. Piston moves from BDC to TDC. Since both valves are closed, the air in the cylinder is compressed by the upward movement of piston. The compression is adiabatic. Pressure and temperature of air increases with decrease in volume. This is represented by line 1-2 on p-V diagram.

   At the end of the stroke, diesel is injected into the cylinder by the fuel injector. Compression ratio of diesel engine is higher than that of petrol engine, so that the temperature of air at the end of compression is sufficiently high to ignite the fuel. The combustion of fuel is nearly at constant pressure. This is represented by line 2-3 in p-V diagram. Thus, the heat added is at constant pressure.

3. ***Expansion stroke :*** As shown in Fig. 7.4(c), both valves remain closed during this stroke. Piston moves from TDC to BDC. High pressure high temperature gases produced due to combustion of fuel act on the piston and push it downward. So, the work is done during this stroke. Thus this is a working or power stroke. The expansion of gas is adiabatic. Pressure decreases and volume increases. This is represented by line 3-4 on p-V diagram.

   At the end of the stroke, exhaust valve opens and the hot gases are discharged into the atmosphere at constant volume. This is represented by line 4-1 on p-V diagram.

4. ***Exhaust stroke :*** As shown in Fig. 7.4(d), inlet valve remains closed and exhaust valve is opened.

   Piston moves from BDC to TDC. During upward motion of piston, it pushes the remaining exhaust gases out of the cylinder at a pressure slightly higher than atmospheric. Pressure remains constant with decrease in volume. This is represented by line 1-0 on p-V diagram. At the end of the stroke, exhaust valve is closed and inlet valve is opened.

   There will be again a suction stroke and the cycle will be repeated.

7.7 Difference between Otto Cycle and Diesel Cycle Engine

Otto cycle engine		Diesel cycle engine	
1.	Petrol engine works on otto cycle	1.	Diesel engine works on diesel cycle
2.	During suction stroke, air petrol mixture is sucked into the cylinder	2.	During suction stroke only air is sucked into the cylinder
3.	Carburettor is required for proper mixing of air and petrol	3.	No carburetor is required, but fuel pump is required to increase the pressure of diesel
4.	Spark plug is required to produce the spark	4.	Fuel injector is required for fine spray of diesel into the cylinder
5.	Air petrol mixture is ignited by spark, so it is a S.I engine	5.	Diesel is ignited by high temperature compressed air, so it is a C.I. engine
6.	Heat is added at constant volume	6.	Heat is added at constant pressure
7.	Compression ratio is about 6 to 10	7.	Compression ratio is about 16 to 20
8.	Due to lower compression ratio, engine is lighter in weight	8.	Due to higher compression ratio, engine is heavier and stronger than petrol engine for the same power
9.	Initial cost is low	9.	Initial cost is high
10.	Thermal efficiency is lower due to lower CR	10.	Thermal efficiency is higher due to higher CR
11.	Due to lighter engine, it is used in light vehicles like scooter, bike, car etc.	11.	Due to heavy engine, it is used in heavy vehicles like bus, truck, jeep etc.

7.8 Two Stroke Cycle Engine

In two stroke engine, the cycle is completed in two strokes of the piston or in one revolution of the crank shaft. Petrol and diesel both engines can work on two stroke cycle. Fig. 7.6 shows working of two stroke petrol engine while. Fig. 7.8 shows working of two stroke diesel engine. Fig. 7.7 and Fig. 7.9 show their p-V diagrams respectively.

In two stroke engine, ports are used instead of valves. Crank case is provided at the bottom of the cylinder, which includes crank and crank shaft. Inlet port opens into crank case while exhaust port opens into atmosphere. Transfer port is provided connecting crank case with cylinder. The exhaust gases are expelled out from the engine cylinder by fresh charge of fuel entering the cylinder. This is known as scavenging.

Working of two stroke petrol engine and two stroke diesel engine are described below.

(a) Working of two stroke petrol engine:

Refer Fig. 7.6 and Fig. 7.7

1. ***First stroke :*** Assume that at the beginning of stroke, piston is at BDC as shown in Fig. 7.6(a). During this stroke, piston moves from BDC to TDC. The air petrol mixture (charge) has already entered into the cylinder from crank case through transfer port. During upward motion, piston first covers the transfer port and then exhaust port. So, the air-petrol mixture is compressed. Thus, this is a compression stroke. Pressure and temperature of mixture increases and volume decreases. This is represented by line 1-2 on p-V diagram.

   During upward movement, piston uncovers the inlet port. Also, partial vacuum is created in the crank case, so fresh mixture of air-petrol is sucked into the crankcase from carburettor through inlet port, as shown in Fig. 7.6(b).

   At the end of the stroke, the compressed charge is ignited by the spark produced from spark plug. Due to combustion of fuel, pressure and temperature of combustion product (hot gases) increase rapidly, nearly at constant volume. Thus, heat is added at constant volume. This is represented by line 2-3 on p-V diagram.

2. ***Second stroke :*** During this stroke piston moves from TDC to BDC. Hot gases, due to their high pressure and temperature, act on the piston. So, the piston moves downward and does mechanical work as shown in Fig. 7.6(c). This is an expansion or working or power stroke. It is represented by line 3-4 on p-V diagram.

   During downward motion, piston covers the inlet port and fresh charge is compressed in the crank case. Further downward motion of piston uncovers the exhaust port and then transfer port as shown in Fig. 7.6(d). The exhaust gases start escaping through the exhaust port. This is shown by line 4-1 on p-V diagram. At the same time, fresh charge compressed in crank case is forced into the cylinder through transfer port. The charge strikes the deflector on the piston crown, rises to the top of the cylinder and pushes out exhaust gases. When the piston reaches BDC, the cylinder is completely filled with fresh charge. Thus, the cycle is completed.
   The cycle is then repeated.

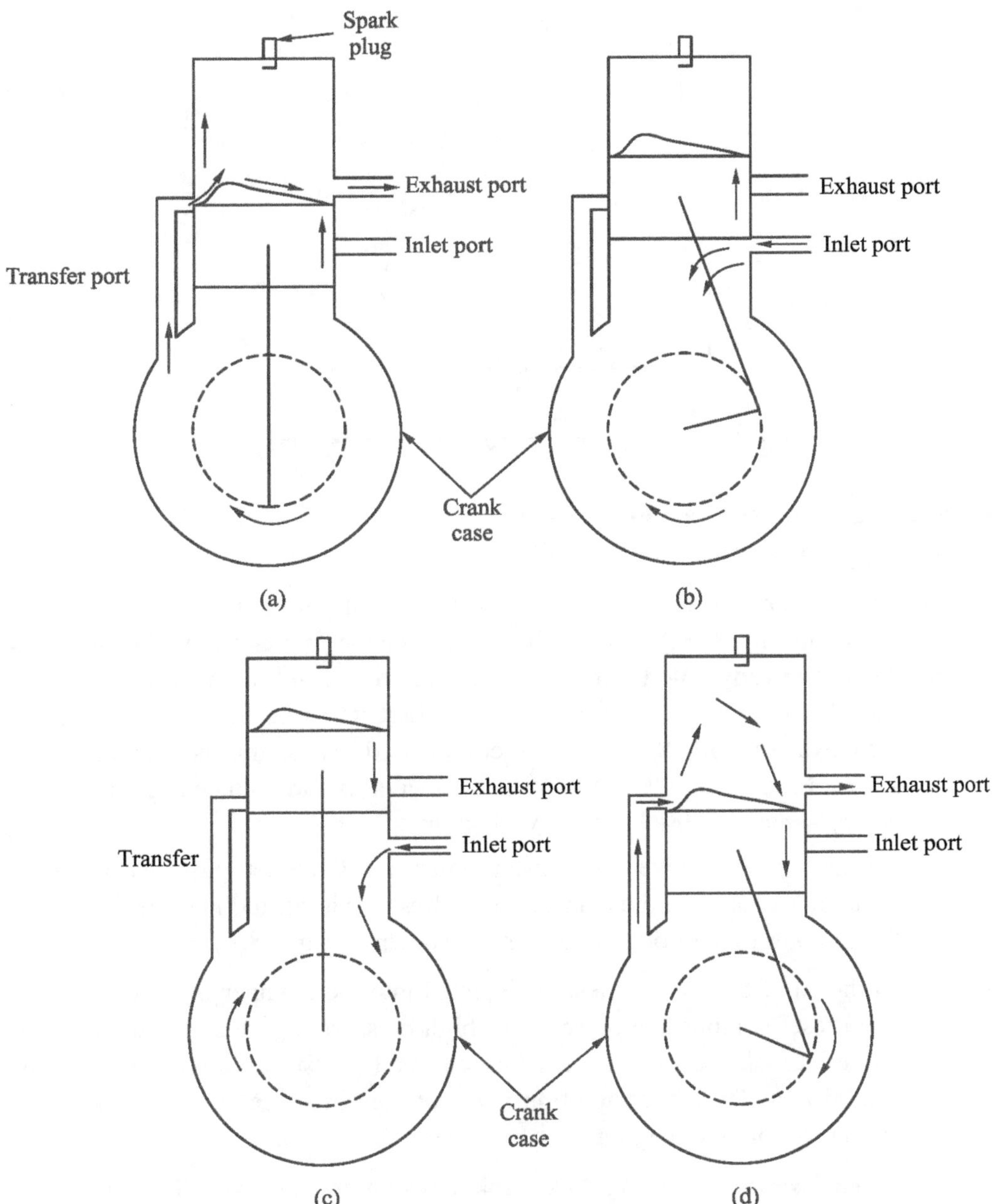

Fig. 7.6 Working of two stroke petrol engine.

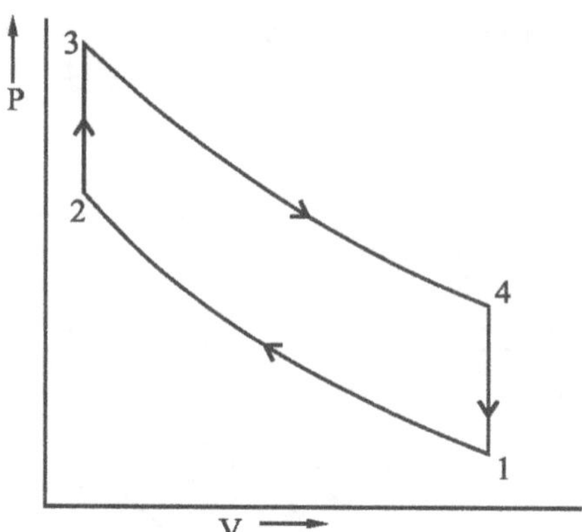

Fig. 7.7 p-V diagram for two stroke petrol engine.

(b) Working of two stroke diesel engine

Refer Fig. 7.8 and Fig. 7.9

1. ***First stroke :*** Assume that at the beginning of stroke, piston is at BDC as shown in Fig. 7.8(a). During this stroke, piston moves from BDC to TDC. Air has already entered into the cylinder from crank case through transfer port. During upward motion of piston, first it covers the transfer port and then exhaust port. So the air is compressed. Thus, this is a compression stroke. Pressure and temperature of air increases and volume decreases. This is represented by line 1-2 on p-V diagram.

   During upward movement, piston uncovers the inlet port. Also, partial vacuum is created in the crank case, so fresh air is sucked into the crank case from atmosphere through inlet port, as shown in Fig. 7.8(b).

   At the en of the stroke, diesel is injected into the cylinder from the injector. The diesel is ignited in presence of high pressure high temperature air. The rate of diesel injection is such that constant pressure is maintained during combustion. Thus, heat addition is at constant pressure. This is represented by line 2-3 on p-V diagram.

2. ***Second stroke :*** During this stroke, piston moves from TDC to BDC. Combustion products (hot gases), due to their high pressure and temperature act on the piston. So the piston moves downwards and does mechanical work as shown in Fig. 7.8(c). This is the expansion or working or power stroke. It is represented by line 3-4 on p-V diagram.

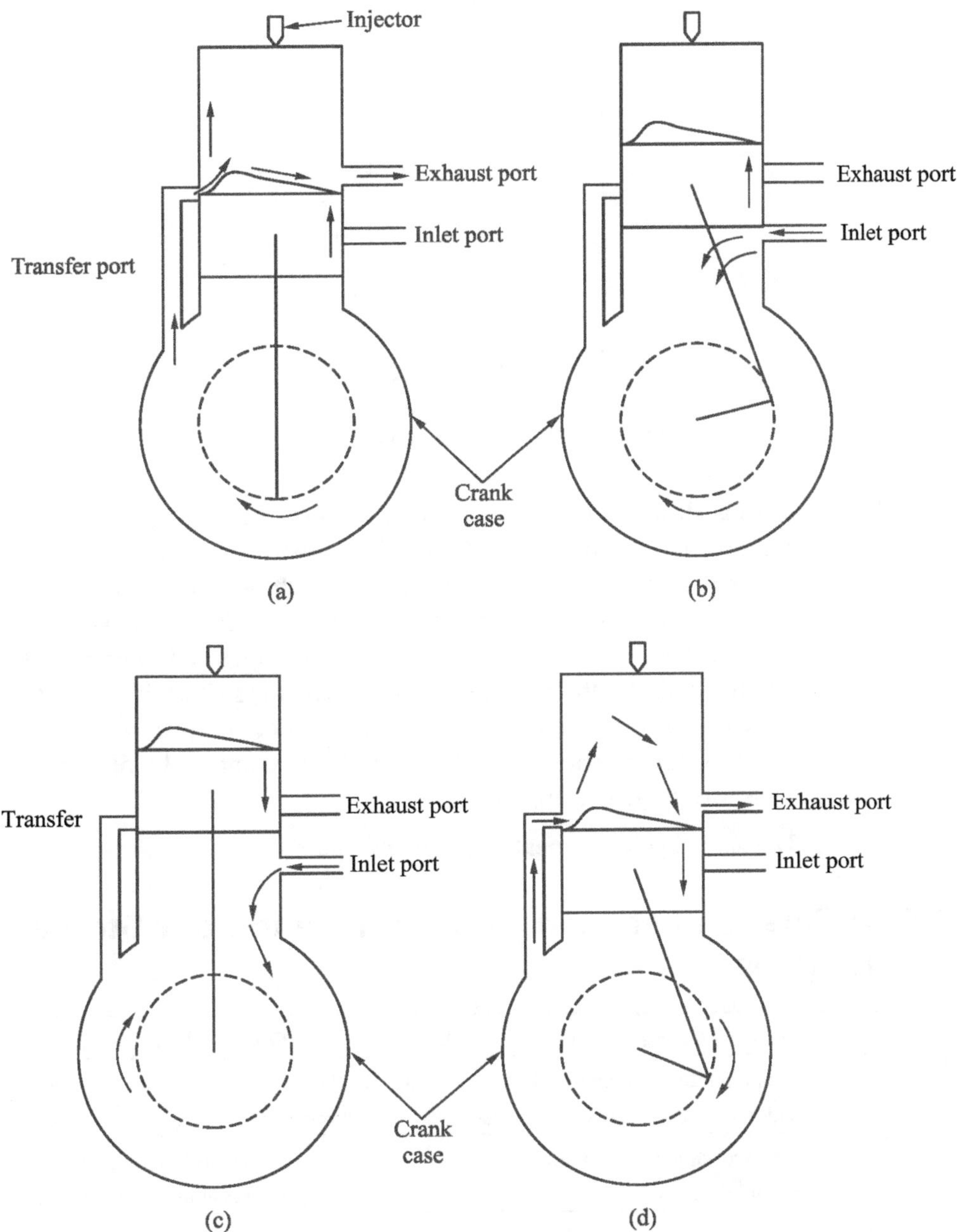

Fig. 7.8 Working of two stroke diesel engine.

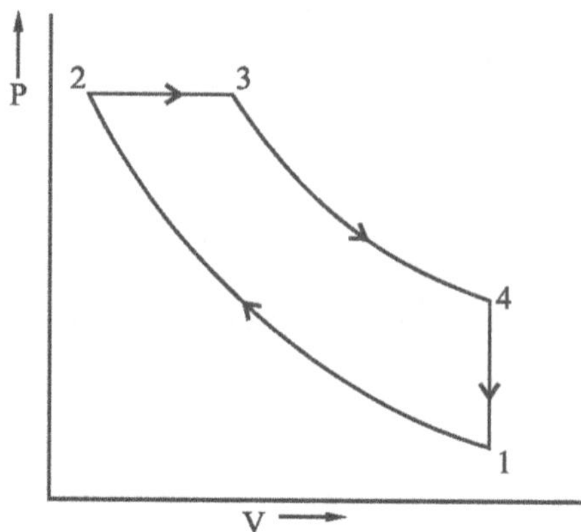

Fig. 7.9 Working of two stroke diesel engine.

During downward motion, piston covers the inlet port and fresh air is compressed in the crank case. Further downward motion of piston uncovers the exhaust port and then transfer port as shown in Fig. 7.8(d). The exhaust gases start escaping through the exhaust port. This is shown by line 4-1 on p-V diagram. At the same time, fresh air compressed in crank case is forced into the cylinder through transfer port. The air strikes the deflector on piston crown. Due to this, air rises up in the cylinder and pushes exhaust gases out. When piston reaches BDC, the cylinder is completely filled with fresh air. Thus, the cycle is completed.

The same cycle is repeated again and again.

7.9 Difference between Two-stroke and Four Stroke Cycle Engines

	Two stroke cycle engine		Four stroke cycle engine
1	The cycle of operation is completed in two strokes and in one revolution of crank shaft.	1.	The cycle of operation is completed in four strokes and in two revolutions of crank shaft.
2.	There is one working or power stroke for every revolution of crank shaft.	2.	There is one working or power stroke for every two revolutions of crank shaft.
3.	Ports are provided instead of valves	3.	Valves are provided
4.	Construction is simple as no valve gear mechanism is required	4.	Construction is complicated because of valve gear mechanism
5.	Some fuel is wasted in petrol two stroke engine during scavenging	5.	No such wastage of fuel

Contd...

Two stroke cycle engine		Four stroke cycle engine	
6.	Thermal efficiency is low due to fuel requirement in each revolution	6.	Thermal efficiency is high
7.	Mechanical efficiency is high due to simple construction	7.	Mechanical efficiency is low
8.	For the same power developed the engine is lighter and occupies less space	8.	The engine is heavier and occupies more space
9.	There is less variation in torque, requiring lighter and smaller flywheel	9.	There is more variation in torque, requiring heavier and larger flywheel
10.	Balancing is better, less vibrations	10.	Balancing is not that good, more vibrations
11.	Engine consumes more lubricating oil due to one power stroke in each revolution	11.	Engine consumes less lubricating oil
12.	As the engine is lighter, it is generally used for scooter, bike etc	12.	Generally used in car, bus, truck etc.

7.10 Indicated Power (ip)

Power developed in the cylinder of an engine is known as indicated power.

If p_m = indicated mean effective pressure, N/m^2 or Pa

A = cross sectional area of piston, m^2

n = number of working stroke per minute

Then

Force acting on piston by hot gases $= p_m \times A$ N

$\therefore$ work done during the stroke $= p_m \times A \times L$ J

$\therefore$ work done per minute $= p_m \times A \times L \times n$ J/min

$\therefore$ work done per second i.e. power

$$ip = \frac{p_m A L n}{60} \text{ w} \quad .....(7.6)$$

If p_m is in kPa, ip will be in kW

For four stroke engine $n = \dfrac{N}{2}$

where N = speed of engine in rpm

$$\therefore \quad ip = \frac{p_m A L N}{60 \times 2} \quad .....(7.7)$$

Fro two stroke engine n = N

$$\therefore \quad ip = \frac{p_m A L N}{60} \qquad .....(7.8)$$

7.11 Brake Power (bp)

Power available at engine shaft is known as brake power. It is less then indicated power because some power is lost in friction of moving parts. Generally, brake arrangement is used to measure it, hence it is named as brake power

$$bp = \frac{p_{mb} A L n}{60} = \frac{2 \pi N T}{60} \text{ W} \qquad .....(7.9)$$

where p_{mb} = brake mean effective pressure, N/m^2 or Pa

T = torque available at engine shaft, N-m .....(7.10)

Friction power Fp = ip – bp

7.12 Efficiencies

Various efficiency terms related with I.C. engines are define below.

1. **Mechanical efficiency η_m**

   The ratio of power available at crank shaft (bp) to the power developed by the engine cylinder (ip) is known as mechanical efficiency.

$$\eta_m = \frac{bp}{ip} \qquad .....(7.11)$$

2. **Thermal efficiency**

   The ratio of work output of the engine to the heat supplied by the fuel is known as thermal efficiency. It may be of two types

   (a) Indicated thermal efficiency

   (b) Brake thermal efficiency

   (a) ***Indicated thermal efficiency η_i***

   It is the ratio of indicated power to the heat supplied by fuel per sec

$$\eta_i = \frac{\text{Indicated power, kw}}{\text{Fuel supplied, kg/s} \times \text{calorific value of fuel, kJ/kg}}$$

$$= \frac{ip}{m_f \times CV} \qquad .....(7.12)$$

   Calorific value of fuel is defined as the amount of heat supplied by combustion of 1 kg of fuel.

(b) ***Brake thermal efficiency η_b***

It is the ratio of brake power to the heat supplied by fuel per sec

$$\eta_b = \frac{\text{Brake power, kw}}{\text{Fuel supplied, kg/s} \times \text{calorific value of fuel, kJ/kg}}$$

$$= \frac{bp}{m_f \times CV} \quad .....(7.13)$$

From equations (7.11) and (7.12), it can be written as

$$\eta_m = \frac{\eta_b}{\eta_i} \quad .....(7.14)$$

Also, from equations (7.6) and (7.9)

$$\eta_m = \frac{p_{mb}}{p_m}$$

3. Relative efficiency η_r

It is the ratio of actual thermal efficiency to its corresponding air standard efficiency

$$\eta_r = \frac{\text{Actual thermal efficiency}}{\text{Air standard efficiency}} \quad .....(7.15)$$

4. Brake specific fuel consumption (bsfc)

The amount of fuel consumed by the engine in one hour to generate one kW brake power is known as brake specific fuel consumption. Its unit is kg/kW–hr i.e. kg/kWh. It indicates the performance of the engine.

$$\text{bsfc} = \frac{m_f \text{ kg/s} \times 3600}{bp} \quad .....(7.16)$$

Examples

1. A two stroke petrol engine has piston diameter 10 cm and stroke length 15 cm. The mean effective pressure is 500 kN/m^2. It runs at a speed of 900 rpm. Mechanical efficiency of the engine is 80%. Find the brake power of the engine.

Solution:

Given

$D = 10$ cm, $L = 15$ cm, $p_m = 500$ kN/m^2, $N = 900$ rpm, $\eta_m = 0.8$

$$ip = \frac{p_m \, A \, L \, N}{60} = \frac{p_m \times \frac{\pi}{4} D^2 \, L \, N}{60}$$

$$= \frac{500 \times \frac{3.14}{4} \times (0.10)^2 \times 0.15 \times 900}{60}$$

$$= 8.83 \text{ kW}$$

$$\eta_m = \frac{bp}{ip}$$

$\therefore$ $$bp = \eta_m \times ip = 0.8 \times 8.83 = \mathbf{7.065 \text{ kW}}$$

2. A four stroke six cylinder engine with stroke volume of 1.75 litre develops 26.25 kW. The m.e.p. is 6 bar. Find speed of the engine.

Solution:

Given

$V_s = 1.75 \times 10^{-3}$ m^3, ip = 26.25 kW, $p_m = 600$ kN/m^2

$$ip = \frac{p_m \times A \, L \, N}{60 \times 2} \times \text{number of cylinders}$$

$$= \frac{p_m \times V_s \, N}{60 \times 2} \times 6 \, \ldots\ldots\ldots (\because \, V_s = AL)$$

$\therefore$ $$N = \frac{ip \times 60 \times 2}{p_m \times V_s \times 6}$$

$$= \frac{26.25 \times 60 \times 2}{600 \times 1.75 \times 10^{-3} \times 6}$$

$$= \mathbf{500 \text{ rpm}}$$

3. A gas engine has a piston diameter of 150 mm and stroke length 250 mm. The speed of the engine is 250 rpm. Average number of explosions are 90 per minute. mep is 7 bar. If average torque of the brake is 140 N-m, Find ip, bp and η_m.

Solution:

Given

D = 150 mm, L = 250 mm, N = 250 rpm, n = 90, p_m = 700 kPa, T = 140 N-m

$$ip = \frac{p_m A L n}{60}$$

$$= \frac{700 \times \frac{\pi}{4} \times (0.15)^2 \times 0.25 \times 90}{60}$$

$$= \mathbf{4.64\ kW}$$

$$bp = \frac{2\pi N T}{60}$$

$$= \frac{2\pi \times 250 \times 140}{60} = 3663\ W$$

$$= \mathbf{3.66\ kW}$$

$$\eta_m = \frac{bp}{ip}$$

$$= \frac{3.66}{4.64}$$

$$= 0.789$$

$$= \mathbf{78.9\%}$$

4. A six cylinder 4-stroke petrol engine is to develop 100 kW ip at 800 rpm. Stroke to bore ratio is 1.25. Mechanical efficiency is 80%. bmep is 5 bar. Find bore and stroke of the engine.

Solution:

Given

$ip = 100\ kW, N = 800\ rpm, L/D = 1.25, \eta_m = 0.8, p_{mb} = 500\ kN/m^2$

$$\eta_m = \frac{p_{mb}}{p_m}$$

$$\therefore \quad p_m = \frac{p_{mb}}{\eta_m} = \frac{500}{0.8} = 625\ kN/m^2$$

$$ip = \frac{p_m A L N}{60 \times 2} \times \text{no. of cylinders}$$

$$= \frac{p_m \times \frac{\pi}{4} D^2 \times 1.25D \times N \times 6}{60 \times 2}$$

$$\therefore \quad 100 = \frac{625 \times \frac{\pi}{4} \times D^3 \times 1.25 \times 800 \times 6}{60 \times 2}$$

$$\therefore \quad 100 = 24531.25\ D^3$$

$$\therefore \quad D^3 = 0.004076$$

$$\therefore \quad D = 0.1597\ m$$

$$= 16\ cm$$

$$\therefore \quad L = 16 \times 1.25$$

$$= \mathbf{20\ cm}$$

5. A petrol engine with a stroke length of 200 mm and diameter of 150 mm has a clearance volume of 5×10^5 mm^3. η_i of the engine is 30%. Find the relative efficiency of this engine. If p_m is 5 bar and the engine speed is 1000 rpm, find ip of the engine.

Solution:

Given

$L = 20$ cm, $D = 15$ cm, $V_c = 5 \times 10^5 \times 10^{-3}$ cm^3, $\eta_i = 0.3$, $p_m = 500$ kPa, $N = 1000$ rpm

$$V_s = \frac{\pi}{4} D^2 L = \frac{\pi}{4} \times 15^2 \times 20 = 3532.5\ cm^3$$

$$r = \frac{V_c + V_s}{V_c}$$

$$= \frac{500 + 3532.5}{500}$$

$$= \mathbf{8.1}$$

$$\eta_a = 1 - \frac{1}{r^{\gamma-1}} = 1 - \frac{1}{(8.1)^{0.4}} = 1 - 0.433 = 0.567$$

$$\eta_r = \frac{\eta_i}{\eta_a} = \frac{0.3}{0.567} = 0.53 = 53\%$$

$$ip = \frac{p_m\, A\, L\, N}{60}$$ assuming two stroke engine

$$= \frac{500 \times \frac{\pi}{4} \times (0.15)^2 \times 0.20 \times 1000}{60}$$

$$= \mathbf{29.4\ kW}$$

6. A C.I engine at rated condition develops 7.5 kW bp. The mechanical losses are 1.5 kW. If η_i is 42% and CV of fuel is 43260 kJ/kg, determine (i) fuel consumption in kg/hr, (ii) η_b.

Solution:

Given

bp = 7.5 kW, fp = 1.5 kW, $\eta_i = 0.42$, CV = 43260 kJ/kg

ip = bp + fp = 7.5 + 1.5 = 9 kW

$$\eta_i = \frac{ip}{m_f \times CV}$$

$$\therefore \quad m_f = \frac{ip}{\eta_i \times CV} = \frac{9 \times 3600}{0.42 \times 43260} = 1.783 \text{ kg/hr}$$

$$\eta_b = \eta_i \times \eta_m = \eta_i \times \frac{bp}{ip}$$

$$= 0.42 \times \frac{7.5}{9} = 0.35 = \mathbf{35\%}$$

7. A petrol engine having a compression ratio of 8 has a brake thermal efficiency which is 40% of the ideal air standard efficiency. The calorific value of petrol is 41500 kJ/kg. Find fuel consumption in kg/hr if brake power of the engine is 16 kW.

Solution:

Given

r = 8, $\eta_b = 0.4\ \eta_a$, CV = 41500 kJ/kg, bp = 16 kW

$$\eta_a = 1 - \frac{1}{r^{\gamma-1}} = 1 - \frac{1}{(8)^{0.4}} = 1 - 0.435 = 0.565$$

$$\eta_b = 0.4 \times 0.565$$

$$= 0.226$$

$$\eta_b = \frac{bp}{m_f \times CV}$$

$$\therefore \quad m_f = \frac{bp}{\eta_b \times CV} = \frac{16}{0.226 \times 41500} = 0.0017 \text{ kg/s}$$

$$= 0.0017 \times 3600 \text{ kg/hr}$$

$$= \mathbf{6.14\ kg/hr}$$

8. Following data refers to a petrol engine

ip = 30 kW, bp = 26 kW, bsfc = 0.35 kg/kWh, CV of fuel = 44000 kJ/kg.

Find (i) η_m (ii) η_i (iii) η_b

Solution:

Given

ip = 30 kW, bp = 26 kW, bsfc = 0.35 kg/kWh, CV = 44000 kJ/kg

$$\eta_m = \frac{bp}{ip} = \frac{26}{30} = 0.8667 = 86.67\%$$

$$bsfc = \frac{m_f \text{ kg/hr}}{bp \text{ kW}}$$

$$\therefore \quad m_f = bsfc \times bp$$

$$= 0.35 \times 26$$

$$= 9.1 \text{ kg/hr}$$

$$\eta_i = \frac{ip}{m_f \times CV} = \frac{30}{\frac{9.1}{3600} \times 44000}$$

$$= 0.2697 = 26.97\%$$

$\eta_b = \eta_i \times \eta_m = 0.2697 \times 0.8667 = 0.2337 =$ **23.37%**

9. The ip of a six cylinder 4-stroke engine is 150 kW at an average piston speed of 320 m/min. Stroke bore ratio is 1.2. If mep is 650 kN/m^2, determine crank shaft speed.

Solution:

Given

ip = 150 kW, piston speed S = 320 m/min, $L/D = 1.2$, $p_m = 650$ kN/m^2

$$S = 2\,LN \qquad \therefore\ LN = S/2$$

$$ip = \frac{p_m\,A\,L\,N}{60 \times 2} \times 6$$

$$\therefore \quad 150 = \frac{650 \times A \times 320/2}{60 \times 2} \times 6$$

$$\therefore \quad A = 0.0288 \ \text{m}^2 = \frac{\pi}{4} D^2$$

$\therefore \quad D^2 = 0.0367 \quad \therefore \quad D = 0.191$ m

$L = 1.2\,D = 1.2 \times 0.191 = 0.229$ m

$S = 2\,L\,N$

$$\therefore \quad N = \frac{S}{2L} = \frac{320}{2 \times 0.229} = \mathbf{698.7\ rpm}$$

10. A four cylinder two stroke cycle petrol engine develops 30 kW bp at 2500 rpm. The indicated mep in each cylinder is 8 bar and mechanical efficiency is 80%. Stroke to bore ratio is 1.5, brake thermal efficiency is 28% and CV of fuel is 44000 kJ/kg. Find (a) bore and stroke of each cylinder (b) brake specific fuel consumption.

Solution:

Given

bp = 30 kW, N = 2500 rpm, p_m = 8 bar, η_m = 0.8, L/D = 1.5, η_b = 0.28, CV = 44000 kJ/kg

$$\eta_m = \frac{p_{mb}}{p_m}$$

$$\therefore \quad p_{mb} = \eta_m \times p_m = 0.8 \times 8 = 6.4 \text{ bar} = 640 \text{ kPa}$$

$$bp = \frac{p_{mb} \times A\,L\,N}{60} \times 4$$

$$\therefore \quad 30 = \frac{640 \times \frac{\pi}{4} \times D^2 \times 1.5D \times 2500}{60} \times 4$$

$$\therefore \quad D^3 = 0.000239 \text{ m}^3$$

$$\therefore \quad D = 0.062 \text{ m} = 62 \text{ mm}$$

$$\therefore \quad L = 93 \text{ mm}$$

$$\eta_b = \frac{bp}{m_f \times CV}$$

$$\therefore \quad m_f = \frac{bp}{\eta_b \times CV}$$

$$= \frac{30}{0.28 \times 44000} \times 3600 \text{ kg/hr}$$

$$= 8.766 \text{ kg/hr}$$

$\therefore$ $$\text{bsfc} = \frac{m_f}{bp}$$

$$= \frac{8.766}{30}$$

$$= \mathbf{0.2922\ kg/kWh}$$

11. A four stroke four cylinder petrol engine has bore 65 mm and stroke 95 mm. It develops torque of 64 N-m while running at 3000 rpm. If clearance volume in each cylinder is 63 cm^3, relative efficiency based on brake thermal efficiency is 0.5 and C.V. of petrol is 42000 kJ/kg, calculate fuel consumption in kg/hr and p_{mb}.

Solution:

Given

$D = 6.5$ cm, $L = 9.5$ cm, $T = 64$ N-m, $N = 3000$ rpm $V_c = 63$ cm^3, $\eta_r = 0.5$, $CV = 42000$ kJ/kg

$$bp = \frac{2\pi NT}{60} = \frac{2 \times 3.14 \times 3000 \times 64}{60} = 20096 \text{ W}$$

$$= 20.096 \text{ kW}$$

$$V_s = \frac{\pi}{4} D^2 L = \frac{\pi}{4} \times (6.5)^2 \times 9.5 = 315 \text{ cm}^3$$

$$r = \frac{V_c + V_s}{V_c} = \frac{63 + 315}{63} = 6$$

Air standard efficiency $\eta_a = 1 - \frac{1}{r^{\gamma-1}} = 1 - \frac{1}{6^{0.4}}$

$$= 1 - 0.488 = 0.512$$

$$\eta_r = \frac{\eta_b}{\eta_a}$$

$\therefore$ $$\eta_b = \eta_r \times \eta_a = 0.5 \times 0.512$$

$$= 0.256$$

Now, $$\eta_b = \frac{bp}{m_f \times CV}$$

$\therefore \quad m_f = \dfrac{bp}{\eta_b \times CV} = \dfrac{20.096 \times 3600}{0.256 \times 42000} = 6.73 \text{ kg/hr}$

Now, $\quad bp = \dfrac{p_{mb} \times A\,L\,N}{60 \times 2} \times 4$

$\therefore \quad p_{mb} = \dfrac{bp \times 60 \times 2}{A\,L\,N \times 4} \qquad AL = V_s = 315 \times 10^{-6}\ m^3$

$$= \frac{20.096 \times 60 \times 2}{315 \times 10^{-6} \times 3000 \times 4 \times 10^2} \text{ bar} \quad \text{as } 1 \text{ bar} = 10^2 \text{ kPa}$$

$$= \mathbf{6.38\ bar}$$

12. The following data is available for 2 stroke diesel engine:

Bore = 10 cm, stroke = 15 cm, piston speed = 300 m/min, torque developed = 58 N-m, η_m = 80%, indicated thermal efficiency = 40%, CV of fuel = 44000 kJ/kg. Calculate (a) indicated power (b) p_m (c) bsfc

Solution:

Given

D = 10 cm, L = 15 cm, piston speed = 300 m/min, T = 58 N-m, $\eta_m = 0.8$, $\eta_i = 0.4$, CV = 44000 kJ/kg

Piston speed = 2 L N

$\therefore \quad 300 = 2 \times 0.15 \times N$

$\therefore \quad N = 1000$ rpm

$$bp = \frac{2\pi N T}{60} = \frac{2 \times 3.14 \times 1000 \times 58}{60}$$

$$= 6071 \text{ W} = 6.071 \text{ kW}$$

$$\eta_m = \frac{bp}{ip}$$

$\therefore \quad ip = \dfrac{bp}{\eta_m} = \dfrac{6.071}{0.8} = 7.59 \text{ kW}$

Now, $\quad ip = \dfrac{p_m\,A\,L\,N}{60}$

$$= \frac{p_m \times \frac{\pi}{4} D^2\,L\,N}{60}$$

$$\therefore \quad 7.59 = \frac{p_m \times \frac{\pi}{4} \times (0.10)^2 \times 0.15 \times 1000}{60}$$

$$\therefore \quad p_m = 387 \text{ kPa} = 3.87 \text{ bar}$$

Now $$\eta_i = \frac{ip}{m_f \times CV}$$

$$\therefore \quad m_f = \frac{ip}{\eta_i \times CV} = \frac{7.59}{0.4 \times 44000} \times 3600$$

$$= 1.5525 \text{ kg/hr}$$

$$\therefore \quad bsfc = \frac{m_f}{bp} = \frac{1.5525}{6.071} = \mathbf{0.2557 \text{ kg/kWh}}$$

Objective Type Questions

1. Two stroke cycle is completed in __________ revolution of crank shaft.
   (a) Four (b) One (c) Two (d) Three
2. Otto cycle is
   (a) Constant volume cycle (b) Constant pressure cycle
   (c) Constant temperature cycle (d) Constant enthalpy cycle
3. In diesel cycle, heat is added at
   (a) Constant volume (b) Constant pressure
   (c) Constant temperature (d) Constant enthalpy
4. Spark plug is provided in
   (a) Petrol engine (b) Diesel engine
   (c) Steam engine (d) Steam turbine
5. Injector is provided in
   (a) Steam engine (b) Steam turbine
   (c) Petrol engine (d) Diesel engine
6. Compression ratio in diesel engine is in the range of
   (a) 6 to 10 (b) 2 to 5 (c) 16 to 20 (d) 30 to 40
7. Thermal efficiency of 2-stroke cycle engine is ____________ a 4-stroke cycle
   (a) Equal to (b) Less than (c) Greater than (d) Half than

8. Carburettor is used in supply

   (a) Air only (b) Petrol only

   (c) Mixture of air and diesel (d) Mixture of air and petrol

Answers

1.	(b)	2.	(a)	3.	(b)	4.	(a)	5.	(d)	6.	(c)
7.	(b)	8.	(d)								

Short Questions

1. Define

   (a) I.C. engine (b) 4-stroke engine (c) 2-stroke engine (d) S.I engine (e) C.I. engine (f) TDC (g) BDC (h) IDC (i) ODC (j) bore of the cylinder (k) stroke (l) swept volume (m) clearance volume (n) compression ratio (o) ip (p) bp (q) fp (r) mep (s) η_m (t) η_i (u) η_b (v) η_r (w) bsfc.

2. Which are the main components of I.C. engine?
3. Is petrol engine a spark ignition engine?
4. Write the function of following components?

   (a) connecting rod (b) piston rings (c) cam (d) fly wheel (e) carburattor (f) spark plug (g) fuel pump (h) injector (i) transfer port.

5. Why piston rings are provided around the piston?
6. Which components are included in valve gear?
7. How valves are operated?
8. Why the speed of cam shaft is kept half of that of crank shaft?
9. How diesel is ignited in diesel engine?
10. Name the four strokes of 4-stroke engine.
11. What is scavenging?
12. Why a deflector is provided on piston head in 2-stroke engine?
13. Why cooling of I.C engine is required?

Long Questions

1. Classify I.C engines
2. Explain, with sketch, working of
   (a) 4-stroke petrol engine
   (b) 4-stroke diesel engine
   (c) 2-stroke petrol engine
   (d) 2-stroke diesel engine
3. Write differences between petrol engine and diesel engine.
4. Write differences between two stroke cycle engine and four stroke cycle engine.

Exercises

1. A four stroke petrol engine has a stroke volume of 10 litres. Its means effective pressure is 600 kPa and speed is 300 rpm. Calculate indicated power of the engine. [15 kW]
2. A two cylinder four stroke diesel engine has stroke of 150 mm and stroke to bore ratio is 1.25. It mep is 7 bar and engine runs at 900 rpm, find ip of the engine.

   [17.81 kW]
3. Two cylinder four stroke diesel has total swept volume of 870 cm^3. Engine speed is 300 rpm. Brake torque us 50 N-m. Mean effective pressure is 10 bar. Find mechanical efficiency of the engine. [72.23%]
4. A four stroke cycle I.C. engine has a piston diameter of 150 mm and average piston speed is 200 m/min. If the mep is 7 bar, find ip of the engine. [10.3 kW]
5. A six cylinder 4-stroke diesel engine develops ip of 90 kW at 800 rpm. Stroke to bore ratio is 1.5. η_m is 0.85. bmep is 5 bar. Find bore and stroke of the engine.

   [D = 269 mm, L = 405 mm]
6. A four cylinder 4-stroke petrol engine develops 185 kW bore at 35 rps. Stroke to bore ratio is 1.5. η_m is 75%. imep is 830 kPa. Determine bore and stroke.

   [D = 153 mm, L = 230 mm]
7. A four cylinder 4-stroke petrol engine has 100 mm bore and 150 mm stroke. It consumes 3.7 kg of fuel per hour having a CV of 41000 kJ/kg. η_i is 41%. mep is 7 bar. Find crank shaft speed [786 rpm]

8. A 4 cylinder 4-stroke petrol engine develops 200 kW bp at 2500 rpm. Stroke to bore ratio is 1.2. If mep is 10 bar and η_m is 81%, find bore and stroke of the engine. Also find η_i and η_b if 65 kg/hr of petrol is consumed having CV = 42000 kJ/kg. [D = 147 mm, L = 176 mm, η_i = 32.56% η_b = 26.37%]

9. A petrol engine with a stroke length of 200 mm and diameter of 150 mm has a clearance volume of 5×10^5 mm^3. η_i of the engine is 30%. Find relative efficiency of the engine. If mep is 5 bar and engine speed is 1000 rpm, find ip is the engine. Assume 4-stroke engine [53.12%, 14.73 kW]

10. A 2-cylinder, 4-stroke petrol engine has 100 mm bore and 140 mm stroke. It runs at 900 rpm. It consumes 2.1 kg/hr of petrol having CV of 42000 kJ/kg. mep in both cylinders is 7.5 bar. Find η_i and η_r if clearance volume is 15% of stroke volume. [50.5%, 90.62%]

11. An engine uses 6.5 kg of oil per hour having CV 30000 kJ/kg. bp of engine is 22 kW. η_m is 85%. Find η_i, η_b and bsfc. [47.8%, 40.6% 0.3 kg/kWh]

12. In a trial on single cylinder oil engine, following observations were made.

Oil consumption	:	10.2 kg/hr
CV of fuel	:	43890 kJ/kg
Speed	:	1900 rpm
Torque on brake drum	:	186 N-m

Find (i) bp (ii) bsfc (iii) ηb [37 kW, 0.2756 kg/kWh, 29.75%]

CHAPTER 8

Speed Control

8.1 Introduction

The variation in speed of an engine can occur in two ways:

1. Variation within one revolution (cyclic variation)
2. Variation in number of revolutions per minute, i.e. variation in rpm.

In first case, the variation in speed is due to variation of pressure of working fluid on piston during the cycle. This variation in speed can be regulated by mounting a suitable flywheel on the shaft.

In second case, variation in speed is due to variation of load on the engine. This requires a governor to maintain a constant speed. It is necessary to control the variation of speed within certain desired limits.

If the load on the shaft increases, the speed of the engine decreases for the same power. So supply of fuel should be increased by opening the throttle valve to increase power supply and maintain the speed. On the other hand, if load on the shaft decreases, the speed of the engine increases. So fuel supply should be decreased by closing the throttle valve to bring the engine back to the original speed. Governor adjusts the supply of fuel by operating throttle valve according to the load and thus maintains the speed of the engine constant.

Function of governor is to regulate the variation of speed of the engine due to change of load.

Fly wheel cannot control the variation of speed due to change of load.

8.2 Governors

Governor adjusts the supply of fuel to the engine according to the load to maintain the speed constant. This control is achieved by the action of centrifugal force of rotating masses of the governor.

8.2.1 Types of Governors

Governors are broadly classified as (i) centrifugal governor (ii) inertia governor

(i) ***Centrifugal governor:*** This type of governor has a pair of masses called 'balls', which rotate at a distance from the axis of rotation. The centrifugal force on the balls is proportional to the speed. So, when the speed of the engine increases, due to higher centrifugal force, the balls tend to rotate at a larger radius from the axis. This movement is utilized through proper linkages to close the throttle valve. Reverse of this will happen when the speed of the engine decreases. Examples of centrifugal governor are Watt governor, Porter governor, Proell governor, Hartnell governor, etc.

(ii) ***Inertia governor:*** In this type of governor, the position of balls is affected by the forces set up by an angular acceleration of the given spindle in addition to centrifugal force on the balls. The response of inertia governor is faster.

Here we will discuss some centrifugal governors.

8.2.2 Watt Governor

It is also known as simple pendulum type governor. It was first used by James Watt, the inventor of steam engine and hence it is known as Watt governor. Fig. 8.1 shows the Watt governor. Two upper arms are hinged at the top of the spindle and two balls are fitted on the other ends of these arms. Two lower arms are hinged with the upper arms on one side and with sleeve on the other side. This sleeve can move up and down over the spindle.

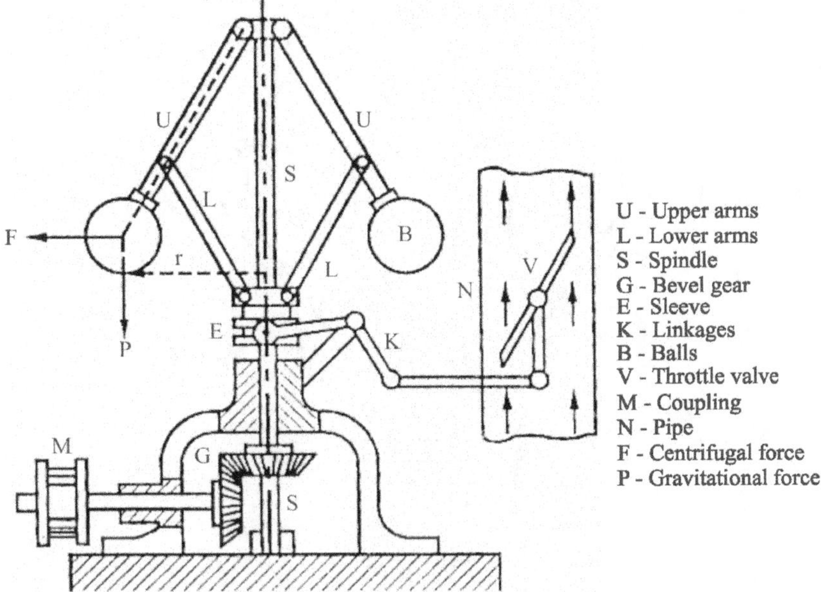

Fig. 8.1 Watt governor.

The speed of the crank shaft of the engine is transmitted to the spindle through a pair of bevel gears when the spindle rotates at a particular speed, the balls rotate at a particular radius.

When the speed of the engine increases, the upper and lower arms open due to extra centrifugal force. The balls move away from spindle axis, causing the sleeve to move upward. This movement is transmitted by levers and links to the throttle valve which causes it to close partially. This causes reduction in fuel supply and the engine speed decreases.

When the engine speed decreases, sleeve moves downward, partially opening the throttle valve. This increases the fuel supply and the engine speed increases.

Watt governor is normally used for low speed engines and has now become practically obsolete.

8.2.3 Porter Governor

This governor is a modification of Watt governor where a central weight is placed over the sleeve as shown in Fig. 8.2. This weight reduces the movement of the sleeve for a given change in speed. So, this governor can be used at higher speeds also.

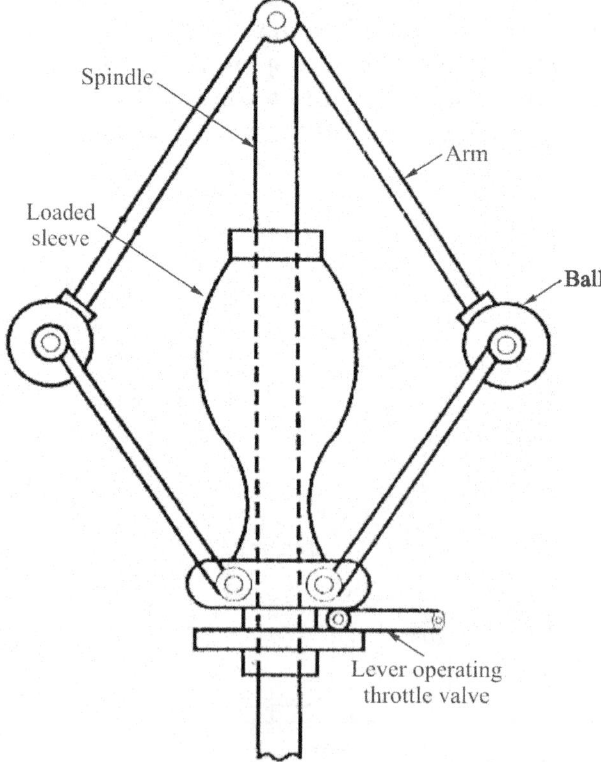

Fig. 8.2 Porter governor

8.2.4 Proell Governor

When the balls of a Porter governor are placed on the upward extension of lower arms as shown in Fig. 8.3, it becomes a Proell governor.

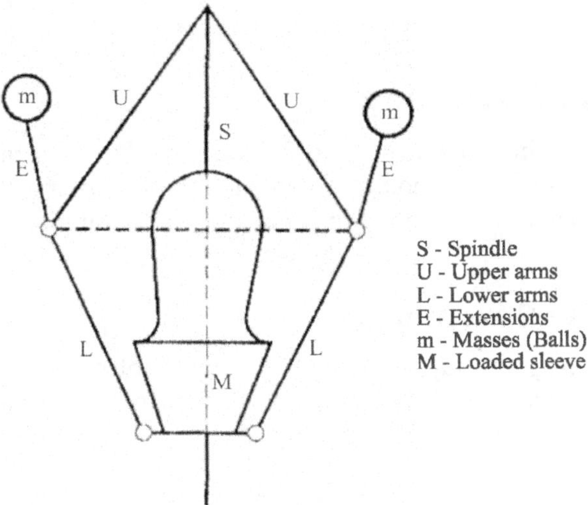

Fig. 8.3 Proell governor.

8.2.5 Hartnell Governor

Hartnell governor is a spring loaded governor as shown in Fig. 8.4. Initially, thc spring is fitted in compression so that the downward force is applied to the sleeve. Two bell crank levers are pivoted on the frame. Each lever carries a ball at one end and a roller at the other end. The roller fits into a groove in the sleeve.

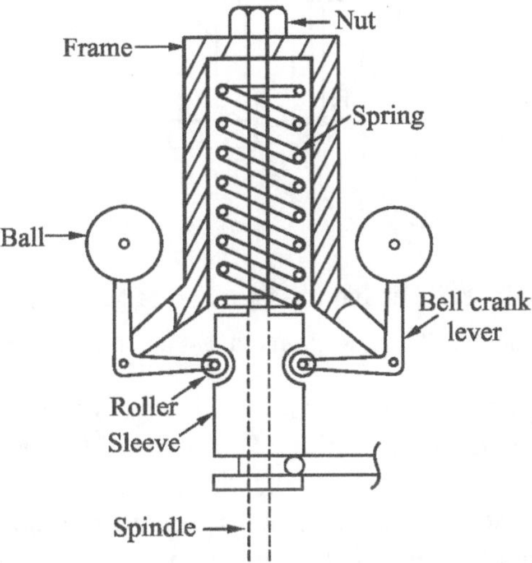

Fig. 8.4 Hatnell governor.

As the speed increases, balls move away from the spindle axis, bell crank levers rotate about the pivot and lift the sleeve against the spring force. If the speed decreases, the sleeve moves downwards. The movement of the sleeve is transmitted to the throttle valve. The spring force can be adjusted by screwing or unscrewing a nut provided on the upper end of the sleeve.

8.3 I.C. Engine Governing

The function of I.C. engine governing is to maintain the speed of the engine constant. It achieves this by controlling the amount of fuel supplied for combustion or the quantity of mixture of air and fuel vapour. Following methods are commonly used for I.C. engine governing.

8.3.1 Hit and Miss Governing

In this type of governing, the supply of fuel is stopped for one or more cycles when the speed of the engine increases. This makes the speed of the crank shaft normal. The arrangement for this governing is shown in Fig. 8.5.

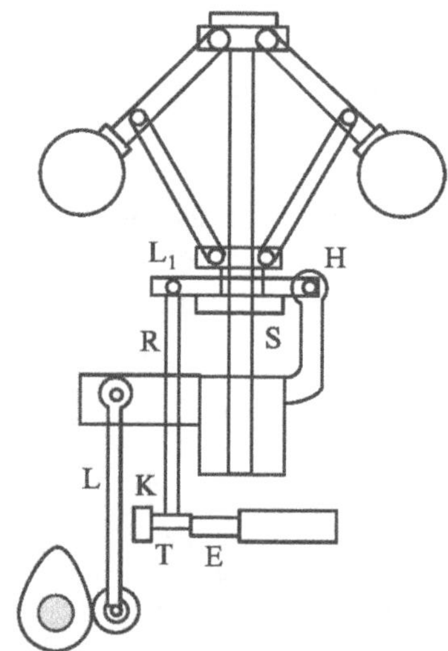

Fig. 8.5 Hit and miss method of governing.

L, L_1-Lever, H-Hinged point, S-Sleeve, R-Drop rod, E-End of the plunger. T-Steel die, K-Striker

A simple pendulum type governor is shown in Fig. 8.5. Liver L_1 rests on the sleeve, one end of which is hinged at the point H and the other end is connected with the governor drop rod R. The lower end of the drop rod has a steel die T. The Lever L carriers a striker K which strikes the steel die at the normal speed due to the rotation of cam so the steel die strikes the end of the fuel pump plunger or the end of the gas valve which is denoted by E. When the engine speed increases, lever L_1 with the drop rod R moves in the upward direction due to the upward movement of sleeve S. So the striker will not strike the plunger end E through the steel die and the fuel will not be supplied during this cycle. The power is, therefore, decreased and the speed is reduced. As a result, the sleeve moves in the downward direction with the drop rod R, again the striker hits the steel die with the supply of fuel.

This method is suitable for small gas or oil engines. In this method, the quality and quantity of mixture burnt inside the cylinder remains the same for each working cycle. So, an ideal proportion of fuel and air may be maintained. It is simple and there is little chance of getting out of order.

The disadvantage of this method is that a large fluctuation of speed takes place in the idle cycle as no fuel is supplied to the engine at this time. To eliminate this disadvantage, a heavy fly wheel is needed. But with a heavy fly wheel, bearing friction will be increased and mechanical efficiency will be decreased.

8.3.2 Quality Governing

In this method, the amount of air taken inside the cylinder remains the same while the amount of fuel is varied. This varies the air-fuel ratio i.e. quality of mixture. To increase the speed of the engine, fuel amount is increased. To decrease the speed, fuel amount is decreased. This method of governing is used in gas and diesel engines.

The fuel quantity can be varied by putting a throttle valve in supply line of gas in gas engines and by changing the length of plunger of fuel pump in diesel engine. The disadvantage of this method is that at low fuel supply at small loads, mixture is very lean and will not burn properly. This will decrease the thermal efficiency of engine.

8.3.3 Quantity Governing

This method is preferred for petrol and gas engines. In this method, the quantity of mixture of air and fuel is varied to control the speed.

In petrol engine, this is achieved by putting a throttle valve between the carburetor and intake manifold. It controls the supply of mixture of air and petrol. In a gas engine, throttle valve controls the mixture of gas and air.

The advantage of this method is its simplicity and constant air-fuel ratio. But when the quantity of mixture is less, compression pressure is low. This affects combustion of fuel and the thermal efficiency is thereby decreased.

The disadvantage of both quality and quantity governing can be eliminated by combining these two methods known as combination method.

8.4 Fly Wheel

In the engine cylinder, the pressure of gas on the piston does not remain constant during working cycle. This causes the fluctuation of speed of the crank shaft in a cycle. A fly wheel is used to control this fluctuation of speed during the cycle. A fly wheel of suitable dimension attached to the crankshaft by a key as shown in Fig. 8.6 makes the moment of inertia of the rotating parts quite large and thus acts as a reservoir of energy. When the speed of the crank shaft tends to increase, fly wheel absorbs certain amount of energy, but at the time of decrease of speed, fly wheel gives up the amount of energy it has absorbed, thus the speed of the crank shaft tends to become normal. So the function of fly wheel is to regulate the fluctuation of speed in a cycle.

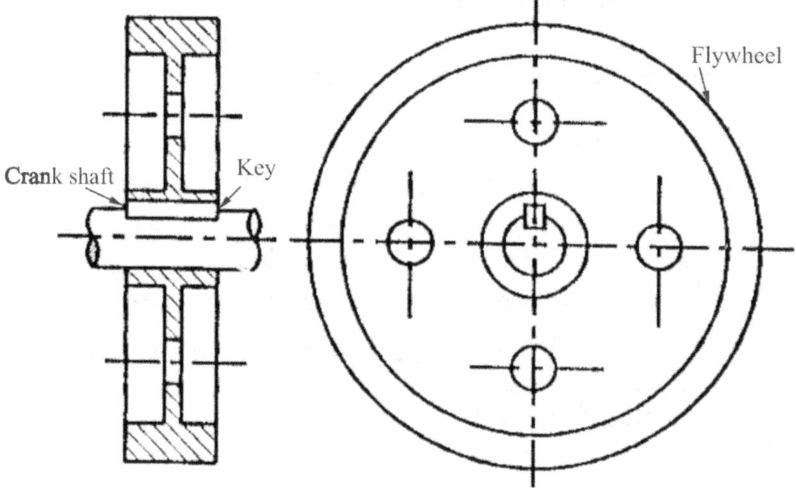

Fig. 8.6 Fly wheel.

The fly wheel is in the form of disc or wheel made generally of cast iron and its mass depends upon the nature of variation of pressure during the cycle in the cylinder. Fly wheel for a two stroke cycle engine is lighter than a fly wheel used for four stroke cycle engine. For multi cylinder engine also, lighter fly wheel is used.

The object of both the fly wheel and the governor is to control the speed of the crank shaft, but there is one fundamental difference between them. The variation of speed, due to the variation of pressure on piston during a cycle is controlled by a fly wheel, while the variation of speed due to the variation of load is controlled by a governor.

Objective Type Questions

1. When load on the engine increases, its speed
   (a) remains constant (b) increases
   (c) decreases (d) becomes zero
2. Which is the spring loaded governor?
   (a) Hartnell governor (b) Watt governor
   (c) Porter governor (d) Proell governor
3. Air fuel ratio varies in
   (a) quantity governing (b) quality governing
   (c) hit and miss governing (d) spark time governing

Answers

1. (c) 2. (a) 3. (b)

Short Questions

1. Why is it necessary to maintain a constant speed in a heat engine?
2. What is a governor?
3. What is the function of governor?
4. What are the various types of governor?
5. What is the limitation of Watt governor? How is it overcome in Porter governor?
6. Name various governing methods used in I.C. engines.
7. What are the advantages and disadvantages of hit and miss governing?
8. What is quality governing as applied to I.C. engine?
9. What is the disadvantage of quality governing?
10. What are the advantages and disadvantages of quantity governing?
11. Explain the difference between quality method and quantity method of governing.
12. Which method of governing is preferred for (i) diesel engine (ii) petrol engine? Why?
13. What is a fly wheel? How does it work?
14. What is the function of a flywheel?

15. How does fly wheel maintain the speed constant?
16. Why fly wheel of two stroke cycle engine is lighter than that of four stroke cycle?
17. What is the fundamental difference between fly wheel and governor?

Long Questions

1. Explain (i) Centrifugal governor (ii) Inertia governor
2. Explain with neat sketch, construction and working of (a) Watt governor (b) Porter governor (c) Hartnell governor
3. Explain following methods of governing used in I.C. engines

   (a) Hit and miss governing (b) Quality governing (c) Quantity governing
4. Explain construction and working of fly wheel.

CHAPTER 9

Pumps

9.1 Introduction

The Pump is mechanical device used to lift or convey the liquid from one place to another place. Pumping means adding of energy to the liquid so that liquid can be raised to higher levels. Centrifugal pumps are all of outward flow type. It is necessary that pump has to be full of water when starting it and it should not be left to drain. Generally centrifugal pumps are made up of radial type. But there are also axial flow or propeller pumps which are particularly adopted for low heads. The basic difference between turbine and pump is that turbine converts pressure energy into mechanical energy and pump converts mechanical energy into pressure energy.

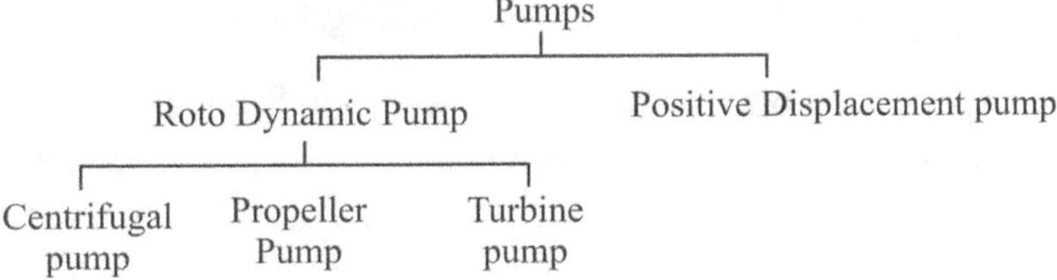

9.2 Construction of C.F. Pump (as shown in Fig. 9.1)

Following are various components of centrifugal Pump

1. Impeller
2. Casing
3. Suction pipe
4. Delivery pipe

1. ***Impeller***

(i) It is a wheel or rotor, series of curved vane are fitted on its periphery.

(ii) Impeller is keyed on the shaft. It is rotated inside casing by means of prime mover.

(iii) Rotation of impeller produces vacuum at the centre due to centrifugal action, liquid enters the impeller through an eye of the impeller and travels towards the periphery of impeller and pressurized liquid is discharged into casing of pump.

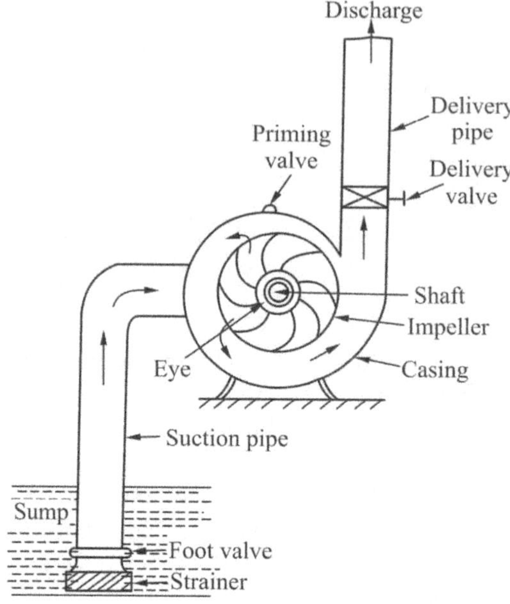

Fig. 9.1 Components of a centrifugal pump.

There are generally three types of impeller (as shown in Fig. 9.2).

(i) Closed impeller (ii) Semi-open impeller (iii) Open impeller

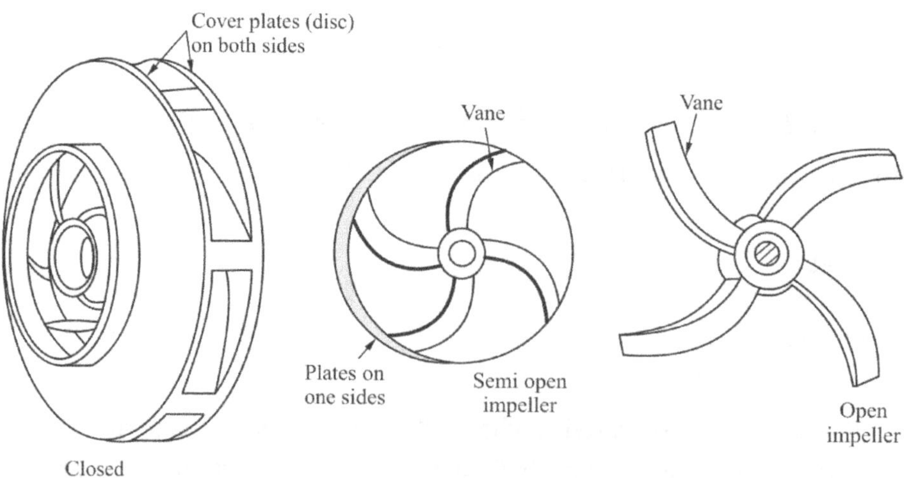

Fig. 9.2 Closed, Semi-open and open impellers.

(i) ***Closed Impeller***: It consists of two shrouds, (plates) curved vanes are inserted between them. This type of impeller handles pure liquid, water, oil etc.

(ii) ***Semi open Impeller***: It consists of only one cover plate (shroud) on which curved vanes is mounted. This type of impeller has less curved vanes compare to closed impeller. It handles viscous liquid, muddy water, paper pulp, sewage water etc.

(iii) ***Open Impeller***: This type of impeller consists of only curved vanes mounted on the hub, both cover plates or shrouds are absent here. This is used to pump water consists of solid impurities such as debris, small solid, paper pulp, muddy water, sand and molasses etc. There is no chance of clogging of impeller.

2. *Casing*

(i) Shaft of impeller is supported in the bearings provided in casing. Casing is closed enclosure in which impeller rotates.

(ii) To avoid leakage of high pressure liquid the casing is provided with gland and stuffing box at the shaft.

(iii) Inlet and outlet pipes are fitted to casing.

(iv) Types of casing are volute casing, vortex casing and diffuser casing. According to this pumps are classified as volute type, vortex type and diffuser type pumps.

(v) Material of casing is generally cast iron or cast steel.

(vi) It should be shaped in such a way that head loss due to formation of eddies is reduced.

3. *Suction Pipe*

(i) It is fitted to the casing at the eye of the impeller and serves as an inlet pipe for liquid.

(ii) It carries water from the sump or well or channel to impeller inlet of pump.

(iii) *Foot valve*: Bottom end of suction pipe is submerged in to liquid. Foot valve is fitted to this end. It is non-return type valve. It opens inside the suction pipe to allow the liquid to enter in suction pipe.

(iv) The foot valve is covered by a strainer which keeps away the floating particles, debris, sand etc., and prevents them to enter in suction pipe.

(v) Vacuum gauge is mounted on the suction pipe near the casing.

(vi) To avoid the cavitations to occur inside suction pipe, the length of pipe and losses in inlet pipe should be minimized.

(vii) To keep low velocity, the diameter of suction pipe must be greater than delivery pipe.

4. ***Delivery Pipe***

(i) It is fitted to the casing at circumference and serves as an outlet pipe for liquid.

(ii) *Delivery valve:* It is the regulating valve fitted to delivery pipe to vary the discharge of the liquid. It should be closed before pump is stopped, otherwise the delivery head appears on the suction pipe which is dangerous.

(iii) Pressure gauge is mounted on the delivery pipe before delivery valve to read pressure of liquid leaving the casing of pump.

9.3 Types of Centrifugal Pump or various Methods to Increase the Efficiency of C.F. Pump or What are various methods to convert K.E into P.E

A. According type of casing (as shown in Fig. 9.3, Fig. 9.4, Fig. 9.5)

(i) Volute type

(ii) Vortex type

(iii) Diffuser type (Guide blade type) or Turbine pump

B. According to number of Stages

(i) Single Stage Pump

(ii) Multi Stage Pump

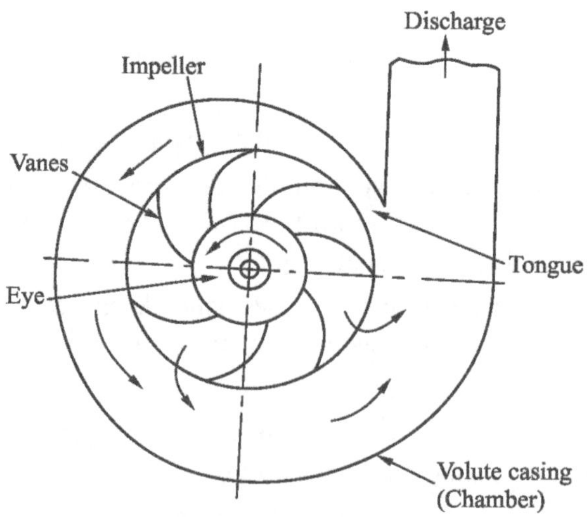

Fig. 9.3 Volute casing.

Volute Type Centrifugal Pump

1. In this pump, the impeller is surrounded by spiral casing known as volute chamber. The area of volute chamber increases gradually towards the delivery side.
2. The water which leaves the vanes is directed to move in the volute chamber circumferentially.
3. Large velocity of liquid entering the casing is reduced.
4. Observations have shown that in this arrangement, the efficiency of the pump is increased slightly, this means that eddies are not completely avoided and some loss of energy takes place.
5. Used for medium head 15 m to 45 m.

Vortex Type Centrifugal Pump

1. This is an improvement over the ordinary volute chamber.
2. In this pump, a vortex or whirl pool chamber or circular ring is inserted between impeller and the volute casing. It is gradually and smoothly converts velocity head into pressure head. Thus action of volute chamber to convert velocity head in to pressure head is improved by vortex.
3. In this arrangement the efficiency is increased considerably.
4. Parallel sided passage around impeller based on free vortex principle that is the velocity drops inversely with radius, used on most of the pumps before volute casing.

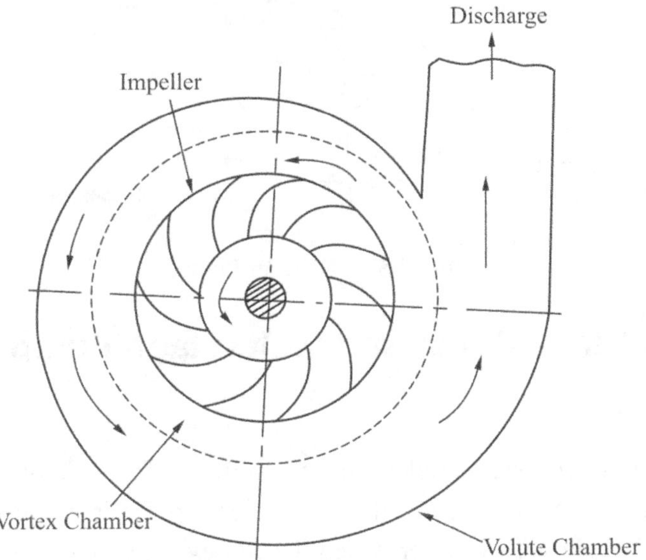

Fig. 9.4 Vortex chamber.

Diffuser Type Centrifugal Pump

1. In this pump, the fixed guide blades are provided in vortex chamber to improve effectiveness. This ring of guide blades is known as the diffuser.
2. These are fixed vanes which receive water, leaving the moving vanes. The angles of guide vanes at their receiving tips are such that the water enters the guide blades without shock. Hydraulic losses are reduced.
3. Guide blades have gradually increasing area, there is better guidance to flow which reduces eddy losses and increases the efficiency of pump.
4. The diffuser converts velocity into pressure, most efficiently and hence this pump is known as turbine pump.
5. Used in centrifugal compressors, super-chargers and turbine type of centrifugal pump.

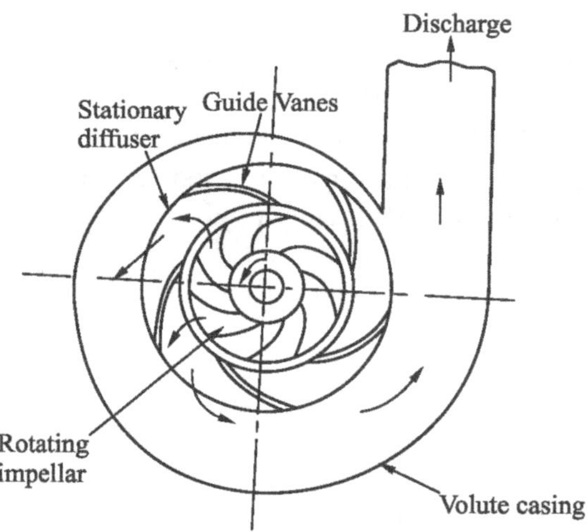

Fig. 9.5 A Diffuser Pump.

9.4 Single Stage Pump and Multi-stage Pump

Single Stage Pump

In single stage pump only one impeller is fitted on the shaft. The required pressure rise is developed in single impeller surrounded by volute casing. To increase its efficiency whirl pool chamber and guide blades are provided. Head developed by single stage centrifugal pump hardly exceed 40 meters.

Multi Stage Pump

In multistage pump two or more impellers are mounted in series on common shaft in a single casing. The discharge of one impeller is guided to inlet of the next impeller and so on. Very high pressure is developed. In order to produce high pressure head and efficiency, multi stage pumps are used as shown in Fig. 9.9.

9.5 Direction of Flow through Impellers (as shown in Fig. 9.6)

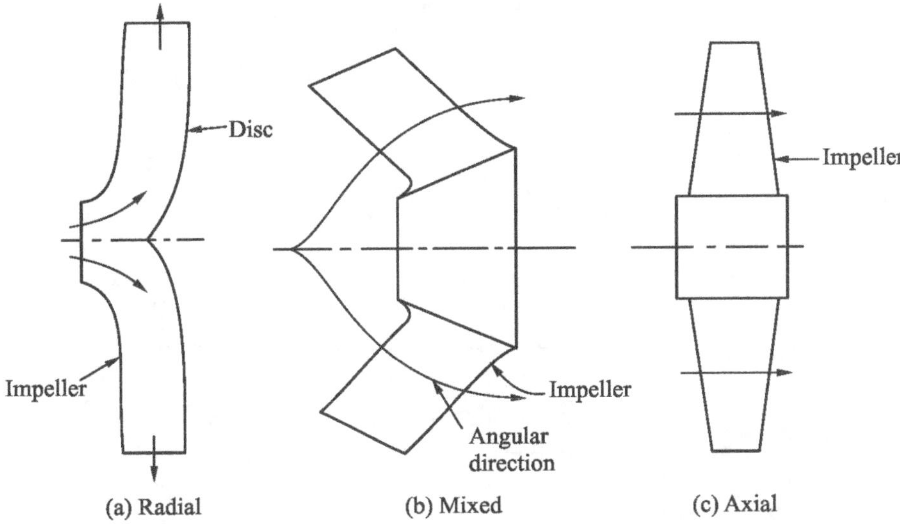

Fig. 9.6 Different impeller shapes or Direction of flow through Impeller.

There are three types of pumps depending upon the direction of flow through impeller.

1. Radial flow
2. Mixed flow
3. Axial flow

***Radial Pump*:** Water enters axially and flows through impeller in radial direction as shown in Fig. 9.6(a). Head is developed by centrifugal action on the liquid. The pumps are used where high head and low discharge is required.

***Mixed flow pump*:** In this type of pump liquid enters axially at the eye of impeller and discharges in angular direction through the impeller. The head is developed partly due to centrifugal action and partly due to propulsive force. It is used for medium discharge and medium head.

***Axial Flow Pumps*:** In this type of pump liquid flows parallel to axis of shaft. In this case head is developed by lift action of vanes on fluid which enters axially. Axial pump have a very high discharge at low head. Such pumps are used for irrigation purpose.

9.6 Types of Vanes (as shown in Fig. 9.7)

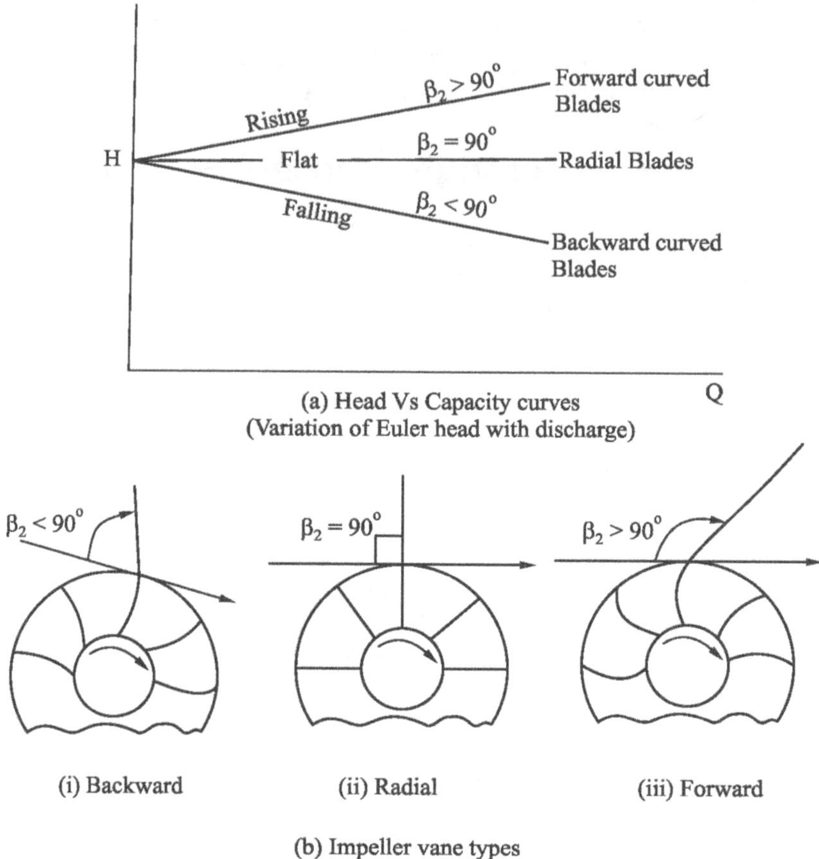

Fig. 9.7 Effect of the Shape of the Blades.

1. ***Backward Curved Vanes*:** Vanes are curved back from direction of rotation of the impeller. H-Q curve is straight line, in this type also. Efficiency is high, about 85 to 95%, slop of line is negative and vane angle at outlet is less than 90^0.
2. ***Radial Vanes*:** These vanes have out tips in radial direction i.e., $\beta_2 = 90^0$. Efficiency obtained by this type, is generally medium, about 80 top 85%.
3. ***Forward Curved Blade*:** In this type the vanes are curved forward in the direction of rotation of impeller (pump). Liquid flows on the concave surface of vanes. H-Q curve is straight line and blade angle at outlet β_2 is greater than 90^0 and slope of blade is positive. These type of blades give low efficiency, about 75^0% and generally not used in practice.

9.7 Single and Double Suction Entry to the C.F. Pump (as shown in Fig. 9.8)

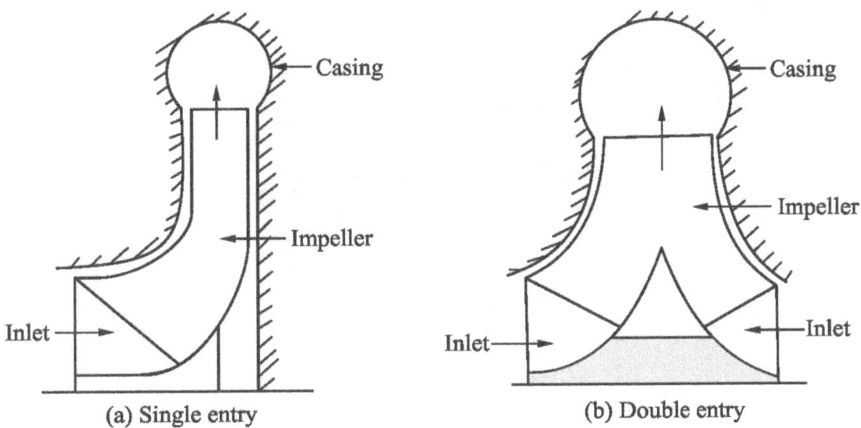

Fig. 9.8 Single and Double-Suction Pumps.

1. Impeller may be single suction impeller or double suction impellers as shown in Fig. 9.8. In a single suction impeller water enters at one end of the casing. But in double suction, water enters the impeller on both side of casing.
2. In case of single suction impeller, due to entry of water on one side of the casing only, a lateral thrust is transmitted to the impeller. But in double suction impeller, thrust acts on both side of the impeller and are balanced.
3. Double suction pump accommodates a large rate of flow compare to single suction pump.

9.8 Pumps Connected in Series and Parallel

Pumps in Series (Fig. 9.9)

1. When two or more impellers are connected in series, the head developed by pump will be the sum of head produced by each impeller.
2. In this arrangement water from one impeller enters into another impeller where head is further increased but discharge passing through one impeller, enters into second impeller. Thus discharge remains constant.

Suppose there are 'n' impellers connected in series and Q is discharge passing through one impeller.

The head developed by pump = n × head developed by one impeller and

Q discharged by pump = Discharge passing through one impeller.

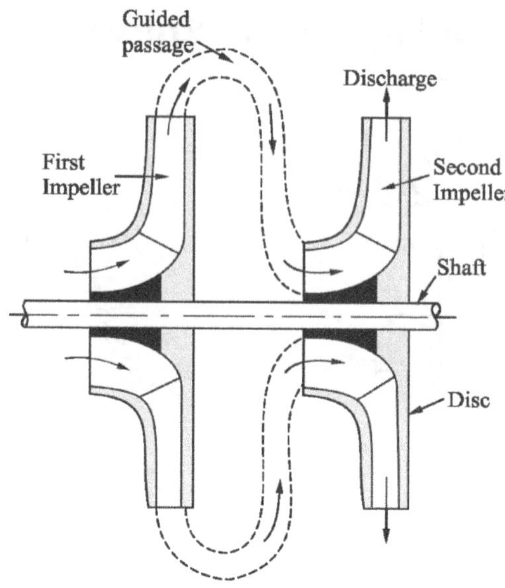

Fig. 9.9 Multistage centrifugal pump in series.

***Pumps in Parallel* (Fig. 9.10):** Head developed by one impeller pump is same when they are connected in parallel but discharge increases as more and more pumps are connected in parallel.

Total discharge when pumps are connected in parallel will be equal to number of pumps multiplied by discharge from one pump.

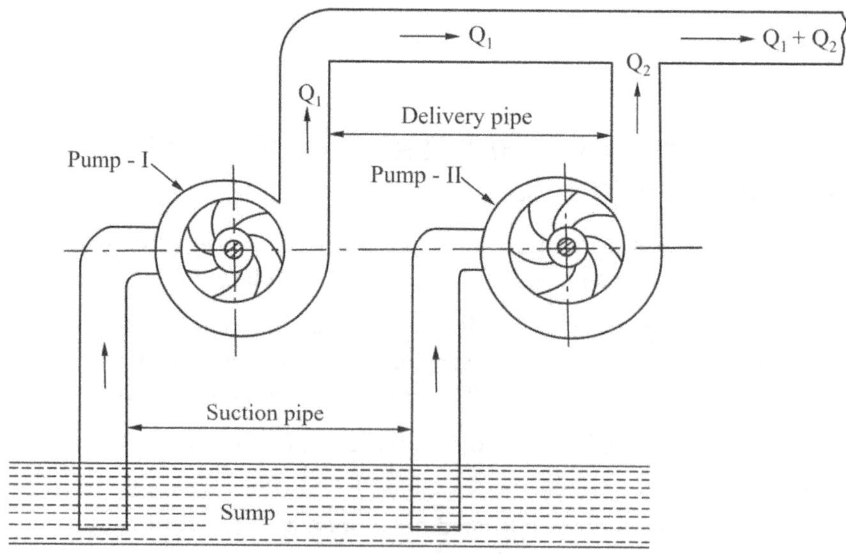

Fig. 9.10 Pumps in parallel.

9.9 Terms used in C.F. Pumps

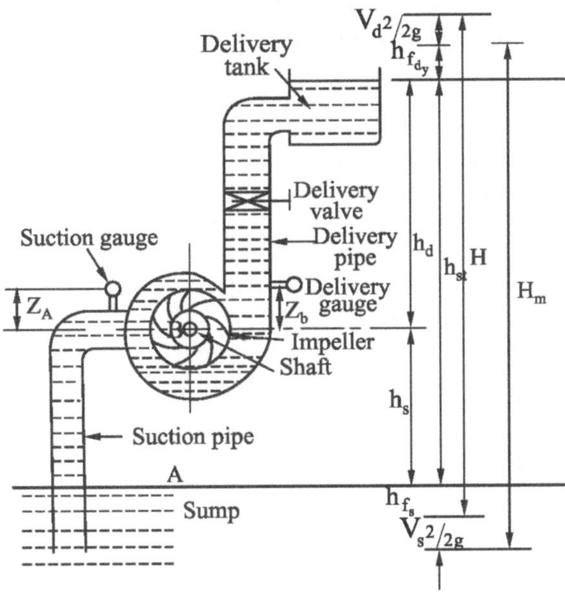

Fig. 9.11(a) Heads of a Pump.

Suction Head (h_s): It is vertical distance between centre line of pump shaft and water level in sump. It is expressed in meter. It is denoted by h_s.

Delivery Head (h_d): It is vertical height between centre line of pump shaft and place where the liquid is discharged. Unit of delivery head is meter. It is denoted by h_d.

Static Head (H_{st}): It is sum of suction head and delivery head. It is expressed in meter. It is denoted by H_{st}.

Manometric Head: It is head against which pump has to work. It is denoted by H_m.

$$\text{The manometric head} = H_m = h_s + h_d + \frac{v_s^2}{2g} + \frac{v_d^2}{2g} + h_{f_s} + h_{f_d}$$

where $\frac{v_s^2}{2g}$ and $\frac{v_d^2}{2g}$ is kinetic head in suction and delivery pipe, are very small and can be neglected in above equation.

Hence $$H_m = h_s + h_d + h_{f_s} + h_{f_d} \qquad .....(1)$$

where h_{f_s} = head lost in friction in suction pipe (in meter)

h_{f_d} = head lost in friction in delivery pipe (in meter)

The Manometric head is ultimately the head developed by the pump. The manometric head is slightly less than the head developed by impeller, but due to some losses in pump. Manometer head is less than impeller power.

$$\therefore \qquad H_m = \frac{V_{w_2} u_2}{g} - \text{impeller loss} \qquad \text{..... (2)}$$

Manometeric head may also be calculated, by difference of energy head, at outlet of pump and energy head at inlet of pump.

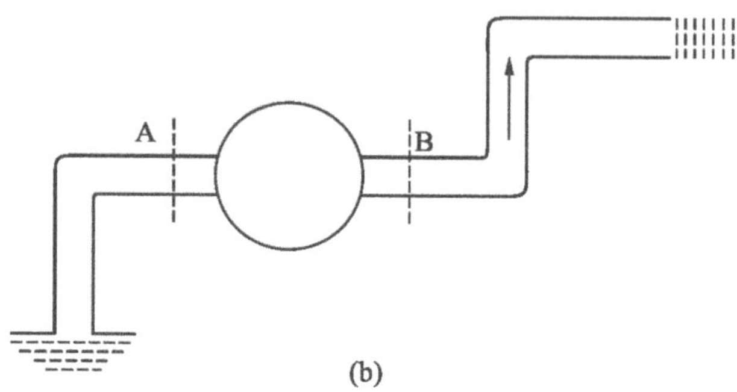

(b)

Fig. 9.11(b)

Let 'A' is section on the suction pipe, close to the pump and 'B' is section on delivery pipe close to the pump (not shown in figure).

Then, Manometric head = H_m = Energy head at B – Energy head at A

$$= \left[Z_B + \frac{p_B}{\rho g} + \frac{V_d^2}{2g}\right] - \left[Z_A + \frac{P_A}{\rho g} + \frac{V_s^2}{2g}\right] \qquad \text{..... (3)}$$

Z_B = Vertical height from centre of pump to centre line of pressure gauge.

Z_A = Vertical height from centre of pump to centre line of suction gauge.

V_d = Velocity of water in delivery pipe.

V_s = Velocity of water in suction pipe

Expressing H_m in terms of total head H,

$$H_m = H + \frac{V_s^2}{2g} - \frac{V_d^2}{2g} \qquad \text{......(4)}$$

If suction and delivery of same diameter, then $H_m = H$. ($\because V_s = V_d$)

Water Power: It is power required by pump to handle liquid against manometric head.

$$\text{Water power} = \rho g\, Q\, H_m$$

It is denoted by W.P, its unit is kw

***Shaft Power*:** It is the power available to the shaft of pump. It is denoted by S.P. Its unit is also kW.

$$\frac{\text{Shaft power}}{\text{Motor power}} = \text{Motor efficiency}$$

***Manometric Efficiency*:** It is the ratio of the manometric head actually generated by impeller. It is denoted by η_{mano}

$$\therefore \qquad \eta_{mano} = \frac{\text{Manometric head}}{\text{head actually generated by impeller}} = \frac{H_m}{\frac{V_{w_2} u_2}{g}} = \frac{gH_m}{V_{w_2} u_2}$$

***Impeller Power*:** This is the power of the impeller determined in terms of work doen by it on the water.

$$\text{Impeller power} = \frac{\rho Q\, V_{w_2} u_2}{1000} \text{ kW}$$

where $Q = \pi\, d_1\, b_1\, V_{f_1} = \pi\, d_2\, b_2\, V_{f_2}$

***Mechanical Efficiency*:** It is ratio of impeller power to power of prime mover (Motor). It is denoted by η_m.

$$\therefore \qquad \eta_m = \frac{\text{Impeller power}}{\text{Power of prime mover}}$$

***Over all Efficiency*:** This is ratio of the work done by pump in lifting the water against gravity and friction in pipe to energy supplied by prime mover (motor).

$$\eta_o = \frac{\text{Water power}}{\text{Shaft Power (power of prime mover)}}$$

$$= \frac{\rho g Q\, H_m}{\text{shaft power}}$$

$$\eta_o = \frac{\text{Water power}}{\text{Impeller power}} \times \frac{\text{Impeller power}}{\text{power of prime mover}}$$

$$\eta_o = \eta_{mano} \times \eta_m$$

***Note*:** H_m represents only pressure energy of liquid where as total head H represents total energy of liquid.

9.10 Specific Speed of Centrifugal Pump

The specific speed of a centrifugal pump may be defined as speed of an imaginary pump, identical with given pump, which will discharge one liter of water (unit mass) while it is being raised through a head of 1 meter (unit head).

It is denoted by N_s

$$N_s = \frac{N\sqrt{Q}}{H^{\frac{3}{4}}} = \frac{N\sqrt{1}}{(1)^{3/4}} = N$$

Let us prove the above formula

N_s = specific speed of the pump

N = Speed of impeller (pump) in RPM

D_2 = Diameter of impeller at outlet

H = Lift of pump

V_2 = Tangential velocity of impeller at outlet

The tangential velocity of impeller varies as head developed by pump $V_2 \propto \sqrt{H}$

Also tangential velocity of the impeller

$$V_2 = \frac{\pi D_2 N}{60} \qquad \left[\because V_2 \propto \sqrt{H}\right]$$

$$\therefore \quad D_2 \propto \frac{\sqrt{H}}{N} \qquad \text{...(i)}$$

Let Q_1 = Discharge of Pump

b_2 = Width of impeller at outlet

V_{f_2} = velocity of flow at outlet

$$\therefore \quad Q = \pi\, D_2\, b_2\, V_{f_2}$$

But $b_2 \propto D_2$ and $V_{f_2} \propto \sqrt{H}$

$$\therefore \quad Q \propto D_2^2\sqrt{H} \qquad \text{...(ii)}$$

Put value of D_2 from equation (ii) we get

$$Q \propto \left(\frac{\sqrt{H}}{N}\right)^2 .\sqrt{H}$$

$$\therefore \quad Q \propto \frac{H^{\frac{3}{2}}}{N^2}$$

$$\therefore \quad H^2 \propto \frac{H^{\frac{3}{2}}}{Q}$$

$$\therefore \quad N \propto \frac{H^{\frac{3}{4}}}{\sqrt{Q}}$$

$$\therefore \quad N = \frac{N_s H^{\frac{3}{4}}}{\sqrt{Q}}$$ (where N_s is constant of proportionality)

$$\therefore \quad N_s = \frac{N\sqrt{Q}}{H^{\frac{3}{4}}}$$

Specific speed of centrifugal pump helps us in selecting the type of centrifugal pump. Following table gives the type of centrifugal pump for the corresponding value of specific speed.

S. No.	Specific Speed (in RPM)	Type
1.	10 to 30	Slow speed with radial flow at outlet.
2.	30 to 50	Medium speed with radial flow at outlet
3.	50 to 80	High speed pump with radial flow at outlet
4.	80 to 160	High speed pump with mixed flow at outlet
5.	160 to 500	High speed pump with axial flow at outlet

9.11 Net Positive Suction Head

It is defined as the total head developed at inlet (suction inlet) above the vapour pressure. It can be calculated by adding gauge reading in meters of liquid at suction side of pump and velocity head at centre line of pump and then subtracting gauge vapour pressure in meters of liquid at pumping temperature.

NPSH = Absolute pressure at pump suction (inlet) + velocity head –vapour pressure corresponding to temperature of liquid at inlet.

$$\text{NPSH} = \frac{P_s}{\rho g} + \frac{V_s^2}{2g} - \frac{P_v}{\rho g} \qquad \text{.....(i)}$$

But absolute pressure at pump suction $= \frac{P_s}{\rho g} = \frac{P_a}{\rho g} - h_s - h_{f_s} - \frac{V_s^2}{2g}$

where P_a = atmospheric pressure on liquid in sump

Ps = absolute pressure at pump suction

By putting value of $\frac{P_s}{\rho g}$ in (i) we get

$$\therefore \quad \text{NPSH} = \left(\frac{P_a}{\rho g} - h_s - h_{f_s} - \frac{v_s^2}{2g}\right) + \frac{v_s^2}{2g} - \frac{P_v}{\rho g}$$

$$\text{NPSH} = \frac{P_a}{\rho g} - \frac{P_v}{\rho g} - h_s - h_{f_s}$$

- Cavitations may occur if insufficient NPSH
- NPSH depend upon suction lift i.e., central line of pump above water level in sump and friction losses in suction pipe.

9.12 Priming of C.F. Pump

1. Removal of air, gas, vapour out of suction pipe, casing of pump and portion of delivery pipe up to delivery valve by filling the liquid in above space, from outside is known as Priming of C.F. Pump.
2. It may be noted that when pump is run on air without priming, head generated is same meters of air as that of water or other liquid. But since density of air is very low, the equivalent meters of water head produced is negligible and hence pump will not deliver any liquid.
3. Non-return valve is provides at the end of suction pipe, to prevent water going back to the sump, when pump is stopped.

Methods of Priming

(i) Manually

(ii) Priming with vacuum pump

(iii) Priming with jet pump

(iv) Priming with separator

(i) ***Manually priming of C.F Pump*:** when priming is being done, an air vent valve is provided in pump casing. When priming liquid is poured in to a pump, with help of funnel, air escapes through air-vent valve. When all air is exhausted from casing of pump, suction pipe and portion of delivery pipe, the primary liquid will flow out of vent valve, indicating that priming is completed. After air is removed from casing, the air-vent valve is closed.

(ii) ***Priming with Vacuum pump***: Since C.F. pump cannot be placed more than about 7 m above sump level, it has to be erected much below ground level inside the well. In such cases, it will be difficult to prime the pump manually. A manually operated reciprocating pump in delivery line is provided, on ground level. The operation of this pump lifts the water from sump to height of 8 to 9 m i.e., slightly above the level of C.F pumps centre line. If now C.F. pump is started, it will be able to delivery the water to desired level. This method can work only for C.F. pump having delivery pipe size say not exceeding 10 cm diameter. Cost of above arrangement will be less than other special type pumps (vacuum pump) to serve the same purpose.

(iii) ***Priming with jet pump***: Water available at high head is allowed to flow through nozzle 'A' as shown in Fig. 9.12. The nozzle is so designed that at the jet outside the nozzle is at the pressure less than the atmospheric pressure thus it is possible to suck the water from the surrounding chamber i.e., sump. The two streams, one through high pressure nozzle and second through low pressure chamber (sump) are mixed together and carried to casing of C.F. pump for priming i.e., removal of air from pump.

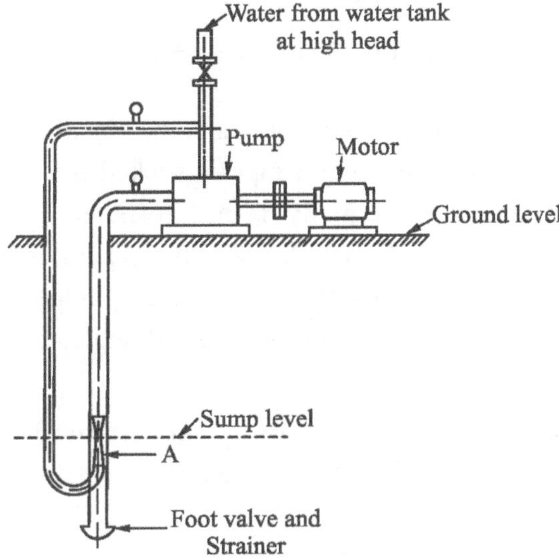

Fig. 9.12 Priming by Jet.

The C.F. Pump can be placed on ground level. The jet pump is generally fixed in suction pipe near the sump level. Jet pump is supplied water initially through high head reservoir or tank. The operation of jet pump lifts the water from sump to the ground level and thus does the priming work for centrifugal pump.

Priming with Separator (Fig. 9.13): It consists of air separator 'A' on delivery side of pump and bent portion of suction pipe 'B', which goes above the centre line of pump. Pump is not provided with any foot valve as it is used for pumping petroleum products, which are almost free from impurities.

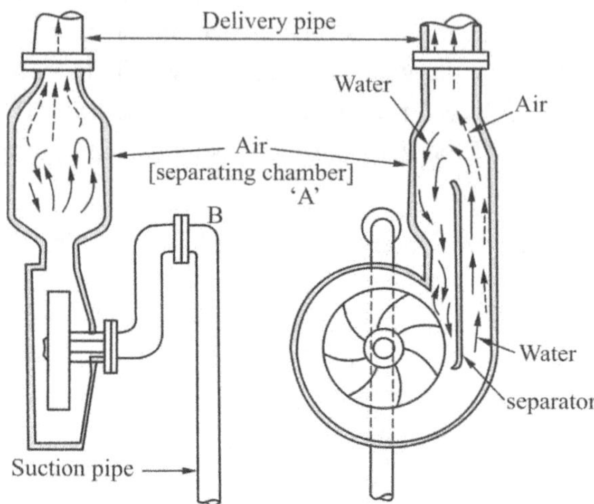

Fig. 9.13 Priming with Separator.

Bent suction pipe portion (B) always contain some liquid. When pump is started this liquid along with air from rest of suction pipe is sucked into impeller and delivered to separator. Since head developed initially is not sufficient, no liquid is delivered at delivery end. The liquid being heavier falls back into separator, leaving air on the upper side of the separator. Thus separation of liquid and air pumped by impeller takes place, the liquid return back to impeller, from other side of separator and air is pushed to the delivery side.

The return of liquid to the impeller makes the mixture sucked richer and its continued working slowly pushes all the air from suction side and thus completes the priming action.

Starting the C.F. Pump

(i) First carryout priming of pump

(ii) Close the delivery valve of C.F. Pump

(iii) If pump is connected to electric-motor, operating switch is made on and pump will be started.

(iv) Pump is started by prime mover i.e., Electric motor or diesel engine and when its normal speed is achieved, the delivery valve is gradually opened to obtain the required discharge of liquid.

9.13 Pump Losses (Fig. 9.14)

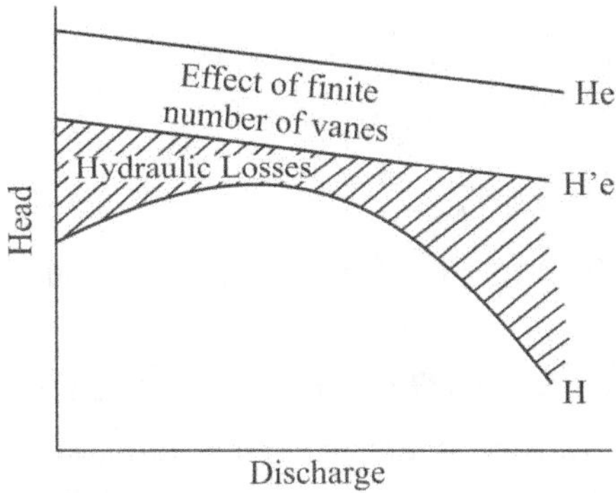

Fig. 9.14 Losses and Head-discharge Relationship.

Hydraulic losses are:

1. Eddy losses
2. Shock at the entrance and exit of the impeller. This is due to shock entry at operating point other than design point.
3. Friction losses in suction and delivery pipes.
4. Friction losses in impeller.
5. Friction and eddy losses in the diffuser and casing.
6. Secondary flow losses.
7. Increase turbulence inflow
8. Viscous motion over the blades.

Mechanical losses are:

1. Friction losses in main bearing and glands.
2. There are losses because of disc friction between impeller and the liquid which fills the gap between impeller and casing

The real flow energy losses are depicted (shown) in Fig. 9.15 below.

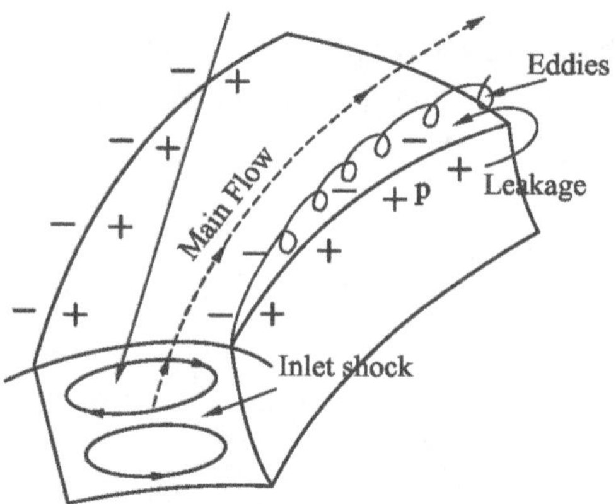

Fig. 9.15 Flow Between Blade Channels.

9.14 Characteristic Curves for C.F. Pump

These curves are necessary to predict the behaviour and performance of pump, when the pump is working under different flow rate, head and speed. The following are the important characteristic curves for C.F. Pumps.

1. Operating characteristic curves
2. Main characteristic curves

Operating Characteristic Curves (as shown in Fig. 9.16): If speed is kept constant, the variation of manometric head (H_m), power and efficiency with respect to discharge, is known as operating characteristic curves of pump.

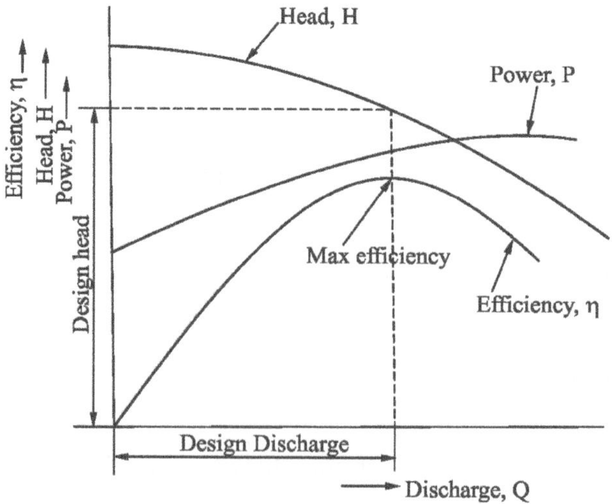

Fig. 9.16 Operating Characteristics.

- Looking to graph of Head and discharge, it is straight line sloping down [$\because$ H $\propto$ (constant – θ)]. In practice there are real flow effects such as friction, viscosity, eddying etc., loss at inlet of blades due to shock entry at operating point, other than the design point and in accordance with turbulent flow concept Head loss = kQ^2 where k = constant, Q = discharge therefore graph of H Vs Q, will be curve and not straight line as shown in Fig. 9.16.
- The power curve does not pass through the origin showing that power is consumed even when delivery valve is closed (i.e., there is no discharge from pump). This is due to whirling or forced vertex motion imparted to confined fluid in casing. The power input increases non-linearly in accordance with head and discharge.
- Efficiency of a centrifugal pump is maximum at design point of speed and discharge. It drops at the off-design values due to real flow effects such as frictional passage, viscous motion over the blades, increased turbulence in flow, undesirable eddying in motion, shock entry, shock exit and secondary flow loss as shown in Fig. 9.16.
- The vertical line in Fig. 9.16, corresponding to point of maximum efficiency. The intersection of this line with the other curves drawn, we can determine the best condition of running the pump.
- A typical set of characteristic is shown in figure where the arrows indicate the trend for the speed increasing above the design speed.

Characteristic Curves of C.F Pump for Speed or Main characteristic curves as shown in Fig. 9.17.

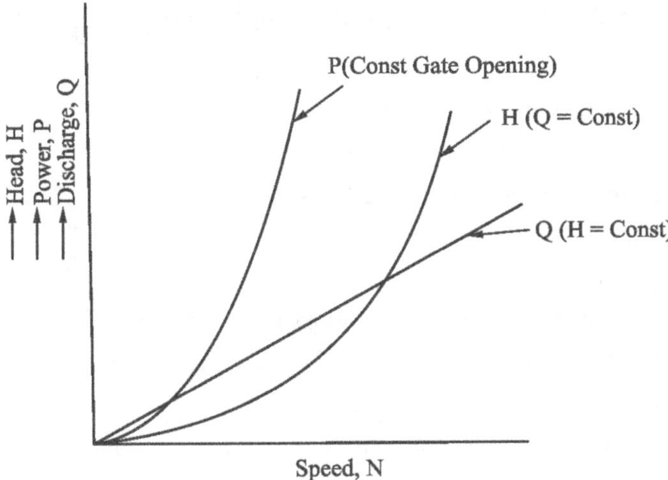

Fig. 9.17 Main characteristic Curve or Constant Head and Constant Discharge Curves.

The main characteristic curve of centrifugal pump consists of variation of head (manometric head H_m), power and discharge with respect to speed.

***Speed versus Discharge*:** Fig. 9.17 show the performance of C.F. Pump under constant head. It is a straight line which shows that discharge increases with the speed.

$$Q = \text{area} \times \text{velocity}$$

$\therefore$ $Q \propto$ velocity for constant area

$$Q \propto \frac{\pi DN}{60}$$

$\therefore$ Curve is straight line

***Speed versus power*:** Fig. 9.17 show the performance of C.F. pump under constant head and constant discharge. It is parabolic curve. It shows power increases with the speed.

$$\text{Power} = \rho\, g\, Q\, H$$

$$\text{Power} \propto QH \qquad \text{But } Q \propto N \text{ for given diameter}$$

$$\text{Power} \propto N.N^2 \qquad H \propto N^2$$

$$\therefore \qquad \text{Power} \propto N^3$$

$\therefore$ Power curve is parabolic

***Speed versus head*:** Fig. 9.17 show the performance of centrifugal pump under constant discharge. It is parabolic curve, which shows that the head increases with speed.

9.15 Advantages of C.F. Pump over other type of Pumps

Following are advantages of Centrifugal Pump:

1. Its discharge is continuous, smooth and free from shocks.
2. It is lighter in weight and requires less space.
3. Discharge is high because it runs at high speed.
4. Uniform torque and less vibration.
5. It is easy to install C.F. pump.
6. Lubrication is easy and liquid to be pumped remain pure.
7. It does not require elaborate foundation because pump is free from vibration.
8. It is less costly than other types of pumps.
9. At low heads efficiency is higher than other type of pumps.
10. Its reliability, simplicity and ease of operation.
11. Make less noise than other type of pumps.

Disadvantages

1. It has limited suction lift.
2. Priming is essential.
3. Low efficiency over range of head and discharge.
4. Single stage centrifugal pump can lift water up to 40 m.

9.16 Uses of Pumps

1. Supplying water for irrigation purposes.
2. For circulation, lifting and storing of liquid at high reservoirs.
3. Used for feed water in boilers.
4. To circulate water in condenser for condensation of steam.
5. Used for chemical, petroleum and textile industries, for circulation of water, chemicals and coolants.
6. To force lubricating oil in lubricating space.
7. It is used to pump sewage water.
8. To remove water or liquid from underground space.
9. To carry fuel from reservoir to the I.C. engines.

9.17 Rotary Pumps

Principle of Operation

1. Rotary pump works on the principle of rotation.
2. It is positive displacement pump. So that liquid is discharged by a positive pressure similar to that in reciprocating pump.
3. It consists of rotating elements like gear, screws, cams, lobes vane etc which are fitted in casing. These elements are rotated by the prime mover. Rotating element of pump generates low pressure in suction line, liquid is sucked, trapped and forcing it through the discharge port at a higher pressure.

9.18 Types of Rotary Pumps

The Rotary pumps are classified according to the rotating elements enclosed in casing as follows:

1. Gear Pump
2. Vane Pump

3. Lobe Pump
4. Screw Pump
1. ***Gear Pumps*** (as shown in Fig. 9.18).

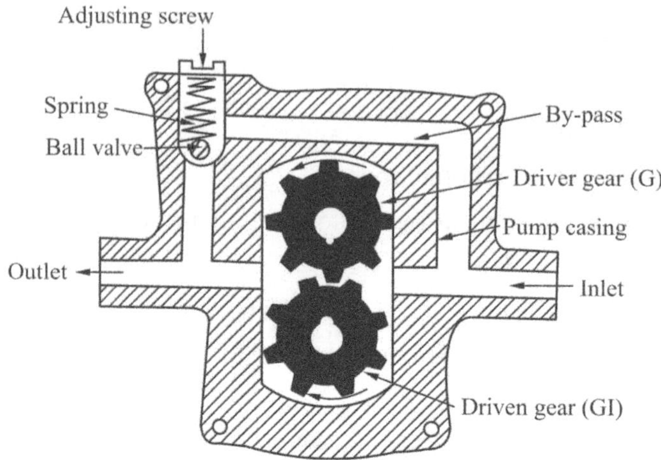

Fig. 9.18 Details of a Gear type Oil Pump.

1. Fig. 9.18 shows gear pump in which spur gears are used. The other gears such as spiral gears can be used.
2. The driving gear G is engaged with driven gear 'G1' inside the casing. Out of two gears one is rotated by prime mover (motor) and other gear rotates in mesh. Two gears generally revolve in opposite direction.
3. The suction side space is filled with liquid due to priming and the liquid entrapped between teeth and the casing is delivered to delivery side due to rotation of gears. The closed fit of meshed gear teeth and close fit between gears and pump casing, prevent the oil from returning to inlet side of pump.
4. It is small in size, simple in operation, and operates with high suction lift.
5. It is used only for small discharge and moderate heads.
6. In order to relieve pressure which might have become excessive due to high engine speeds or clogged oil lines, a pressure relief valve is also provided in certain oil pumps. For relieving of excessive pressure, the spring compressed ball is lifted off its seat with the oil passing back to the inlet side through by pass provided. A screw is provided to adjust the pressure for valve.
7. Oil pressure delivered by pump is about 3-4 bar.

2. ***Vane Pumps*** (as shown in fig. 9.19)

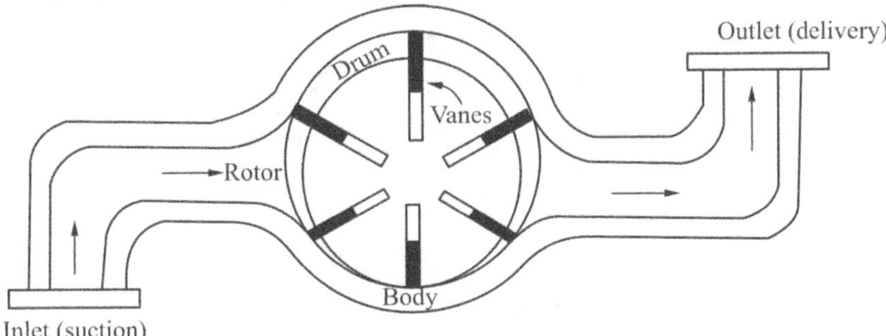

Fig. 9.19 Vane Type Pump.

1. It consists of rotor mounted eccentrically in circular casing. Rotor touches the casing at the top and keeps gap at the bottom.
2. The sliding vanes are placed in radial slots of rotor by spring. These vanes move outwards from centre of the rotor due to spring, gravity and centrifugal action.
3. The rotation of the rotor carries the liquid entrapped between the vanes and casing from the suction side towards the delivery side.
4. When the number of vanes is more in number, the leakage losses are less because pressure difference between adjacent space is small.
5. The vanes are usually made of steel.
6. Pump gives a very uniform torque, so that there is no need of any flywheel, requires work input.
7. It has low weight.
8. It used for pressure ratio up to 8.5 : 1.
9. Speed is limited to 3000 RPM.

3. ***Screw Pump*** (as shown in Fig. 9.20 and 9.21)

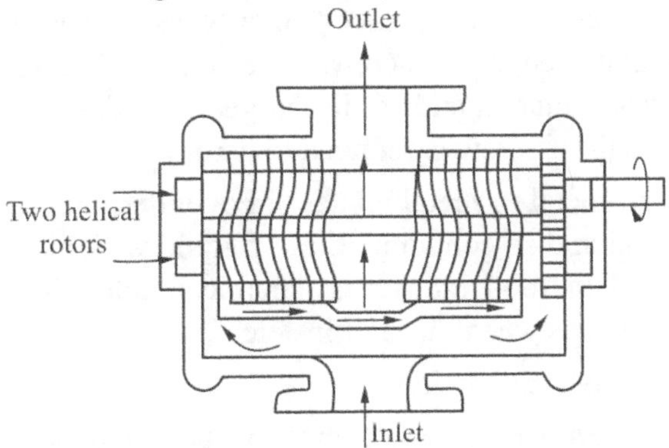

Fig. 9.20 Screw Pump.

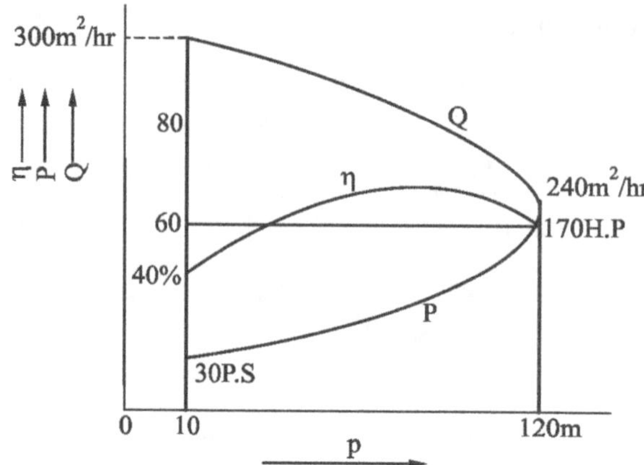

Fig. 9.21 Screw Pump Characteristics.

1. Screw pump is positive displacement pump in which flow is accomplished by the enmeshing of two mating helically grooved rotors suitably housed in cylinder provided with suitable inlet and discharge ports.
2. Rotors are generally made from carbon steel and required perfect machining. Rotors are dynamically balanced after machining. Helical screw mounted on one shaft is driven by motor; other shaft is geared to first shaft. Both spindles are enclosed in a casing.
3. Bearings are sleeve type. Bearings also work as shaft and seals within the pump.
4. When pair of screws which have been already in mesh commence to disengage liquid, enters through inlet port into the enlarging space between them. As the rotor continues to rotate the liquid is moved past the suction port and sealed in space between pair of screws. From the suction end of machine remeshing commences and continues along the rotor, so that liquid trapped in space is gradually compressed and displaced axially as screws rotate. At fixed point where the dealing edge of screws and edge of discharge port co-inside, compression ceases and liquid is discharged into delivery line until volume of liquid between pair of screw is reduced to zero.
5. Since there is no metal to metal contact in cylinder, the rotor need no (very very small) lubrication and so contamination of liquid with oil is avoided. It is for this reason it is used in food industries, chemical industries and transportation of loose material such as wheat, iron-chips etc.
6. It requires very small foundation.
7. Screw pumps have advantages of both the turbo pump and reciprocating pump.
8. The main disadvantage is noise which is of high frequency, caused by the discontinuous suction and discharge of the liquid.

9. The pump is self-priming and is suitable for pumping water, juices and oils. Its speed is same to that of gear pumps and is dependent on viscosity of pumped liquid.
10. Discharge is regulated with help of change in speed and this is done by change of poles. Between 1000 RPM and 1500 RPM, the discharge is proportional to speed.
11. Casing is made of grey C.I, bronze or cast steel. The spindle teeth are made of bronze or cast steel. The maximum discharge of this type of pump is 400 m^3/hr; for pressure equal to 24.525 bar (25 kg/cm^2).
12. The efficiency of pump is greatly influenced by leakage losses and viscosity of fluid to be pumped. Low viscosity necessitates very narrow gaps and viscous fluids, gap required is large. This will increase the leakages losses but the force required to overcome friction decreases.
13. Each spindle is provided with portion of right hand screw and left hand screw, hence the pump is double flow pump in which suction takes place in axial direction from both sides and carries to centre of pump. In this way the axial thrust gets balanced on both sides and hence axial thrust on spindles is zero. In spite of this the pump is provided with bearings to take axial thrust if any.
14. The spindles are to be so arranged that metallic contact between the teeth is avoided.

4. *Lobe Pumps* (as shown in Fig. 9.22 and 9.23)

Fig. 9.22 Lobe pump.

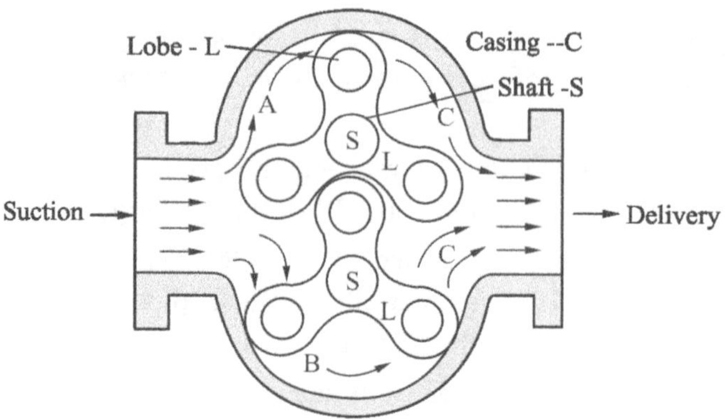

Fig. 9.23 Lobe type pump.

1. As shown in figure 9.22 it consists of two rotor (lobe type). They are mounted on separate axes, which rotate in opposite direction.
2. One of the rotor are connected to electric motor and other rotor is geared from the first for correct mating the lobes of rotors are of cycloidal or involute form. The lobes mesh like teeth of gear pump and rotate about 15000 RPM in opposite direction.
3. Through intake port liquid enters inside and trapped between lobe and casing and final discharge into receiver.
4. Water in space A and B is at atmospheric pressure P_1 and water in space C increases in pressure to the receiver pressure.
5. The efficiency of lobe pump decreases with increase in pressure ratio.
6. It can develop a pressure ratio up to 2 in single stage and up to 3 in two stage.
7. The rotors of lobe pump are made of steel and ground exactly to required shape.

9.19 Application of Rotary Pumps

1. Rotary pump combines the advantage of reciprocating and centrifugal pumps and hence it is used in many applications.
2. For pumping thick and viscous liquids such as tar, grease, heavy lubricating oil, vegetable oil etc.
3. For pumping lighter liquids with high vapour-pressure such as alcohol, gasolene, benzene etc.

9.20 Reciprocating Pump

Single Acting Reciprocating pump (as shown in Fig 9.24)

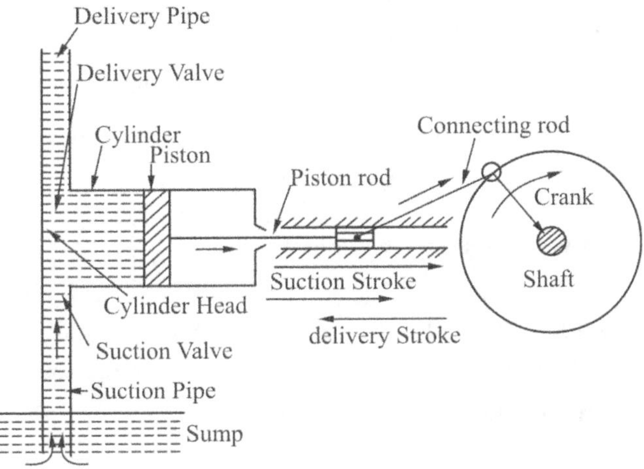

Fig. 9.24 Single Acting Reciprocating Pump.

1. It is positive displacement pump.
2. It consists of a piston and cylinder with suction valve and delivery valve fitted with suction and delivery pipes as shown in Fig. 9.24.
3. In single acting reciprocating pump, liquid acts on one side of the piston. Suction and delivery of liquid takes place on same side of piston or cylinder end.
4. During suction stroke, the piston moves away from the cylinder head side and expansion of air inside the cylinder creates the vacuum, so suction valve opens inwards and liquid is sucked into the cylinder from the sump through suction pipe and suction valve. Delivery valve is closed due to pressure of spring.
5. During return (delivery) stroke piston moves towards cylinder head side. This pushes the liquid to close the suction valve and to open the delivery valve in delivery pipe. Thus liquid rushes through the delivery pipe due to positive pressure, created by piston. So pump is known as a positive displacement pump.
6. The cycle is repeated to discharge the liquid at high pressure. Here there is discharge during delivery stroke and no discharge during suction stroke so it is known as single acting pump. If pump discharge is obtained during both the forward and backward strokes, the pump is known as double acting pump.
7. For continuous and constant pressure of discharge, they are provided with air vessel.

Consider Single Acting Reciprocating Pump

A = Cross sectional area of piston in m^2

L = length of stroke in m

r = Radius of crank in m

N = Rotational speed of the pump in RPM

H_s = height of the cylinder axis above the sump water level in m

H_d = height of delivery outlet above the axis of the cylinder in m

Total height to which water is lifted = $H_s + H_d$

Kinetic head of water leaving the delivery pipe = $\frac{V_d^2}{2g}$

where V_d = velocity at outlet. Since V_d is generally small and subjected to variation during stroke, it can be neglected, except in cases where the total lift is small.

volume delivered in one revolution = AL

volume delivered per minute = ALN $\frac{m^3}{min}$

$$= \frac{ALN}{60} \frac{m^3}{sec}$$

Total mass discharged per second $= \frac{\rho ALN}{60}$ kg/sec

Total weight (force) discharged per second $= \frac{\rho g ALN}{60}$ N/sec

work done = ρALN ($H_s + H_d$)

$$\text{Theoretical power} = \frac{\rho ALN(H_s + H_d)}{60 \times 1000} \text{ kw}$$

Actual power required is greater than above value due to frictional losses, mechanical losses and leakage losses.

***Slip of Pump*:** Actual average discharge of the pump is less than theoretical average discharge due to leakage of liquid.

The Difference between theoretical average discharge and the actual average discharge is known as slip of pump.

$$\text{Slip} = \text{Theoretical average discharge} - \text{Actual average discharge}$$

$$\text{Slip percentage} = \frac{\text{slip}}{\text{Theoretical discharge}} \times 100$$

$$= \frac{Q_{th} - Q_{act}}{Q_{th}} \times 100$$

$$= \left(1 - \frac{Q_{act}}{Q_{th}}\right) \times 100$$

$$= (1 - Cd) \times 100 \qquad \left(\because \frac{Q_{act}}{Q_{th}} = Cd\right)$$

where

Q_{th} = Theoretical discharge

Q_{act} = Actual discharge

Cd = co-efficient of discharge

Negative Slip

Slip is equal to difference between theoretical discharge and actual discharge. If actual discharge is more than theoretical discharge the slip of pump will be negative. In this case the slip of pump is known as negative slip.

Negative slip occurs when suction pipe is long, delivery pipe is short and pump runs at high speed.

Consider single acting pump, the absolute pressure head in cylinder at end of suction stroke

$$= H_{atm} - H_s + \frac{l_s}{g}\frac{A}{a}\omega^2 r$$

$$= H_{atm} + \left(\frac{l_s}{g}\frac{A}{a}\omega^2 r - H_s\right)$$

Suppose in an installation, H_s is small

l_s is large

H_d is small

Under above conditions, in such cases the quantity $\left(\frac{l_s}{g}.\frac{A}{a}\omega^2 r - H_s\right)$ is greater than delivery head H_d, the pressure of water in the cylinder will be large enough to force open the delivery valve. Therefore there will be some discharge into delivery pipe near the close of suction stroke. The discharge is in addition to usual discharge that takes place during the delivery stroke. Therefore actual discharge is more than theoretical discharge and slip is negative.

9.21 Air Vessel or Air Chamber (as shown in Fig. 9.25)

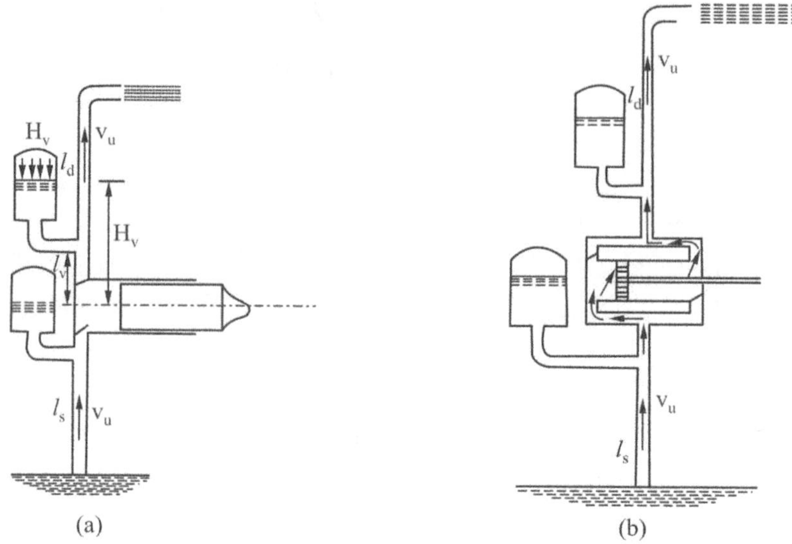

Fig. 9.25 (a) Single Acting Pump with Air Vessels. **(b)** Double Acting Pump.

1. For a single acting pump, the flow of liquid in suction and delivery pipe is fluctuating. For double acting pump, flow is continuous but velocity in suction and delivery pipe changes as piston moves from IDC to ODC or from ODC to IDC. To obtain continuous flow and uniform velocity in suction and delivery pipe with constant pressure, a air vessel (chamber) is provide on suction and delivery pipe, near to the cylinder. This vessel has an opening at bottom through which liquid may flow in or out of it. When vessel is empty it contains atmospheric air in it.
2. During delivery stroke, delivery valve is open and water enters into air vessel, which compresses the air in air-vessel and absorb most of the pressure fluctuations and air expands when liquid flows out.
3. During suction stroke when delivery valve is closed, the compressed air in vessel on delivery pipe expands and flow to discharge tank is continuously maintained.
4. The air vessels store water when piston accelerates and delivery water when piston (water) decelerates in pipe.

9.22 Double Acting Reciprocating Pump (Piston Pump)

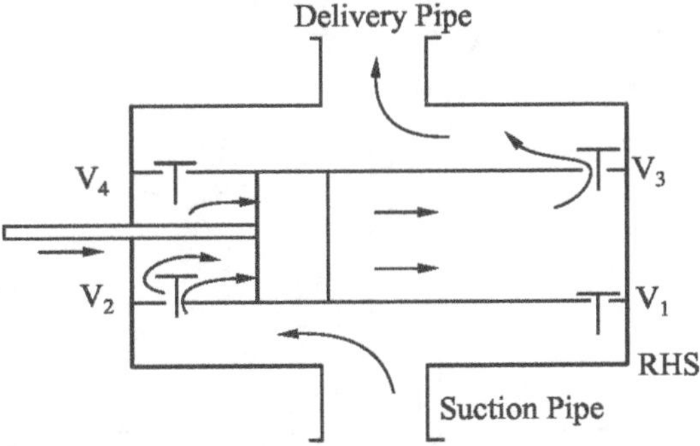

Fig. 9.26 Double Acting Piston Pump.

1. Here suction and delivery takes place simultaneously.
2. During forward stroke piston moves toward right hand side RHS of cylinder. The suction valve V_1 is closed and delivery valve V_3 is opened due to increase in water pressure on that side in the cylinder as shown in Fig. 9.26. So water is forced through valve V_3 to delivery side at the same time suction valve V_2 is opened and delivery valve V_4 is closed due to decrease in pressure on that side in the cylinder, so that water is sucked through valve V_2 into the cylinder.
3. During backward stroke the valve operations are interchanged. The valves V_2 and V_3 are closed and valve V_1 is opened and water is sucked in to cylinder and valve V_4 is opened to deliver water from cylinder to delivery side.
4. The cycle is repeated and water is continuously delivered and constant pressure is obtained.
5. The double acting pump gives more uniform discharge than single acting pump as fluid is delivered in both strokes.

9.23 Single Acting Plunger Pump

(as shown in Fig. 9.27 and fig. 9.28)

1. The pump- is operated by handle (Rod)
2. It consists of inlet valve, delivery valve, plunger, operating handle known as Rod, pump chamber and air vessel.
3. During upward stroke of plunger, the pressure in the pump chamber falls below atmosphere, the suction valve or inlet valve opens. So water or oil enters the pump chamber.
4. During downward stroke, suction valve is closed and delivery valve is opened, water or oil through delivery valve enters into delivery pipe.

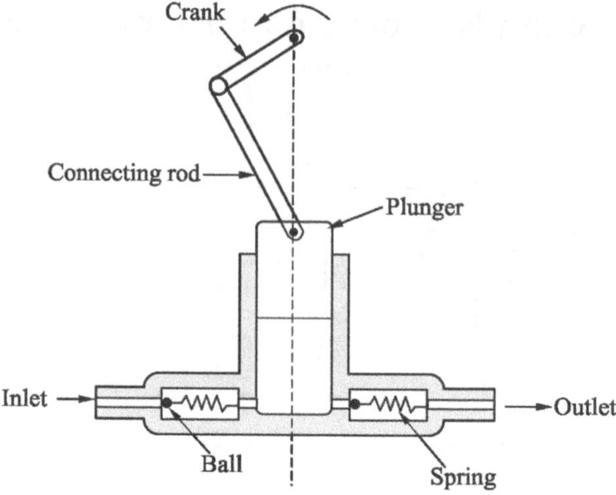

Fig. 9.27 Constructional Details of a Plunger type oil pump.

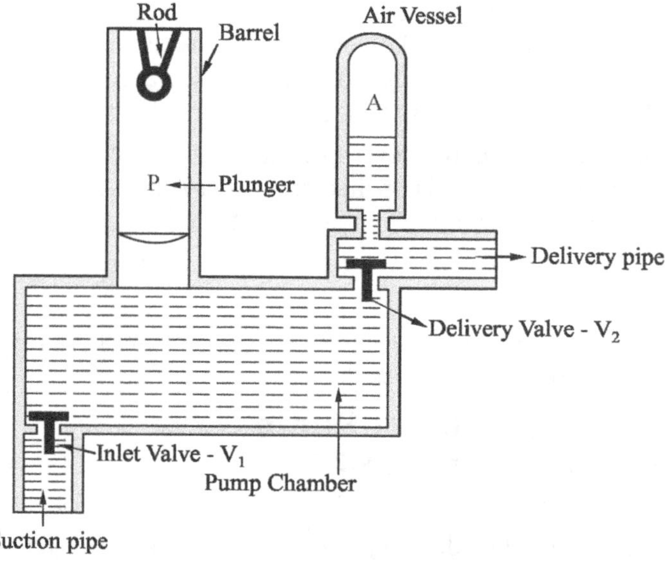

Fig. 9.28 Single Acting Plunger Pump (with air-vessel).

5. Plunger pump is single acting pump, the discharge of water or oil is not continuous and its pressure is also not constant. This produces shock on delivery valve. To avoid these shocks and to obtain continuous and constant pressure

discharge, the air vessel is fitted over and near the delivery valve as shown in figure 9.28.

6. Plunger pumps are used in oil and diesel engines as fuel pump.
7. Plunger pump has relatively long length and small diameter compare to piston.
8. Cylinder having small diameter is known as barrel.

9.24 Bucket Pump (as shown in Fig. 9.29)

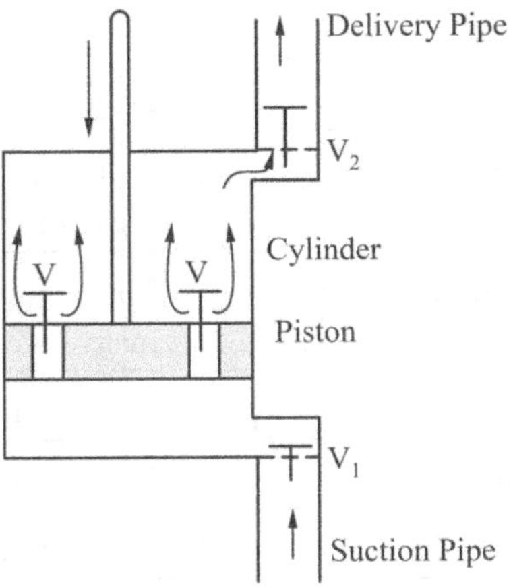

Fig. 9.29 Bucket Pump.

1. Piston is fitted with two non-return type valves on the piston as shown in Fig. 9.29.
2. During down stroke, suction valve V_1 and delivery valve V_2 are closed and piston valves V are opened, so the air and water in lower portion of cylinder rushes towards upward portion of piston through valves "V".
3. During upward stroke, the piston valve "V" is closed and suction valve V_1 and delivery valve "V_2" are opened so the sump water is sucked through valve V_1 in lower part of piston and water in the upper part is lifted to delivery side, through delivery valve V_2.
4. The cycle is repeated. This is single acting pump so discharge is during upward stroke only. Continuous discharge is not obtained.
5. When pump is operated by hand, it is known as hand pump.

9.25 Comparison between Centrifugal Pump and Reciprocating Pump

Centrifugal Pump	Reciprocating Pump
1. This is high speed pump.	1. This is slow speed pump
2. This pump is meant for large discharge	2. This pump is meant for low discharge
3. Low pressure head	3. High pressure head
4. Discharge is smooth and even	4. Discharge is fluctuating
5. Priming is essential	5. Priming is not required
6. At low heads efficiency is high	6. At low heads efficiency is low, cause of greater losses due to friction and valves
7. This pump is compact and occupies less space.	7. This pump occupies large space and has many components.
8. Weight of pump is less	8. Weight of pump is more
9. It is easy to install this pump	9. Installation is comparatively difficult
10. Uniform torque	10. Fluctuating torque
11. Less vibration	11. More vibration
12. Small noise	12. More noise
13. Low cost	13. High cost
14. Not sensitive to impurities, not affected by dirty water and solid bodies in liquid	14. Sensitive to impurities, valves are affected by solid impurities in the liquid
15. This is known as Dynamic pressure pump	15. This is known as positive displacement pump
16. Multi stage pump is used to increase the pressure head	16. Multi-stage method is not used.
17. Lubrication is easy and liquid remain pure.	17. Lubrication is difficult and liquid become impure due to lubricant.
18. This involves less maintenance cost	18. This involves more maintenance cost.

Solved Problems

1. Single acting reciprocating pump, running at 50 RPM delivers 0.01793 m^3/sec; diameter of suction pipe is 25 cm and stroke length 45 cm. Determine.

   (i) Theoretical discharge of pump

   (ii) Co-efficient of discharge

   (iii) Slip and percentage of slip

***Solution*:**

Given

N = 50 RPM

Q_{act} = Actual discharge = 0.01793

D_s = Diameter of suction pipe = 25 cm = 0.25 m

L = 45 cm = 0.45 m

$$A = \text{Area of suction pipe } = \frac{\pi}{4}(0.25)^2 = 0.0491$$

(i) ∴ Q_{th} = Theoretical discharge

$$= \frac{ALN}{60} = 0.0491 \times \frac{0.45 \times 50}{60} = 0.0184 \text{ m}^3/\text{sec}$$

∴ $Q_{th} - Q_{act} = 0.0184 - 0.01793 = 0.00047$

(ii) ∴ $$\text{Coefficient of discharge } = \frac{Q_a}{Q_{th}} = \frac{0.01793}{.0184} = 0.974$$

(iii) Slip = 0.0184 – 0.01793 = 0.00047

∴ $$\text{percentage slip } = \frac{0.00047}{0.0184} \times 100 = 2.554\%$$

2. A double acting reciprocating pump, running at 40 RPM is discharging 0.01666 m^3 of water per second. The pump has stroke of 40 cm, diameter of piston is 20 cm. The delivery and suction heads are 20 m and 5 m respectively. Find slip and power required to drive pump.

***Solution*:**

Given

N = 40 RPM

Q_{act} = 0.01666 m^3/sec

L = 40 cm = 0.40 m

D = Piston diameter = 20 cm = 0.2 m

$$\therefore \qquad A = \frac{\pi}{4}(0.2)^2 = 0.031416 \text{ m}^2$$

Suction head = h_s = 5m

Delivery head = h_d = 20 m

$$\text{Theoretical Discharge} = Q_{th} = \frac{2ALN}{60} = \frac{2 \times 0.031416 \times 0.4 \times 40}{60}$$

$$= 0.01675 \text{ m}^3/\text{sec ('2' stands for double acting)}$$

(i)
$$\text{Slip} = Q_{th} - Q_a$$
$$= 0.01675 - 0.01666$$
$$= 0.00009 \text{ m}^3/\text{sec}$$

(ii)
$$\text{The power required to drive the pump} = \frac{2\rho g\, ALN(h_s + h_d)}{1000 \times 60}$$

$$= \frac{2 \times 1000 \times 9.81 \times 0.031416 \times 0.4 \times 40 \times (5+20)}{1000 \times 60}$$

$$= \frac{2 \times 9.81 \times 0.031416 \times 0.4 \times 40 \times 25}{60}$$

$$= 4.109 \text{ kw}$$

3: A Centrifugal pump delivers 0.03 m^3/sec of water to a height of 18.5 m through 10 cm diameter pipe 90 meters long. If overall efficiency of pump is 75%, find power required to drive the pump. Take f = 0.01.

***Solution*:**

$$Q = 0.03 \text{ m}^3/\text{sec}$$
$$H_s + H_d = 18.5 \text{ m}$$
$$d = ds = 10 \text{ cm} = 0.1 \text{ m}$$
$$\eta_0 = 75\%$$
$$L = L_s + L_d = 90 \text{ m}$$
$$f = 0.01$$

velocity of water in suction and delivery pipe

$$= v_s = v_d = \frac{0.03}{\frac{\pi}{4}(0.1)^2} = 3.82 \text{ m/sec}$$

$$\text{Loss of head in pipe line} = h_{fs} + h_{fd} = \frac{4fLv^2}{2gd} = \frac{4 \times 0.01 \times 90 \times (3.82)^2}{2 \times 9.81 \times 0.1} = 26.78 \text{ m}$$

$\therefore$ Manometric head $= H_m = (H_s + H_d + h_{fs} + h_{fd})$

$$= 18.5 + 26.78$$

$$= 45.28 \text{ m}$$

Power required to drive the pump $= \dfrac{\rho g Q H_m}{1000\eta_0}$

$$= \frac{1000 \times 9.81 \times 0.03 \times 45.28}{1000 \times 0.75}$$

$$= 17.768 \text{ kw}$$

4. Find the number of pumps required to take water from a deep pump under a head of 89 m. All pumps are identical and are running at 800 RPM. The specific speed of pump is 25 while the rated capacity of each pump is 0.16 m^3/sec.

***Solution*:**

Given

Head = 89 m

Speed = 800 RPM

$N_s = 25$

$Q = 0.16 \text{ m}^3/\text{sec}$

Let H_m = Head developed by each pump

(i) So first calculate H_m from $N_s = \dfrac{H\sqrt{Q}}{Hm^{\frac{3}{4}}}$

(ii) Then Number of pumps = Total head developed head developed by one pump

Now we know $N_s = \dfrac{H\sqrt{Q}}{Hm^{\frac{3}{4}}}$

$$25 = \frac{800\sqrt{0.16}}{H_m^{\frac{3}{4}}}$$

$$H_m^{\frac{3}{4}} = \frac{800 \times \sqrt{0.16}}{25}$$

$H_m = 29.94$ m

∴ Number of pumps required

$$= \frac{\text{Total head}}{\text{Head developed by one pump}}$$

$$= \frac{89}{29.94}$$

$$= 2.973$$

i.e., 3 pumps

5. Centrifugal pump is required to lift water to a head of 40 meters at a rate of 50 lit/sec. Find power required for the pump, if the overall efficiency is 62%.

***Solution*:**

Given

$$H_m = 40 \text{ m}$$

$$Q = 50 \text{ lit/sec} = \frac{50}{1000} = 0.05 \text{ m}^3\text{/sec}$$

$$\eta_0 = 62$$

Find power = p

$$P = \frac{\rho g Q H_m}{1000 \times \eta_0} = \frac{1000 \times 9.81 \times 0.05 \times 40}{1000 \times 0.62} \text{ kw}$$

$$= 31.6452 \text{ kW}$$

6. Calculate the brake power to pump 1500 kg of water per minute against a head of 70 metre. Assume efficiency of pump to be 70%.

***Solution*:**

Brake power $= \dfrac{Pv}{1000\eta_p}$ where P = Pressure of water N/m^2

V = Volume in m^3/kg

$$= \frac{\rho g H V}{1000\eta_p} \quad (\because P = \rho g H)$$

$$= \frac{gH \times \text{mass}}{1000\eta_p} \quad \text{where } \rho.V = \text{mass}$$

$$= \frac{g\,mH}{1000\eta_p}$$

$$= \frac{1500 \times 9.81 \times 70}{60 \times 1000 \times 0.7} \quad \text{where m = kg/sec}$$

$$= 24.52 \text{ kw}$$

7. Following parameters are measure for C.F. Pump

(i) Power consumed = 18 kW

(ii) Delivery head = 20 m

(iii) Suction head = 5 m

(iv) Water collected = $4m^3$/min

Find overall efficiency of pump.

***Solution*:**

Total head developed by pump = 20 + 5 = 25 m

Power consumed = 18 kw

$$Q = \frac{4}{60} = 0.06667 \text{ m}^3/\text{sec}$$

$$\therefore \quad \text{Water power} = \frac{\rho g QH}{1000}$$

$$\therefore \quad \eta_0 = \frac{\text{Water Power}}{\text{Power consumed}} = \frac{1000 \times 9.81 \times 0.06667 \times 25}{1000 \times 18}$$

$$= 0.9083$$

$$= 90.83\%$$

8. Pump is directly coupled with electric motor. It raises 1500 kg of water/min through a height of 30 m. The efficiency of pump is 65%. What should be the power of pump shaft (impeller power). Also calculate power of motor in kW, if Mechanical efficiency of motor is 85%.

***Solution*:**

$$\text{Power required to drive pump} = \frac{\rho g Q H}{1000 \times \eta_{pump}}$$

$$= \frac{g m H}{1000 \eta_{pump}} = \frac{9.81 \times 1500 \times 30}{60 \times 1000 \times 0.65} = 11.319 \text{ kw}$$

$$\ldots(\because m = \rho Q)$$

ρ = density of water

m = Mass of water pumped/sec

H = height through which water is raised

$$\text{Power of motor} = \frac{\text{pump Power}}{\text{Motor efficiency}}$$

$$= \frac{11.319}{0.85} = 13.3165 \text{ kw}$$

Questions

1. Why Reciprocating pump is called a positive displacement pump?
2. Classify Reciprocating pumps.
3. Explain with neat sketch the working of double acting reciprocating pump.
4. Explain with neat sketch the working of single acting reciprocating pump.
5. Why air vessel (chamber) is provided on reciprocating pump?
6. Should be a limit to suction lift? How this is limited?
7. Define slip in reciprocating pump. Why it occurs?
8. What do you mean by negative slip? Under what circumstances can this occur?
9. What is difference between piston and plunger pumps?
10. Explain with diagram the working of vane pump.
11. Explain with neat diagram working of gear pump.
12. Why safety valve is provided on gear pump?
13. What are functions of positive displacement pumps?
14. Principle of working of reciprocating pump.
15. Distinguish between co-efficient of discharge and slip of reciprocating pump.
16. Define centrifugal pump. Explain working of a single stage centrifugal pump with sketch.
17. Define the terms (i) suction head; (ii) delivery head; (iii) static head;

    (iv) Manometric head related to C.F. Pump.
18. What do you mean by (i) Manometric efficiency; (ii) Mechanical efficiency; and (iii) Over all efficiency of centrifugal pump?
19. What is difference between single stage and multistage centrifugal pump?
20. Describe Multi-stage pumps when connected

    (i) Impellers are in parallel,

    (ii) Impellers are in series.
21. What is priming? Why it is necessary? What are the various methods of priming?
22. Define specific speed of C.F. Pump and Develop expression for the same.
23. Is there any limit to the suction lift of the pump?
24. What are various types of centrifugal pump?
25. What are various methods to increase efficiency of centrifugal pump?

26. Explain with sketch the volute and diffusion pump.
27. What is role of volute casing of C.F. Pump?
28. What are various expressions of Manometric head?
29. What are various types of impeller and where they are used?
30. What do you understand by characteristic curve of C.F. Pump?

Exercises

1. A single acting reciprocating pump has plunger diameter 0.250 m and stroke of 0.180 m. If pump runs at 40 RPM, calculate discharge of pump.

   [0.00589 m^3/sec]

2. A double acting reciprocating pump has a bore of 30 cm and stroke of 40 cm and runs at 40 RPM. Calculate discharge of pump.

   [6.0377 m^3/sec]

3. A single acting reciprocating pump has piston diameter of 40 cm and stroke of 30 cm. The Pump lifts water to a total height of 1200 cm. The speed of pump is 50 RPM and actual discharge of pump is 0.030 m^3/sec. Calculate,

   (i) Slip of pump and percentage slip

   (ii) Theoretical power required to drive pump

   [(i) 0.0014 m^3/sec, 4.46%

   (ii) 3.53 kw]

4. Find power required to drive a centrifugal pump which delivers 0.02 m^3/sec of water to a height of 30 m through 10 cm diameter pipe and 90 m long. The overall efficiency of pump is 70% and f = 0.009 in following formula.

   $h_f = \frac{4flv^2}{2gd}$ with usual notions

CHAPTER 10

Air Compressors

10.1 Introduction

The function of compressor is to take a definite quantity of fluid (gas or air) and deliver it at a required pressure.

Air compressor sucks air from atmosphere, compresses it and delivers the high pressure air to the storage tank. From storage tank it is conveyed by the pipe-lines where the compressed air is needed.

A compressor has to be driven by some prime-mover (diesel engine or electric motor) as compression of air requires that work should be done upon air. The energy received by compressor from prime mover, some will be absorbed in work done against friction, some will be lost to radiation and some will be carried away by coolant which is employed to cool the machine and rest will be utilized to increase the pressure of air. The Prime-mover converts only a fraction of the heat, which it receives from source into work and so far as compressor is concerned it receives energy which is available at the shaft of the prime-mover.

10.2 Air Compressor Terminology

1. ***Free air delivered (F. A. D)*:** Free air delivered is the volume of air delivered under the condition of temperature and pressure existing at the compressor intake.

   In the absence of any given free air conditions they are generally taken as 1.01325 bar and 15 °C. (standard conditions).
2. ***Capacity of Compressor*:** It is volume of air delivered by compressor, expressed in m^3/min.
3. ***Pressure ratio*:** It is ratio of discharge pressure to inlet pressure both expressed in absolute pressure.
4. ***Displacement of the compressor*:** The swept volume of the piston in the first cylinder of multi-cylinder compressor is known as displacement of compressor. It is given by $\frac{\pi}{4}D^2L$ for single acting compressor.

where D = diameter of cylinder
L = stroke length

5. ***Inlet pressure*:** This is absolute pressure of air at inlet to a compressor.
6. ***Discharge pressure*:** This is absolute pressure of air at outlet of a compressor.
7. ***Compression ratio:*** This is the ratio of the absolute discharge pressure to absolute inlet pressure.
8. ***Single stage and multi-stage compressors*:** In single stage compressor, the compression of the air from initial pressure to discharge pressure (final pressure) is carried out in one cylinder.

   In multi stage compressor, air is compressed from initial (atmospheric) pressure to desired discharge pressure is carried out in more than one cylinder, in a sequence, delivery of one cylinder is supplied to next cylinder.
9. ***Single acting and double acting compressors*:** In single acting reciprocating compressor, the suction, compression and delivery of air take place on one side of piston only.

   In double acting reciprocating compressor suction, compression and delivery are carried out on both sides of piston.

   Single acting unit gives one delivery stroke per revolution of the crank where as double acting unit gives two delivery strokes per revolution of crank.

   The double acting unit requires a piston rod and cross head in addition to the connecting rod for single acting compressor.
10. ***Reciprocating compressor:*** In this type of compressor the air is compressed by means of piston, reciprocating inside the cylinder.
11. ***Rotary compressor:*** In this type of compressor, the compression of air is affected by rotating vane or impeller that imparts velocity to flowing air to give desired pressure (by using diffuser vanes).
12. ***Clearance volume*:** When crank is at TDC (in case of vertical compressor) or IDC (In case of horizontal compressor) the volume between top of piston and cylinder head is known as clearance volume. It is measured in m^3. It is denoted by v_c.

   where TDC = Top dead centre
   IDC = Inner dead centre
13. ***Clearance ratio*:** It is defined as ratio of clearance volume to swept volume. It is denoted by k and is equal to $\frac{V_c}{V_s}$. In percentage, this ratio will have a value in general, between 4% to 10% of V_s, where V_s = stroke volume.
14. ***Theoretical power:*** It is power required to compress the air adiabatically up to desired delivery pressure, without any provision for loss of energy.

15. ***Brake power***: Brake power is actual power input required by compressor to compress the required quantity of air and at required delivery pressure taking to all losses into consideration.
16. ***Volumetric efficiency***: This is the ratio of capacity of a compressor to piston displacement of compressor.

10.3 Classification of Air Compressors

1. ***According to number of stages***
   - (a) single stage
   - (b) Multi stage
     - (i) Single stage – delivery pressure up to 5 bar
     - (ii) Two stage – delivery pressure up to 35 bar
     - (iii) Three stage – delivery pressure up to 85 bar
     - (iv) Four stage – delivery pressure above 85 bar
2. ***According to moving parts***
   - (a) Reciprocating (b) Centrifugal (c) Rotary
3. ***According to method of cooling***
   - (a) Air cooled (b) Water cooled
4. ***According to power drive***
   - (a) Direct drive (b) Belt drive (c) Chain drive
5. ***According to number of air cylinder***
   - (a) Simplex (single cylinder)
   - (b) Duplex (double cylinder)
   - (c) Triplex (three cylinder)
6. ***According to nature of installation***
   - (a) Portable
   - (b) Semi fixed (semi-stationary)
   - (c) Fixed (stationary)
7. ***According to medium to be compressed***
   - (a) Air
   - (b) O_2, N_2, CO_2 etc.
8. ***According to application***
   - (a) Rock drill compressor
   - (b) Pneumatic hand tool compressor
   - (c) Road building compressor

(d) Quarrying compressor

(e) Spray painting compressor

(f) Sand blasting compressor

(g) Mine prospecting compressor

(h) According to arrangement of cylinder

   (i) horizontal (ii) vertical

   (iii) V-type (iv) radial

9. ***According to Principle of Operation***

Pressure of air may be increased either by decreasing its volume or by increasing the kinetic energy and then convert it to pressure energy by providing diffuser.

Accordingly Compressors are classified as:

(a) Positive displacement compressors

(b) Dynamic compressors

In positive displacement compressors the volume of air is reduced there by pressure is increased. Prominent types of such compressors are :

(i) Reciprocating compressor

(ii) Rotary compressors

In dynamic compressors K.E is imparted to air by multi bladed rotor and then it is converted into pressure energy in diffuser.

Compressors belongs to this category are sub-divided into:

(i) Centrifugal compressors

(ii) axial compressors

10. ***According pressure rise***

(a) Fan develops few cm of water column or 0.07 bar gauge pressure or pressure ratio less than 1.1

(b) Blower pressure ratio lies in between 1.1 to 2.5

(c) Compressor develops pressure above 2.5 bar gauge and up to 980 bar (Pressure ratio > 2.5).

11. ***According to volume of air delivered***

Low capacity – volume delivered 0.2 m^3/s or less,

Medium capacity – volume delivered up to 5 m^3/s,

High capacity – volume delivered 5 m^3/s to 50 m^3/sec.

Note: In Reciprocating compressor volume delivered up to 5m^3/sec.

In Rotary compressor volume delivered up to 50 m^3/sec.

10.4 Working of a Single Stage Reciprocating Compressor

Fig 10.1(a) shows the reciprocating compressor consists of a piston, connecting rod, crank and crank shaft. Both suction and discharge valves are disc type. They are opened and closed by difference in air pressure acting on their two sides. When the pressures are equal on both sides, they are kept closed by springs. The crank shaft is rotated by electric motor. The crank converts rotary motion into reciprocating motion of connecting rod and piston. A Fly wheel is fitted on the main shaft of compressor to ensure uniform turning moment to be supplied throughout the cycle of operation.

During outward or suction stroke, piston moves downward, the pressure inside the cylinder falls below atmospheric pressure as a result of which intake valve (suction valve) opens. The atmospheric air is sucked into cylinder until piston completes down-ward or suction stroke.

During inward (or compression) stroke piston will move upward, a slight increase in cylinder pressure will first close the inlet valve. Since both the inlet and delivery valves are closed further compression of air takes place as piston moves towards the top of its stroke. When pressure of air in cylinder exceeds, that in receiver, the delivery valve opens and air is delivered to receiver for remaining of stroke. At the end of delivery stroke, a small quantity of air at high pressure is left in clearance space as piston again starts its suction stroke, the air contained in clearance space expands and pressure inside the cylinder falls below atmospheric pressure, suction valve opens and air is sucked in to the cylinder. Cycle is repeated to get high pressure air, which is stored in receiver.

The receiver is a simple vessel which acts as storage tank. The compressor is driven by electric motor or diesel engine. When compressor is to be started against receiver pressure, the prime-mover will have to supply very high starting torque which will over load the electric motor. To avoid this, hand unloader or automatic unloader is used for releasing pressure from compressor cylinder when compressor is stopped.

It may be noted that there is intermittent flow of air in single acting reciprocating air compressor. Reciprocating compressor may be of double acting type in which the compression takes place on both sides of the piston and flow of air is continuous.

Single stage air compressor are used for delivery pressure up to 10 bar

Two stage compressor are used for pressure between 10 bar 200 bar

Three stage compressor are used 200 – 250 bar

Average piston speed of reciprocating compressor is limited to about 300 – 400 m/sec, to reduce friction wear.

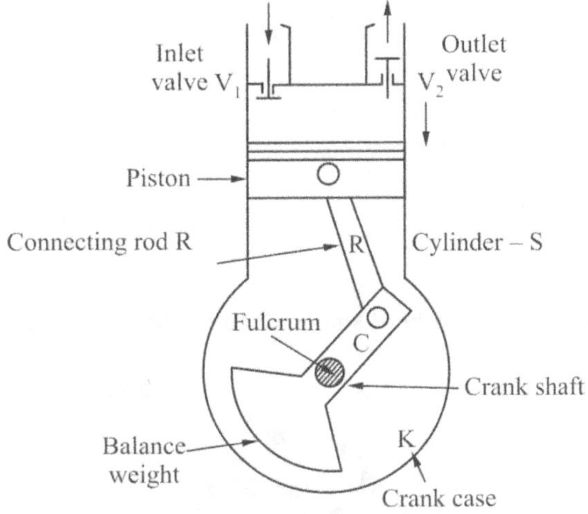

Fig. 10.1 (a) Single acting reciprocating compressor.

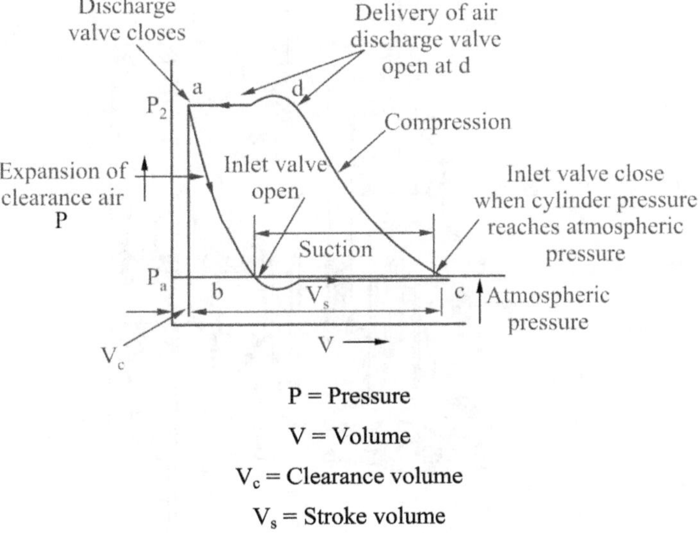

P = Pressure

V = Volume

V_c = Clearance volume

V_s = Stroke volume

Fig. 10.1 (b) Typical indicator card.

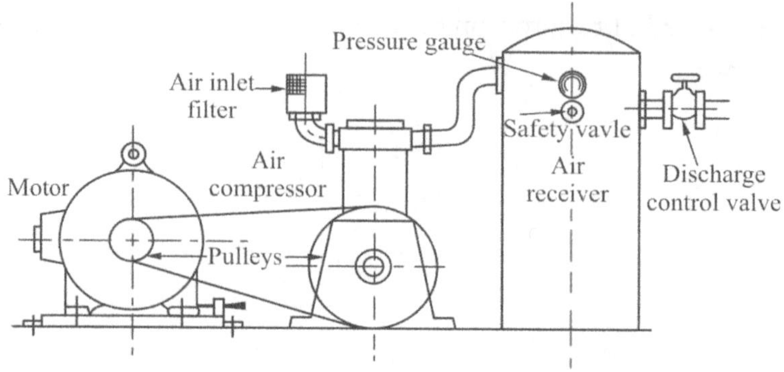

Fig. 10.1 (c) Arrangement of single-stage reciprocating air compressor.

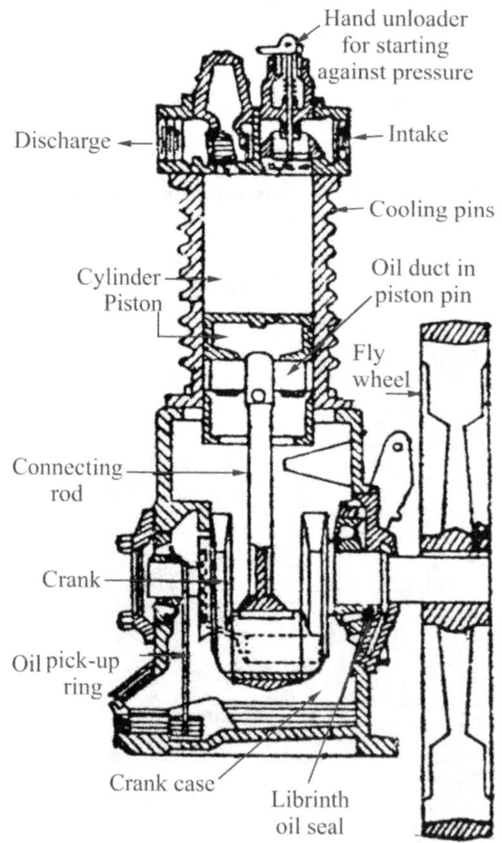

Fig. 10.1 (d) Sectional view of a single-acting air cooled, reciprocating air compressor.

10.5 Indicated Diagram for Compressor without Clearance

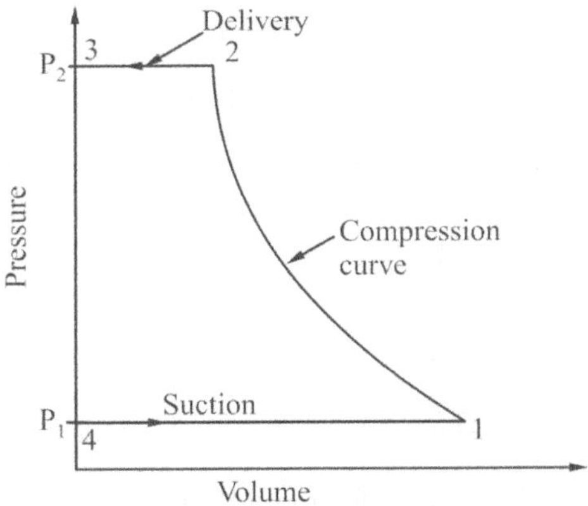

Fig. 10.2 Theoretical indicator diagram of single stage air compressor without clearance.

During suction stroke air is drawn into the cylinder along 4 - 1 at constant pressure p_1 which is slightly below that of atmosphere. At point 1, the piston completes the suction stroke, compressor stroke starts from 1 and ends at point 2. During compression both valves being closed, the air is compressed along compression curve 1 - 2. At point 2, pressure p_2 is reached which is slightly higher than the pressure in the receiver. The discharge valve opens at point 2 and delivery of compressed air takes place along line 2 - 3. At pressure P_2, Piston has reached at top of cylinder and again starts on its suction stroke. The pressure in the cylinder will be lowered again to p_1. The cycle of operation will be repeated. The network done required for compression and delivery of air per cycle is represented by area 1 - 2 - 3 - 4 - 1.

10.5.1 Actual Indicator Diagram of a Single-Stage Air Compressor

Actual air compressor differs from the ideal compressor in many respects, because the cylinder, the valves and the piston, store and release energy in the form of heat and further because air is throttled through the valves. An indicator diagram of an actual air compressor is shown in Fig 10.3 superimposed upon an ideal one (shown dotted) for the same compressor. The air pressure is to be raised from atmospheric pressure p_1 to a receiver pressure p_2.

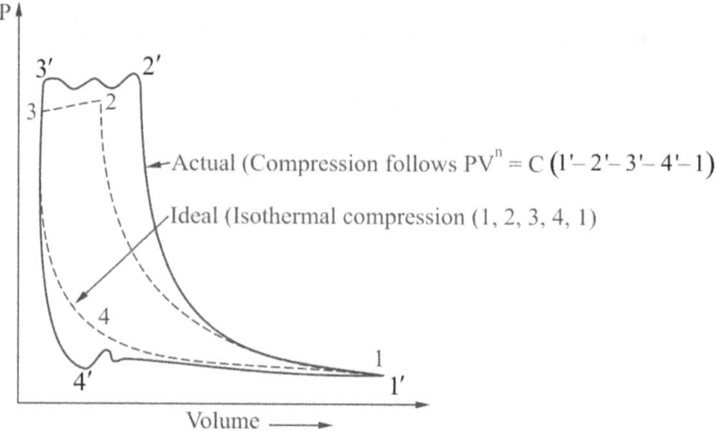

Fig. 10.3 Actual indicator diagram of single-stage air compressor

The ideal indicator diagram is drawn for isothermal compression of air in the cylinder and for isothermal expansion of the clearance air. In actual compressor, isothermal compression is not achieved, instead polytrophic compression is obtained i.e. the compression falls between the adiabatic (isentropic) and the isothermal, but it is closer to adiabatic as indicated by compression line $1' - 2'$ in Fig. 10.3. The value of index n varies from 1.25 to 1.35. In actual compressor, the clearance air will become cooler than the cylinder walls with which it is in contact hence it will receive heat during the latter part of the expansion process. The expansion line in actual compressor lies between an adiabatic and an isothermal, but generally it is closer to isothermal. The actual expansion line obtained will have shape and position similar to line $3' - 4'$ instead of line 3 - 4.

In ideal air compressor, the intake valve will open as soon as the clearance air pressure has decreased to atmospheric but actually the intake valve does not open until the pressure drops, a little below atmospheric. After the valve is open and air is in motion, there are generally several oscillations of the valve and the air column as indicated by the wavy portion on the suction line $4' - 1'$. The discharge valve does not open till the pressure attained is slightly above that in the receiver; and after opening, it behaves much like the intake valve for similar reason. The discharge line actually obtained will have a shape and position similar to line $2' - 3'$ instead of line 2 - 3.

There is always some leakage past the valves, piston rings and piston rod packing in the actual compressor; and greater the leakage the larger will be the work required for any given mass of air delivered.

10.6 Effect of Nature of Compression (i.e.,) n Curve on work Required for Compression

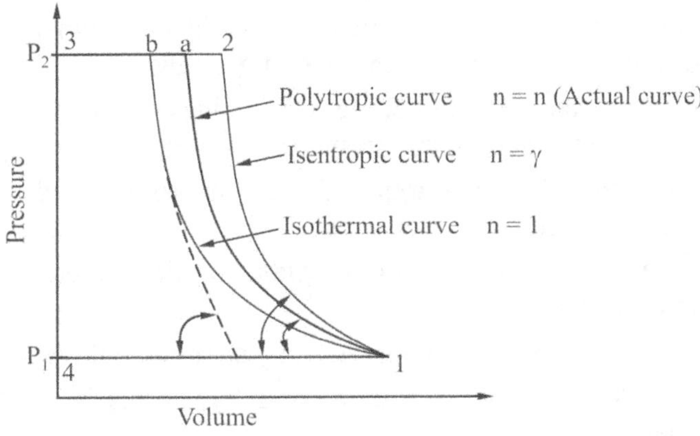

Fig. 10.4 P-V Diagram

1. Slope of compression curve represented by $pv^{\gamma} = c$ depends upon the value of index n. The larger the values of index n say $n = \gamma$ will be comparatively a steeper curve i.e., the tangent to curve $pv^{\gamma} = c$ makes larger angle with line 4 – 1.
2. The value of n = 1 i.e., Isothermal compression (1 – b) will be less steeper than isentropic curve (1 – 2) and polytrophic curve i.e., the tangent to curve pv = c makes smaller angle with 4 – 1 compare to isentropic and polytrophic curve.
3. The middle curve (1 – a) shows the curve which is obtained in actual practice. The curve is polytrophic ($pv^n = c$) having value of n = 1.3 for the water cooled cylinder.
4. The isentropic work required per cycle on air is represented by area 1 – 2 – 3 – 4 – 1 and work required per cycle on air during isothermal compression is represented by area 1 – b – 3 – 4 – 1 which is less than isentropic work. It is clearly seen that an isothermal compression is economical and efficient and requires less work to be done on air for same pressure ratio.
5. In practice, Isothermal compression can not be achieved. To have isothermal compression,
   (i) The compressor must run very slow, but in practice compressors run at high speed, so that as maximum possible air is compressed in given time.
   (ii) perfect water cooling should be provided so that temperature during compression process remain constant.

6. Isentropic compression also can not be achieved in practice because:
   (i) Compressor has to run very very fast, so that time for heat transfer is reduce.
   (ii) The cylinder should be insulated to avoid heat transfer from cylinder to atmosphere. This increases the size of machine.
7. So in practice, the compression curve, with values of index n between 1 and 1.4 will fall within the isothermal and isentropic curves.

   It will be seen that the work required for compression and delivery of air per cycle decreases as value of n decreases.
8. To reduce the value of n i.e., compressing the air isothermally approximately, three methods are used.
   (i) Cooling the air during compression by spraying cold water into cylinder.
   (ii) Cooling the air during compression by circulating cold water through cylinder jacket.
   (iii) Adopting multi-stage compression with inter stage cooling.
   (i) ***Cold Water Spray***: In this method water is sprayed into cylinder during compression. This reduces the temperature of air and the compression curve will be approximately $pv^{1.2} = c$. This shows that compression is brought nearer to isothermal compression which results in saving of work.
   (ii) ***Water Jackets***: In this method water is circulated in cylinder jacket which absorbs heat from compressed air there by keeping the temperature rise as small as possible and compression of air is carried out nearly isothermally and there will be saving in work for same pressure rise.
   (iii) ***Multi stage compression***: In this method compression of air is carried out in two or more than two cylinders. Inter cooler is provided in between the cylinders of multistage compressor. Air is compressed first in L.P cylinder then it is cooled in intercooler before air enters in H.P. cylinder. This cooling between the cylinders that keeps the compression very near to isothermal and there is saving in work.

10.7 Application of Compressed Air or Air Compressors

The air compressor delivers high pressure compressed air which can be stored in tanks (receiver) and used in various applications as listed below.

1. For driving air motors in mines where electric-motor or I.C engines are not used due to risk of fire.
2. For producing air blast in blast furnace.
3. For starting I.C engines.
4. For spray fuel oil in cylinder of large diesel engines.

5. For operating pneumatic tools such as drills, hammers, riveters, screw driver etc.
6. Operating pneumatic brakes in the locomotives and automobiles.
7. For operating of gas-turbine plants.
8. Scavenging and super-charging of I.C engine
9. For spraying painting
10. For clearing machines in workshop.
11. use for refrigeration and air conditioning.
12. Use in chemical factories for manufacturing acids, soda etc.
13. For pumping water from tube wells.
14. For operating chipping and caulking cools.
15. For filling air in automobile tires.

10.8 Equation for Work Done for Single Stage Reciprocating Air Compressor (without clearance)

Assumptions

1. There is no pressure drop through suction and delivery valves.
2. Compressor has no clearance volume.
3. Pressure in suction and delivery lines remains constant.
4. Entire compression is carried out in single cylinder.

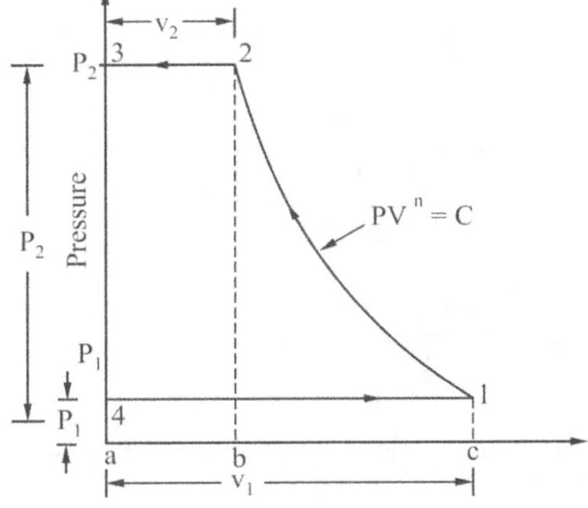

Fig. 10.5 P-V Diagram

(A) ***Without Clearance***: p-V diagram for single stage and single acting reciprocating air compressor is shown in Fig. 10.5. It is assumed that there is no pressure drop through suction and delivery valves and friction losses are neglected.

The sequence of operations as shown on P-V diagram are as follows:

Process 4 – 1 represents suction stroke in which volume of air v_1 is taken into compressor at constant pressure P_1 and temperature T_1.

Process 1 – 2 represents compression of air. During this process air is compressed according to Law $pv^n = c$ from pressure P_1 to Pressure P_2. Volume decreases from v_1 to v_2. Temperature increases from T_1 to T_2 (as temperature is directly proportion to pressure)

Process 2 – 3 The compression continues till the pressure p_2 in cylinder is sufficient to force open delivery valve at "2", after that no more compression takes place with the inward movement (or upward movement) of the piston. Now during remaining part of compression stroke, the compressed air is delivered into receiver or tank till piston reaches cylinder's head end side.

Work done during polytrophic compression $pv^n = c$

Work done on air/cycle or total shaft work done/cycle = area 4 – 1 – 2 – 3 – 4

But area 4–1–2–3–4 = (area a–b–2–3–a) + (area b–2–1–c–b–) – (area a–4–1–c–a)

$$= p_2 v_2 + \frac{p_2 v_2 - p_1 v_1}{n-1} - p_1 v_1$$

$$= \frac{p_2 v_2 (n-1) + p_2 v_2 - p_1 v_1 - (n-1) p_1 v_1}{n-1}$$

$$= \frac{n p_2 v_2 - p_2 v_2 + p_2 v_2 - p_1 v_1 - n p_1 v_1 + p_1 v_1}{n-1}$$

$$= \frac{n p_2 v_2 - n p_1 v_1}{n-1}$$

$$= \frac{n (p_2 v_2 - p_1 v_1)}{n-1}$$

W.D on air/cycle $= \dfrac{n}{n-1} (p_2 v_2 - p_1 v_1)$ .....(1)

Note

(i) Cycle in case of a reciprocating compressor consists of two strokes. In case of single acting air compressor there is only one working stroke. (i.e. delivery stroke). Therefore work done per cycle is also known as W.D per working stroke also known as W.D. per revolution because in one revolution there is one working stroke.

(ii) Since process 1 – 2 is a compression process. The work is expressed $\frac{p_2v_2 - p_1v_1}{n-1}$ and not as $\frac{(p_1v_1 - p_2v_2)}{n-1}$ because work is done on air is taken as –ve i.e., $\frac{-(p_1v_1 - p_2v_2)}{n-1} = \frac{p_2v_2 - p_1v_1}{n-1}$

Equation (1) can be expressed in different form as shown below

Second form of equation for W.D on air/cycle

$$\text{W.D on air/cycle} = \frac{n}{n-1}(p_2v_2 - p_1v_1)$$

$$= \frac{n}{n-1}p_1v_1\left[\frac{p_2v_2}{p_1v_1} - 1\right] \quad .....(2)$$

Now apply pressure-volume relationship for polytrophic process

i.e., $p_1v_1^n = p_2v_2^n$

$\therefore \quad \frac{v_2^n}{v_1^n} = \frac{p_1}{p_2}$

$\therefore \quad \left(\frac{v_2}{v_1}\right) = \left(\frac{p_1}{p_2}\right)^{\frac{1}{n}}$

Put this value of $\frac{v_2}{v_1}$ in equation (2), we get

$$\therefore \quad \text{W.D on air/cycle} = \frac{n}{n-1}\ p_1v_1\left[\left(\frac{p_2}{p_1}\right)^{\frac{n-1}{n}} - 1\right] \quad \text{second form}$$

Third form of Equation in which W.D on air/min is represented

$$p_1v_1 = \text{ma } RT_1$$

where m_a = mass of air in kg/min.

$$\therefore \quad \text{W.D on air} = \frac{n}{n-1} p_1 v_1 \left[\left(\frac{p_2}{p_1} \right)^{\frac{n-1}{n}} - 1 \right]$$

$$\therefore \quad = \frac{n}{n-1} m_a RT_1 \left[\left(\frac{p_2}{p_1} \right)^{\frac{n-1}{n}} - 1 \right] \frac{kJ}{min} \qquad .....(3)$$

$$\left(\text{unit of maRT}_1 = \frac{kg}{min} \times \frac{kJ}{kgk} .k = \frac{kJ}{min}\right)$$

Fourth form of equation in which W.D on air is expressed

$$\text{W.D on air} = \frac{n}{n-1} m_a \; RT_1 \left[\left(\frac{p_2}{p_1} \right)^{\frac{n-1}{n}} - 1 \right]$$

But we know $\frac{T_2}{T_1} = \left(\frac{p_2}{p_1} \right)^{\frac{n-1}{n}}$

Put this value in equation (3)

$$\therefore \quad \text{W.D on air} = \frac{n}{n-1} m_a \; RT_1 \left(\frac{T_2}{T_1} - 1 \right)$$

$$= \frac{n}{n-1} m_a \; RT_1 \left(\frac{T_2 - T_1}{T_1} \right)$$

$$= \frac{n}{n-1} \; m_a \, R \, (T_2 - T_1) \; \frac{kJ}{min} \qquad ..... (4)$$

Note

If m_a is taken in $\frac{kg}{min}$ the unit of W.D will be $\frac{kJ}{min}$ because $\frac{n}{n-1}$ is constant

$$\text{unit of } m_a = \frac{kg}{min}$$

$$\text{unit of R} = \frac{kJ}{kgk}$$

$$\text{unit of } (T_2 - T_1) = k$$

$$\therefore \quad \frac{kg}{min} \times \frac{kJ}{kgk} \times k = \frac{kJ}{min}$$

B. Work done during Isothermal Compression

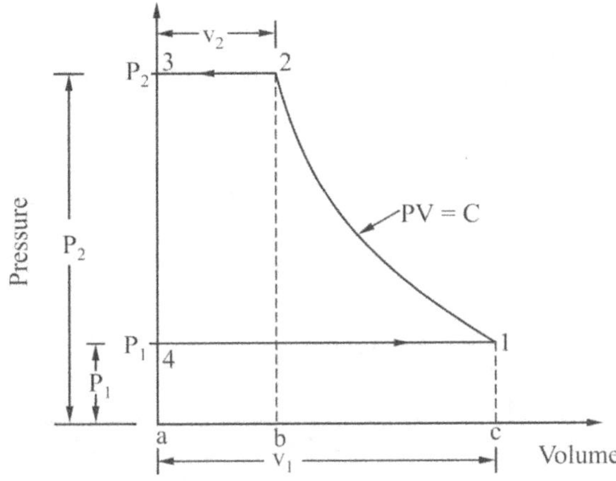

P – V Diagram

W.D on air/cycle = area (4–1–2–3–4) = area (a–b–2–3–a) + area (b–2–1–c–b) – area (a–c–1–4–a)

$$\text{W.D on air/cycle} = p_2v_2 + p_1v_1 \log \frac{v_1}{v_2} - p_1 v_1$$

But $p_1 v_1 = p_2 v_2$ isothermal compression

$$\therefore \quad \text{W.D on air/cycle} = p_1 v_1 \log_e \frac{v_1}{v_2} \quad \text{(i) Form}$$

$$= p_2 v_2 \log_e \frac{v_1}{v_2} \quad \text{(ii) Form}$$

$$= p_2 v_2 \log_e \frac{p_2}{p_1} \quad \text{(iii) Form}$$

$$= m_a R T_1 \log_e \frac{v_1}{v_2} \quad \text{(iv) Form}$$

$$p_1 v_1 = p_2 v_2$$

$$\therefore \quad \frac{v_1}{v_2} = \frac{p_2}{p_1}$$

***Note*:** Units in each equation is already discuss in (A)

C. Work done during isentropic (Reversible adiabatic)

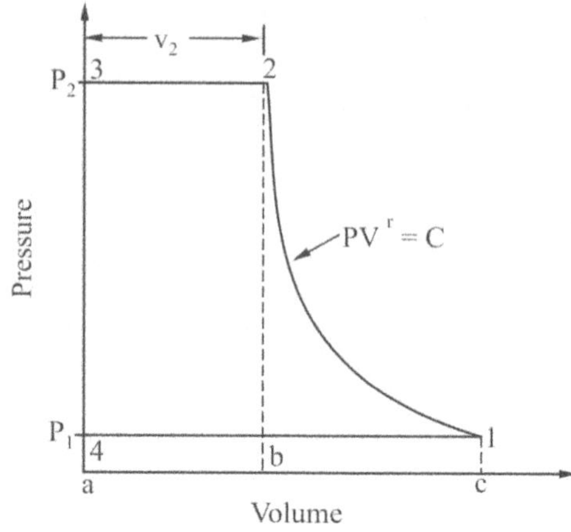

P – V Diagram

W.D on air/cycle = area 4 – 1 – 2 – 3 – 4 = area (a – b – 2 – 3 – a) + area (b – 2 – 1 – c – b) – area (a – 4 – 1 – c – a)

W.D is represented in different forms (ii, iii)

$$= p_2v_2 + \frac{p_2v_2 - p_1v_1}{\gamma - 1} - p_1v_1$$

$$= (p_2 v_2 - p_1 v_1) + \frac{p_2v_2 - p_1v_1}{\gamma - 1}$$

$$= (p_2 v_2 - p_1 v_1) + \left(1 + \frac{1}{\gamma - 1}\right)$$

$$= \frac{\gamma - 1 + 1}{\gamma - 1}(p_2v_2 - p_1v_1)$$

$$= \frac{\gamma}{\gamma - 1}(p_2v_2 - p_1v_1) \quad \text{.....(i)}$$

W.D on air/cycle $= \frac{\gamma}{\gamma - 1} m_a R(T_2 - T_1)$ .....(ii)

W.D on air/cycle $= \frac{\gamma}{\gamma - 1} p_1v_1\left[\left(\frac{p_2}{p_1}\right)^{\frac{\gamma-1}{\gamma}} - 1\right]$ .....(iii)

We know that $\gamma = \frac{c_p}{c_v}$ and $R = (c_p - c_v)$

From this we get $R = c_p\left(1 - \frac{1}{\gamma}\right) = c_p\left(\frac{\gamma - 1}{\gamma}\right)$

Put value of R in equation (iii) we get

$$\text{W.D on air/cycle} = \frac{\gamma}{\gamma - 1}.m_a.c_p\left(\frac{\gamma - 1}{\gamma}\right)(T_2 - T_1)$$

$$= m_a.c_p\,(T_2 - T_1) \qquad \text{.....(iv)}$$

***Note*:** Unit of work done per cycle in each form of equation is discussed in case (A).

Power required to compress the Air

We have already discussed the expressions for work-done per cycle during isothermal, polytrophic and isentropic compression. Power required (theoretical) to drive the compressor may be obtained as follows.

$$\frac{WN}{60} = \text{walt}$$

W = work required to compress air, in J/cycle or J/working stroke

($\because$ For single acting W.D/cycle = W.D/working stroke)

N = Number of working cycles/min i.e., RPM

***Note*:** In Reciprocating air compressor cycle consists of two stroke – (suction and delivery). But 1 revolution is equal to two strokes. i.e., For single acting = W.D/cycle = W.D/revolution = w.D/working stroke

For double acting in one revolution there will be two working strokes i.e., double acting air compressor delivers nearly double quantity of air.

In short $P = \frac{W.N_w}{60 \times 1000}$ kw

$$N_w = 2N$$

Let N = RPM of compressor

$N = N_w =$ for single acting $\qquad \therefore\ P = \frac{W.N}{1000 \times 60}$ for single acting

$N_w = 2$ N for double acting $\qquad P = \frac{W \times 2N}{60 \times 100}$ for double acting

Since the compression takes place in three different ways therefore power obtained from different work done will be different. In generally the three values of power obtained.

1. Isothermal power $= \dfrac{\text{W (in Isothermal compression) N}}{60}$ watt

2. Isentropic power $= \dfrac{\text{W (in Isentropic compression) N}}{60}$ watt

3. Indicated power $= \dfrac{\text{W (in Polytropic compression) N}}{60}$ watt

Indicated power is also known as air power of compressor.

10.9 Equation for Work done with Clearance

$v_1 - v_4$ = Effective swept volume

$v_1 - v_3$ = swept volume = swept volume = $v_s = v_1 - v_c$ $\qquad (\because v_3 = v_c)$

$v_3 = v_c$ = clearance volume

v_1 = Total volume

1. In the previous article 10.8, we have assumed that there is no clearance volume in the compressor cylinder and entire volume of air in cylinder is compressed by inward stroke (or upward stroke in case of vertical compressor).
2. In practice it is not possible to reduce clearance volume to zero, because it is not desirable to allow piston to come in contact with cylinder head and valves. If piston strikes the cylinder head and valves 600 times in one minute (assuming RPM of compressor = 600 RPM). It will reduce the life of cylinder head and valves. It may also cause noise.
3. Now consider a reciprocating compressor with clearance volume as shown in Fig. 10.6.

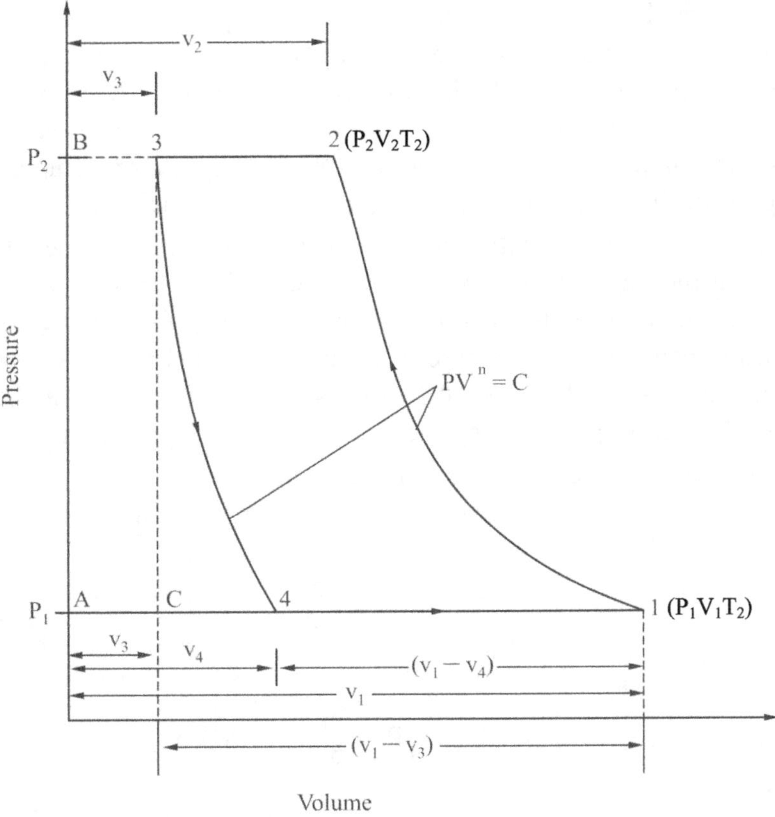

Fig. 10.6 P – V Diagram

Let p_1 = Initial pressure of air before compression

v_1 = Initial volume of air before compression

T_1 = Initial temperature of air before compression.

P_2, v_2, T_2 = corresponding values of pressure, volume and temperature for final conditions respectively.

$$r = \text{pressure ratio} = \frac{p_2}{p_1}$$

$$v_s = \text{swept volume} = (v_1 - v_3) = (v_1 - v_c) \qquad (\because v_3 = v_c)$$

v_c = Clearance volume at point 3 or v_3

n = Polytrophic index for compression and expansion

4. At point 1 cylinder is full of intake air, having pressure p_1 and v_1 and piston is about to start its compression stroke. Air is compressed according to law $pv^n = c$ till the pressure p_2 in cylinder is sufficient to force, the delivery valve to open at point 2. After that no more compression takes place with inward or (upward) movement of piston. Now during the remaining part of compression stroke, compressed air is delivered to receiver till the piston reaches point 3.
5. At this stage some air is left in the clearance space of the cylinder at pressure P_2. As piston moves outward (down ward), the air in clearance space will expand during some part of outward stroke of the piston i.e., expansion stroke 3-4. This expansion continues till pressure in cylinder P_1 is below atmospheric pressure. Due to this pressure difference the valve is opened at point 4. After that the air is sucked from atmosphere during suction stroke 4 – 1 at pressure p_1.
6. W.D by compressor/cycle = area 1 – 2- 3- 4 – 1

$$= (\text{area } 1 - 2 - B - A - 1) - \text{area } (A - 4 - 3 - B - A)$$

$$= \frac{n}{n-1} p_1 v_1 \left[\left(\frac{p_2}{p_1} \right)^{\frac{n-1}{n}} - 1 \right] - \frac{n}{n-1} p_1 v_4 \left[\left(\frac{p_2}{p_1} \right)^{\frac{n-1}{n}} - 1 \right]$$

$$= \frac{n}{n-1} p_1 \left[\left(\frac{p_2}{p_1} \right)^{\frac{n-1}{n}} - 1 \right] (v_1 - v_4)$$

$$= \frac{n}{n-1} p_1 [v_1 - v_4] \left[\left(\frac{p_2}{p_1} \right)^{\frac{n-1}{n}} - 1 \right] \quad \ldots\ldots (1)$$

$$P_1(v_1 - v_4) = m_a RT_1$$

where $v_1 - v_4$ is actual volume of air taking during suction stroke.

Replace $p_1(v_1 - v_4)$ in equation (1) By m_aRT_1 we get

$$\therefore \quad \text{W.D on air/cycle} = \frac{n}{n-1} m_a RT_1 \left[\left(\frac{p_2}{p_1} \right)^{\frac{n-1}{n}} - 1 \right] \quad \ldots\ldots (2)$$

Note: It will be noted that work done on air delivered is not seriously affected by clearance volume. Because the expanding air from 3 to 4 develops positive work equal to that done on it during compression. However there is considerable reduction in the volumetric efficiency due to clearance volume. Due to clearance actual volume aspired is reduced.

10.10 Efficiency of Reciprocating Air Compressor

1. ***Isothermal efficiency (Compressor efficiency)***

It is ratio of work (power) required to compress the air isothermally to actual work required to compress the air for same pressure ratio. It is denoted by η_c

$$\therefore \quad \eta_c = \frac{\text{Isothermal power}}{\text{Indicated power (Actual power)}} \text{ or } \frac{\text{Isothermal work}}{\text{Actual work}}$$

$$= \frac{\text{Isothermal work}}{\text{Indicated work}}$$

Isothermal work done (on air to increases the pressure from p_1 to p_2 and to decrease the volume v_1 to v_2) = $10^2\, P_1V_1, \log_e P_2/P_1$ kJ/sec.

If p is in bar, and V_1 is in m^3/sec then unit of W.D will be KJ/sec and if V_1 is m^3/stroke, the unit of W.D will be KJ/stroke.

Indicated work done by compressor = W.D during polytrophic compression.

$$= \frac{n}{n-1} 10^2 p_1 v_1 \left[\left(\frac{p_2}{p_1} \right)^{\frac{n-1}{n}} - 1 \right] \quad \text{KJ/cycle or (KJ/stroke)}$$

$$\therefore \quad \eta_c = \frac{10^2 p_1 v_1 \log e \frac{p_2}{p_1}}{\frac{n}{n-1} 10^2 p_1 v_1 \left[\left(\frac{p_2}{p_1} \right)^{\frac{n-1}{n}} - 1 \right]}$$

$$= \frac{\log \frac{p_2}{p_1}}{\frac{n}{n-1} \left[\left(\frac{p_2}{p_1} \right)^{\frac{n-1}{n}} - 1 \right]} \quad \text{..... (1)}$$

2. ***Over-all efficiency (Isothermal)***

It is ratio of isothermal power to shaft power or brake power of the motor or engine, required to drive the compressor. It is denoted by η_0.

$$\eta_o = \frac{\text{Isothermal power}}{\text{Shaft power or Brake power of motor}}$$

3. ***Mechanical Efficiency*:** It is ratio of the indicated power (knows as W.D during polytrophic compression) to shaft power or brake power of the motor or engine to drive the compressor

It is represented by η_m

$$\therefore \qquad \eta_m = \frac{\text{Indicated power}}{\text{Shaft power or Brake power of motor}}$$

***Note*:** In engine $\eta_m = \frac{BP}{I.P}$ but in air compressor B.P of motor is input power and I.P is out power (i.e., air indicated power).

$$\therefore \qquad \eta_m = \frac{IP}{BP}$$

4. ***Isentropic Efficiency*:** It is ratio of isentropic power to the brake power required to drive the compressor. It is denoted by η_i,

$$\therefore \qquad \eta_i = \frac{\text{Isentropic power}}{\text{B.P required to drive compressor}}$$

5. ***Volumetric Efficiency*:** It is ratio of free air delivered (F.A.D) per stroke to swept volume of piston. Volumetric efficiency of a reciprocating air compressor is different when it is with or without clearance volume.

Explain the volumetric Efficiency of Reciprocating compressor

Volumetric efficiency of single stage reciprocating compressor is defined as the volume of air entering the compressor per minute to piston displacement/min.

Therefore for single stage compressor,

$$\eta_{VOL} = \frac{\text{volume of air entering compressor per minute}}{\text{Piston Displacement per minute}}$$

The volumetric efficiency of multistage compressor is defined in the same way except that the piston displacement is taken as that of the low pressure cylinder.

$$\therefore \quad \eta_{voL} \text{ of Multi} - \text{compressor} = \frac{\text{Volume of air entering compressor per minute}}{\text{Displacement of low pressure per minute}}$$

Various factors affecting the volumetric efficiency

$$\text{We know } \eta_{voL} = 1 + k - k\left(\frac{p_2}{p_1}\right)^{\frac{1}{n}}$$

$$= 1 - k\left(\frac{p_2}{p_1}\right)^{\frac{1}{n}} - 1 \qquad \text{Rearranging and taken K common}$$

where k = clearance ratio

1. If $p_1 = p_2$ the volumetric efficiency will be 100%. It shows that no compression takes place.
2. $p_2 > p_1$ i.e., pressure ratio increases, volumetric efficiency decreases.
3. Volumetric efficiency decreases with increase in clearance ratio. If clearance is very large then no air is discharged by compressor.

   As delivery pressure of a single stage reciprocating air compressor is increased, the mass flow through compressor decreases.
4. Following are some other factors which decrease the volumetric efficiency.
   (i) Valve leakage
   (ii) Obstruction at inlet valve
   (iii) Piston ring leakage which allows air to past the Piston.
   (iv) Heating of air by contact with hot cylinder walls
   (v) Very high speed of rotation

   With the decreases of volumetric efficiency, the capacity (quantity of free air delivery) of compressor decreases.

Expression for Volumetric Efficiency of Reciprocating Air Compressor with Clearance (First method)

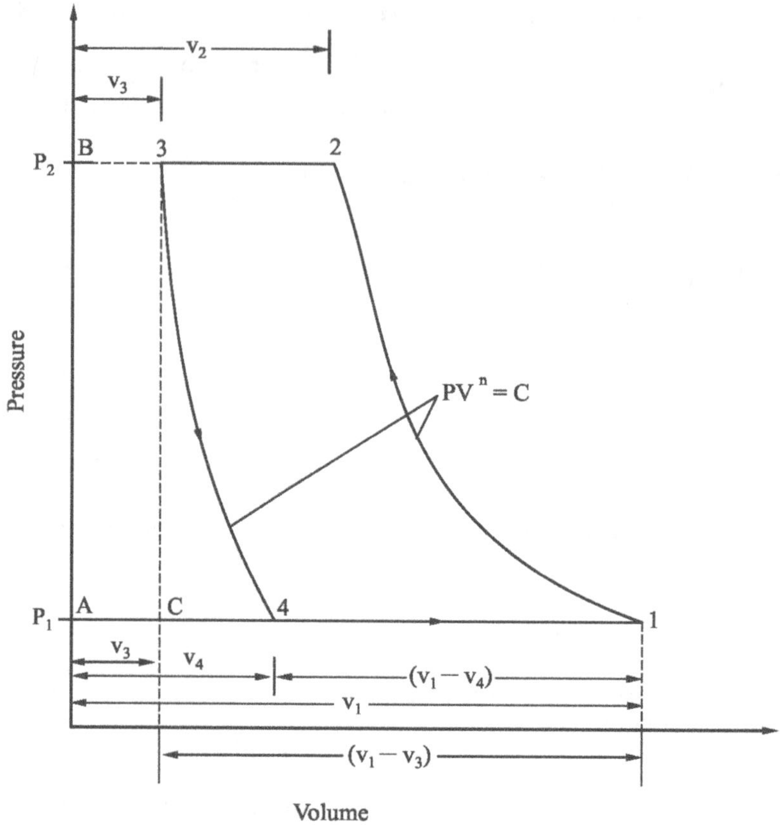

Fig. 10.7 P – V Diagram

$$\text{Volumetric efficiency} = \frac{\text{actual swept volume}}{\text{swept volume or stroke volume}}$$

$$= \frac{v_1 - v_4}{v_1 - v_3}$$

$$= \frac{v_1 + v_3 - v_3 - v_4}{v_1 - v_3} \quad \text{(add and subtract } v_3\text{)}$$

$$= \frac{v_1 - v_3}{v_1 - v_3} + \frac{v_3}{v_1 - v_3} - \frac{v_4}{v_1 - v_3}$$

$$= 1 + \frac{v_3}{v_1 - v_3} - \frac{v_4}{v_1 - v_3} \quad .....(1)$$

Let us Define Clearance Ratio: It is ratio of clearance volume to swept volume. It is denoted by k.

$$k = \frac{V_3}{V_1 - V_3} = \frac{V_c}{V_s}$$

Replacing $\frac{V_3}{V_1 - V_3}$ by k in equation (1) we get

$$\eta_{voL} = 1 + k - \frac{V_4}{V_1 - V_3}$$

$$= 1 + k - \frac{V_3}{V_1 - V_3} \times \frac{V_4}{V_3} \qquad \text{(Multiplying and dividing by } v_3\text{)}$$

$$= 1 + k - \frac{V_3}{V_1 - V_3} \times \frac{V_4}{V_3}$$

$$= 1 + k - k\left(\frac{V_4}{V_3}\right) \qquad .....(2) \qquad \because \ k = \frac{V_3}{V_1 - V_3}$$

$$= 1 + k - k\left(\frac{p_2}{p_1}\right)^{1/n} \qquad .....(3) \qquad P_3V_3^n = P_4V_4^n$$

$$\therefore \ \frac{P_3}{P_4} = \left(\frac{V_4}{V_3}\right)^n$$

$$\therefore \ \frac{V_4}{V_3} = \left(\frac{P_3}{P_4}\right)^{\frac{1}{n}}$$

$$\frac{V_4}{V_3} = \left(\frac{P_2}{P_1}\right)^{\frac{1}{n}}$$

$\because \ P_1 = P_4, P_3 = P_2$

Second Method (with same p-V diagram)

$$V_4 = V_3\left(\frac{p_2}{p_1}\right)^{\frac{1}{n}} = v_c\left(\frac{p_2}{p_1}\right)^{\frac{1}{n}} \qquad [\because \ v_3 = v_c]$$

The volume of fresh air taken during suction = $V_a = (V_1 - V_4)$

$\therefore \quad V_a = (V_1 - V_4)$

$$= V_s + V_c - V_c\left(\frac{P_2}{P_1}\right)^{\frac{1}{n}} \left(\because \ V_1 = V_s + V_c \text{ and } V_4 = V_3\left(\frac{P_2}{p_1}\right)^{\frac{1}{n}}\right.$$

$$= v_c \left(\frac{p_2}{p_1}\right)^{\frac{1}{n}}) \quad = v_s - v_c\left[\left(\frac{p_2}{p_1}\right)^{\frac{1}{n}} - 1\right] \text{ (Taking } V_c \text{ common)}$$

$$\therefore \text{ volumetric efficiency} = \frac{\text{Volume of free air delivered per stroke}}{\text{volume swept by piston per stroke}}$$

$$= \frac{v_1 - v_4}{v_s}$$

$$= \frac{v_s - v_c\left[\left(\frac{p_2}{p_1}\right)^{\frac{1}{n}} - 1\right]}{v_s}$$

$$= \frac{v_s}{v_s} - \frac{v_c}{v_s}\left[\left(\frac{p_2}{p_1}\right)^{\frac{1}{n}} - 1\right] = 1 - k\left[\left(\frac{p_2}{p_1}\right)^{\frac{1}{n}} - 1\right]$$

Where k = pressure ratio. It is also denoted by c

$$\therefore \text{ volumetric efficiency } = 1 - k\left[\left(\frac{p_2}{p_1}\right)^{\frac{1}{n}} - 1\right]$$

Expression for Volumetric Efficiency (Third Method)

When atmospheric temperature and pressure is not same to that of at the end of suction stroke i.e., point '1', in practice, the temperature and pressure at end of suction i.e., at point 1 is not ambient because fresh air, due to its passage over hot valves, cylinder surface and mixes with residual air of clearance. There are also obstructions in suction to fresh air.

Applying general gas equation to atmospheric condition of air and condition of air before compression (or end of suction):

Let p_1 = Initial pressure of air (at end of suction i.e., point 1)

v_1 = Initial volume of air (at end of suction or before compression)

T_1 = Initial temperature of air (at end of suction or before compression)

P_a, v_a T_a = corresponding values of ambient (N.T.P) condition.

$$\frac{P_a v_a}{T_a} = \frac{P_1(V_1 - V_4)}{T_1}$$

$\therefore$ Volume of Air sucked referred to ambient conditions = V_a

$$\therefore \quad V_a = \frac{P_1 T_a}{P_a . T_1} (V_1 - V_a)$$

We know that

volumetric efficiency = $\dfrac{\text{actual volume of air sucked referred to ambient condition}}{\text{Swept volume of Piston}}$

$$= \frac{V_a}{V_s} \qquad \text{Replacing } V_a \text{ by } \frac{P_1 T_a}{P_a T_1}(V_1 - V_4)$$

$$= \frac{P_1 T_a}{P_a T_1} \frac{(V_1 - V_4)}{V_s}$$

$$= \frac{P_1 T_a}{P_a T_1} \frac{(V_s + V_c - V_4)}{V_s} \qquad \because v_1 = v_s + v_c$$

$$= \frac{P_1 T_a}{P_a T_1} \left(\frac{V_s}{V_s} + \frac{V_c}{V_s} - \frac{V_4}{V_s} \right)$$

$$= \frac{P_1 T_a}{P_a T_1} \left(1 + k - \frac{V_4}{V_s} \times \frac{V_c}{V_c} \right) \qquad \text{multiply and divide by } v_c$$

$$= \frac{P_1 T_a}{P_a T_1} \left(1 + k - \frac{V_4}{V_c} . k \right)$$

$$= \frac{P_1 T_a}{P_a T_1} \left(1 + k - k \frac{V_4}{V_c} \right) \qquad \text{by rearranging}$$

Consider air trapped in clearance is expanded from point 3 where pressure is P_2 and volume is V_c to point 4 where pressure is P_1 and volume is V_4.

$$p_2 V_c^n = p_1 . V_4^n$$

$$\therefore \quad \frac{V_4}{V_c} = \left(\frac{P_2}{P_1} \right)^{\frac{1}{n}} \qquad .....(B)$$

Put this value of $\dfrac{V_4}{V_c}$ in equation A, we get

$$\eta_{vol} = \frac{P_1 T_a}{P_a T_1} \left[1 + k - k \left(\frac{P_2}{P_1} \right)^{\frac{1}{n}} \right]$$

10.11 Comparison between Reciprocating and Rotary compressor

The advantages and disadvantages of Reciprocating and Rotary compressors will be clear from the following comparison.

Reciprocating Compressor	Rotary Compressor
1. They are used when low rate of flow with high pressure rise are required	1. They are used when large rate of flow and medium rise of pressure are required
2. Low discharge – 30 m^3/min	2. Large discharge – 3000 m^3/min
3. High pressure – 980 bar	3. Low pressure – 10 bar
4. Low speed 50 to 650 RPM	4. High speed up to 40,000 RPM
5. As discharge is not continuous i.e., it is intermittent discharge, receiver for compressed air is essential.	5. Continuous discharge as uniform delivery of air. Air receiver is not must.
6. Large size	6. Small size because of high speed of rotation
7. Balancing is not problem because of low speed.	7. Perfect balancing is required due to high speed.
8. Air discharged contains particles of lubricating oil therefore after cooler is required.	8. Clear air due to freedom from lubricating oil contact.
9. Complicated lubrication system	9. Simple lubricating system because of, less number of sliding parts.
10. Cannot be directly coupled to turbine or electric motor due to low speed.	10. It can be directly coupled to turbine and electric motor due to high speed.
11. 'Thermodynamic efficiency of reciprocating compressor is isothermal. Due to low speed slow speed of compressor and by providing sufficient cooling, the compression may be achieved isothermally	11. While that of Rotary compressor the efficiency is adiabatic. It is due to reason that if no inter cooler is provided in case of Rotary compressor, the compression index may reach as high as 1.6 to 1.7. It has high value of compression index (1.7) the large amount of work is required. However the value of compression index can be greatly reduced by providing sufficient cooling. But it is not possible to bring the value of compression index to that of reciprocating compressor.

10.11.1 Comparison Between Axial and Centrifugal Compressor

Axial flow Compressors		Centrifugal Compressors	
1.	The flow of air is parallel to axis of shaft of compressor	1.	Flow of air is in the perpendicular direction to the axis of shaft of compressor i.e., in radial direction.
2.	Pressure ratio per stage is 1.2 : 1 and for high pressure ratio, more number of stages are required.	2.	Pressure ratio per stage is 4:1 for high pressure ratio multi-stage C.F compressor is used.
3.	Isentropic efficiency is 85 to 88% with modern aero foil blades.	3.	Isentropic efficiency is about 70%.
4.	Less flexibility of operation	4.	More flexibility of operation due to adjustable prewhirl and diffuser vanes
5.	Frontal area is less therefore they are adopted for aircraft	5.	Frontal area is more cannot be adopted for aircraft.
6.	Performance of axial compressor is affected by formation of deposits on surface of impeller	6.	Performance is not affected as it is centrifugal compressor
7.	Starting torque required is high therefore used universally with large gas turbines.	7.	Comparatively starting torque is low, suitable for super-charging of I.C Engines and compression of refrigerants and industrial-gases.
8.	Efficiency vs speed curve is less flat, less flat curve indicates that efficiency changes as speed changes.	8.	Efficiency vs speed curve is more flat. Flat curve indicates that efficiency is nearly constant at various speeds.

10.12 Advantages of Multistage Compressions

The following are the advantages of Multistage Compression

1. ***Better mechanical balance:*** Due to the phasing of operation into two or more stages, a better mechanical balance for air compressor is attained, thus needing smaller fly wheel.
2. ***Improved overall volumetric efficiency*:** If all compression is done in one cylinder the air in the clearance would expand to great extent before the new intake of fresh air could begin. By cooling the air between stages a higher efficiency can be obtained.

   In other words volumetric efficiency increases due to lower temperature and pressure range in multistage compression.
3. ***Less Power*:** Air can be cooled perfectly at pressure intermediate between suction pressure and delivery pressure. This would result in less power requirement to run multistage compression as compare to a single stage compression for same delivery pressure and same quantity of free air delivered.

4. Less loss of air due to leakage past the piston of cylinder due to lower pressure range in each cylinder and better lubrication.
5. ***Better Lubrication*:** Better lubrication is possible due to lower working temperature.
6. Since maximum temperature reached during the compression process is greatly reduced by cooling the air in inter-cooler before entering the H.P cylinder. This reduces lubrication difficulties and explosion hazards.
7. The cylinder of single stage compressor must be strong to withstand delivery pressure where as the low pressure cylinder of multistage compressor can be lighter.
8. The air gets chance to cool at a pressure between inlet and delivery pressures which reduces size of H.P cylinder.
9. Due to lower pressure range, low pressure cylinder are made light where as high pressure cylinder of third stage is made robust.

Disadvantages

1. Multi-stage compressor requires more than one cylinder, intercoolers with pipe connections and supply of air continuously.
2. Initial cost of multistage compressor is more.
3. Its construction is more complications.

10.13 Working of Two stage Air Compressor with Inter-cooler

The air is first drawn into low pressure cylinder at point 1 with pressure P_1, Volume V_1. The air is compressed according law $pv^n = c$ to point $2'$ having intermediate pressure P_2 and temperature T_2, is passed on to inter cooler where it is cooled at constant pressure P_2 and temperature of air is reduced from T' to T_1. This is represented by constant pressure $2' - 2$ and volume $2 - p_2$ to $2' - p_3$. After leaving the intercooler the air enters the H.P (High Pressure) cylinder for further compression along line $2 - 3$. Pressure of air is increased from P_2 to P_3 and finally air is discharged into the receiver at P_3.

Complete or perfect inter cooling means that existing air temperature (T_2) from inter cooler will be equal to T_1. Similarly perfect after cooling makes the air to leave the H.P. cylinder at temperature T_1. This after cooling reduces the volume of the air leaving the compressor and entering the receiver.

If compression of air is carried out in a single stage it will follow polytrophic curve $1-3''$. The saving in work input (power) by use of multi-stage compression with inter cooler is shown by shaded area $2'-3''-3-2-2'$.

Both the L.P and H.P cylinders are mounted on the same crank shaft and are driven by Electric motor or prime mover.

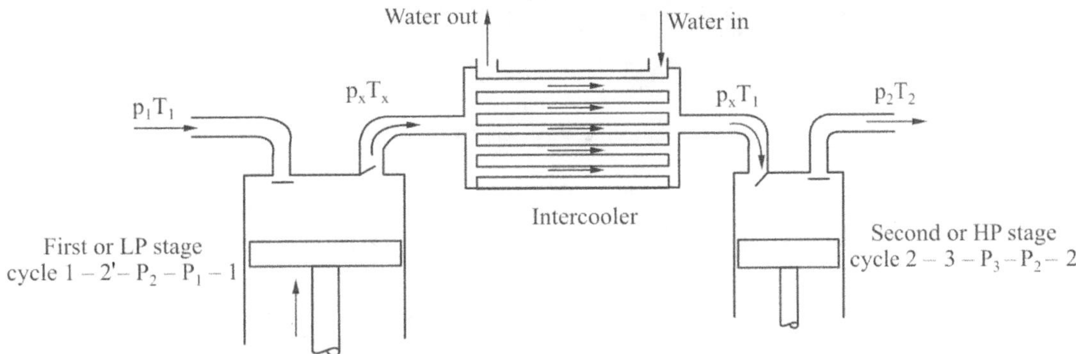

Fig. 10.8 Plan showing inter-cooling between compressor stages.

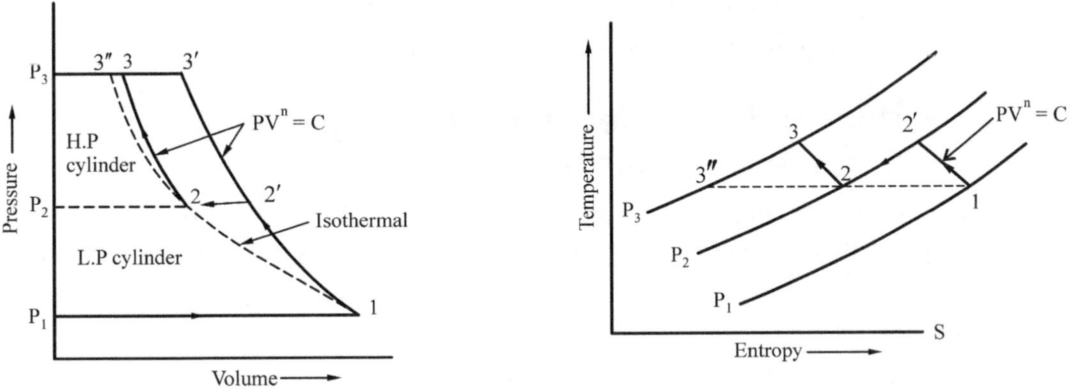

Fig. 10.9 & 10.10 P-V and T-S diagrams for two stage compression with perfect inter-cooling showing the work saved.

Similarly, perfect after-cooling makes the air, leaving the H.P. compressor, is cooled also to the inlet temperature T_1. This after-cooling reduces the volume of the gas leaving, and thus the size of the receiver becomes smaller. The clearance volume in both the cylinders here has been assumed to be the same.

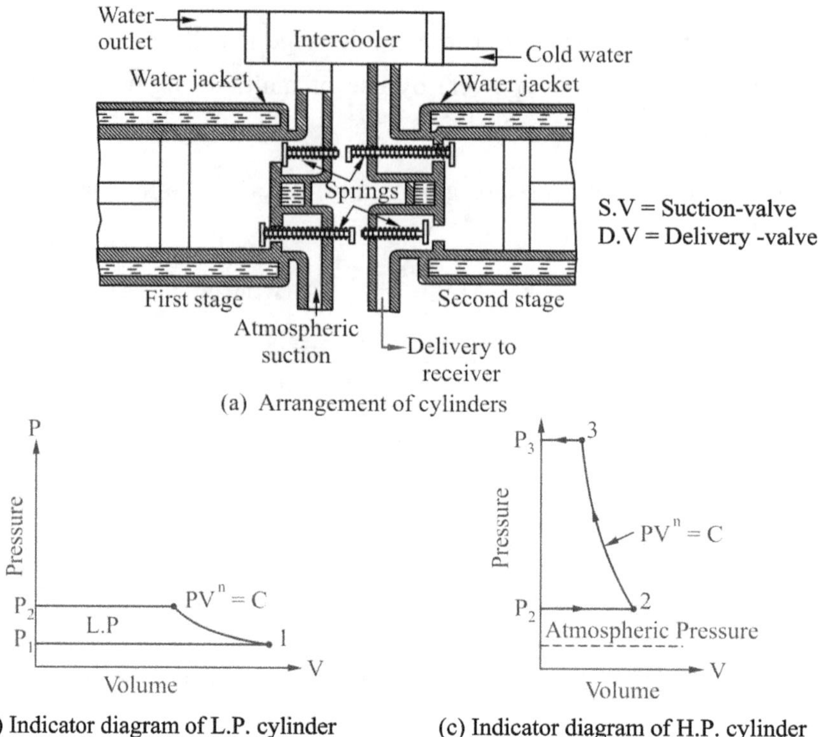

Fig. 10.11 Two-stage air compressor with intercoolers.

10.14 Perfect and Imperfect Inter-cooling in Two stage Air compressor

If temperature of air leaving the intercooler is equal to intake temperature of air T_1, the inter-cooling is called perfect or complete inter-cooling. The point 2 lies on Isothermal curve as shown in Fig. 10.12.

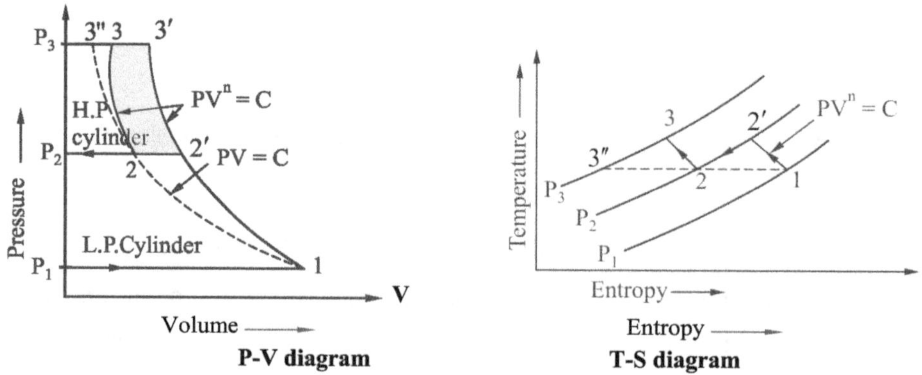

Fig. 10.12 Complete (perfect) inter-cooling of air

If air leaving the intercooler is not cooled to intake air temperature T_1 that is point 2 will not lies on isothermal curve. The inter-cooling is called imperfect or incomplete inter-cooling (Fig. 1013).

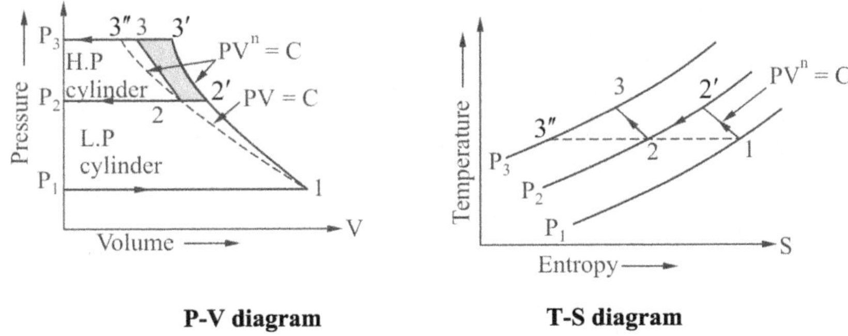

P-V diagram **T-S diagram**

Fig. 10.13 Incomplete inter-cooling of air

Three stage air compressor with intercoolers (See Fig. 10.14(a) also)

The indicated diagram for three stage compressor with perfect inter-cooling is shown in Fig. 10.14(b). The air delivered from the first stage at C is cooled to point 'd' on the Isothermal curve PV = C (shown dotted) before admission to second cylinder. Again the

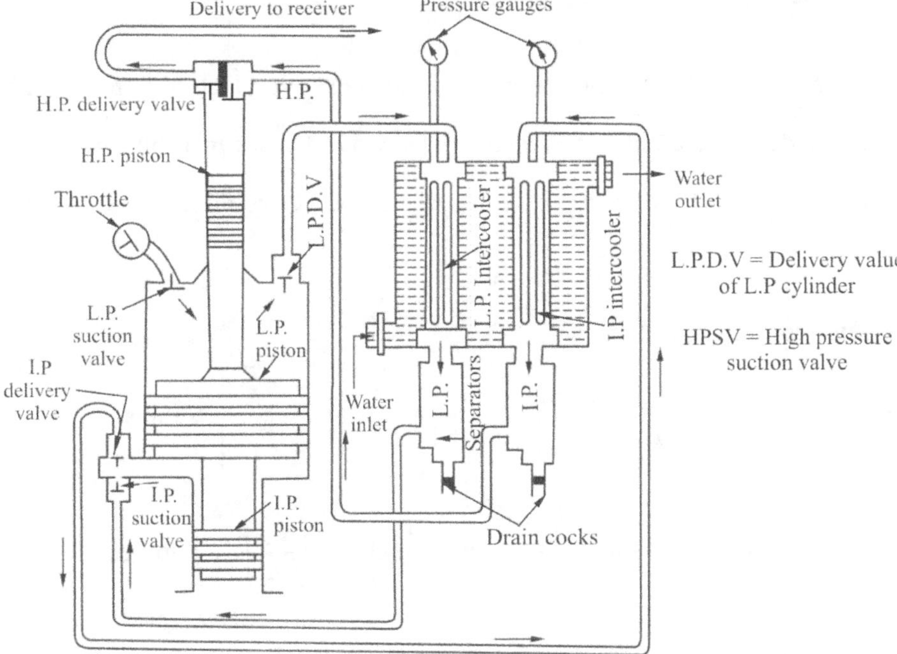

Fig. 10.14(a) Arrangement of three-stage air compressor with intercoolers.

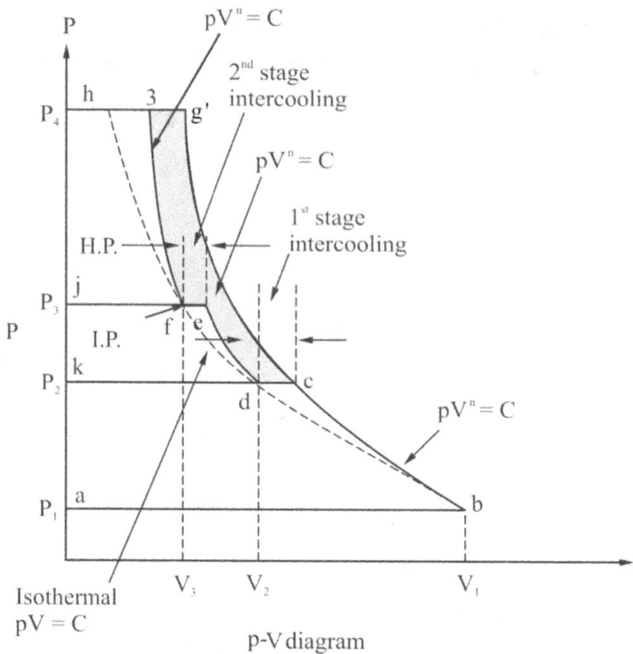

Fig. 10.14(b) Three-stage compression with perfect inter-cooling.

air is cooled from etof after compression in second stage cylinder and before admission to third stage cylinder. Shaded area represents the saving of work resulting from such inter-cooling since without inter-cooling the compression curve would have followed the path bcg'.

When inter-cooling is perfect $P_1V_1 = P_2V_2 = P_3V_3$

(a) $\therefore$ W.D $= \dfrac{n}{n-1}P_1V_1\left[\left(\dfrac{P_2}{P_1}\right)^{\frac{n-1}{n}} + \left(\dfrac{P_3}{P_2}\right)^{\frac{n-1}{n}} + \left(\dfrac{P_4}{P_3}\right)^{\frac{n-1}{n}}\right]$ Joules

(b) Also W.D $= \dfrac{n}{n-1}RT_1\left[\left(\dfrac{P_2}{P_1}\right)^{\frac{n-1}{n}} + \left(\dfrac{P_3}{P_2}\right)^{\frac{n-1}{n}} + \left(\dfrac{P_4}{P_3}\right)^{\frac{n-1}{n}}\right]$ Joules

(c) Work required is minimum when $\dfrac{P_2}{P_1} = \dfrac{P_3}{P_2} = \dfrac{P_4}{P_3}$

10.15 Maintenance of Reciprocating Compressors

Maintenance means to keep the compressors in state of good order and efficient operation.

Maintenance broadly comprises of the six factors:

1. Inspection	2. Cleaning	3. Care
4. Adjusting	5. Renewing	6. Repair

1. ***Inspection*:** It is necessary to inspect the compressors regularly at definite intervals in order to ensure necessary corrections such as replacement of working parts, adjustment for wear of working parts and cleaning of air valves, air filters and crank case, oil level in crank case must be checked everyday.
2. ***Cleaning*:** The machinery must be kept absolutely clean so that dust and other foreign matter containing abrasive particles may not result in causing wear of the working part.
3. ***Care*:** When the compressors have to run continuously, it must be taken care of supplies such as oils, packing and gaskets etc. are kept in hand and applied and adjusted at right time and with right technique.
4. ***Adjusting, Renewing and Repair*:** There is no such machine on the earth which will operate without adjustments or occasional repair. Whatsoever best type of lubricant we may use but the chances of wear due to friction can never be eliminated altogether. There are many other causes of break downs of machines. It is essential that a needed minor repair must be at once attended to, so that it does not develop into major repair, putting off till tomorrow what should be done today, may cause great damage to the machine and may reduce its working life.

Important Points about General Service

1. Compressor must be inspected regularly. Time spent on inspection and care taken will be repaid in long life and trouble free operation.
2. Always keep the compressor in perfect working condition.
3. Keep the unit always clean.
4. The water temperature must be checked from time to time.
5. In winter season, if the unit is shut down in an unheated location, the water jacket and water piping must be drained completely.
6. The oil level in crank case and force feed lubricator must be checked at regular intervals.
7. The condensate from inter cooler must be drained at regular intervals.

8. Air intake filter must be regularly cleaned as per manufacturer's instruction.
9. Safety valves must be checked from time to time to ensure that they will blow-off at required pressure settings.
10. Leaks in discharge pipe must be checked from time to time.
11. The valve assemblies must be inspected regularly.
12. Use only the highest grade of lubricating oil manufactured by reliable refinery.

10.16 Rotary Air Compressor

Whenever large quantity of air required at relatively low pressure, rotary compressors are employed.

Types of Rotary Air Compressor

There are many type of Rotary Air Compressors, yet the following are important form subject point of view.

1. Root Blower Compressor
2. Vane blower compressor

} Known as positive displacement compressors

3. Centrifugal blower compressor
4. Axial flow compressor

} Known as turbo compressors

5. Fan
6. Turbo Blower

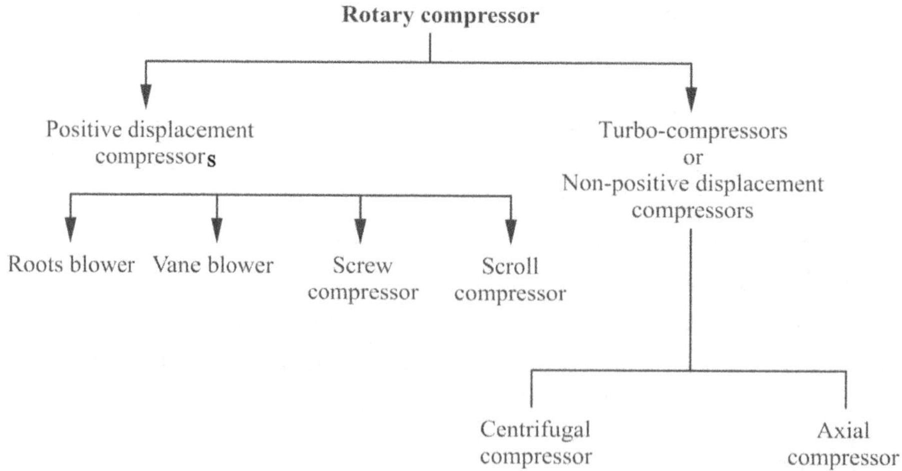

Positive Displacement Compressors have two sets of engaging surfaces or lobes. The air is trapped between the lobes and pressure rise takes place by squeezing action and back flow of air.

Non-Positive Compressor or Turbo Compressor

In this type of compressor air is compressed by centrifugal force to final pressure above 2.5 bar. Dynamic compressors are subdivided into:

(i) Centrifugal compressors (ii) Axial flow compressors.

If the air enters the centre and flows radially outward towards the circumference, (flow of air through impeller) then compressor is known as centrifugal compressor. If the air flows parallel to the shaft, it is called axial flow compressor.

Note: Positive displacement compressor such as root blower and vane blower are not so popular regarding practical point of view however they have some academic importance. The only important rotary compressor is centrifugal blower compressor.

10.16.1 Root Blower Compressor or only Root Blower

It has two rotors revolving in opposite direction by pair of spur gears (not shown in Fig. 10.15) of equal size. One of the rotor is connected to the drive (Electric motor) and second rotor is gear driven from the first. Both the rotor rotates at same speed.

A small clearance is provided between the rotor and the outer casing and between the rotor themselves. The profile of lobes is of cycloidal or involute form giving correct mating of the lobes to seal the delivery side from inlet side. In order to reduce wear, a small clearance of 0.1 to 0.2 mm is allowed between the lobes resulting in leakage through the clearance which decreases the efficiency as pressure ratio increases.

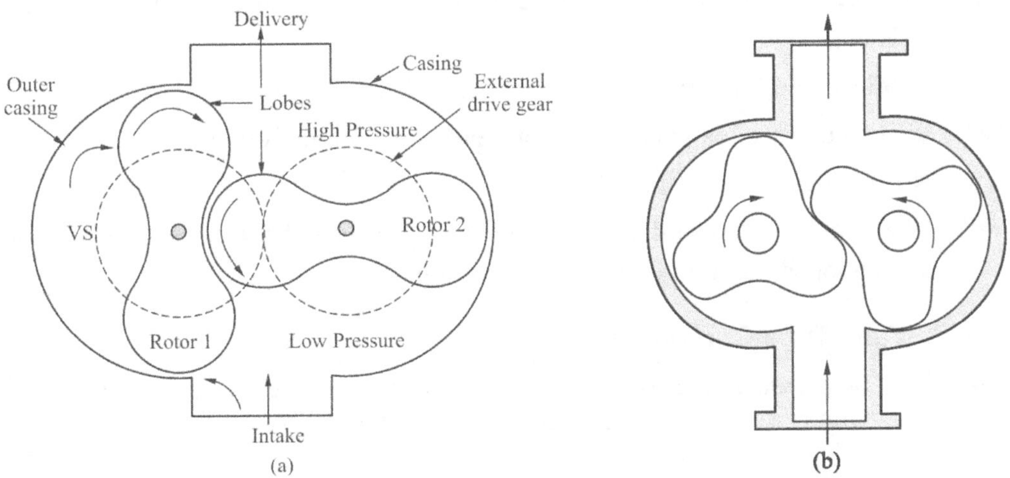

Fig. 10.15 Root Blower.

As rotor rotates, the air at atmospheric pressure P_1 and volume V_1 are trapped between the lobe and casing. The rotary motion of the lobes delivers the entrapped air into the receiver. Thus more and more flow of air into receiver, increases the pressure in receiver. Finally the air at high pressure is discharge from receiver for utilization. Theoretically there should be no pressure rise because flow area does not decrease from inlet to exit. But when outlet port is uncovered by rotating lobe, some high pressure air from receiver blows back to delivery of blower discharge. This is known as backflow process. Due to back flow, air from receiver gets mixed up with entrapped air. Thus the pressure of air entrapped in the clearance is increased at constant volume entirely by the back flow of air. The air is delivered by the rotation of the lobes. This process is carried out four times per revolution of driving shaft.

Root blowers are suitable for capacities varying from 0.13 m^3/min to 1400 m^3/min and pressure ratio of 2 : 1 for single stage and 3 : 1 for two stage machines. The rotors are made of steel tube with required shape.

(i) These machines are used for super charging and scavenging of I.C. engines.

(ii) To furnish draught furnaces, basemer, cupolas.

Advantages

(i) They run at very high speed, therefore they can be directly coupled to steam or gas turbine.

(ii) It can supply large quantity of air up to 3000 m^3/minute.

(iii) Simple in construction.

(iv) Maintenance requirements are minimum.

(v) Operation is dynamically balanced as rotors are symmetrical about their centre of rotation.

(vi) Lubricating oil required is less as all parts (Rotors) are rotary.

(vii) Wear and tear of parts is less therefore life of machine is more.

(viii) No cooling of air is required.

(xi) Whatever air is taken during intake some amount is delivered to receiver and no air is trapped (or stored) in clearance volume.

10.16.2 Vane Type Compressors

***Vane Type Compressor*:** As shown in Fig. 10.16, vane compressor is also known as the sliding vane compressor. It consists of spring loaded vanes mounted in slot on the rotor. The rotor is eccentrically mounted inside circular casing. The rotor is mounted on the shaft which is supported in bearing in end cover of body. The vanes are made up of non-metallic material usually synthetic fiber or carbon. The vanes can slide in or out against the force of spring because of this arrangement, vanes are always in contact with inner surface of the body. The construction of vane compressor is such that the suction side opens to the larger area of vanes and delivery opens to smaller area between two vanes. It has intake and delivery ports.

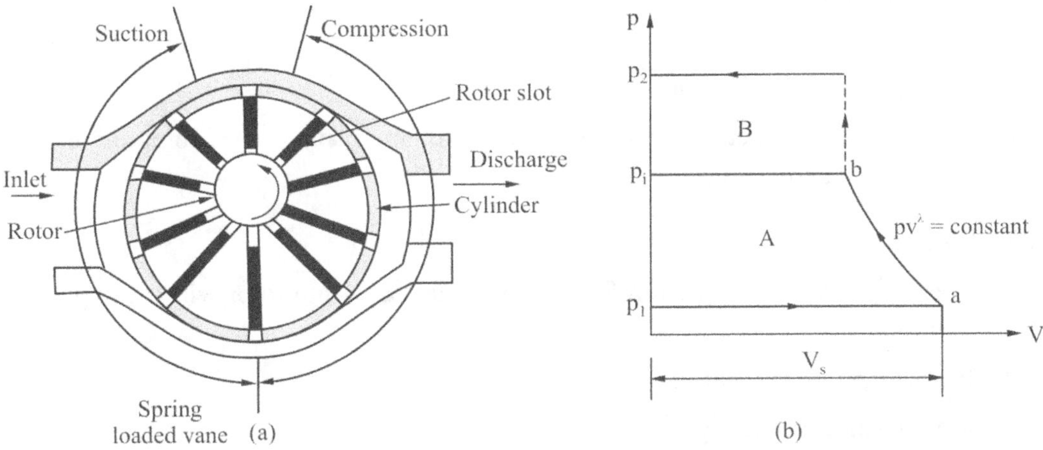

Fig. 10.16 Vane-type positive displacement compressor with p-V diagram

When rotor rotates the centrifugal force drives the vanes out towards the cylinder wall. The space between the inner surface of body and drum decreases from inlet to out side. In this way the quantity of air which enters at inlet, decreases in volume, because of which pressure of air increases as it reaches the exit. The compressor is so designed the half the total pressure rise takes place through internal reversible compression before the entrapped air is released to delivery port. Further compression can be obtained by the back-flow of air from the receiver, which occur in an irreversible manner.

Fig. 10.16 (b) shows the p-V diagram v_s is the induced volume at pressure P_1 and temperature T_1 compresses to pressure Pi. At this pressure the displaced air is open to receiver and pressurized air from receiver flowing back rise the pressure irreversibly to P_2.

Advantages

(i) The vane blower (compressor) requires less power compare to root-blower (compressor) for same capacity and pressure rise.

(ii) They are commonly used to deliver up to 150 m^3 of air per minute at pressure ratio of 8.5.

(iii) No valves are required

(iv) No clearance volume therefore discharge is complete (i.e., no air is stored in clearance)

(v) Compressor gives a very uniform torque so that there is no need of flywheel. It has low weight.

(vi) It is compact

(vii) It possesses high efficiency

(viii) They have limited capacity as compared with non displacement compressors i.e., (centrifugal and axial compressors).

Disadvantages

(i) Pressure which is required in forcing the vanes back into rotor will causing some vibration.

(ii) It has high rate of wear of sliding vanes.

(iii) Maintenance cost is high.

(iv) Speed limit of vane compressor is 3000 RPM.

10.16.3 Centrifugal Compressor

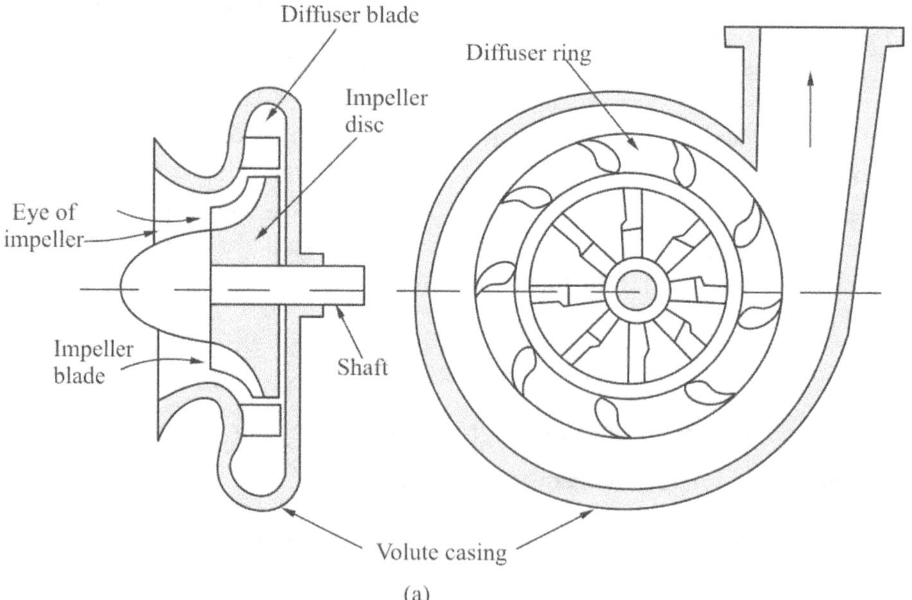

(a)

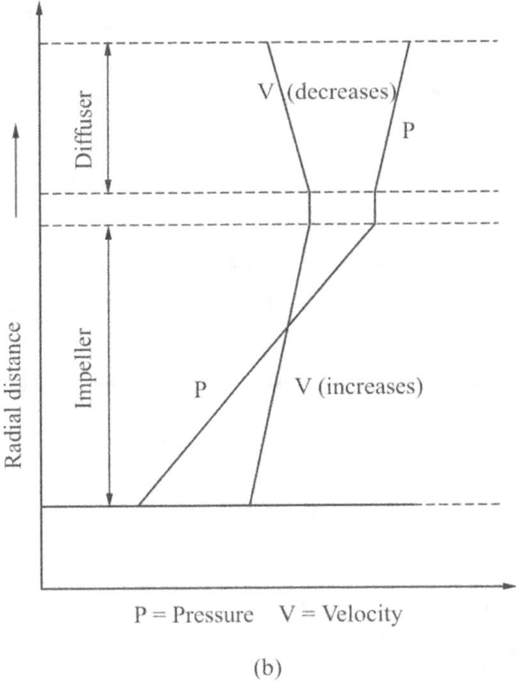

(b)

Fig. 10.17 Centrifugal compressor.

Fig. 10.17 shows the main components of centrifugal Compressor

(i) It consists of impeller having large number of blades. It may be single or double sided. It is mounted on shaft inside stationary casing.

(ii) Diffuser

(iii) Casing

(iv) inlet and outlet pipes.

When impeller is rotated by electric motor or engine or turbine which produces a centrifugal head of air within the casing, which causes air to flow rapidly outward from casing. The fresh air enters at eye of impeller as vacuum is created in casing. During flow of air from eye to tip of impeller, there is an increase in pressure and velocity, due to increased momentum of air. From impeller air enters into fixed diffuser or diffuser blades which convert most of K.E of the air into pressure energy. Air is then collected in volute (scroll) casing and taken out from outlet pipe.

In this type of compressor the air enters axially but leaves in radial direction. This arrangement is known as single stage compression and is suitable for small pressure ratio say up to 4 : 1. In order to have high pressure ratios, number of such compressor are used in series. Blades of impeller may be radial, backward and forward turned. Material of impeller of centrifugal compressor is generally forged or die-casted of low silicon aluminium alloys.

The centrifugal compressor has the following advantages as compare with axial flow compressor.

1. It is simple.
2. It is compact and smaller in length
3. It attains higher pressure ratio per stage
4. It is able to operate efficiently over a wide range of mass flow at any particular speed.
5. It is not liable to loss of performance when working in a contaminated atmosphere

Disadvantages

1. larger frontal area
2. Lower maximum efficiency
3. Less adaptability to multi-staging.

10.16.4 Axial Flow Compressor

***Construction*:** Axial Flow compressor consists of a rotor, fixed blades and casing (also known as stator). Rotor blades are mounted to the Rotating rotor (drum) and fixed blades are attached to casing. The rotor as well as the stator blade channels are of diverging type i.e., area increases from inlet to outlet.

A row of moving blades and row of fixed blades is called one stage of axial flow compressor. Fixed blades converts K.E of air into pressure energy (fixed blades acts as diffuser. The stator blades also redirect the air to an angle suitable for entry to the succeeding set of moving blades to reduce eddy losses.

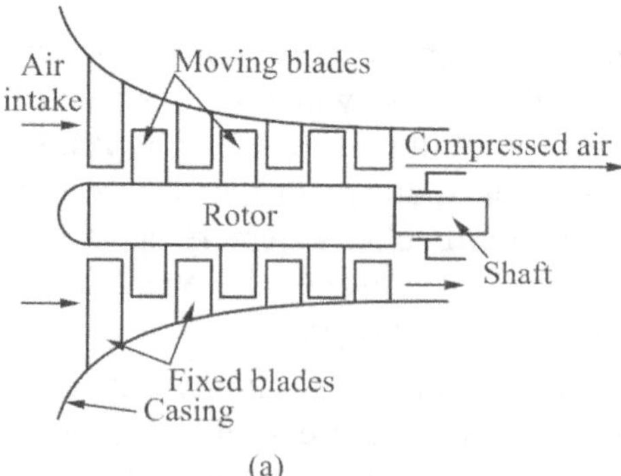

(a)

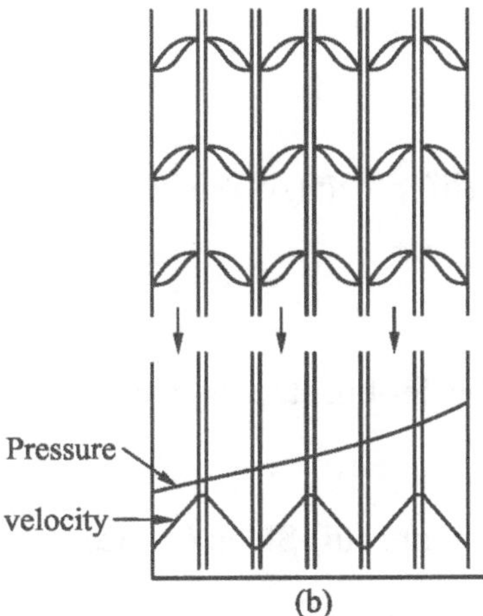

Fig. 10.18 Axial compressor

The axial flow compressors are mostly multi-stage compressors. Pressure ratio which can be produced by single stage of the axial flow compressor is 1.12 to 1.2 and multistage axial compressor produces pressure ratio of 8.

At the inlet of the compressor an extra row of fixed vanes called inlet guide vanes are fitted. These guide vanes redirect the air at a correct angle to first row of moving blades. The rotor and stator blade rings (wheel) must be as close as possible for smooth and efficient flow. There should be a very small clearance at the end of the rotor and stator blades. Height of moving blades and fixed blades is reduced to compensate for reduction in volume, resulting from increasing pressure from stage to stage.

There are two types of rotors – drum and disc. Disc type is used in aircraft application due to low weight. The drum type is more suitable for industrial application.

Working

The shaft of axial flow compressor is rotated by mechanical energy which rotates the drum. The air enters from left side of compressor and the air flows in an axial direction from the intake to the delivery. The air enters the moving blades, the rotor increases the velocity of air and blades of rotor also acts as diffuser to convert the K.E into pressure energy and pressure of air is increased. Air from rotor enters the fixed blades which acts as diffuser and remaining K.E of air is again converted into pressure energy. Thus increase in pressure also occurs during is passage through stator.

Uses

They are used in application of aircraft and industrial gas turbine power plant due to its high efficiency less, frontal area and is capable of producing higher compression ratio.

Materials

Rotor blades

(i) Aluminium; (ii) Titanium; (iii) Steel; (iv) Nickel alloy

Shaft: steel

Disc: titanium

Stator blades: Same as Rotar blades but steel is more common.

Castings

(i) Cast magnesium; (ii) Aluminium; (iii) Steel; (iv) Iron

Solved Problems

1. A single acting air compressor is required to deliver 15 kg of air per minute at 6 bar. At the beginning of compression stroke, the pressure and temperature are 0.9 bar and 27^0 C. Assuming law of compression being $pv^{1.32} = C$, calculate power required to drive compressor, if the mechanical efficiency is 80%. Find also final temperature. Take R for air 0.287 KJ/kg. K.

***Solution*:** Given $m = 15 \text{ kg/min} = \frac{15}{60} \text{ kg/sec}$

$P_1 = 0.9$ bar

$T_1 = 273 + 27 = 300$ k

$PV^{1.32} = C$ i.e. $n = 1.32$

$P_2 = 6$ bar

Mechanical efficiency = 80%

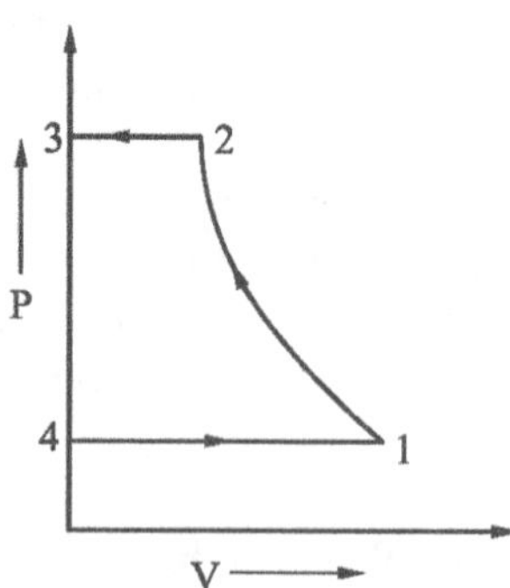

$$\text{Work done on air/sec} = \frac{n}{n-1} P_1 V_1 \left[\left(\frac{P_2}{P_1} \right)^{\frac{n-1}{n}} - 1 \right]$$

***Note*:** In the problem, mass of air delivered per min is given so we have to rewrite above formula in terms of mass i.e.,

$$10^2\, p_1 v_1 = mRT_1 \qquad \text{(mass is taken kg/sec)}$$

∴ W.D on air/sec (also known as indicated power)

$$= \frac{n}{n-1} mRT_1 \left[\left(\frac{P_2}{P_1} \right)^{\frac{n-1}{n}} - 1 \right]$$

$$= \frac{1.32}{1.32-1} \times \frac{15}{60} \times 0.287 \times 300 \left[\left(\frac{6}{9} \right)^{\frac{1.32-1}{1.32}} - 1 \right]$$

$$= 4.125 \times 0.25 \times 0.237 \times 300 \times 0.584$$

$$= 51.854 \text{ KW}$$

$$\therefore \text{ power require to drive the compressor} = \frac{51.854}{0.8}$$

$$= 64.82 \text{ kw} \quad \textbf{Ans.}$$

$$\text{Temperature at end of compression} = T_1 \left(\frac{p_2}{p_1} \right)^{\frac{n-1}{n}}$$

$$= 300 \times \left(\frac{6}{0.9} \right)^{\frac{0.32}{1.32}}$$

$$= 300 \times 1.584$$

$$= 475.2 \text{ k}$$

$$= \mathbf{202.2^0\ C} \quad \textbf{Ans..}$$

2. A single acting reciprocating compressor takes in 0.1 m^3/sec at 1 bar and 15° C and compresses to delivery pressure of 6 bar according to $PV^{1.25} = C$. Calculate.
   1. Temperature at end of compression
   2. Work done during suction by air
   3. W.D on air during compression
   4. W.D on air during delivery

5. Net W.D on air
6. Heat transfer
7. Change in internal energy
8. Brake power if mechanical efficiency is 86%.

Take R for air as 0.287 kJ/kg K.

Solution: Given $v_1 = 0.1$ m3/sec

$P_1 = 1$ bar

$T_1 = 273 + 15 = 280$ K

$P_2 = 6$ bar

$R = 0.287$ kJ/kg K

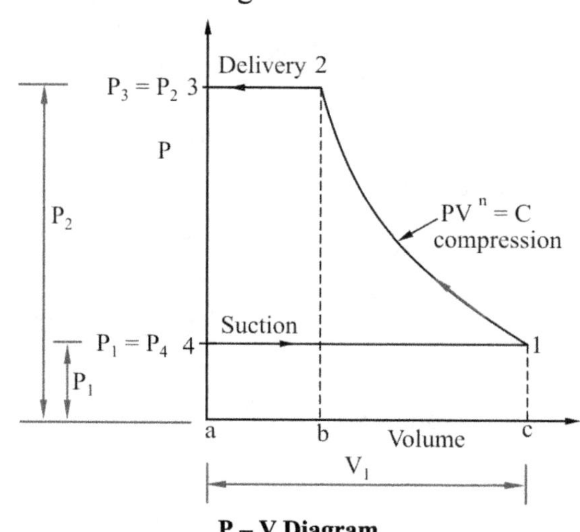

P – V Diagram

Step

(i) W.D by air during suction = area 1-4-a-c-1

$$= p_1 v_1$$

(ii) W.D on air during delivery = area 2 – 3 – a – b – 2

$$= p_2 v_2$$

(iii) W.D during compression $= \dfrac{p_2v_2 - p_1v_1}{n-1}$

$$= \text{area } 1 - 2 - b - c - 1$$

1. $$T_2 = T_1\left(\frac{p_2}{p_1}\right)^{\frac{n-1}{n}}$$

$$= 288\left(\frac{6}{1}\right)^{\frac{0.25}{1.25}}$$

$$= 412.1 \text{ K} = \mathbf{139.1\ ^\circ C} \quad \textbf{Ans.}$$

2. W.D by air during suction $= p_1\, v_1$

$$= 10^2\, p_1\, v_1 \qquad \left[\frac{KN}{m^2}\cdot\frac{m^3}{sec}\right],$$

$$= 10^2 \times 1 \times 0.1 \left[\frac{KN-m}{sec} = \frac{KJ}{sec} = kw\right]$$

$$= \mathbf{10\ kw} \quad \textbf{Ans.}$$

3. W.D on air during compression $= \dfrac{p_2 v_2 - p_1 v_1}{n-1}$

Note in this formula v_2 is to be calculated. To avoid further calculation for v_2, above formula should be rewritten in terms of temperature because we have already calculated T_2, only calculate mass 'm'.

$$10^2\, p_1\, v_1 = mRT_1$$

$$\therefore \quad m = \frac{10^2 \times 1 \times 0.1}{0.287 \times 288} = 0.121$$

$$\therefore \quad \text{W.D on air during compression} = \frac{mR}{n-1}(T_2 - T_1)$$

$$= \frac{0.121 \times 0.287}{0.25}(412.1 - 288)$$

$$= \mathbf{17.236\ kw} \quad \textbf{Ans.}$$

4. Heat transfer to cylinder walls $= \dfrac{\gamma - n}{\gamma - 1} \times$ W.D during polytropic compression.

$$= \frac{1.4 - 1.25}{1.4 - 1} \times 17.236 = \mathbf{6.464\ kw} \quad \textbf{Ans.}$$

5. W.D during delivery $= 10^2\, p_2\, v_2$

$$= m\, RT_2$$

$$= 0.121 \times 0.287 \times 412.1 = \mathbf{14.311\ kw}$$

6. Net W.D on air = 17.236 + 14.311 – 10 = 21.547

= 21.547 kw **Ans.**

or

Net W.D on air may also be found from equation

$$\text{W.D on air/sec} = \frac{n}{n-1} 10^2 p_1 v_1 \left[\left(\frac{p_2}{p_1} \right)^{\frac{n-1}{n}} - 1 \right]$$

$$= \frac{1.25}{0.25} \times 10^2 \times 1 \times 0.1 \left[(6)^{\frac{0.25}{1.25}} - 1 \right] = \mathbf{21.548\ kw} \quad \textbf{Ans.}$$

7. Change in internal energy takes place during compression of air i.e, during 1 – 2 process because during this temperature is increased from T_1 to T_2. During suction 4 – 1 and 2 – 3 pressure is constant therefore temperature is assumed to be constant.

∴ Change in internal energy = (Heat transfer during compression) (W.D during compression)

$$\Delta u = Q - W \qquad \text{.....(i)}$$

In this case Q is rejected ∴ $Q = -Q$ (– ve sign indicates that heat is rejected)

Work is done on the air ∴ $W = -W$ (– ve sign because work is done on air i.e on system)

Put this value in equation $\Delta = -Q - (-W)$

$$= W - Q$$

$$= 17.236 - 6.464$$

$$= \mathbf{10.772\ kw} \quad \textit{Ans.}$$

Brake power or motor shaft power $= \dfrac{\text{Net W.D.R}}{\eta_m}$ indicated air work

$$= \frac{21.548}{0.86} = \mathbf{25.056\ kw} \ \ \textbf{Ans.}$$

Note (For information onlty)

(i) During isothermal compression of air

W.D during suction = $p_1 v_1$

W.D during compression = $p_1 v_1 \log \frac{p_2}{p_1}$

W.D during delivery = $p_2 v_2$

During isothermal $\pi_2 = T_2$

$\therefore$ Change in internal energy = 0

$\therefore$ W.D = heat transfer

(ii) If air is compressed adiabatically or isentropically

W.D during suction = p_1v_1

W.D during compression $= \frac{p_2v_2 - p_1v_1}{\gamma - 1}$

W.D during delivery = $p_2 v_2$

During isentropic compression heat transfer is zero

$Q = W.D + \Delta_u$

$\therefore$ $W.D = Q - \Delta_u$

$W.D = 0 - \Delta_u$

$W.D = -\Delta_u$

As work is done by the system on air, therefore work will be – ve.

$\therefore$ $-W.D = -\Delta_u$

$\therefore$ $W.D = \Delta_u$

3. What is the power required to drive an air compressor which takes in 15 kg of air per minute at 27 °C and 1 bar. The compression is carried out in one stage to 8 bar, air is delivered to receiver at this pressure. Compression curve follows $PV^{1.3} = C$ and mechanical efficiency of compressor being 80%. Neglecting losses due to clearance, cooling and leakage, find also the heat lost through the cylinder walls.

***Sol*:**

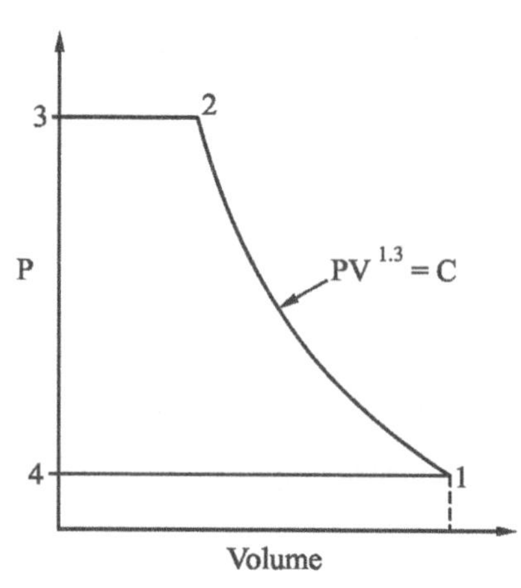

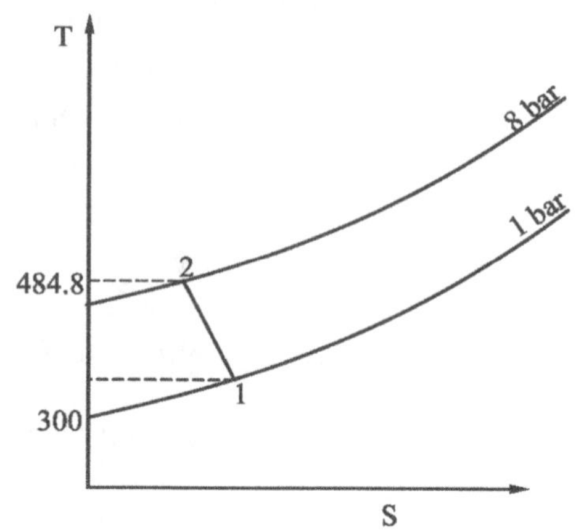

P – V Diagram

T – S Diagram

***Note*:**

(i) To represent process 1 – 2 on T – S diagram first find T_2 i.e.,

temperature of air at delivery $= T_1 \left(\dfrac{P_2}{P_1} \right)^{\frac{n-1}{n}}$

(ii) Point 1 can be located on T – S diagram as T_1 and P_1 is known

(iii) Point 2 can be located on T – S diagram as P_2 and T_2 both are known. Then join point 1 and point 2.

Given

$$m = 15 \text{ kg/min} = \frac{15}{60} \text{ kg/sec} = 0.25 \text{ kg/sec}$$

$P_1 = 1$ bar, $T_1 = 273 + 27 = 300$ k

$P_2 = 8$ bar,

$n = 1.3$

$\eta_m = 80\%$

R is not given in problem, assume 0.287 kJ/KgK

(i) W.D on air/min $= \frac{n}{n-1} 10^2 P_1 v_1 \left[\left(\frac{P_2}{P_1}\right)^{\frac{n-1}{n}} - 1\right]$

where 10^2 standard to convert bar in KP_a

$$= \frac{n}{n-1} mRT_1 \left[\left(\frac{P_2}{P_1}\right)^{\frac{n-1}{n}} - 1\right]$$

$$= \frac{1.3}{1.3-1} \times 0.25 \times 0.287 \times 300 \left[(8)^{\frac{0.3}{1.3}} - 1\right]$$

$$= \frac{1.3}{0.3} \times 0.25 \times 0.287 \times 300 \, [1.615866\text{-}1]$$

= 57.445 kw

$\therefore$ Power required to drive compressor $= \frac{57.445}{0.8}$

= 71.81 kw **Ans.**

(ii) Temperature of air at delivery $= T_1 \left(\frac{p_2}{p_1}\right)^{\frac{n-1}{n}}$

$$= 300 \, (8)^{\frac{1.3-1}{1.3}}$$

$= 300 \times 1.616 = 484.8^\circ$ K **Ans.**

(iii) From fig $Q = C_P T_2 = C_P T_1 + w$

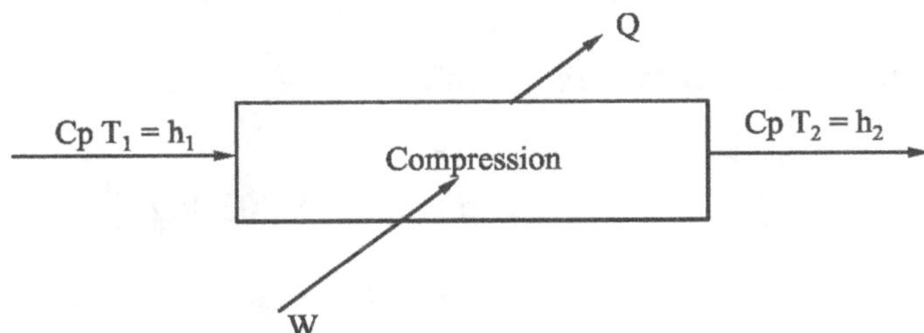

Q = heat lost per kg air

$= m\,C_pT_1 + W.D - m\,C_pT_2$

(Apply total energy in = Total energy out)

$= (0.25 \times 1.005 \times 300) + 57.445 - (0.25 \times 1.005 \times 484.8)$

$= 75.3624 + 57.445 - 121.806$

= 11 kJ/sec

assuming Cp = 1.005 kJ/kg K

4. Single Stage reciprocating air compressor is required to compress 5 kg of air from 1 bar and initial temperature of 30^0C to 6 bar. Compare the work required in following cases.

1. Isothermal compression
2. Isentropic compression
3. Polytrophic compression with index n = 1.2

Take R for air as 0.287 kJ/kg K

Solution:

Mass = 5 kg

$P_1 = 1$ bar, $T_1 = 273 + 30 = 303$ k

R = 0.287 kJ/kg K

n = 1.2

For isentropic compression $\gamma = 1.4$

1. W.D on air or work required by compressor

$= 10^2\, p_1\, v_1 \log_e \frac{P_2}{P_1}$ (for isothermal compression)

But $10^2\, P_1\, v_1 = mRT_1$

$\therefore$ W.D required by compressor $= mRT_1 \log_e \frac{P_2}{P_1}$

$= 5 \times 0.287 \times 303 \log_e \frac{6}{1}$

= 779.1 kJ **Ans.**

(because mass is not given Kg/min or kg/s cannot be written in terms of KW)

2. W.D on air or work required by compressor

$$= \frac{\gamma}{\gamma - 1} 10^2 p_1 v_1 \left[\left(\frac{P_2}{P_1} \right)^{\frac{\gamma-1}{\gamma}} - 1 \right]$$

Equation can be rewritten in mRT_1 terms

$$= \frac{\gamma}{\gamma - 1} mRT_1 \left[\left(\frac{P_2}{P_1} \right)^{\frac{\gamma-1}{\gamma}} - 1 \right]$$

$$= \frac{1.4}{1{:}4 - 1} \times 5 \times 0.287 \times 303 \left[(6)^{\frac{1.4-1}{1.4}} - 1 \right]$$

$= 1017.35$ KJ *Ans.*

3. W.D on air $= \frac{n}{n-1} mRT_1 \left[\left(\frac{p_2}{p_1} \right)^{\frac{n-1}{n}} - 1 \right]$

$$= \frac{1.2}{1.2 - 1} \times 5 \times 0.287 \times 303 \left[(6)^{\frac{0.2}{1.2}} - 1 \right] = \mathbf{907.89\ kJ}$$ **Ans.**

5. A single stage, single acting air compressor is belt driven from an electric motor at 400 RPM. The cylinder diameter is 15 cm and stroke 18 cm. The air is compressed from 1.05 bar to 7.5 bar and law of compression is $pv^{1.3} = C$. Find the power of motor if transmission efficiency is 97% and mechanical efficiency of compressor is 90%, neglecting clearance.

Solution: Given

Single acting single stage $N = 400$ RPM

$d = 15$ cm

$l = 18$ cm

$P_1 = 1.05$ bar

$P_2 = 7.5$ bar

$n = 1.3$

$\eta_{trans} = 97\%$

$\eta_{Mech} = 90\%$

Step I : Find indicated work of air per working stroke (or per cycle)

by using formula $\frac{n}{n-1}10^2 P_1 v_1\left[\left(\frac{P_2}{P_1}\right)^{\frac{n-1}{n}}-1\right]$

Note $\left[\left(\frac{P_2}{P_1}\right)^{\frac{n-1}{n}}-1\right]$ is constant, $\frac{n}{n-1}$ = constant, then

If P_1 is given in bar, then 10^2 P_1 will have unit as $\frac{KN}{m^2}$ and

V_1 is swept volume in $\frac{m^3}{cycle}$, then unit of 10^2 P_1 V_1 $=\frac{KN}{m^2}.\frac{m^3}{cycle}$

$$\therefore \quad \text{unit of work done } \frac{kN-m}{cycle}=\frac{kJ}{cycle} \qquad (\because N-m=J$$

$$\therefore kN\text{–}m = KJ)$$

Step II : Power required by compressor $=\frac{\frac{W.D}{cycle}\times \text{No of working stroke/min}}{60}$

For single acting, number of working stroke/min = RPM

$$\therefore \quad \text{Power required by compressor } =\frac{W.D/cycle\times N}{60}$$

$$\therefore \quad \text{Power of motor } =\frac{\text{Power of Compressor Required}}{\text{Transmission Efficiency}\times \text{Motor efficiency}}$$

Step III: If clearance is neglected

$$V_s = V_1$$

Step I:

$$V_1 = \frac{\pi}{4}.d^2.l$$

$$=\frac{\pi}{4}\left(\frac{15}{100}\right)^2\frac{18}{100}$$

$$= 0.003179 \text{ m}^3\text{/stroke}$$

Step II: W.D on air (or W.D required by compressor or indicated work of air)

$$=\frac{n}{n-1}10^2 p_1 v_1\left[\left(\frac{p_2}{p_1}\right)^{\frac{n-1}{n}}-1\right]$$

$$= \frac{1.3}{1.3-1} \times 10^2 \times 1.05 \times 0.003179 \left[\left(\frac{7.5}{1.05} \right)^{\frac{1.3-1}{1.3}} - 1 \right]$$

$$= \frac{1.3}{.3} \times 10^2 \times 1.5 \times 0.003179 \, [1.5742-1]$$

$= 0.8305$ kJ/cycle

Step III: Power required by compressor/sec $= \dfrac{\text{W.D/cycle} \times \text{N}}{60} = \dfrac{\text{W.D/cycle} \times 400}{60}$

$= 5.537$ kw

$$\text{Power of motor} = \frac{5.537}{\eta_t \times \eta_{Mech}} = \frac{5.537}{0.97 \times 0.9} = 6.342 \text{ Kw}$$

6. Find the volumetric efficiency of each of the three ideal air compressor having 6% clearance volume and operating with junction pressure of 1 bar and discharge pressure 5 bar if exponents of expansion process in three compressors are. (i) 1.4 (ii) 1.2 (iii) 1

Solution: Volumetric efficiency $= 1 - k \left[r_p^{\frac{1}{n}} - 1 \right]$

where k = clearance ratio = 0.06

r_p = pressure ratio = 5 = P_2/P_1

(i) For n = 1.4

$$\eta_{VOL} = 1 - 0.06 \left[5^{\frac{1}{1.4}} - 1 \right]$$

$= 1 - 0.1296$

$= 87.04\%$

(ii) For n = 1.2

$$\eta_{VOL} = 1 - 0.06 \left[5^{\frac{1}{1.2}} - 1 \right]$$

$= (1 - 0.1694)$

$= 83.05\%$

(iii) for n = 1

$$\eta_{VOL} = 1 - 0.06 [5^1 - 1]$$

$= 1 - 0.24 = 76\%$

For this problem it is clear that as value of 'n' increases volumetric efficiency also increases

If n = 1, $\eta_{VOL} = 76\%$

If n = 1.2, $\eta_{VOL} = 83.05\%$

If n = 1.4 $\eta_{VOL} = 87.04\%$

7. A single acting single stage reciprocating air compressor of 15 kw at 240 RPM takes in air at 1 bar and deliver it at 8 bar. Index of compression curve is 1.35. Find diameter of cylinder and stroke length at an average piston speed of 180 m/min neglecting clearance.

***Solution*:**

First draw p.v diagram

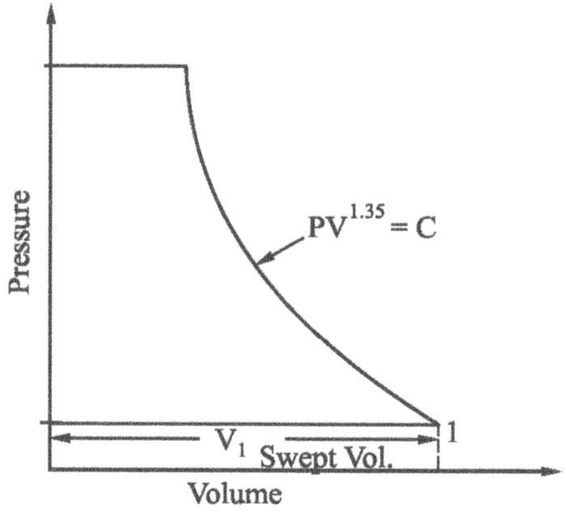

p-V diagram

Given

Single acting, Single stage

Power of compressor = 15 kw

Mean piston speed = 2 LN = 180 m/min

N = 240 RPM

$P_1 = 1$ bar

$P_2 = 8$ bar

n = 1.35

Step I: From mean piston speed (Piston always has linear speed) in one revolution of crank the distance travel by piston will be 2L where L = stroke length

Mean speed of piston = 2 LN

$$180 = 2L \times 240$$

$$L = \frac{180}{480} = 0.375 \text{ m or } 37.5 \text{ cm}$$

Step II: Compressor power is given. From formula swept volume in single acting and single stage will be v_1 as shown in figure.

$$15 \text{ kw} = \frac{n}{n-1} 10^2 p_1 V_1 \left[\left(\frac{P_2}{P_1} \right)^{\frac{n-1}{n}} - 1 \right] \qquad P_1 = \frac{kN}{m^2}, V_1 \frac{m^3}{sec}$$

$$\therefore P_1 V_1 = \frac{kNm}{sec} \text{ kJ} = \text{kw}$$

(If power is K.W than V_1 will be in m^3/sec)

$$\therefore \quad 15 = \frac{1.35}{1.35-1} . 10^2 \times 1 \times v_1 \left[(8)^{\frac{0.35}{1.35}} - 1 \right]$$

$$= \frac{1.35}{0.35} \times 10^2 \times 1 \times v_1 [\ 1.7145 - 1]$$

$$= 275.59 \ V_1$$

$$\therefore \quad v_1 = \frac{15}{275.59} = 0.05443 \text{ m}^3/\text{sec}$$

But $$v_1 = \frac{\pi}{4} d^2 L \frac{N}{60}$$

$$0.05443 = \frac{\pi}{4} d^2 . \frac{240}{60} \times 0.375$$

$$d^2 = .04623$$

$$d = 0.215 \text{ m}$$

$$\therefore \quad d = 21.5 \text{ cm}$$

Diameter of piston 21.5 cm

Length of stroke = 37.5 cm

8. The clearance volume in a single cylinder single acting reciprocating type compressor is 4% of swept volume. Air is drawn in at 1 bar and temperature 38° C. Compression and expansion curves follow the law $p.v^{1.2} = C$ delivery, pressure is 7 bar and the atmospheric pressure and temperature are 1.01325 bar and 15°C respectively. Determine.

(a) Volumetric efficiency

(b) Volumetric efficiency referred to atmospheric condition

(c) The required work input per kg of air

Solution:

Given $v_c = 0.04\ Vs$

$T_1 = 273 + 38 = 311\ k$

$P_1 = 1\ bar$

$n = 1.2$

$P_2 = 7\ bar$

Atmospheric conditions $P = 1.01325\ Bar$

$T = 288\ k$

Step I Draw necessary required figure

Step II Find value of $v_a = v - v_4$ in terms of V_s

Step I Apply formula for volumetric efficiency i.e.,

$$\eta_{vol} = \frac{v_1 - v_4}{v_s} = \frac{\text{actual value of taken in}}{\text{stroke volume (swept volume)}}$$

From fig;

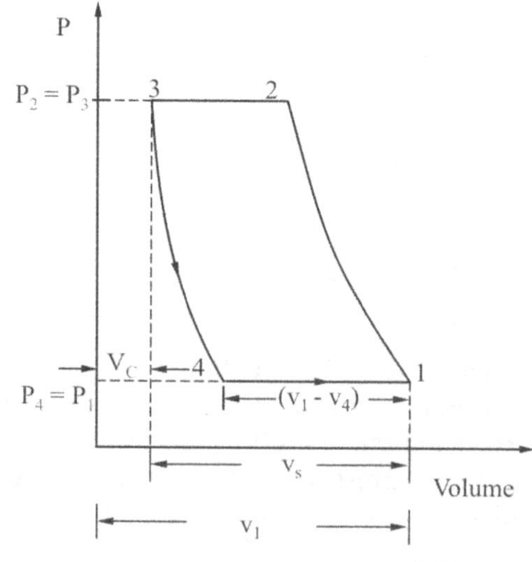

Single stage Compressor with Clearance

$$v_3 = v_c = 0.04\ v_s$$

Always calculate v_4 by applying polytrophic law between point 3 and 4 i.e.,

$$v_4 = v_3 \left(\frac{P_3}{P_4}\right)^{\frac{1}{n}}$$

$$v_4 = 0.04\ v_s \left(\frac{P_2}{P_1}\right)^{\frac{1}{n}} \qquad \therefore\ P_3 = P_2, \quad \because\ P_1 = P_4$$

$$v_4 = 0.04\ v_s \left(\frac{7}{1}\right)^{\frac{1}{1.2}}$$

$$v_4 = 0.04\ v_s \times 5.06$$

$$v_4 = 0.2024\ v_s \qquad \text{.... (i)}$$

$$v_1 = v_s + 0.04\ v_s = 1.04\ v_s$$

$$\therefore \qquad v_a = v_1 - v_4 = 1.04v_s - 0.2024\ v_s = 0.8376\ v_s \qquad \text{....(ii)}$$

But $$\eta_v = \frac{V_a}{V_s} = \frac{0.8376V_s}{V_s} = 0.8376$$

i.e. **83.76%**

Alternatively, we can find volumetric efficiency by applying equation for volumetric efficiency

$$\text{Volumetric efficiency} = 1 - \frac{V_c}{V_s}\left[\left(\frac{P_2}{P_1}\right)^{\frac{1}{n}} - 1\right]$$

$$= 1 - \frac{0.04v_s}{v_s}\left[\left(\frac{7}{1}\right)^{\frac{1}{1.2}} - 1\right]$$

$$= 1 - 0.04\ (5.06\text{-}1)$$

$$= 1 - 0.1624$$

$$= \mathbf{0.8376\ or\ 83.76\%} \quad \text{Same as before}$$

(b) In this problem the intake conditions at point 1 differ from atmospheric conditions. The volumetric efficiency is to be calculated refer to atmospheric conditions.

$$\therefore\ \eta_v = \frac{\text{volume of free air drawn in}}{V_s}$$

If P_1 = Initial pressure of air at end of suction i.e., point 1

T_1 = Initial temperature of air at end of suction

V_1 = Initial volume of air at end of suction

P_a, V_a, T_a = corresponding values of ambient conditions

$$\therefore \quad P_a = 1.01325 \text{ bar}$$

$$T_a = 273 + 15 = 288 \text{ k}$$

$$V_a = ?$$

Apply general characteristic law between point "1" and atmosphere

$$\frac{P_a v_a}{T_a} = \frac{P_1(v_1 - v_4)}{T_1} \quad \text{where } v_1 - v_4 = \text{value of air at '1'}$$

$$\therefore \quad V_a = \frac{P_1}{T_1} \cdot \frac{T_a}{P_a} (v_1 - v_4)$$

$$V_a = \frac{P_1}{P_a} \cdot \frac{T_a}{T_1} (v_1 - v_4)$$

$$V_a = \frac{1}{1.0125} \times \frac{288}{311} (v_1 - v_4)$$

But $v_1 - v_4 = 0.8376$ from equation (ii)

$$V_a = \frac{1}{1.01325} \times \frac{288}{311} \times 0.8376 \ v_s$$

$$\eta_v = \frac{v_1 - v_4}{v_s} \quad \therefore \ v_1 - v_4 = 0.837 \ v_s$$

$$V_a = 0.7761 \ v_s$$

$\therefore$ η_v referred to atmospheric condition

$$= \frac{\text{actual volume at atmospheric conditions}}{\text{swept volume}}$$

$$= \frac{0.7761 v_s}{v_s}$$

$$= 0.7761 \text{ i.e., } 77.61\%$$

(c)
$$w = \frac{n}{n-1} p_1 (v_1 - v_4) \left[\left(\frac{p_2}{p_1} \right)^{\frac{n-1}{n}} - 1 \right]$$

But $\quad P_1 (V_1 - V_4) = mRT_1, \qquad P_1 v_1 = RT_1; \quad m = 1 \text{ kg}$

$$\therefore \qquad \text{W.D/kg of air} = \frac{n}{n-1} RT_1 \left[\left(\frac{p_2}{p_1} \right)^{\frac{n-1}{n}} - 1 \right]$$

$$= \frac{1.2}{0.2} \times 0.287 \times 311 \left[\left(\frac{7}{1} \right)^{\frac{1.2-1}{1.2}} - 1 \right]$$

$$= \mathbf{205.159\ kJ/kg\ of\ air}$$

Note work input/kg of air is not affected by clearance

Questions

1. What are uses of compressed air?
2. What is object of compressing air in to stages?
3. How are air compressor classified?
4. Describe briefly the working of single acting single stage reciprocating air compressor.
5. What is function of receiver in case of reciprocating air compressor?
6. Estimate work required for compression of 1 kg of air for single stage, single acting air compressor.
7. What are advantages of multi-stage compression over single stage compression?
8. Explain what to mean by single acting and double acting air compressor.
9. Define volumetric efficiency of compressor. What factor affects the volumetric efficiency?
10. Prove that clearance does not affect the theoretical work needed to compress 1 kg of air.
11. State the methods to improve the volumetric efficiency of air compressor.
12. Draw P-V and T-S diagram for single acting reciprocating air compressor.
13. Derive the expression for work done in case of single acting single stage air compressor, when compression is (a) Isothermal (b) Polytrophic.
14. When multi-stage compression is used?
15. Explain the effect of inter cooling in a multistage reciprocating air compressor.
16. Describe briefly the working of two stage air compressor with inter-cooler. Draw the ideal P-V diagram.
17. Explain with P-V diagram the perfect and imperfect inter-cooling in two stage reciprocating air compressor.

18. Why cooling of air compressor during compression is essential?
19. What do you understand by isothermal efficiency of a compressor? Find expression for the same?
20. Explain the working of centrifugal compressor with neat diagram.
21. Explain with neat sketch the working of axial flow compressor.
22. Distinguish between:
    (i) reciprocating and rotary compressor
    (ii) centrifugal and axial flow compressor
23. Differentiate between a fan, a blower and a compressor.
24. Describe with neat sketch, the working of vane blower compressor and show its P-V diagram. For what application it is used?

Objective Type Question, Choose the Correct Answer

1. Which compression law gives minimum work input to compressor

   (a) $pv = c$ (b) $pv^n = c$

   (c) $pv^\gamma = c$ (d) none of above

2. For reciprocating air compressor, the law of compression desired is adiabatic when it is possible

   (a) very low sped

   (b) very high speed

   (c) at any speed as speed does not affect the compression law

   (d) none of the above

3. Work input of air compressor with 'n' as index of compression

   (a) increases with increase in index "n"

   (b) decreases with increase in value of "n"

   (c) remain same what ever the value of "n"

   (d) none of above

4. With increase in clearance volume, the ideal work of compression per kg of air

   (a) increases (b) decreases

   (c) remain same (d) none of above

5. Clearance ratio in case of reciprocating air compressor is given by

(a) $\frac{\text{clearance volume}}{\text{swept volume}}$ (b) $\frac{\text{Total volume of cylinder}}{\text{clearnace volume}}$

(c) $\frac{\text{clearance volume}}{\text{Total volume of cylinder}}$ (d) $\frac{\text{swept volume of cylinder}}{\text{clearance volume}}$

6. For some overall pressure ratio, the leakage of air past the piston, for multistage compression as compared to single stage compressor is

(a) more (b) less

(c) constant (d) none of the above

7. Why the inter cooling in multistage compression is done

(a) to minimize the work (b) to cool the air during compression

(c) to cool the air at delivery (d) none of the above

8. In air compressor installation where are separators generally used

(a) before inter cooler (b) after inter cooling

(c) between after cooler and receiver (d) before first stage suction

9. Why is an after cooler is used

(a) to reduce the volume of air

(b) to cool the air

(c) to remove impurities from air

(d) to cause moisture and oil-vapour to drop out

10. Which is a non-positive displacement compressor

(a) axial flow compressor (b) reciprocating compressor

(c) root blower (d) vane blower

11. In compressor installations the function of receiver is to

(a) collect compressed air

(b) to reduced temperature of compressed air

(c) to reduce moisture content of compressed air

(d) dampen pulsations or pressure wave in compressor discharge.

12. Why intake air filters are provided on compressors
    (a) to reduce the temperature of suction air
    (b) to reduce the pressure of suction air
    (c) to remove dirt and dust from suction air
    (d) none of the above
13. Capacity of compressor is expressed in
    (a) $\frac{m^3}{kg}$ (b) $\frac{kg}{m^3}$
    (c) $\frac{m^3}{min}$ (d) none of the above
14. The volume of the air delivered by compressor is called
    (a) compressor capacity (b) free air delivered
    (c) swept volume (d) none of above
15. At high altitude the compressor will draw
    (a) more power (b) less power
    (c) same power (d) none of the above
16. Which is a positive displacement compressor
    (a) axial flow compressor (b) centrifugal compressor
    (c) root blower (d) none of above
17. In order to check moisture troubles, the branch connection from compressor air line should be taken from which side of main
    (a) left (b) right
    (c) top (d) bottom
18. Even when the compressor is running at high speed, Isothermal compression efficiency can by approached by using
    (a) cold water spray (b) multi-stage compression
    (c) both (a) and (b) (d) none of the above
19. Volume of air delivered by compressor is called
    (a) compressor capacity (b) free air delivered
    (c) swept volume (d) discharge pressure

20. Increase in speed at a given pressure ratio in centrifugal compressor causes
    (a) decrease in efficiency (b) decrease in flow
    (c) increase in flow (d) increase in flow and decrease in efficiency
21. In two stage compressor what is optimum intermediate pressure
    (a) average of suction and delivery pressure
    (b) geometric mean
    (c) 45 percent difference of two
    (d) none of the above
22. Reciprocating compressors are employed to compress the air upto pressure of _______ bar
    (a) 20 bar (b) 40 bar
    (c) 80 bar (d) more than 100 bar
23. Regarding air receivers which of the following statement is incorrect
    (a) they are used to dampen pulsations
    (b) they improve the performance of compressor
    (c) they act as reservoir to take care of sudden demand
    (d) none of the above
24. In a centrifugal compressor the main loss is due to
    (a) impeller channel loss (b) diffuser losses
    (c) inlet losses (d) all of the above
25. Isothermal compression in multi-stage compressor is achieved by
    (a) running the compressor at very slow speed
    (b) constantly cooling the cylinder
    (c) employing inter cooling
    (d) insulating the cylinder

Fill in the Blanks

1. In case of reciprocating air compressor the value of index 'n' lies between ______ and ______
2. Mechanical efficiency of reciprocating air compressor is ratio of ______ to ______

3. In a reciprocating air compressor B.P is ______ than IP
4. In centrifugal air compressor flow is ______ to its axis
5. The clearance volume is provided in reciprocating compressor to______
6. Isothermal compression in reciprocating air compressor is possible when compressor runs______
7. ______ type of air compressor is used in aeroplanes
8. ______ is most efficient method of compressing air
9. The volume delivered by the compressor is called______
10. Rotary compressor is suited for ______ quantity of air at ______ pressure
11. When the intake temperature of compressor is ______ its capacity will be highest
12. Maximum work is required to compress the air when compression is______
13. Ratio of discharge pressure to inlet pressure in air reciprocating compressor is called ______
14. In ______ compressor the flow of air is perpendicular to the axis.
15. In centrifugal compressor ______ is the usual value of power input factor
16. In rotary compressor the ideal compression is______

Answer the objective type Questions

10.1	(a)	10.2	(b)	10.3	(a)	10.4	(c)
10.5	(a)	10.6	(b)	10.7	(a)	10.8	(d)
10.9	(b)	10.10	(a)	10.11	(d)	10.12	(c)
10.13	(c)	10.14	(a)	10.15	(a)	10.16	(c)
10.17	(c)	10.18	(c)	10.19	(a)	10.20	(c)
10.21	(c)	10.22	(b)	10.23	(d)	10.24	(b)
10.25	(c)						

Answers

Fill up the Blanks

1. 1 to 1.4
2. IP to BP
3. Greater
4. Perpendicular (Radial)
5. Allow for wear and mechanical freedom
6. very slow
7. Axial
8. Isothermal

9. Capacity of compressor
10. Large, low
11. Lower
12. Adiabatic
13. Compression ratio
14. Centrifugal
15. 1.3
16. Adiabatic

Exercises

1. A single stage, single acting air compressor deals with $85 m^3$ of free air delivered per hour at 1.01325 bar and 15°C. The pressure and temperature in the cylinder during the suction stroke remain constant at 1 bar and 40°C respectively. The index of compression is 1.22 and the delivery pressure is 5.5 bar. If mechanical efficiency is 85%, neglecting clearance volume, find the power required to drive the compressor.

   [6.107 kw]

2. A single acting single stage air compressor is belt driven from an electric motor at 300 R.P.M. The cylinder diameter is 20 cm and stroke is 24 cm. Air is compressed from one atmosphere to 8 atmosphere and law of compression is $PV^{1.25} = c$. Find power of electric motor if transmission efficiency is 96 percent and mechanical efficiency of compressor is 85 percent. Neglect clearance volume.

   [12.05 kw]

3. Single acting, single cylinder air compressor run at 100 RPM. Air is compressed from 1 bar to 8 bar. The stroke volume is $0.125 m^3$ and law of compression and expansion is $PV^{1.3} = c$. If clearance volume is 5% of stroke volume, calculate

   (i) volumetric efficiency

   (ii) volume of air taken per hour

   (iii) indicated power of the compressor

   [(i) 80.245% (ii) 10.031 m^3/min (iii) 44.55 kw]

4. In single stage, single acting air compressor 15 m^3 of air per minute at 100 kP_a and 297°C is compressed to 700 kP_a. Calculate the power required to drive the compressor. If the compression is

   (a) Isothermal

   (b) Polytropic with index 1.3

   (c) Isentropic

   Take $\gamma = 1.4$ for air neglecting clearance

   [(a) 48.648 kw (b) 61.533 kw (c) 65.1 kw]

5. Determine the size of the cylinder for a double acting air compressor of 40 kw indicated power, in which air is drawn in at 1 bar and 15°C and compressed according to law $PV^{1.2} = C$ to 6 bar. The compressor runs at 100 RPM with average piston speed of 152.5 m/min neglecting clearance. Assume stroke to diameter ratio of 1.5.

[D = 0.31m, L = 0.465 m]

6. A reciprocating air compressor delivers air at a constant pressure of 8 bar. The condition of air at inlet is one bar and 27°C. If index of compression is 1.3. calculate the W.D and heat-transferred during compression. Also find the work done during delivery. Assume mass flow rate as unity

[W.D = W_c = 176.75 kJ/kg of air, 44.19 kJ/kg, W.D during delivery = 139.12 kJ/kg of air]

7. Determine the kg of air dealt with per kwh by reciprocating compressor when initial temperature is 15°C. The index of compression is 1.3 and pressure ratio is 5. What will be the mass of air if compression is isothermal?

[mass of air/kwh = 22.35 kg/kwh, for isothermal m = 27.06 kg/kwh]

8. A single stage double acting air compressor takes air in at a pressure of one bar and deliver at 8 bar. The clearance is 5% of stroke. Assume compression and expansion to follow law $PV^{1.25} = C$. Find

   (i) volumetric efficiency of compressor

   (ii) value of free air delivered per kwh [(i) 78.6% (ii) 13.96 m^3/kwh]

9. A single stage, single acting reciprocating air compressor has bore of 200 mm and stroke of 300 mm. It receives air at 1 bar and 20°C and delivers it at 5.5 bar. If compression follows the law $PV^{1.3} = c$ and clearance volume is 5% of the stroke volume, determine

   (i) the mean effective pressure

   (ii) power required to drive the compressor, if it runs at 500 RPM.

[Pm = 1.807 bar, 14.19 kw]

10. An air compressor takes in air at 0.98 bar and 20°C and compresses it according to law $PV^{1.2} = c$. It then delivers to receiver at constant pressure of 9.8 bar.

    Determine

    (i) Temperature at end of compression

    (ii) W.D per kg of air

    (iii) The heat transfer during the compression

    (iv) W.D during delivery

    Take R = 287 J/kg K and $\gamma = 1.4$

[(i) 150°C (ii) 236 kJ/kg (iii) 118 kJ (iv) 123.7 kJ]

CHAPTER 11

Refrigeration and Air Conditioning

11.1 Refrigeration

(i) A Refrigeration is the technique of producing cold by artificial abstraction of heat so that the temperature below that of surrounding is produced in a substance as within space.

(ii) The Refrigeration may also be defined as process of removal of heat from a body colder than that of its surrounding with help of external work.

11.2 Refrigeration by Non-Cyclic Processes

Refrigeration is cooling of a system below the temperature of surrounding.

(i) ***Melting of ice:*** The melting of ice or snow was one of the easiest methods of refrigeration and is still employed. Ice melts at 0°C. So when ice is placed in given space warmer than 0°C, heat flows into ice and space is cooled or refrigerated. The latent heat of fusion of ice is supplied from surroundings and the ice changes its state from solid to liquid.

(ii) ***Solid Carbon dioxide:*** An other medium of refrigeration is solid carbon dioxide or dry ice. At atmospheric pressure CO_2 cannot exist in liquid state and when solid CO_2 is exposed to atmosphere, it sublimates i.e., it goes directly from solid to vapour by absorbing the latent heat of sublimation (620 kJ/kg at atm, –78.5 °C) from the surrounding. Thus dry ice is suitable for low temperature refrigeration.

In these two examples it is observed that refrigeration effect has been accomplished by non cyclic processes. In this method cooling substance is consumed and discarded. But in thermodynamic cycle the cooling substance is used again and again.

11.3 Refrigerator

(i) A Refrigerator is device used to remove heat from a body to produce low temperatures for useful purposes.

(ii) It is machine for producing cold. It is the reversed heat engine and it operates on the reversed heat engine cycles.

(iii) The amount of heat energy removed by refrigeration process in refrigerator is called refrigerating effect.

(iv) The basic principle of Mechanical refrigerator is that the latent heat is absorbed by fluid while changing its state from liquid to vapour and the heat is given up while condensing.

11.4 Unit of Refrigeration

1. We know that capacity of engine is given in horse power or kW. Similarly the capacity of refrigeration system is given in tones of refrigeration.

   A ton of refrigeration is defined as refrigerating effect (cooling) produced by melting of one ton of ice at 0 °C in 24 hours when temperature of water supplied is 0 °C.

   Taking one short ton = 900 kg where latent heat of ice = 335 kJ/kg

   $$\text{One ton of Refrigeration} = \frac{900 \times 335}{24 \times 3600} = 3.5 \text{ kJ/sec}$$

   $$= 3.5 \text{ kW}$$

   Sometimes the kW rating of motor for refrigerator is expressed in kW/ton of refrigeration $= \dfrac{3.5}{\text{C.O.P}}$.

11.5 Refrigerating Effect

(i) It may be defined as the amount of heat absorbed in evaporator to evaporate circulating refrigerant.

(ii) It may also be defined as amount of heat removed from the space to be cooled.

11.6 Co-efficient of Performance of Refrigerator (C.O.P)

(i) The C.O.P is defined as the ratio of refrigerating effect (N) or heat extracted at low temperature to external work done by refrigerator.

$$\therefore \quad \text{C.O.P} = \frac{\text{Refrigerating effect}}{\text{External work done}} = \frac{N}{W}$$

$$= \frac{\text{Heat extracted (Removed)}}{\text{External work done}} = \frac{Q}{W}$$

where the Q is quantity of heat removed by the refrigeration system and W is the work supplied to system.

(ii) ***Relative C.O.P*:** It is defined as the ratio of actual C.O.P to theoretical C.O.P.

$$\text{Relative C.O.P} = \frac{\text{Actual C.O.P}}{\text{Theoretical C.O.P}}$$

Refrigerant	C.O.P	Motor Horse Power Per ton
Carnot cycle	5.74 (highest)	0.82 (lowest)
R_{11}	5.09	0.93
R_{113}	4.92	0.96
Ammonia	4.76	0.99
R_{12}	4.7	1.0
R_{22}	4.66	1.01
Carbon dioxide	2.56 (lowest)	1.84 (highest)

11.7 Refrigerator and Heat Pump

Refrigerator is used to remove heat continuously from the space and maintain the temperature below surrounding temperature and reject the heat extracted from space into surrounding atmosphere.

Heat pump is used to supply the heat continuously to the space for maintaining the temperature above surrounding by absorbing the heat from surrounding which is at low temperature.

The Refrigerator and heat pump are shown in Fig. 11.1.

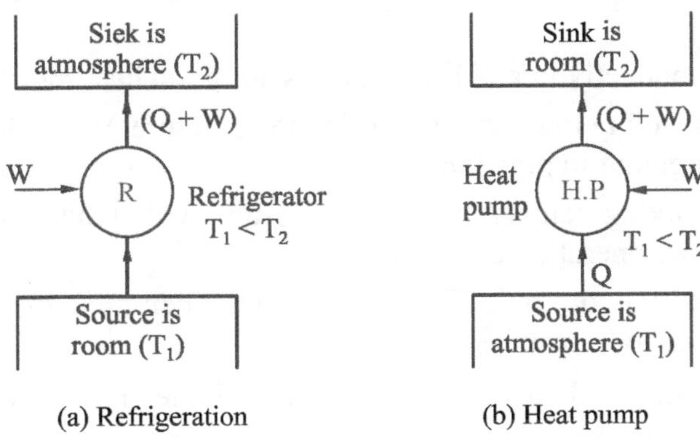

Fig. 11.1

$$\text{C.O.P} = \frac{\text{N}}{\text{W}} \text{ or } \frac{\text{Q}}{\text{W}} \quad \text{for refrigeration system}$$

$$\text{C.O.P} = \frac{\text{Q}+\text{W}}{\text{W}} \quad \text{for heat pump}$$

$$= \frac{\text{Q}}{\text{W}} + 1$$

(C.O.P) Heat pump = 1 + C.O.P of refrigerator

11.8 Refrigerants

It may be defined as substance (liquid form) which is used for heat transfer in refrigeration system. It absorbs heat by evaporation at low temperature and pressure (evaporator or space) and gives up this heat in condenser by condensing at higher temperature and pressure.

Properties and Selection of Refrigerant

The working fluid used in refrigerator is known as the refrigerant.

A good refrigerant should possess the following properties. The selection of the refrigerant is based on these properties.

1. High latent heat of evaporation, to minimize the flow of refrigerant. This reduces initial cost on refrigerant and size of the system will be small which further reduces the cost.
2. Critical pressure well above maximum working pressure.
3. Critical temperature well above maximum working temperature.
4. Low specific volume
5. High vapour and liquid densities.
6. High C.O.P
7. High thermal conductivity and low viscosity for good heat transfer.
8. Freezing point well below lowest working temperature, this will avoid freezing and logging of refrigerant in pipe line.
9. Stability at working temperature and pressure with tendency to breakdown or decompose under operating conditions.
10. Non-Poisonous and non toxic. Toxic refrigerant when mixed with small percentage of air is injurious to human being.
11. Non-inflammables and non-explosive to avoid the danger of explosion or fire hazard during high compression and over heated conditions.
12. Reaction with material of construction. Certain metals are attacked by the refrigerant. Ammonia reacts with copper, brass or other cuprous alloys in the

pressure of water. Therefore in ammonia system the common metals used are iron and steel.

Freon group does not react with steel, copper, brass, zinc, tin and aluminum. But it tends to dissolve natural rubber in packing and gaskets. But synthetic rubber such as neoprene is entirely suitable.

13. Leak tendency should be low and detection of leak should be easy. The tendency of refrigerant to leak should be as low as possible and detection of leaks should be quite easy so as to avoid loss of refrigerant from the system. The ammonia leak is easily detected by odour. Burning sulphur in presence of ammonia will form white fumes which will pin out the leak.

    For odour less refrigerants, usual method of detecting leaks is to observe the flame of an alcohol or prestolite gas torch, if there is no leakage of refrigerant flame is light blue, but in presence of refrigerant it blazes in to blue-green colour.

    An electronic detector is available which is many times more sensitive than torch method.

14. ***Action with Oil*:** There should not be any chemical reaction between refrigerant and oil of the compressor. Secondly miscibility of the oil i.e., the ability of refrigerant to dissolve with oil and vice-versa is quite important, as some oil should be carried out of compressor crank-case with hot refrigerant vapour to lubricate piston and discharge valves properly.

15. Cost of refrigerants should be reasonable and it should be available in sufficient quantity.

    The relative cost of three most common refrigerants ammonia, refrigerant 12 and refrigerants 22 is given below.

Ammonia	Cheapest
F_{12}	More expensive
F_{22}	Most expensive

16. Damage to refrigerated products

    In case of leakage of the refrigerant from the system, it will come in contact with products. Some products contain moisture and we knew ammonia dissolve in water, spoil the test and may cause toxicity. The alogenated hydrocarbons have no effect upon food and fabrics.

From Table it may seen that:

(i) CO_2 operates at extremely high pressure thus requiring strong metal for compressor, piping and vessels. On other hand R_{11} and R_{113} operating below atmospheric pressure where some equipment is required to purge air from the system.

(ii) The high value of flow of refrigerant in m^3 per hour per ton for R_{11} and R_{113} indicate why they are used predominantly in centrifugal compressor system.

Refrigerant	Freezing temperature in °C	Volume flow per ton per hour at evaporative temperature –15° C	Evaporative pressure at 15° C (in bar)	Condenser pressure at 29 °C (in bar)
Ammonia	–77.8	1.56	2.3	11.3
Carbon dioxide	–56.7	5.58	23.25	69.85
R_{11}	–111	5.9	0.2016	1.26
R_{12}	–157.8	9.45	1.77	7.18
R_{22}	–160	5.58	2.97	12.03
R_{113}	–13	1.64	0.069	0.542
Water	0	-	-	-

11.9 Vapour Compression System Versus Vapour Absorption System

The Main points of difference between a vapour compression system and vapour absorption system are given below.

Vapour Compression System	Vapour Absorption System
1. System has more wear and tear and noise due to moving parts of the compressor.	1. The system is quite in operation i.e., no noise.
2. Energy supplied is high grade energy i.e., mechanical energy is supplied.	2. Energy supplied is mainly heat i.e., low grade energy.
3. Amount of energy supplied is less. It is only ½ to ¼ of refrigeration effect.	3. Energy is supplied is high, about 1½ times the refrigeration effect.
4. System has poor performance at partial loads.	4. Performance of system is not affected by load variation.
5. The system is used where high grade energy is available.	5. The system can also be used at remote places as it can work even with a simple kerosene oil lamp, of course in small capacities.
6. Charging of refrigerant is quite simple.	6. Charging of system is difficult.
7. There are more chances of leakage of refrigerant from the system.	7. There is no compressor or any reciprocating components to cause leakage.
8. Liquid traces in suction line may damage the compressor	8. Liquid traces of refrigerant present in piping at exit of evaporator or in suction line constitute no danger.

11.10 Application of Refrigeration

(a) Industrial

1. Ice plants for making ice.
2. For liquifying gases and vapour, in chemical and pharmaceutical industries.
3. Treatment of metals and machine tools industries for different purposes.
4. Food processing and preservation industries.
5. Cotton mills and textile mills.
6. For cold storage.

(b) Commercial

1. For cooling of storage chamber for hospitals, retail shops, restaurants, hotels, for preserving perishable items such as Food, Vegetables, Milk, Medicines etc.

(c) Transportation and Marine

1. For transportation of perishable goods such as milk, food, fruits, vegetables etc.
2. For cold storage of fish on boats.

(d) Domestic

Home refrigerators are for preserving food and perishable items and water coolers.

(e) Air Conditioning

1. Human comfort, air conditioning of rooms, houses, buildings, schools, offices, shops, factories, buses and cars, theatres etc.
2. Air conditioning for machine comfort such as computer cabin, T.V. shops, telephone houses, process factories etc.

11.11 Classification of Refrigeration System

1. Natural Refrigeration
2. Chemical Refrigeration
3. Mechanical Refrigeration
   - (a) Air Refrigeration
   - (b) Vapour Refrigeration
     - (i) Vapour Compression System
     - (ii) Vapour absorption System
4. Heat energy Refrigeration
   - (a) Steam jet Refrigeration
   - (b) Vapour absorption Refrigeration

These above methods are discussed below.

1. ***Natural Refrigeration***

   (i) In this system, cooling substance like ice or cold water are kept in contact with material for cooling.

   (ii) The latent heat for melting of ice is supplied to the material (substance) and it gets cool.

   (iii) Similarly the material can be cooled by extracting heat for sublimation of solid CO_2.

2. ***Chemical Refrigeration***

   Cooling effect may be obtained by chemical means, in which a chemical reaction is carried out which absorbs heat for completion. This heat required for the purpose is taken from substance or space to be cooled. This method is impracticable on commercial scale.

3. ***Mechanical Refrigeration***

   (i) In this, the cooling effect is achieved by expenditure of mechanical work.

   There are two types of mechanical refrigeration such as:

   (a) Gas or Air refrigeration where gas, generally air is used as working fluid.

   (b) Vapour Refrigeration where ammonia or R_{12}, R_{22} and similar other refrigerants are used as working fluid.

      (i) Vapour compression method in which vapour is first compressed by using compressor.

      (ii) Vapour absorption method in which vapour is first absorbed by the water.

11.12 Domestic Refrigerator

1. Domestic Refrigerator is also known as household Refrigerator, it works on a vapour compression cycle.
2. In Modern domestic refrigerators, we use a capillary tube, in place of expansion valve. It is small diameter tube which carries refrigerant from condenser to the low temperature coil of the evaporator. Tube is in contact with suction line serves a dual function of reducing the pressure of refrigerant by throttling action and sub cooling it by giving its heat to the vapour going to the suction of compressor.

3. Basic components of system are compressor, condenser, capillary and evaporator and receiver (accumulator).

   The Compressor is directly coupled to electric motor and both are scaled in steel box. Therefore it is known as sealed unit. This arrangement helps to reduce the noise of motor, prevents the leakage of refrigerant and motor is cooled by refrigerant vapour as motor is exposed to vapour refrigerant.

   Condenser: It is generally made up of grid of copper tubes provided with fins and mounted behind the unit (cabin-box) and cooled by the natural air convection.

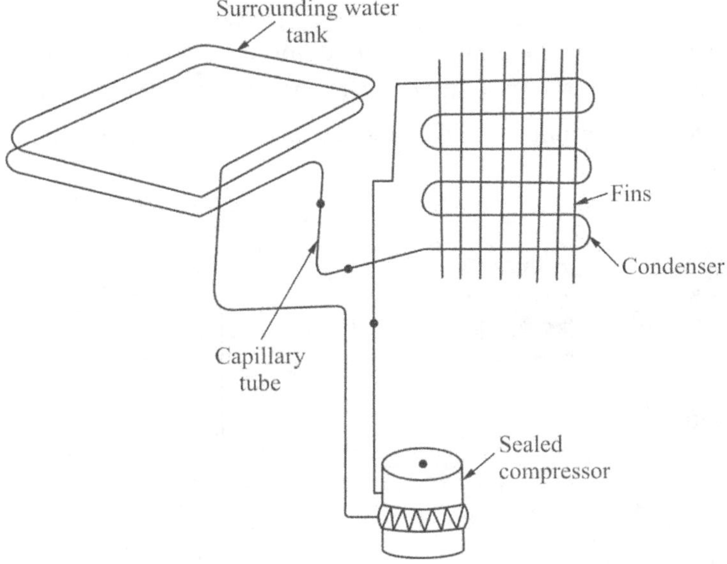

Fig. 11.2

   Capillary: It is used as an expansion or throttling device for reducing temperature and pressure of refrigerant.

4. ***The Refrigerator***: It is a box, forms the evaporator part of the system and it is heavily insulated with thermo Cole to reduce the heat flow from surrounding to the evaporator (Fig. 11.2). This refrigerator is generally divided into three compartments. The top compartment works as freezer where temperature of the refrigerant is minimum (–10 °C to –18 °C) as the refrigerant first enters into this compartment. Other middle and bottom, two compartments are known as cold chambers. These two chambers or compartments are cooled by natural

convection as the cooled air in freezer descends and then goes up after absorbing the heat from the middle and bottom compartments.

5. The top compartment of the refrigerator is used for making ice, ice cream etc. The middle and bottom compartments are used for preserving milk, butter, curd vegetables and fruits.
6. The door of the refrigerator is also provided with space which is used for putting the water bottles, cool drinks, sauce and jam bottles. The top part of door is used for putting the eggs.
7. The control used is on-off control and it is operated automatically by thermostat valve.
8. The refrigerant F_{12} is only used in domestic refrigerator as it is most safe refrigerant. It does not spoil the foods even it comes in contact, due to leak.
9. The power consumption of 165 liters capacity refrigerator is about 60 to 80 kW per month.

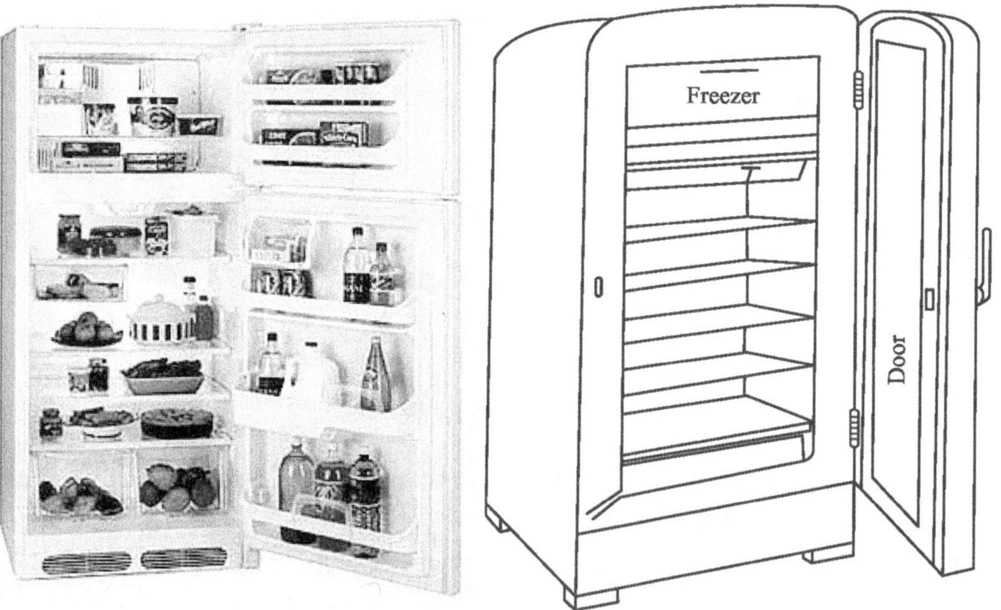

Fig. 11.3　　　Fig. 11.4

Some important points regarding household Refrigerators

1. They are highly durable and reliable, providing satisfactory service for over 15 years.

2. Household refrigerator is actually combination of refrigerator and freezer since it has a freezer compartment to make ice and to store frozen food.
3. Today's refrigerator use much less energy as a result of using smaller and higher efficiency motors and compressors, better insulation materials, larger coil surface areas and better door seals as shown in Fig. 11.3.
4. Typical operating efficiency for a freezer temperature of – 18 °C and ambient temperature of 32 °C.

   Vapour compression C.O.P = 1.3

   Absorption refrigeration C.O.P = 0.4

   Thermo-electric refrigeration C.O.P = 0.1
5. All Full-size refrigerators are designed to maintain the freezer section at –18 °C and the refrigerator section at 3 °C.
6. All refrigerators have large air tight drawer for leafy vegetables and fresh fruits to seal in moisture and protect them from drying effect of cool air circulating in refrigerator. Covered egg compartment in the lid extends the life of eggs by slow down the moisture loss from the eggs. Some models have temperature controlled meat compartment maintained at – 0.5 °C which keeps meat at safe temperature without freezing it and thus extending its storage life.
7. The more expensive models come with an automatic ice maker located in freezer section that is connected to the water line. A typical ice maker can produced 2 to 3 kg of ice per day and store 3 to 5 kg of ice in removable ice storage container.
8. Household refrigerators consume about 90 to 600 w of electrical energy when running and are designed to perform satisfactorily in environments up to 43 °C. Refrigerator runs intermittently, about 30 percent of time under normal use in house at 25 °C.
9. The total food storage volume has been increased over the years without an increase in the external dimensions by using thinner but more effective insulation and minimizing the space occupied by compressor and condenser. Switching from the fiber glass insulation (Thermal conductivity k = 0.032 – 0.040 w/m °C) to expanded in place urethane (k = 0.019 w/m °C) made it to reduce wall thickness from about 98 to 48 mm for freezer section and from about 70 mm to 40 mm for refrigerator section. The entire shell of refrigerator must be carefully sealed to prevent any water leakage or moisture migrated into the insulation since moisture degrades effectiveness of insulation.

10. ***Open the refrigerator Door, the fewest times*** possible for shortest duration. Each time the refrigerator is opened, the cool air inside refrigerator is replaced by the warmer air outside, which will increase the cooling load on refrigerator.
11. ***Cool the hot food to room temperature first***, before putting them into refrigerator. Moving a hot pan from the oven directly into the refrigerator not only wastes energy by making the refrigerator to work longer, but it also causes the near by perishable food to spoil by creating a warm environment in its immediate surrounding.
12. ***Clean the condenser coil*** located behind the refrigerator. The dust and grime that collect on the coil act as insulation that slows down heat dissipation through them. Cleaning coil a couple of times a year with damp cloth or vacuum cleaner will improve cooling ability of refrigerator while cutting down power consumption a few percent. Sometimes a fan is used to cool the condenser of large refrigerator, and the strong air motion keeps the coil clean.
13. ***Check the door gasket for air leakage.*** This can be done by placing a flash light in to the refrigerator, turning off the kitchen lights and looking for light leaks. Heat transfer through the door gasket region and thus any defective door gaskets must be repaired immediately.
14. ***Avoid Unnecessarily Low Temperature*:** Recommended temperature for freezer and refrigerators are –18 °C and 3 °C respectively. Setting the freezer temperature below – 18 °C adds significantly to energy consumption but does not add much to the storage life of frozen foods. Keeping temperature 6°C below recommended levels can increase the energy use as much as 25 percent.
15. ***Avoid excessive ice build up on the interior surfaces of the evaporator:*** The ice layer on surface acts as insulation and slow down the heat transfer from the freezer section to refrigerant. The refrigeration should be defrosted by manually turning off the temperature control switch when the ice thickness exceeds few milli meter. Defrosting is done automatically in no-frost refrigerator by supplying heat or by hot refrigerant gas, for short period. The water is than drained to pan outside, where it is evaporated using heat dissipated by condenser. The no-frost evaporators are basically finned tubes subjected to air flow circulated by a fan; practically all the frost collects on fins, which are the oldest surfaces, leaving the exposed surfaces of the Freezer-section and fronzen food forst-free.

16. ***Use of Power-Saver Switch:*** The switch controls the heating coil and prevents condensation on outside surfaces in humid environments. The low voltage heaters are used to raise the temperature of outer surfaces of refrigerator at critical locations above dew point in order to avoid water droplets, forming on surfaces and sliding down. Condensation is most likely to occur in summer in hot and humid climates in homes without air conditioning. The moisture formation on surfaces is undesirable since it may cause the painted finish of the outer surface to deteriorate and it may wet the kitchen floor. About 10 per cent of total heat-energy consumed by the refrigerator can be saved by turning this heater off and keeping it off unless there is visible condensation on outer surfaces.
17. Do not block the air flow passages to and from the condenser coil of the refrigerator. The heat dissipated by the condenser to the air is carried away by air that enters through the bottom and sides of refrigerator and leaves through the top. Any blockage of this nature-convection air circulation path by large objects such as several cereal boxes on the top of the refrigerator will degrade the performance of the condenser and thus the refrigerator.

11.13 Water Cooler

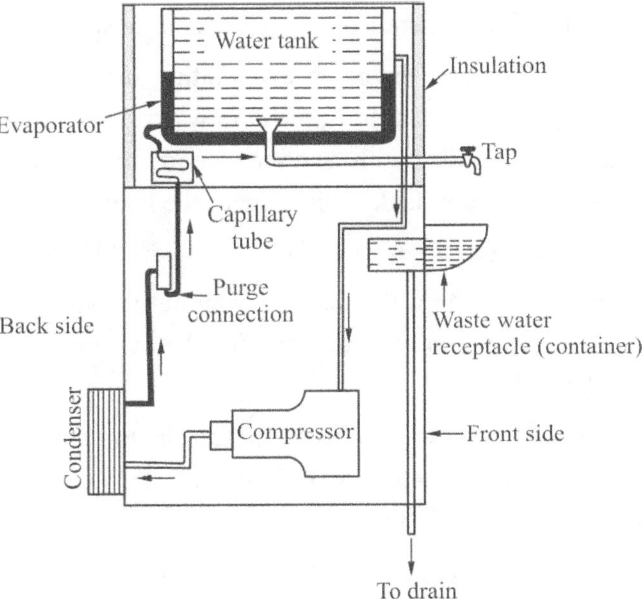

Fig. 11.5

1. Water cooler uses vapour compression refrigeration system.
2. Basic components of water cooler are compressor, condenser, capillary, evaporator, water tank, tap connection with water tank, top cover, M.S cabin, insulator, waste water receptacle (container) etc.
3. The compressor is mounted at bottom and condenser with fins is mounted at backside of cooler. The capillary is near the water tank. Hallow space around the water tank works as evaporator.
4. Refrigerant vapour usually F_{12} is compressed in the compressor. Its temperature and pressure is raised to such a point, so that it may be easily condensed in condenser with normal available condensing media (water or air). It maintains continuous flow of refrigerant through the system.
5. Discharge line conveys the high pressure and temperature vapour refrigerant from compressor to condenser.
6. Condenser is to provide heat transfer surface through which heat passes from the refrigerant to condensing medium which is either water or air.
7. Liquid line carries liquid refrigerant from condenser to capillary tube where its pressure and temperature is reduced considerably so that refrigerant may take sufficient amount of latent heat from the refrigerating space during evaporation.
8. The vapour liquid mixture enters in evaporating tubes around the water tank. Where it absorbs its latent heat from water in water tank and water is cooled. The water box (tank) is provided with heavy insulation to prevent the heat flow from the surroundings into water tank. This reduces the load on the system.

   Generally sealed unit (compressor and motor enclosed in closed vessel is used).
9. The cooled water from the water tank is taken out through a water tap as shown in Fig. 11.5 Water receptacle is provided below tap and water falling in receptacle is drained.
10. Water can be filled in the water tank by lifting the top cover on the cooler or by providing water inlet connection with float control. Water coolers of 100 to 200 liters capacities are available in market.
11. The control used is on-off control and is operated automatically by thermostatic valve.

Vapour Compression Refrigeration System

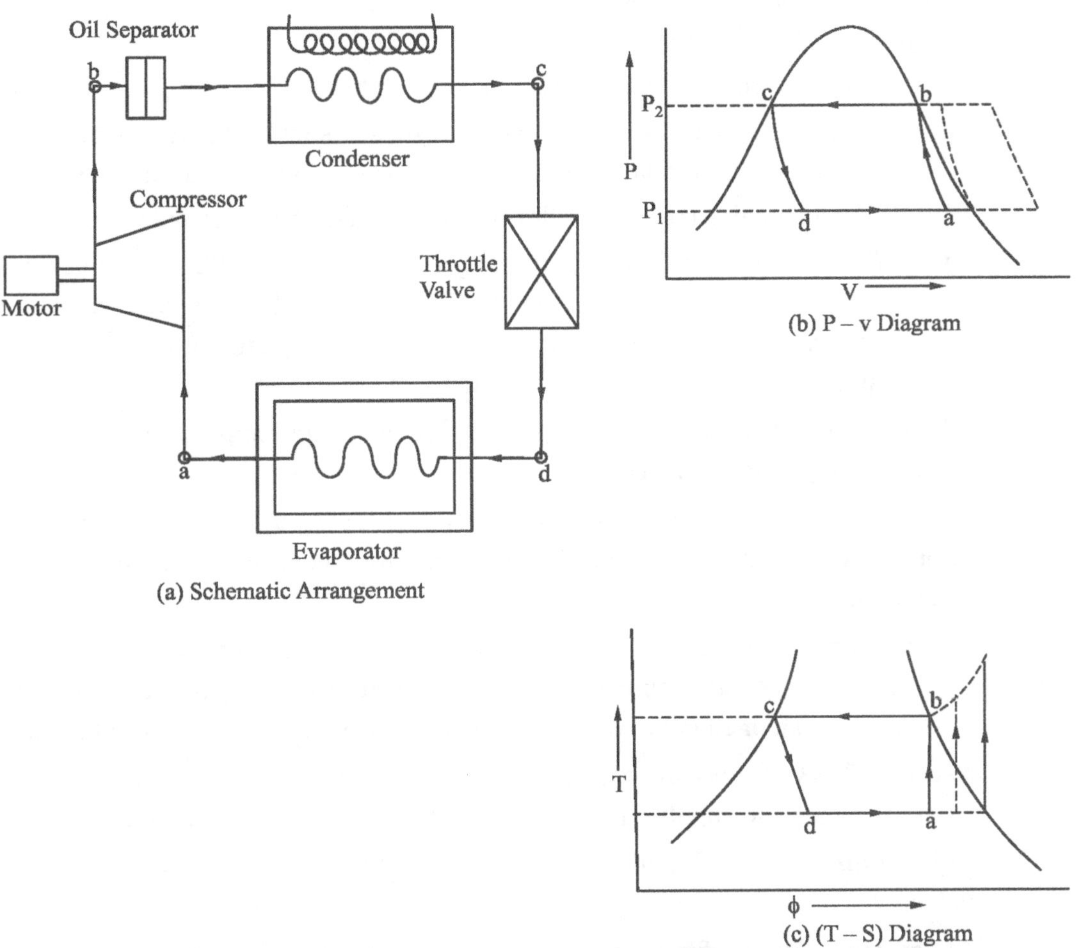

Fig. 11.6(a) Vapour Compression Refrigeration.

The Principal parts of simple vapour refrigeration system and their functions and working.

1. ***Evaporator*:** Its function to provide a heat transfer surface through which heat can pass from the refrigerated space into vaporizing refrigerant. Common type evaporator is coil brazed on to the plate called plate evaporator. In a flooded evaporator the coil is filled only with a liquid refrigerant. In an indirect expansion coil, water (up to 0 °C) or brine (between 0 °C and – 21 °C) may be chilled in evaporator. The chilled water or brine may then be used to cool some other medium.

2. ***Suction Line*:** It carries the low pressure vapour from the evaporator to the suction inlet of the compressor.
3. ***Compressor*:** The function of compressor is:
   (i) To draw refrigerant vapour from the evaporator
   (ii) To raise its temperature and pressure to such a point so that it may be easily condensed with normally available condensing media i.e., air or water.
   (iii) It also maintains continuous flow of the refrigerant through the system.
   (iv) When the volume flow rate of refrigerant is large centrifugal compressors are used. Rotary compressors are used for small units, Reciprocating compressors are used in plant up to 100 ton capacity and for plant of higher capacity centrifugal compressor is used.
4. ***Discharge Line*:** It conveys the high pressure and high temperature refrigerant from the compressor to the condenser.
5. ***Condenser*:** The function of condenser is to provide a heat transfer surface through which heat passes from the refrigerant to condensing medium, is either water or air.

   It must first de-superheat and then condense the compressed refrigerant. Condenser may be either air cooled or water cooled. Air cooled condenser is used in small self contained unit. Water cool condensers are used in large installation.
6. ***Receiver Tank*:** It acts as reservoir which stores the liquid refrigerant coming from condenser and supplies it to the evaporator according to requirements.
7. ***Liquid Line*:** It carries liquid refrigerant from receiver and convey it to expansion valve.
8. ***Expansion valve:*** Its function is to reduce the pressure of liquid refrigerant and also to regulate the flow of refrigerant to evaporator.

   Two widely used types of expansion device are (i) capillary tube (ii) throttle valve (thermostatic expansion valve).

Its function is to reduce the pressure of liquid refrigerant and also to regulate the flow of refrigerant to evaporator.

Two widely used types of expansion devices are

(i) Capillary tube

(ii) Throttle valve (Thermostatic expansion valve).

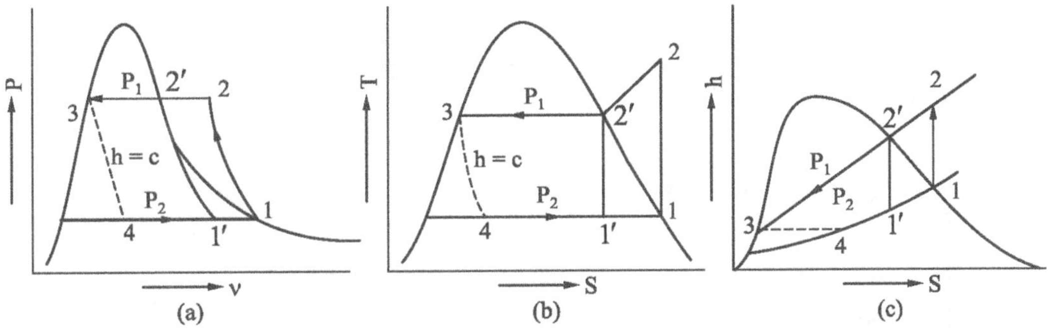

Fig. 11.7 Vapour compression refrigeration cycle – properly diagrams.

Working

The various operations represented are as follows for idealized plant.

1. ***Compression*:** A reversible adiabatic process 1-2 or $1'-2'$ either starting with saturated vapour (state 1) called dry compression or starting with wet vapour (state $1'$) called wet compression. Dry compression 1-2 is always preferred to wet compression $(1'-2')$ because with wet compression there is danger of liquid refrigerant being trapped in head of the cylinder (of compressor) by the rising piston which may damage the valve or the cylinder head and the droplets of liquid refrigerant may wash away the lubricating oil from walls of cylinder thus accelerating wear.
2. ***Cooling and Condensing*:** Reversible constant pressure 2-3, first vapour refrigerant is de-superheated and then condensed by cooling water or by atmospheric air ending with saturated liquid. Heat Q_1 is transferred to or rejected to cooling medium.
3. ***Expansion*:** An adiabatic throttling process 3-4. During throttling enthalpy remain constant. State 3 and 4 are equilibrium points. Process 3-4 is adiabatic then only $h_3 = h_4$ by S.F.E.E but not isentropic. Hence it is irreversible and cannot be shown in property diagram states 3 and 4 have simply joint by dotted line.
4. ***Evaporation*:** Constant pressure reversible process 4-1 which completes the cycle. The refrigerant throttle by the expansion valve to a pressure, the saturation temperature at this pressure is below the temperature of surroundings. Heat flows by virtue of temperature difference from the surroundings which get cooled and refrigerant gets evaporated absorbing the heat of evaporation. The evaporator thus produces the cooling or refrigerant effect, absorbing heat Q_2 from the surroundings by evaporation.

Performance and Capacity of a Vapour Compression Plant

Fig. 11.8 shows the simplified diagram of a vapour compression refrigeration plant.

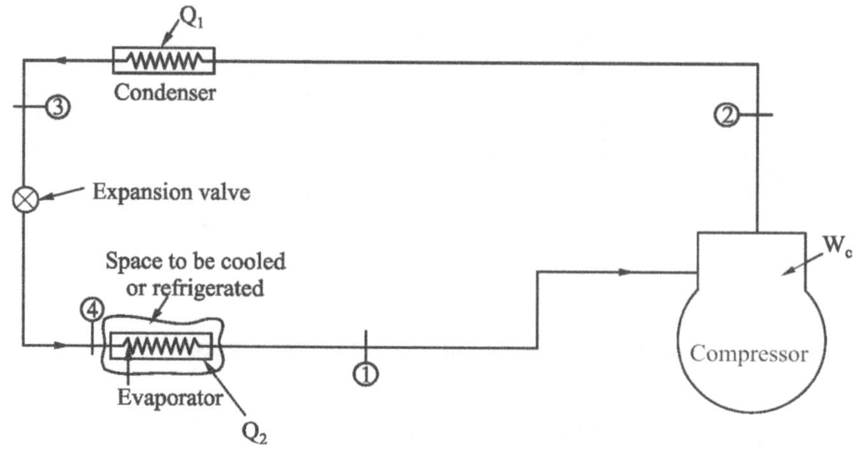

Fig. 11.8 Vapour compression plant.

When steady state has been reached, for 1 kg flow of refrigerant through the cycle, the steady flow energy equations (neglecting K.E. and P.E. changes) may be written for each of the components in the cycle as given below.

Compressor $h_1 + W_c = h_2$

$\therefore \quad W_c = (h_2 - h_1)$ kJ/kg .....(i)

Condenser $h_2 = Q_1 + h_3$

$\therefore \quad Q_1 = (h_2 - h_3)$ kJ/kg .....(ii)

Expansion valve $h_3 = h_4$

or $(h_f)P_1 = (h_f)P_2 + x_4\,(h_{fg})P_2$

where P_1 is condenser pressure and P_2 is evaporator pressure

$$x_4 = \frac{(h_f)_{P_1} - (h_f)_{P_2}}{(h_{fg})_{P_2}} \qquad .....(iii)$$

This is the quality of the refrigerant at the inlet to the evaporator (mass fraction of vapour in liquid-vapour mixture).

$(h_f)_{P_1}$ = enthalpy of liquid at pressure P_1

$(h_f)_{P_2}$ = sensible heat at P_2

$(h_{fg})_{P_2}$ = latent heat at pressure P_2

Evaporator

$$h_4 = Q_2 + h_1$$

$$Q_2 = (h_1 - h_4) \qquad ..... (iv)$$

Q_2 is known as the refrigerant effect

$$\therefore \quad C.O.P = \frac{Q_2}{wc} = \frac{h_1 - h_4}{h_2 - h_1}$$

If m is the mass flow of refrigerant in kg/s, then the rate of heat removed from space

$$= m(h_1 - h_4) \quad KJ/sec$$

$$= m(h_1 - h_4)\ 3600 \quad KJ/hr$$

One ton of refrigerant is defined as the rate of heat removed from surroundings equivalent to the heat required for melting 1 ton of ice in one day.

If latent heat of fusion of ice is taken as 336 KJ/kg

Then 1 ton is equivalent to heat removal at rate of $\frac{(1000 \times 336)}{24} = 14000$ KJ/hr

$$\therefore \text{ Capacity of refrigerant plant } = \frac{m(h_1 - h_4) \times 3600}{14000} \text{ tons} \quad .....(v)$$

The rate of heat removed in condenser $= m(h_2 - h_3)$ KJ/sec .....(vi)

$$Q_1 = m(h_2 - h_3) = m_c\, c_c\, (t_{c_2} - t_{c_1}) \quad .....(vii)$$

where m_c = Mass flow rate of cooling water in condenser in kg/sec

$(t_{c_2} - t_{c_1})$ = rise in temperature of water.

C_c = Specific heat of water

Rate of work input to compressor $= m(h_2 - h_1)$ KJ/sec .....(viii)

11.15 Simple Vapour Absorption System

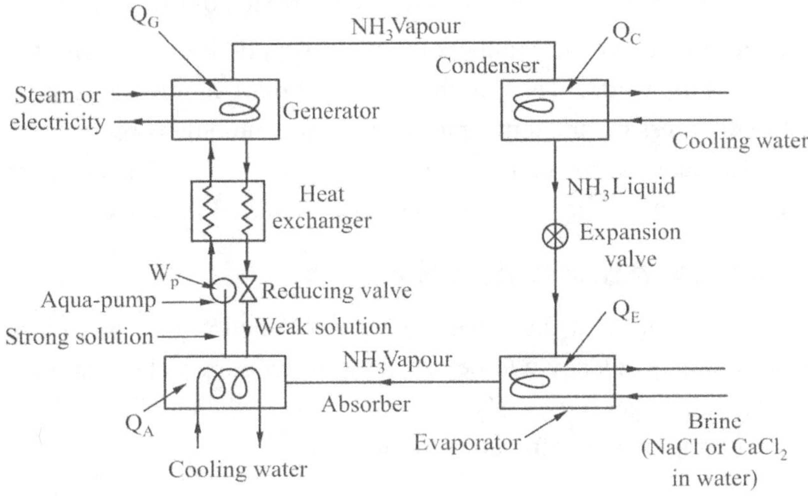

Fig. 11.9 Vapour Absorption Refrigeration Plant–flow Diagram.

If the compressor in vapour compression system were replaced by generator, pump, and absorber assembly the result will be simple absorption system.

Simple vapour absorption system consists of :

1. Evaporator
2. Absorber
3. Pump
4. Generator
5. Condenser
6. Receiver
7. Expansion valve
8. Heat exchanger

The low pressure Ammonia Vapour refrigerant from evaporator is absorbed in absorber by weak solution of refrigerant in water. The absorption of ammonia vapour lower the pressure in absorber, which in turn draws more ammonia vapour from evaporator. Water cooling is employed in absorber to remove heat of solution in absorber this increases the absorption capacity of water because at high temperature, water absorbs less ammonia vapour. The solution in absorber is called strong solution because it is rich in ammonia refrigerant.

The pump draws strong solution from absorber and increases the pressure up to 10 bar and forces the strong solution in the generator.

In the generator, the strong solution of ammonia is heated by some external source such as gas or steam or electricity. In heating process the ammonia vapour is given out of solution because low boiling point of ammonia, leaving behind, in generator a weak solution. The weak solution flows back to absorber through a reducing valve which maintains the pressure difference between high and low side of the system .

From the generator the ammonia vapour is supplied to condenser where it is condensed by cooling water. Then high pressure liquid ammonia is passed through throttle valve where pressure and temperature of liquid ammonia is reduced. From throttle valve liquid refrigerant enters in to evaporator where it absorbs its latent heat, thus producing cooling effect and liquid refrigerant gets converted in vapour refrigerant.

11.16 Practical Absorption System

The simple absorption system discussed above can function and provide refrigerating effect but its operating efficiency will be very low. In order to make it more practical it is fitted with heat exchanger, an analyzer and rectifier as shown in Fig. 11.6 these accessories help to improve the efficiency of the system.

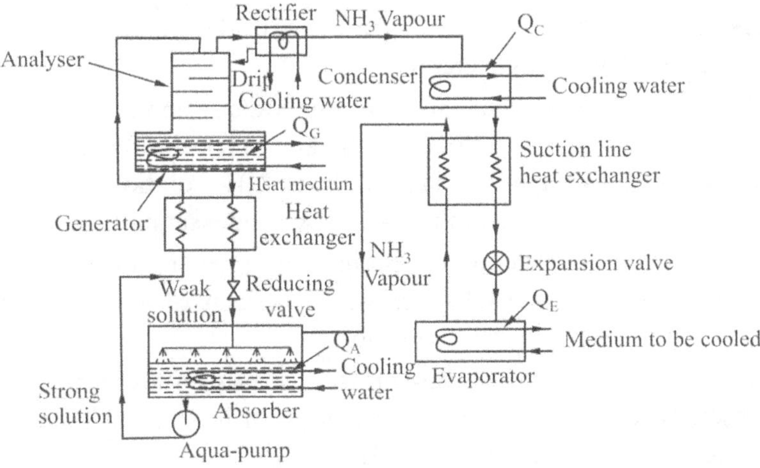

Fig. 11.10 Actual Vapour Absorption Refrigeration Plant with Analyzer and Rectifier.

Working of Practical (Actual) Absorption System

(i) Evaporator: The wet vapour of NH_3 from throttle valve enters into evaporator where it absorbs latent heat and dried up at low pressure to form NH_3 vapour.

(ii) Absorber: The low pressure NH_3 vapour (Dry) coming out from evaporator is absorbed in the absorber by weak solution of refrigerant in water. Absorption of ammonia vapour lower the pressure in absorber which in turn draws more and more NH_3 vapour from evaporator.

The absorption process gives out heat and increases the temperature of solution. But increase in temperature reduces the absorption. Therefore solution is cooled by removing heat of solution by the circulation of cooling water.

(iii) Pump: The strong NH_3 solution leaves the absorber and enters the pump where its pressure is increased to about 10 bar. The compression work is very small because of high density of liquid solution as compare to vapour.

(iv) Heat Exchanger: Location of heat exchanger between the generator and absorber is ideal. The strong solution pumped from absorber to the generator must be heated and hot weak solution from generator to the absorber must be cooled. The heat exchanger between two streams reduces both the cost of heating the generator and cost of cooling the absorber, weak solution is always hot solution from absorber.

(v) Generator: The warm and high pressure solution of ammonia refrigerant enters the generator. It is heated by steam, gas or electricity. In heating process the NH_3 vapour is driven out of solution because of low boiling point of NH_3 than water. The remaining weak NH_3 solution is first throttled and return to the absorber through heat exchanger.

(vi) **Analyser:** The analyser is direct contact heat exchanger, consisting of a series of trays mounted above the generator.

The NH_3 vapour from generator contains water vapour which may freeze in expansion valve and choke the pipe line and block the flow. To avoid this NH_3 vapour is passed through a stream of strong NH_3 solution from absorber in the analyser. This removes most of the water by condensation from vapour.

(vii) **Rectifier:** The final reduction or elimination of water vapour occurs in rectifier. It is water cooled heat exchanger which condenses water vapour (and some ammonia) and returns it to the generator.

(viii) **Condenser:** The pure NH_3 vapour is then passed from the rectifier to the condenser. Latent heat of NH_3 vapour is given to condenser cooling water and NH_3 vapour is condensed to liquid NH_3.

(ix) **Throttle valve:** The high pressure liquid NH_3 is throttled to very low pressure by throttle valve. This forms wet vapour of NH_3 at low temperature and pressure.

This system is simple and low in operating cost but it requires large amount of cooling water and low pressure steam for heating.

The net refrigerating effect of such a machine is the heat extracted in the evaporator. The total energy supplied for operating the machine is the sum of work done by the pump and heat supplied in generator.

$$\therefore \qquad \text{c.o.p} = \frac{\text{Heat extracted from the evaparator}}{\text{W.D by pump} + \text{Heat supplied in generator}}$$

11.17 Air Refrigerator Working on a Bell Coleman Cycle or Reversed Joule or Brayton Cycle

Bell Coleman refrigerator may work on any one of the following two methods.

(i) ***Open System*:** In this system the air comes in direct contact with refrigerated chamber or the product to be cooled in the cold chamber.

(ii) ***Closed System or Dense Air System*:** In this system the air is passed through the coils placed in chamber and thus abstracts heat from cold chamber without coming in direct contact with the product kept therein. In some cases, the coils carrying cold air are immersed in brine (aquous solution of NaCl or $Cacl_2$ in water). For 23% NaCl solution the freezing point is (– 21°C). The cool air from expander extracts heat from brine. The resulting chilled brine is then circulated around the chamber to be cooled.

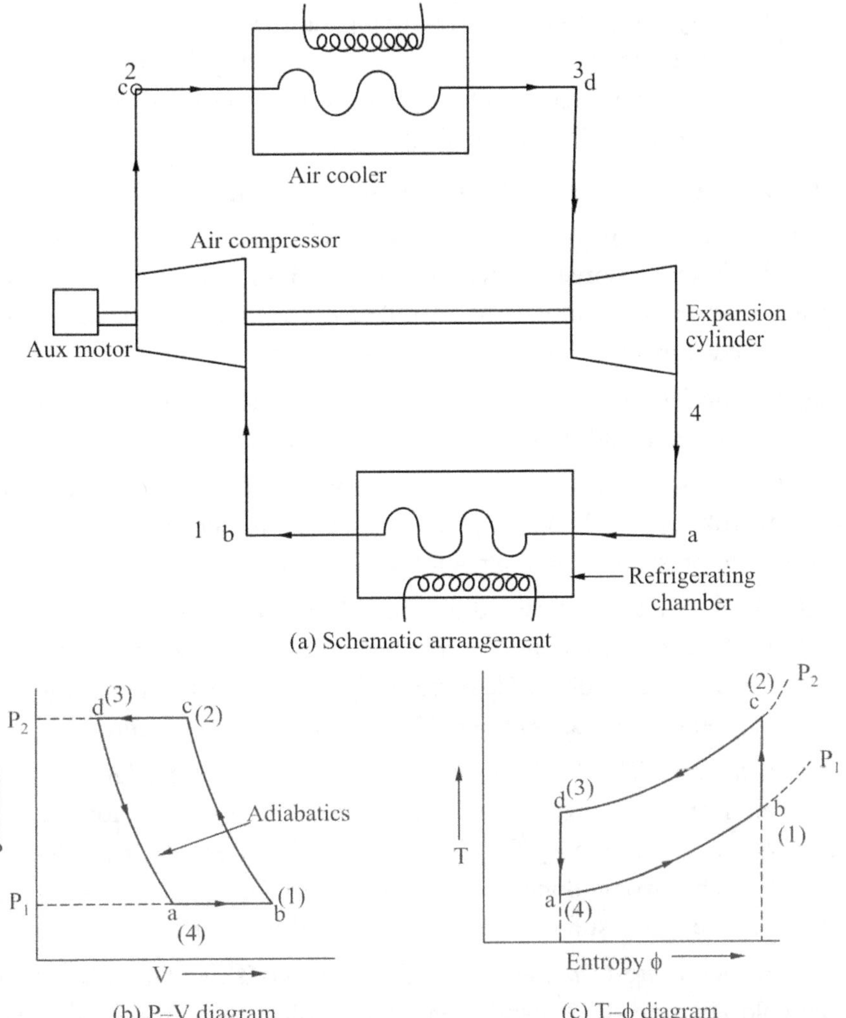

Fig. 11.11 Air Refrigeration working on Bell-Coleman Cycle (Reversed Joule Cycle).

Advantages of Closed System over Open System

(i) In open system air comes in direct contact with products to be cooled and picks up moisture from products kept in refrigerated chamber. Moisture may freeze during expansion and could result in choking of the valve of the expander. In closed system this difficulty does not occur at all.

(ii) In open system the expansion of air is limited by atmospheric pressure prevailing in cold chamber but for closed system there is no such restriction.

(iii) In a closed system, same air is re-circulated, the suction to the compressor may be at high pressure. This results in dense air entering the compressor. The size of expander and compressor may be reduced as compare to open system. In such cases the system is called dense air system.

The Reverse Brayton Cycle or Bell Column on Air Cycle

Actual refrigeration cycle using air as working substance is not based on the reversed carnot cycle because of some difficulties involved in carrying out isothermal heat absorption and isothermal heat rejection. An ideal cycle which can be more easily used in practice is reversed Brayton cycle o Bell-columan air cycle.

The cycle consists of four main components (i) compressor (ii) Air cooler (iii) Expander (Engine or Turbine) (iv) Refrigerating chamber or cold room.

Process 1-2 (compressor): The air leaving the cold chamber enters into air compressor during suction stroke (point 1) at temperature T_1 and Pressure p_1 and air is compressed isentropically up to pressure p_2 and temperature T_2.

Process 2-3 (Cooler): Hot compressed air leaving the compressor is passed through the pipes of cooler. The cooler contains a number of pipes immersed in circulating water and air is passed through these tubes. Heat from air is rejected to circulating water in the cooler. The air is cooled from temperature T_2 to T_3 at constant pressure.

Process 3-4 (Expander): The Air from cooler is drawn to the expander. In this process the expansion of air from high pressure p_2 and temperature T_3 to the low pressure p_1 and lowest temperature T_1 in the cycle. Temperature T_1 is lower than the refrigerating temperature T_2. The work done during expansion process assists in driving the compressor unit. The rest is supplied by the auxiliary drive unit.

Process 4-1 (Refrigerating Chamber or Cold Chamber): This is the place required to be cooled. The cold air from the expander enters the cold room, absorbs the heat from the material (or substance) to be cooled at constant pressure p_1 with increase in temperature from T_4 to T_1. Since the air from expander is colder than the refrigerating temperature T_1, the air from cold chamber enters the compressor and cycle is repeated.

C.O.P of air refrigerator (Bell Coleman Cycle)

Assume 1 kg of air flowing through the cycle

Refrigerating effect $= q_{ab} = c_p(T_1 - T_4)$ KJ/kg

Heat rejected to air cooler $= q_{rej} = c_p\,(T_2 - T_3)$ KJ/kg

$\therefore$ Work input to compressor = Heat rejected – Heat absorbed

$$= c_p(T_2 - T_3) - cp\,(T_4 - T_1)$$

$$C.O.P = \frac{\text{Refrigerating Effect}}{\text{Work Input}} = \frac{Cp(T_1 - T_4)}{C_p(T_2 - T_3) - c_p(T_4 - T_1)}$$

$$= \frac{T_1 - T_4}{(T_2 - T_3) - (T_4 - T_1)}$$

Uses or Application of Air Refrigerator

Due to non toxic, non explosive and non poisonous and non expensive properties of air it is used in ships and air craft for safety.

This plant is bulky due to large amount of air required and it has high power consumption i.e., low C.O.P. It has a disadvantage of freezing the moisture present in air. This may choke up valves of expander of the system.

11.18 Comparison Between Air Refrigeration and Vapour Compression Refrigeration

	Air Refrigeration	Vapour Compression Refrigeration
1.	Working refrigerant is air.	1. Working refrigerant is NH_3, R_{12}, R_{11} etc.
2.	Refrigerant is non poisonous, harmless. There is no danger of explosion due to its leakage.	2. In some of refrigerant used in vapour compression refrigeration, there is danger of explosion due to its leakage.
3.	It consists of compressor, combustion chamber, gas turbine, primary heat exchanger, Refrigerating heat exchanger cooling turbine etc.	3. It consists of motor, compressor, condenser, throttle valve, evaporator etc.
4.	No mechanical or electrical power is required to run the system. Thus system is highly reliable.	4. Electric power is required therefore comparatively less reliable.
5.	Large volume of air is required to be handled for each ton of refrigeration. This results in greater size of compressor and expander.	5. Comparatively small size of compressor condenser, evaporation and other parts. Therefore whole unit very is small.
6.	Low c.o.p (one or less than one). Since air has poor conducting and poor refrigerating effect.	6. c.o.p is high (4 to 5) as the working cycle of this system is near the carnot cycle.
7.	Heat transfer in condenser and evaporator is sensible heat only.	7. In condenser and evaporator heat transfer takes place in form of latent heat.
8.	Here working refrigerant (air) is discharged into surrounding atmosphere.	8. Working refrigerant after cooling and absorbing heat from refrigerated space is re-circulated again and again in closed circulation.

Contd...

Air Refrigeration	Vapour Compression Refrigeration
9. Working fluid air is available free of cost	9. Cost of refrigerant R_{12}, R_{11}, NH_3 is much higher than air refrigerant.
10. Initial investment cost is less	10. Initial investment cost is higher than the air refrigeration.
11. Leakage of refrigerant air can be easily prevented	11. Prevention of leakage of refrigerant in this system is major problem
12. The air refrigeration system requires five times more power than vapour compression refrigeration system of the same capacity.	12. The running cost of this system is less than air refrigeration system.
13. Here instead of expansion valve, Expander is there, so large range of temperature, the required temperature in evaporator can not be achieved of same unit	13. This system can be employed over a large range of temperatures by adjusting the expansion valve of the same unit, the required temperature in evaporator can be achieved.

11.19 Vertical Packaged A.C. Unit known as Unitary System or Self-contained Unit

1. These units are similar to window type but size is large. They have capacity from 5 to 30 tons. Now the units are available in capacities of more than 100 tons.
2. They are factory assembled units having complete system mounted in a cabinet or frame.
3. They are installed near the space to be conditioned.
4. This type of unit may be console or of remote unit type. The console is placed in room below the window sill and has suitable fresh air inlet opening provided in wall.
5. The condenser of this unit is water cooled for which cooling tower or spray tower are used to provide condenser water.
6. The power consumption per ton of refrigeration is lower than that of window A.C.
7. The condensing unit is mounted in the bottom of the console, air blower in middle and the evaporator coil at the top of the cabinet. The return air enters the lower grille while the cooled air is discharged at upper grill as shown in Fig 11.12.
8. Condenser may be air cooled or water cooled. Some condensers are located away from the packaged units.

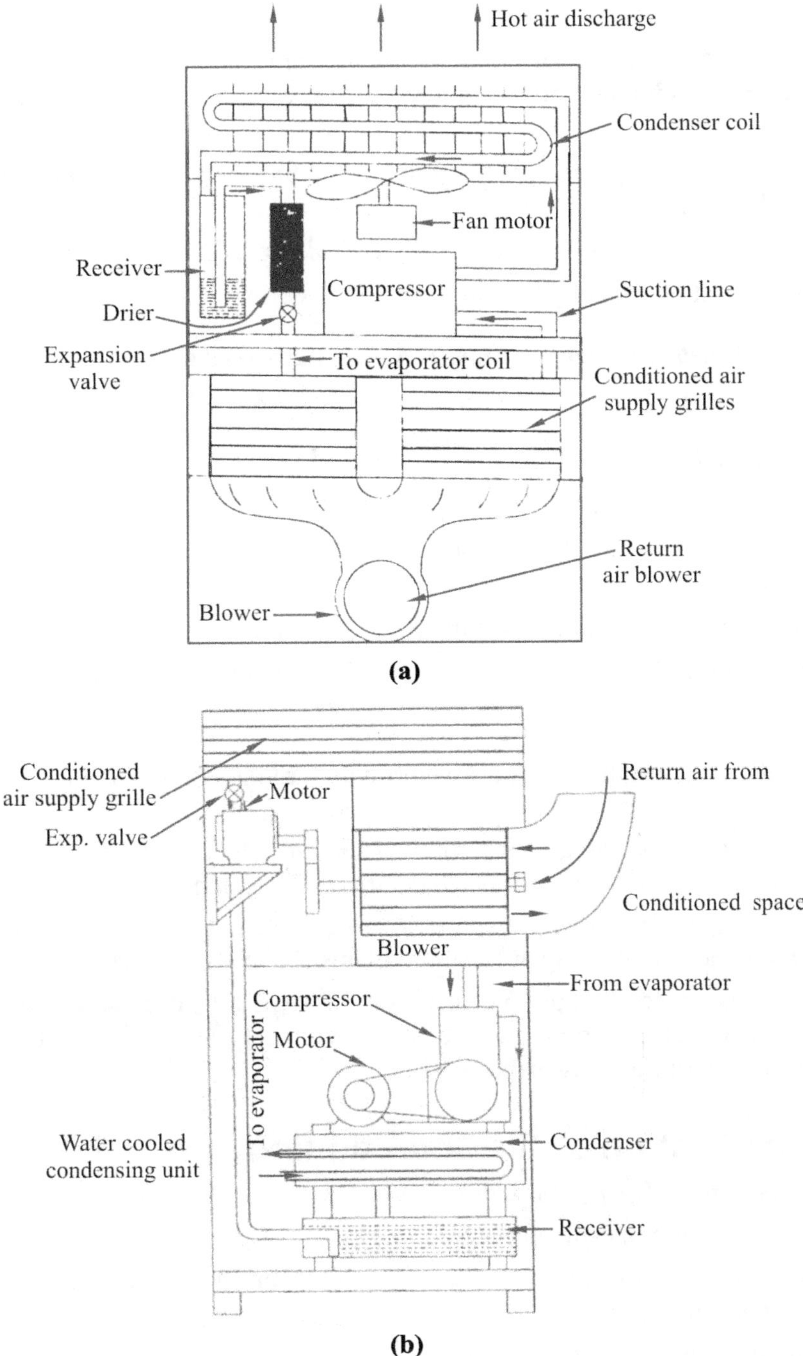

Fig. 11.12 Self contained air conditioning: (a) with water and (b) with air cooled condensing units.

Central A.C. System

1. In this system, all the air conditioning equipments are centralized in a separate room and conditioned air is supplied to the various rooms in the large building through pipes.
2. In this type of system, the various components of system are designed for assembly at site.
3. This system is useful for several rooms requiring individual control of each room.
4. Its refrigerating capacity is more than 25 tons.

Advantages and disadvantages of central A.C system over unitary A.C system

Advantages

1. Low initial cost
2. Noise and vibration are not the problems.
3. It can be installed away from building room at a place which may not be useful otherwise.
4. Maintenance is easy due to separate central plant.
5. Exhaust air can be used again and again thus heating and cooling are reduced.

Disadvantages

1. No saving in installation and assembly as various equipments are assembled at site.
2. In unitary A.C. system, distribution system and duct work for conditioned air is eliminated which is not the case in central A.C. System.
3. In the failure of unit (central A.C system) complete A.C. of all rooms stops. So to continue the conditioning at least two plants are required in central system.
4. Further expansion is possible in unitary system for successive room with new units. But same central A.C system will not be sufficient to meet A.C. requirement of (i.e., new expansion).

11.20 Application of Air Conditioning

1. *Application for Human Comfort*
   (i) A.C. for domestic purpose
   (ii) A.C in shops, offices, theatres, auditorium, hospital, clubs, cars, Railway coaches
2. *Application for Machine Comfort*
   (i) Automatic Telephone Exchanges

(ii) Xerox machines, T.V and Radio centers.

(iii) For electronic equipment manufacturing and operating.

(iv) A.C. in computer centers

3. *Industrial Application*

(i) Paper Industries

(ii) Textile Industries

(iii) Clothe Printing and Dyeing

(iv) Food Processing Industries

(v) High Precision Metal cutting

(vi) Chemical factories

(vii) Electronic equipment manufacturing

4. *Special Application*

(i) Museums for preserving scare collection of various old articles and to prolonging their life.

(ii) Space Research centers

11.21 Window Air Conditioner

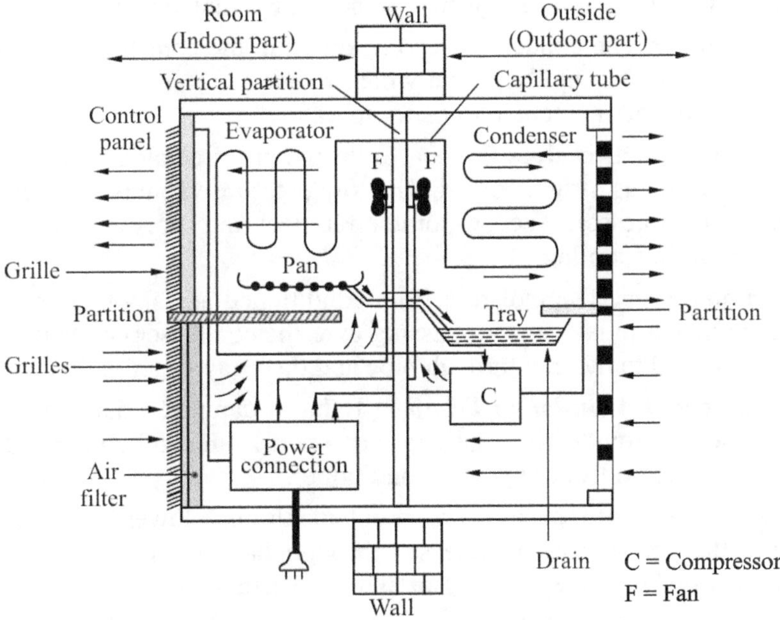

Fig. 11.13(a) Window Air Conditioner Unit.

1. A window type A.C, is a small, compact and self contained unit which is fitted to the window of the room in such a way that condenser and compressor is outside the room.
2. The unit is having capacity from 1 to 3 tons and it is sufficient for office room, bed room and Drawing hall etc.
3. Fig. 11.13(a) shows the schematic diagram of window A.C. unit.
4. The various components of window A.C are divided into two parts by vertical partition as:

   (a) Outdoor parts (b) Indoor parts. The unit is fitted in such a way that outside parts are outside the window and indoor parts are inside the room.
5. Outdoor parts have compressor and tray below the small horizontal partition, a fan and condenser above it.
6. Indoor parts have power connection and air filter below the small horizontal partition, Fan (blower), evaporator, Pan, control panel and remote bulb refrigerant control above it.
7. ***Power Connection*:** It gives electric supply to compressor motor, condenser fan, and evaporator fan (Blower), and control panel.
8. ***Compressor*:** It receives the hot and low pressure vapour refrigerant from the evaporator and compresses it at high pressure.
9. ***Condenser*:** It receives high pressure vapour refrigerant from compressor and condenses it due to heat rejected to the air, circulated by the fan.
10. ***Evaporator*:** The condensed refrigerant from the condenser passes through the filter and to the capillary tube which reduces the pressure and temperature of refrigerant and it enters in to the evaporator.

    The refrigerant receives the heat from the air circulated by the evaporator fan (Blower). Thus air get cooled. The low pressure vapours refrigerant flow back to the compressor. The evaporator removes the moisture of the air to some extent during cooling.
11. ***Pan and Tray*:** The moisture from conditioned air is condensed and collected in the pan. Similarly the air passing over the condenser condenses the tray. Pan water is laid to the tray through pipe and then tray water is drained.
12. ***Exhaust and Ventilation Damper*:** Exhaust damper, exhaust the room air and fresh air is provided by ventilation damper. Both can be operated with help of knob placed on front of the air conditioner.
13. ***Condenser Fan*:** It sucks air from outside through lower half drills and drives it over the condenser and releases through half grills to the outside. The air circulation keeps the compressor and condenser cool.
14. Evaporator fan sucks the room air through air filter evaporator, air gets cooled and conditioned air is released in room through upper half grills.

15. The Control Panel controls the operation automatically. The unit stops when required temperature is reached in the room and restarts it when room temperature rises. This is done with help of thermostat which is placed ahead of filter that will sense the temperature of air entering the unit.

    The compressor motor is protected from damage because of an over current or over temperature by providing over load protector on compressor motor.

    The capacitor is provided on compressor motor to improve starting and running characteristics of motor.

    Starting switch which is located on the front of the air-conditioner is for starting and stopping the unit.

16. Filter is fitted near evaporator coil. Evaporator side is insulated form the condenser side. Dampers are used for appropriate air distribution as shown in Fig. 11.13(b), especially from heat transfer point of view. Chilled air should not be short circulated in order to have efficient air transfer. Damper is also used to regulate the fresh air supply.

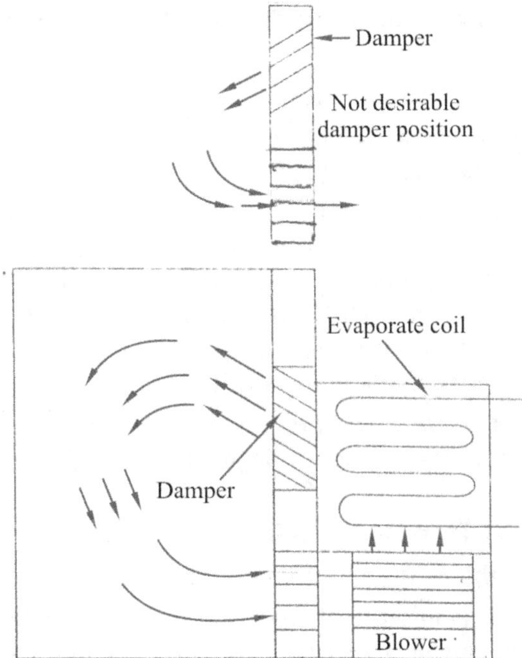

Fig. 11.13(b) Air discharge from air conditioning system.

11.22 Terms in Air Conditioning

1. ***Psychrometry*:** It is a science dealing with properties of moist air (air and water vapour) or it is study of properties of moist air.
2. ***Psychrometric Chart*:** It is the graphical representation of the thermodynamic properties of moist air.
3. ***Psychrometer*:** It is the instrument measuring the relative humidity of air by means of difference between dry bulb and wet bulb temperature.
4. ***Dry Air*:** Dry air is mechanical mixture of oxygen, Nitrogen, CO_2, H_2, argon, Neon, Helium, Ozone and xenon. However oxygen and nitrogen make up major part of the combination.

   Dry air is considered to consist of 21% of oxygen and 79% of nitrogen by volume and 23% of oxygen and 77% of nitrogen by mass. Complete dry air does not exist in nature, water vapour varying amount is diffused through it. Partial pressure P_w of water vapour is very small and the saturation temperature of water vapour at P_w is less than atmospheric temperature. The water vapour in air exists in super heated state and air is said to be unsaturated.
5. ***Sling Psychrometer*:** It has two thermometers dry bulb and wet bulbs are mounted on a frame with handle. The handle is rotated so that there is good air motion.
6. ***Moisture*:** The water vapour present in the air is known as moisture.
7. ***Dry Bulb Temperature (DBT)*:** The temperature of air measured by ordinary thermometer is known as dry bulb temperature of air.
8. ***Wet Bulb Temperature (WBT)*:** The temperature of air measured by thermometer when its bulb covered with wet cloth is known as wet bulb temperature. The Difference between dry bulb and wet bulb is known as wet bulb depression. When DBT and WBT are simultaneously measured by combined thermometer, it is known as psychrometer. This temperature depends up on water vapour content in the air. It is less than DBT.
9. ***Dew Point Temperature (DPT)*:** It is the temperature air recorded by dry bulb thermometer when water vapour begins to condensate. It is the temperature of saturated air.

   $$\text{DPT} = \text{DBT} = \text{WBT}.$$
10. ***Moist Air*:** It is the mixture of dry air with water vapour.
11. ***Saturated Air*:** It is mixture of dry air with the maximum water vapour that air can contain at given pressure and temperature. If temperature is above saturation vapour superheats, when the temperature decreases below saturation temperature, the water vapour is condensed.

12. ***Conditioned Air*:** It is air of controlled temperature, humidity, velocity and purity produced by air-conditioning system.
13. ***Humidity*:** It is mass of moisture (water vapour) contained in air at a given temperature.
14. ***Specific Humidity or Humidity Ratio*:** It is mass of water vapour contained in one kg of dry air. It is generally given in grams per kg of air.
15. ***Absolute Humidity*:** It is mass of water vapour contained in one m^3 of dry air.
16. ***Relative Humidity (R.H)*:** It may be defined as ratio of actual mass of water vapour in a given volume of moist air to the mass of water vapour in the same volume of saturated air at the same pressure and temperature. It is given by P_v / P_{vs}

    where P_v is actual partial pressure of water vapour in air and

    P_{v_s} the partial pressure of water vapour when air is fully saturated.
17. ***Degree of Saturation*:** It is defined as ratio of actual mass of water vapour to the mass of water vapour in saturated air for same mass of dry air under the same condition of pressure and temperature.
18. ***Dew Point Depression*:** It is the difference between dry bulb temperature and dew point temperature of air.
19. ***Sensible Heat of Air*:** The quantity of heat which can be measured by measuring the DBT of air is known as sensible heat.
20. ***Total Heat of Air*:** It is the sum of sensible heat of dry air and sensible and latent heat of water vapour associated with dry air.
21. ***Effective Temperature*:** It is the degree of warmth sensed by human being due to combined effect of temperature, humidity and velocity of air. It cannot be measured directly. It is determined by DBT, WBT and air velocity and referring charts, performing experiments and using imperial formula.

11.23 Air Conditioning (A.C)

***Air Conditioning*:** It is defined as simultaneous control of (i) Temperature (ii) humidity (iii) motion of air (3 to 8 m/min) (iv) purity of air within an enclosed space.

Purpose of Air Conditioning

1. To produce healthy and comfortable atmospheric or weather conditions for people in residential and Commercial buildings, offices, shops, stores, theaters, auditoriums, museums, Public halls, hospitals, railway coaches and cars etc.
2. To increase the efficiency of working staff in factories by providing conditioning air for improving their comfort, health and emotional conditions.

3. To provide controlled atmosphere for complete control of manufacturing processes and quality of product in industries such as textile, paper, printing, drugs and chemicals.
4. To improve efficiency, durability and accuracy of electronic and other precision equipments such as computers, T.Vs.

Factors of Air Conditioning for Human Comfort

The air conditioning system should provide the control of following factors for complete comfort of human beings during summer and winter seasons.

***Purity of Air*:** Clean air is highly essential for breathing and health. Air is purified in air filter to remove dust, dirt and other impurities and germs. The dust and dirt present in air are objectionable in food product, drugs and chemicals etc.

***Temperature Control*:** For maximum comfort a temperature from 20 °C to 25 °C is desirable. So this temperature should be achieved by cooling the air during summer and heating the air in winter.

***Humidity of Air*:** For maximum healthy and comfortable conditions, the humidity or moisture content of air should be controlled. It improves the quality and reduces cost in manufacturing of Automobile, Electric, Textile, Pharmaceutical products. Humidity is controlled by adding the moisture to the air during winter and removing the moisture from air during the summer.

***Motion of Air*:** This is the circulation and distribution of air effectively and evenly in all parts of space to be air conditioned at all times. It provides even temperature through out the building. Motion of air is taken 3m/min to 8 m/min.

***Ozonation or Ionisation of Air*:** It is the process by which odour is eliminated or unpleasant smells are removed from the air by addition of small quantity of ozone. It keeps the air fresh throughout the operation.

11.24 Classification of A.C. Systems

Air conditioning (A. C) systems are classified by different ways as follows.

1. According to the purpose
   - (i) A.C. for human comfort
   - (ii) A.C for machine comfort
   - (ii) Industrial A.C. Systems

2. According to the seasons
    (i) Summer A.C. System
        (a) A.C. for hot and dry weather
        (b) A.C for hot and wet weather
    (ii) Winter A.C. systems
    (iii) Year-round (all season) A.C. systems.
3. According to arrangement of equipments
    (i) Unitary A.C. System
        (a) Window Air Conditioners
        (b) Vertical Package Air Conditioners
    (ii) Central A.C. system

Unitary A.C. System

This system has each room or space provided with its individual A.C. unit. These air conditioner units are factory assembled and installed at the space to be conditioned.

There are two types of unitary A.C. system

(i) Vertical packed air conditioner

(ii) Window type air conditioner

11.25 Split Air Conditioner

All split air conditioners consist of two main components, the indoor unit (IDU) and the out door unit (ODU). Indoor unit is installed inside the air conditioned room and ODU is kept outside because all the heat from inside is rejected outside through outdoor unit. The ODU houses the compressor, condenser and fan for condenser. The condenser coil is coated with flue fins to increase efficiency and life, and to protect it from chemicals and corrosion due to salinity. The all ODU components are capable of standing a 500 hour salt spray test.

I.D.U is mounted on one of the walls of the room to be air conditioned. The outdoor unit can be mounted on a wall, sunshade or skirting out side the room. Two units are connected through refrigerant pipes made of copper which are insulated to reduce loss of refrigerating effect as in Fig. 11.14(a).

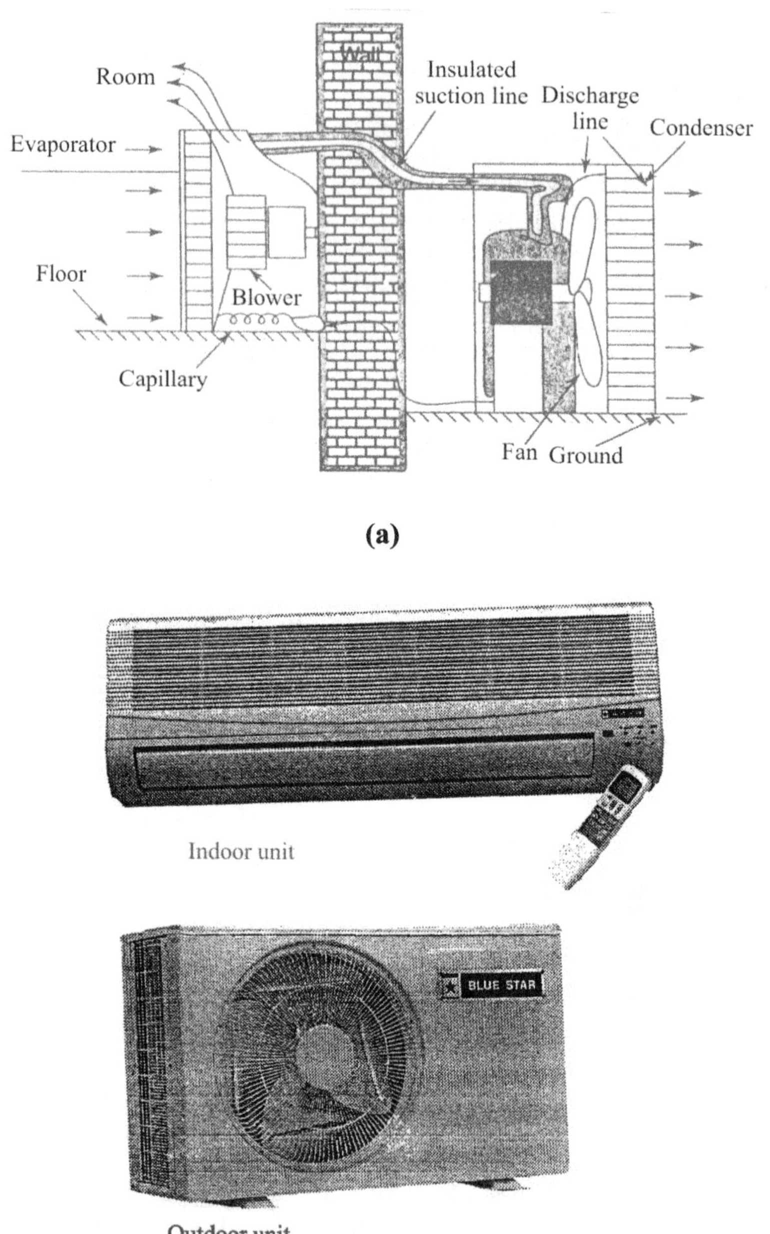

Fig. 11.14 Split air-conditioning system.

Locating the indoor and outdoor units properly, with help optimize the performance of air conditioner.

Locating the Indoor Unit (IDU) Fig. 11.5

1. Locate the IDU for the best cool air condition. There should be no obstruction nearby.
2. Do not locate the unit directly opposite a door which is opened frequently. The cold air will go out of the room each time the door is opened.
3. All IDU will form condensate water. See that the condensate water can be drained out of the room to toilet or pantry etc. The U-tube bend is provided in the drain to prevent insects from coming into room through the drain tube.
4. The IDU of air conditioner should be at an accessible height.

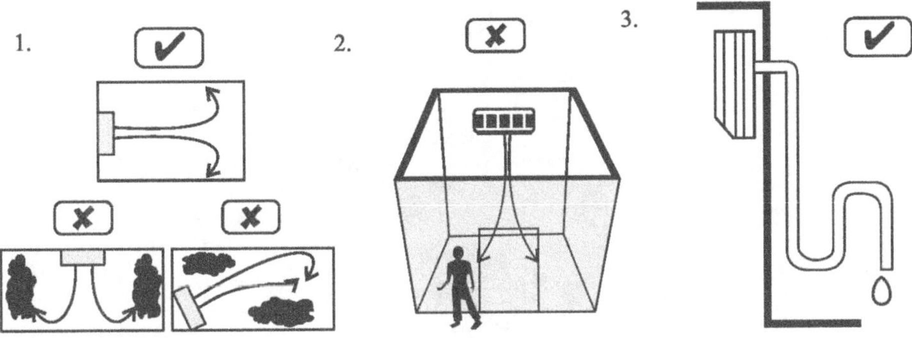

Fig. 11.15

Locating Out Door Unit (Fig. 11.16)

The outdoor unit houses the compressor, condenser, condenser fan and electrical components. Heat removed from the room is expelled to atmosphere through ODU.

1. The outdoor unit can be mounted on a wall, sunshade or skirting.
2. The ODU should not be too far away from the IDU. The recommended distances are marked in Fig. 11.10.
3. Ensure that air can pass freely through the unit. Allow at least 2 m free space on the outlet
4. If more than one ODU is to be installed, they must be properly separated, so that warm air from the outlet of one not enters the inlet of other.

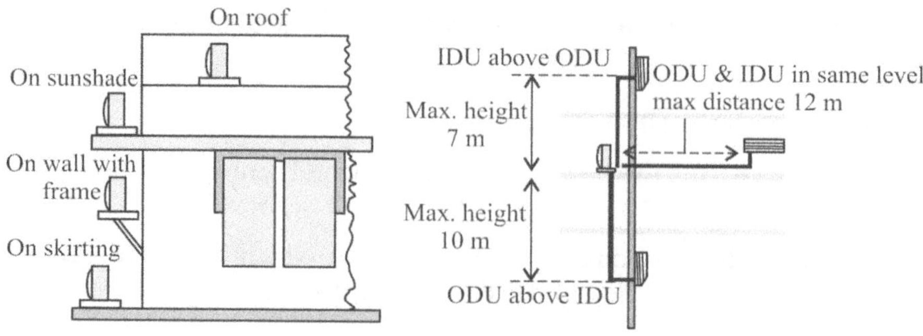

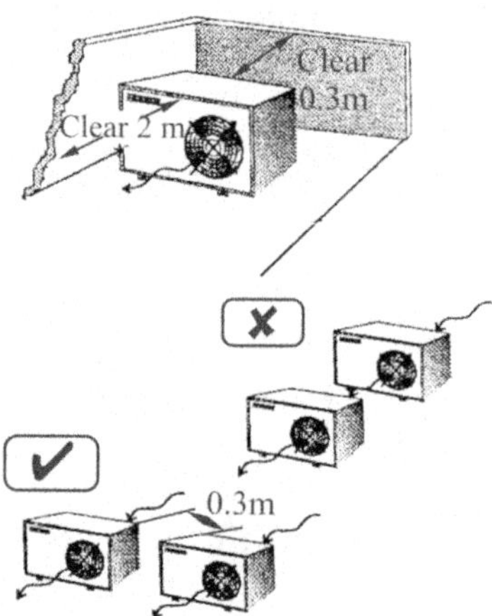

Fig. 11.16

Distance between IDU and ODU (Fig. 11.17)

1. Avoid long and complicated routing between IDU and ODU as shown in Fig. 11.11.

   1-a, too long

   1-b, too high

   1-c, too many bends
2. The ODU must be placed on a strong frame or support. If mounted on wall, ensure that the wall is thick enough and frame is supported adequately.
3. ODU should be easily accessible to the service technician. If the ODU is to be located on high ledge, the sufficient space must be provided for service technician to service the unit. It is good idea to provide a "Cat wall" (a platform) around the unit and an access door or window close to the ODU location.

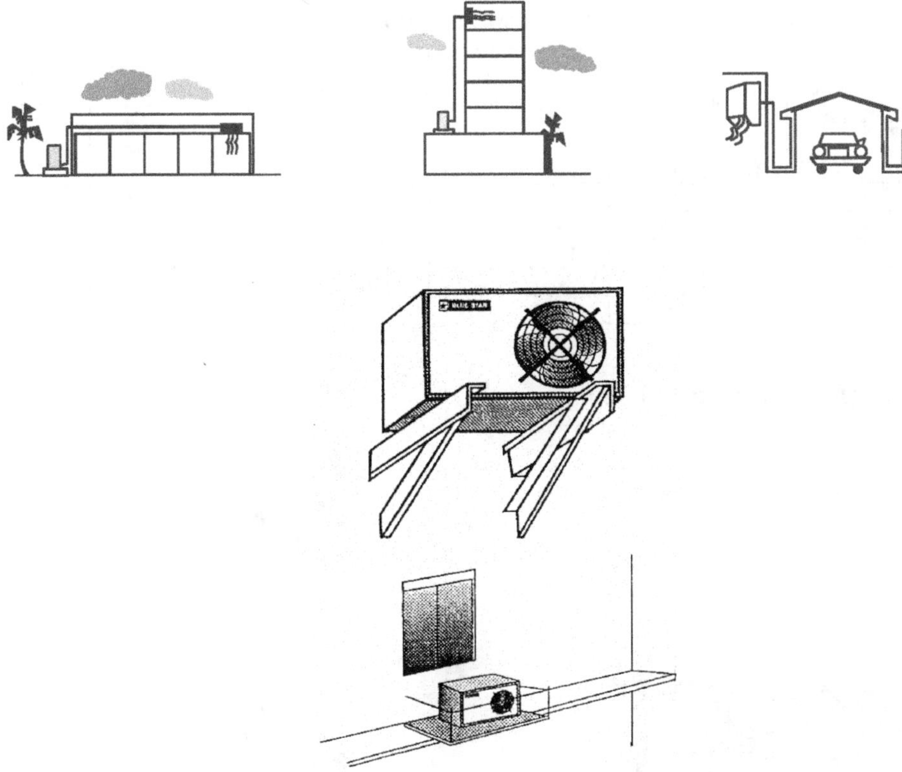

Fig. 11.17

Electrical Connection

After the location of the ODU and IDU is finalised, suitable Electrical connections for IDU and ODU must be provided.

Advantages and Disadvantages of Split Air Conditioner

1. Less or no noise compare to window air conditioner.
2. Less vibration
3. Proper distribution of cold air inside room.
4. Occupies more volume
5. Initial cost is more compare to window air conditioner
6. More power consumption.

Note : Working and various components of split air conditioner are the same as window air conditioner only split air conditioner of IDU and ODU instead of single unit.

Technical specifications

Sr. No.	Cooling Capacity of Refrigerant	No. of IDU's	Dimensions of IDUs			Wt. of IDU in kg.	Dimensions of ODU in mm			Wt. of ODU in kg	Power Consumption Watts
			Width	Depth	Height		Width	Depth	Height		
1	1	1	750	188	250	8.5	850	370	540	50	1400
2	1.5	1	906	235	286	13.5	850	370	540	50	1900
3	2	1	1080	222	330	17	850	370	540	59	2400
4	3	2	906	235	286	2×13.5	1180	464	810	73	3800
5	4	2	1080	222	330	2×17	1180	464	810	87	4800

Maintenance (Fig. 11.18 and 11.19)

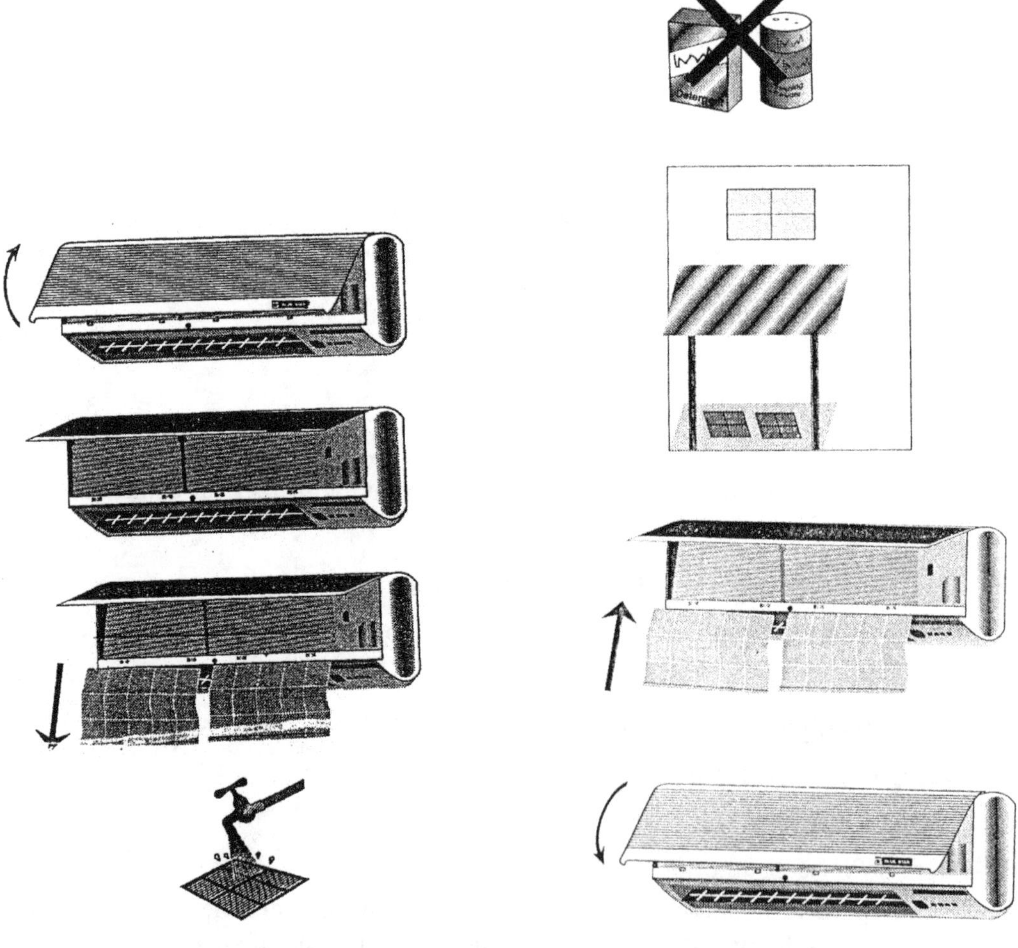

Fig. 11.18

Fig. 11.19

There are some simple maintenance jobs that user should carry out easily on his own way.

1. Lift the cover by applying gentle pressure on both sides and lifting it upward.
2. The filter will be visible
3, Lift the Filter slightly upwards
4. Pull out the filters.
5. Wash the filters under tap until they are clean.
6. Do not use detergents for cleaning of the filter.
7. Dry fitter under shade.
8. Reinstall dry filters
9. Press the cover gently back in place.

Objective Type Questions

1. During a refrigeration cycle, heat is absorbed by the refrigerant in a

   (a) Compressor (b) Condenser

   (c) Evaporator (d) expansion valve.

2. The flow of a refrigerant in a refrigeration cycle is controlled by

   (a) Condenser (b) Expansion Valve

   (c) Evaporator (d) Receiver

3. In a mechanical refrigeration system, the refrigerant occurs as liquid at low pressure (very wet vapour) between.

   (a) Compressor outlet and condenser (b) Condenser and expansion valve

   (c) Expansion valve and evaporator (d) Before and after expansion valve

4. Dry ice is

   (a) Free from dissolved air and gases (b) Does not contain impurities

   (c) solidified form of CO_2 (d) None of the above

5. In a vapour compression system, the highest temperature occurs after

   (a) Condenser (b) Compression

   (c) Expansion (d) Evaporator

6. In vapour compression cycle the lowest temperature, during the cycle occurs after

   (a) Evaporator (b) Expansion valve

   (c) Condenser (d) None of above

7. Refrigerant commonly used in vapour absorption system is

(a) SO_2 (b) Ammonia

(c) Freon (d) Aqua-Ammonia

8. Boiling point of Ammonia is

(a) – 10.5° C (b) – 30° C

(c) – 33.3° C (d) – 77.6° C

9. The most common type vapour absorption system in use for industrial application is

(a) Brine and water (b) Hydrogen and water

(c) Ammonia and water (d) None of above

10. c.o.p of practical vapour compression system as compare to that for vapour absorption system is

(a) less (b) More

(c) May be less or more (d) None of above

11. Shaft work required for vapour refrigeration system to produce 1 ton refrigeration may be only

(a) 40% to 50% of that required for vapour compression system

(b) 15 to 35% of that required for vapour compression system

(c) 1 to 2% of that required for vapour compression system

(d) None of the above

12. Which of the following equipment perform the function of compressor in absorption system.

(a) Generator (b) Absorber

(c) Generator, absorber and liquid pump (d) Absorber and liquid pump

Answers

1. c 2. b
3. c 4. c
5. b 6. b
7. d 8. c
9. c 10. b
11. c 12. c

Questions

1. What is unit of Refrigeration and define it.
2. With a neat sketch explain vapour compression system.
3. State the function of the each component of vapour compression system.
4. Define Net Refrigerating effect and C.O.P.
5. What is refrigerant and what are its important properties?
6. Draw a line diagram of vapour absorption system and explain its working
7. Give function of each component used in vapour absorption system.
8. What is refrigeration and How it is produced.
9. State advantages and disadvantages of vapour absorption system
10. What is principle of absorption system using ammonia as refrigerant?
11. Explain with neat diagram the working of air refrigeration system.
12. Discuss the advantages and disadvantages of vapour compression system over air refrigeration system.
13. What is refrigerant? Mention at least five desirable characteristics of a refrigerant.
14. Why air condition is required in air craft.
15. What refrigerant is used in vapour compression system and why.
16. Draw neat sketch of window air conditioner and explain its working.
17. Give function of various components used in air conditioner.
18. Why air conditioner is fixed on window. Why it is not kept inside the room on Floor.
19. What is split air conditioner, why now-a-days people prefer it.
20. Draw neat sketch of water cooler and explain its working.
21. Draw neat diagram of Domestic refrigerator and explain its working. Why its condenser is always black in colour.
22. What is air conditioning ?
23. Define the following terms related to air conditioning.
    (i) DBT and WBT (ii) Relative humidity
    (iii) Specific humidity (iv) Dew point temperature and Dew point depression.
24. Classification of A.C systems
25. What are various applications of air conditioning?
26. Advantages of central A.C. system over unitary A.C. system.
27. What are advantages of unitary system over central A.C. system?

CHAPTER 12

Production Processes

12.1 Introduction

Turning is a machining process in which a cutting tool, typically a non-rotary tool bit, describes a helix tool path by moving more or less linearly while the workpiece rotates. The tool's axes of movement may be a straight line, or they may be along some set of curves or angles, but they are essentially linear. "Turning" is meant for the generation of *external* surfaces by this cutting action, whereas this same essential cutting action when applied to *internal* surfaces (that is, holes, of one kind or another) is called "boring". Thus both "turning and boring" categorizes the larger family of (similar) processes. The cutting of faces on the work piece (surfaces perpendicular to its rotating axis), whether with a turning or boring tool, is called "facing".

When turning, a piece of relatively rigid material (wood, metal, plastic, or stone) is rotated and a cutting tool is traversed along the axes of motion to produce precise diameters and depths. Turning produces tubular components to various geometries. Turning can be done manually, in a traditional lathe, which frequently requires continuous supervision by the operator, or by using an automated lathe which does not. Today the most common type of such automation is computer numerical control (CNC).

The turning processes are typically carried out on a lathe, considered to be the oldest machine tools, and can be of four different types such as *straight turning*, *taper turning*, *profiling* or *external* grooving. Those types of turning processes can produce various shapes of materials such as *straight*, *conical*, *curved*, or *grooved* workpiece. In fact, turning uses simple *single-point cutting* tools. Each group of workpiece materials has an optimum set of tools angles which have been developed for respective requirements.

12.2 Turning

12.2.1 Turning Operations

This operation is one of the most basic machining processes i.e. the part is rotated while a single point cutting tool is moved parallel to the axis of rotation. Turning can be done on the external surface of the part as well as internally (boring). The starting material is generally a workpiece generated by other processes such as casting, forging, extrusion, or drawing.

Tapered turning

It is a method for generating shallow tapers on a a workpiece.

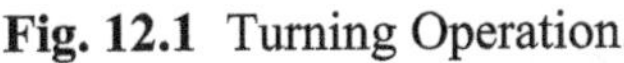

Fig. 12.1 Turning Operation.

Spherical generation

The proper expression for making or turning a shape is to generate a form around a fixed axis of revolution: (a) using hydraulic copy attachment, (b) computerised numerically controlled lathe, (c) using a form tool, (d) using bed jig.

Hard turning

Hard turning is a turning done on materials with a Rockwell C hardness greater than 45. It is typically performed after the workpiece is heat treated. The process is intended to replace or limit traditional grinding operations. However, when it is applied for finishing where form and dimension are critical, grinding is superior. Grinding produces higher dimensional accuracy of roundness and cylindricity. It is used for gears, injection pump components, hydraulic components, among other applications.

Facing

In facing, cutting tool moves at right angles to the axis of rotation of the rotating workpiece. This can be performed by the operation of the cross-slide, as distinct from the longitudinal feed (turning). It is the first operation performed in the production of the workpiece.

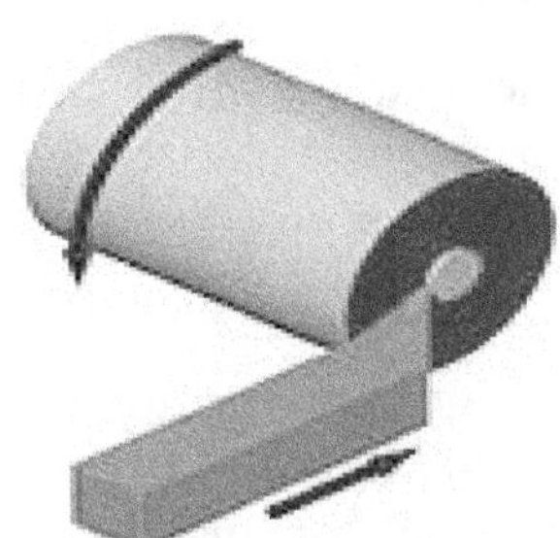

Fig. 12.2 Facing Operation.

Parting

This process, called parting off or cutoff, is used to create deep grooves which will remove a completed or part-complete component from its parent stock.

Grooving

Grooving is like parting, except that grooves are cut to a specific depth instead of severing a completed/part-complete component from the stock. Grooving can be performed on internal and external surfaces, as well as on the face of the part (face grooving or trepanning).

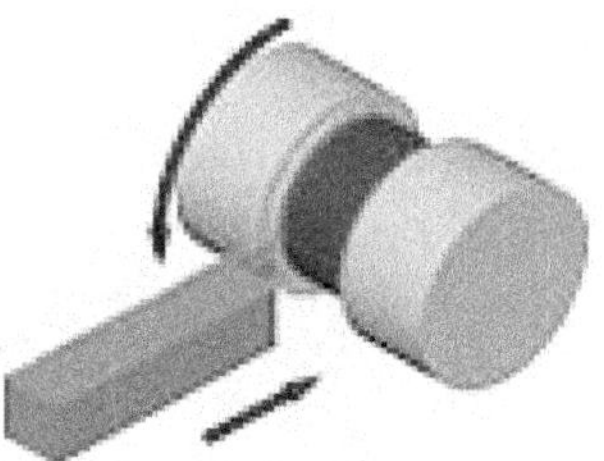

Fig. 12.3 External grooving.

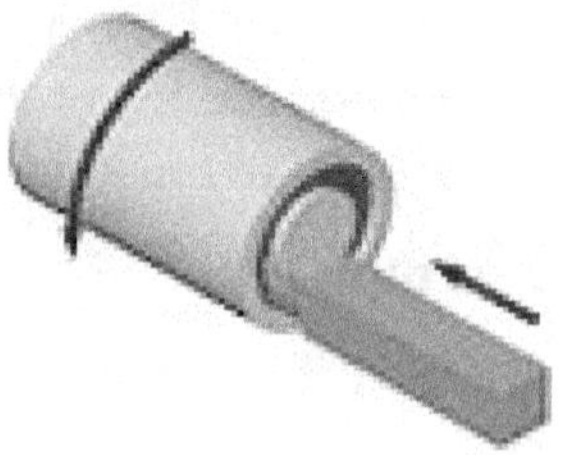

Fig. 12.4 Face grooving

Boring

Enlarging or smoothing an existing hole created by drilling, moulding a) by mounting workpiece to the spindle via a chuck or faceplate, b) by mounting workpiece onto the cross slide and placing cutting tool into the chuck. This work is suitable for castings that are too awkward to mount in the face plate. On long bed lathes large workpiece can be bolted to a fixture on the bed and a shaft passed between two lugs on the workpiece and these lugs can be bored out to size.

Drilling

It is used to remove material from the inside of a workpiece. This process utilizes standard drill bits held stationary in the tail stock or tool turret of the lathe. The process can be done by separately available drilling machines.

Knurling

The cutting of a serrated pattern onto the surface of a part to use as a hand grip, by a special purpose knurling tool.

Fig. 12.5 Knurling Operation.

Reaming

The sizing operation that removes a small amount of metal from a hole already drilled. It is done for making internal holes of very accurate diameters. For example, a 6mm hole is made by drilling with 5.98 mm drill bit and then reamed to accurate dimensions.

Threading

Both standard and non-standard screw threads can be turned on a lathe using an appropriate cutting tool. (Usually having a 60, or 55° nose angle). Tapping of threaded nuts and holes are done (a) using hand taps and tailstock centre, (b) using a tapping device with a slipping clutch to reduce risk of breakage of the tap. Threading operations include (a) all types of external and internal thread forms using a single point tool also taper threads, double start threads, multi start threads, worms, leadscrew with single or multistart threads. (b) by the use of threading boxes fitted with 4 form tools, up to 2" diameter threads.

Polygonal Turning

In this process, non-circular forms are machined without interrupting the rotation of the raw material.

12.2.2 Geometry of Single Point Cutting Tool

Convenient way to specify tool angles by use of a standardized abbreviated system is known as tool signature. It indicates the angles that a tool utilizes during the cut. It specifies the active angles of the tool normal to the cutting edge. This will always be true as long as the tool shank is mounted at right angles to the workpiece axis.

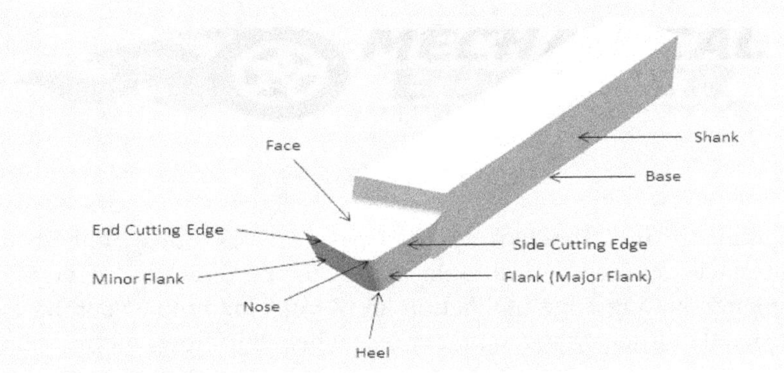

Fig. 12.6 Tool signature

The seven elements that comprise the signature of a single point cutting tool are stated in the following order:

(i) Back rake angle
(ii) Side rake angle
(iii) Clearance or End Relief angle
(iv) Side Relief angle
(v) End cutting edge angle
(vi) Side cutting edge angle
(vii) Nose radius

A typical tool signature of single point cutting tool is 0-7-6-8-15-16-0.8. Here, this tool signature indicates that the tool has 0, 7, 6, 8, 15, 16 degree back rake, side rake, end relief, side relief, end cutting edge, side cutting edge angle and 0.8 mm nose radius.

12.2.3 Orthogonal and Oblique Cutting

It is appears in the giagram that while turning ductile material by a sharp tool, the continuous chip would flow over the tool's rake surface and in the direction apparently perpendicular to the principal cutting edge, i.e., along orthogonal plane which is normal to the cutting plane containing the principal cutting edge. But practically, the chip may not flow along the orthogonal plane for several factors like presence of inclination angle, λ, etc.

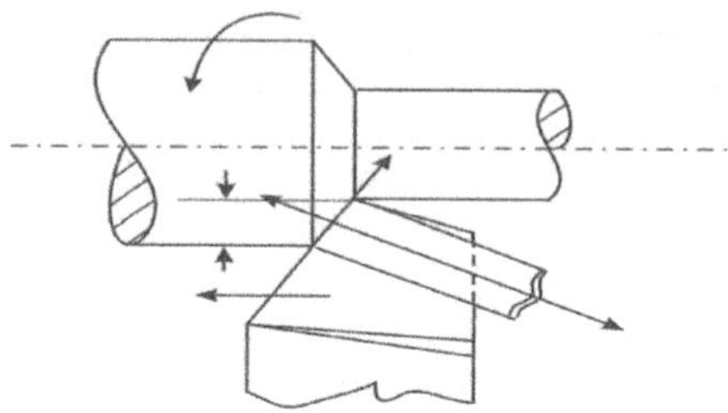

Fig. 12.7 Oblique cutting

In **orthogonal cutting,** the tool approaches the work piece with its cutting edge parallel to the uncut surface and at right angles to the direction of cutting. Thus tool approach angle and cutting edge inclination are Zero. This type of cutting is also known as two-dimensional cutting. *Orthogonal cutting:* when chip flows along orthogonal plane, π_o, i.e., $\rho_c = 0$

In **oblique cutting,** the cutting edge of the tool is inclined at an acute angle with the direction of tool feed or work feed, the chip begin disposed of at a certain angle. This type of cutting is also called three-dimensional cutting. *Oblique cutting*: when chip flow deviates from orthogonal plane, i.e. $\rho_c \neq 0$. But practically ρ_c may be zero even if $\lambda = 0$ and ρ_c may not be exactly equal to λ even if $\lambda \neq 0$. Because there are some other (than λ) factors exist which may cause chip flow deviation.

12.2.4 Lathe

A lathe is a machine tool used principally for shaping pieces of metal, wood, or other materials by causing the workpiece to be held and rotated by the lathe while a tool bit is advanced into the work causing the cutting action. Lathes can be divided into three types: engine lathe, turret lathe, and special purpose lathes. Smaller ones are bench mounted and semi-portable. The larger lathes are floor mounted and may require special transportation if they must be moved. The engine lathe is ideally suited for common purpose. A trained operator can accomplish more machining jobs with the engine lathe than with any other machine tool. Turret lathes and special purpose lathes are usually used in production or job shops for mass production or specialized parts, while basic engine lathes are usually used for any type of lathe work.

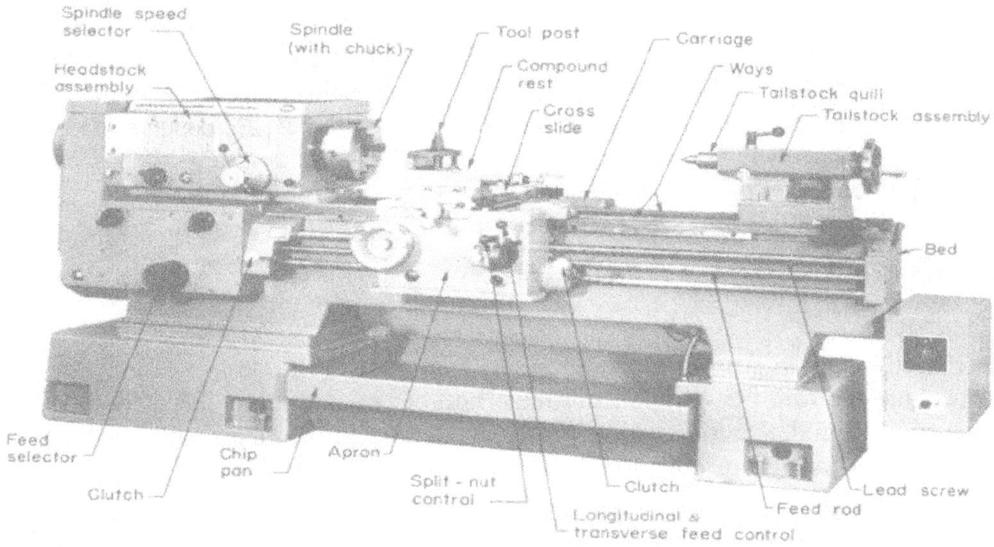

Fig. 12.8 A lathe with nomenclature of its parts

Workholding Methods

- ***Chuck*:** Chucks are a very common workholding method. There are many types, some for round and square stock, and other for irregular shapes.
- ***Collet*:** Primarily used for small round workpieces.
- ***Faceplate*:** A faceplate, drive dog, and mandrel may be used to turn workpieces such as gear blanks.
- ***Drive center*:** Use hydraulic or spring-loaded teeth that "bite" into the end of workpieces and can be used when the entire length of the workpiece must be machined.

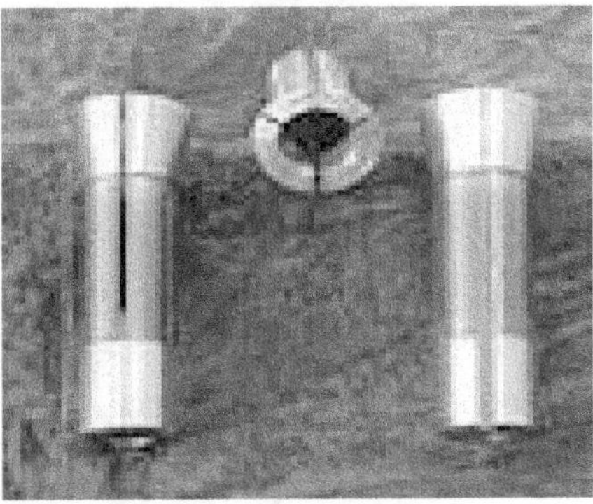

Fig. 12.9 Collets.

Tooling

The various angles, shapes, and sizes of a single-point cutting tool have direct relation to the resulting surface of a workpiece in machining operations. Different types of angle such as rake angle, side rake angle, cutting-edge angle, relief angle, nose radius exist and may be different with respect to the workpiece. Also, there are many shapes of single-point cutting tools, such as V-shaped and Square. Usually, a special toolholder is used to hold the cutting tool firmly during operation.

Dynamics of Turning

Forces

The relative forces in a turning operation are important in the design of machine tools. The machine tool and its components must be able to withstand these forces without causing significant deflections, vibrations, or chatter during the operation. There are three principal forces during a turning process:

- The *cutting or tangential force* acts downward on the tool tip allowing deflection of the workpiece upward. It supplies the energy required for the cutting operation. The specific cutting force required to cut the material is called specific cutting force. Cutting force is depends on the material.
- The *axial or feed force* acts in the longitudinal direction. It is also called the feed force because it is in the feed direction of the tool. This force tends to push the tool away from the chuck.
- The *radial or thrust force* acts in the radial direction and tends to push the tool away from the workpiece.

Speed

Speeds and feeds for turning are chosen based on cutter material, workpiece material, setup rigidity, machine tool rigidity and spindle power, coolant choice, and other factors.

Feed

- The distance the tool advances into the material in one revolution is called "feed". It is specified as mm per revolution (mm/rev).

12.2.5 Classification of Lathe

In a lathe the work piece revolves along with the chuck to which the motor is connected by various gears and shafts. In lathes tools have either longitudinal or transverse motion or cross feed. There are many types of lathes but they are broadly classified in to seven types:

1. Speed lathes
2. Engine lathes
3. Bench lathes

4. Tool room lathes
5. Capstan and turret lathes
6. Special purpose lathes
7. Automatic lathes

Speed Lathe

This lathe is simplest of all the lathes and is easily understandable to everyone. As the name indicates very high speeds are generated. These lathes contain a bed, head stock, tail stock mounted on a adjustable slide. These lathes do not have tool post, feed rod, and lead screw. Very few ranges of speeds are possible by speed lathes. These lathes are in general employed for wood working, centering polishing and spinning.

Engine Lathe

It is the most commonly used lathe and is of high importance in lathe family. The name engine is kept as in the past steam engines are used for running these lathes. This is similar to the speed lathes but the head stock is little robust in construction. Engine lathes have additional mechanisms to produce a wide range of speeds. Many parts like lead screw, feed rod, tool post etc. are also provided in these lathes. Speeds both in anti clockwise and clockwise directions can be generated very effectively.

Bench Lathe

This is a very small lathe and is generally mounted over a bench. This contains all the parts of engine lathe but the major difference arises in size. These lathes are small in size and are generally used for doing small and precision works.

Tool Room Lathe

It has similar features of engine lathe but it is more accurately built and a wide range of speeds ranging from very low to very high speeds up to 2500rpm can be generated. It has many attachments like chuck, taper turning equipment, draw in collet attachment, thread chasing rest, pump for coolant etc. This is used for doing precision works like tools dies etc. tool room lathes are costlier when compared to engine lathes of same size.

Capstan and Turret Lathe

These are the developments made in engine lathe. These lathes found application in production work. In this lathes the tail stock is replaced by hexagonal turret. Based on the way the turret is mounted they will be classified in to capstan and turret lathes. Each face of the turret is mounted with a tool. These tools are arranged in an order. When the operation of one tool gets completed then while returning back of turret it takes a turn of 1200 making the next tool to come in to picture. By employing capstan and turret lathes production rate can be increased to a larger extent. These lathes found application where a lot of identical pieces needed to be generated.

Special Purpose Lathes

As the name indicates these lathes are designed for a specific purpose. These lathes are not so generally used as they have a unique job production. These type of lathes are used when there is requirement of mass production. Some of the special purpose lathes are wheel lathe which is used for finishing of journals, T-lathes, Gap bed lathes, Duplicating lathes, missile lathes etc.

Automatic Lathes

These are latest king of lathes where every operation is automatically done by employing special computers. These lathes have high speed, heavy duty and are used for mass production. This lathe requires no flour men. A properly trained operator can efficiently run 7 to 10 machines at a time as all the work starting from loading till obtaining of finished goods every operation requires no man power. These lathes produce jobs with minimum tolerances and of very high accuracy. These lathes generally have very high cost. But the cost per piece is very less for this machine when compared to any other lathes.

12.2.6 Specifications of a Lathe

The various specifications are:

- Swing the largest work diameter that can be swung over the lathe bed.
- The shape of bed ways and horse power of the driving motor sometimes taken into specifications.
- Distance between head stock and tail stock centre.
- Some manufacturers designate the lathes by the swing and length of the bed.

12.2.7 Parts of Lathe Machine

Various parts of lathe machine are discussed as follows:

1. **Bed:** It is the main body of the machine. All main components are bolted on it. It is usually made by cast iron due to its high compressive strength and high lubrication quality. It is made by casting process and bolted on floor space.
2. **Tool post:** It is bolted on the carriage. It is used to hold the tool at correct position. Tool holder mounted on it.
3. **Chuck:** Chuck is used to hold the workspace. It is bolted on the spindle which rotates the chuck and work piece. It is four jaw and three jaw according to the requirement of machine.
4. **Head stock:** Head stock is the main body parts which are placed at left side of bed. It is serve as holding device for the gear chain, spindle, driving pulley etc. It is also made by cast iron.

5. **Tail stock:** Tail stock is situated on bed. It is placed at right hand side of the bed. The main function of tail stock is to support the job when required. It is also used to perform drilling operation.
6. **Lead screw:** Lead screw is situated at the bottom side of bed which is used to move the carriage automatically during thread cutting.
7. **Legs:** Legs are used to carry all the loads of the machine. They are bolted on the floor which prevents vibration.
8. **Carriage:** It is situated between the head stock and tail stock. It is used to hold and move the tool post on the bed vertically and horizontally. It slides on the guide ways. Carriage is made by cast iron.
9. **Apron:** It is situated on the carriage. It consist all controlling and moving mechanism of carriage.
10. **Chips pan:** Chips pan is placed lower side of bed. The main function of it to carries all chips removed by the work piece.
11. **Guide ways:** Guide ways take care of movement of tail stock and carriage on bed.
12. **Speed controller:** Speed controller switch is situated on head stock which controls the speed of spindle.
13. **Spindle:** It is the main part of lathe which holds and rotates the chuck.

12.2.8 Types of Operations of Lathe Machine

Lathe machine is used to cut the metal from cylindrical work piece, and convert it into desire shape. It turns the cylindrical work piece, and during turning a sharp edge cutting tool introduce, which cuts the metal. Its working can be described as follow.

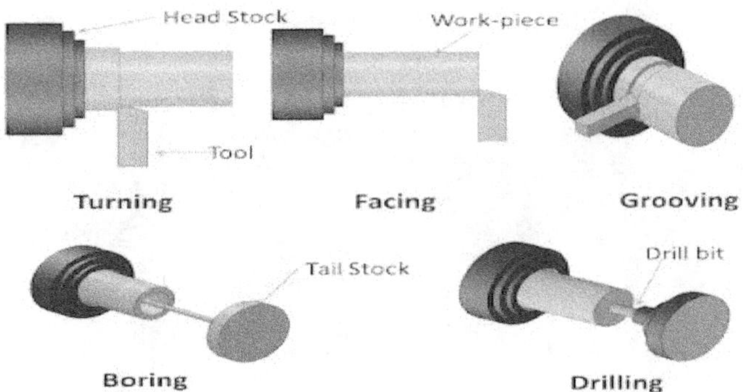

Fig. 12.10 Lathe machine operation

1. A cylindrical work piece fixed to the chuck. A chuck may have three jaw or for jaw according to the requirement. The work piece is at the centre or some eccentric according to the process perform.

2. The spindle starts to rotate and set it at desire speed. The spindle speed plays a huge role during cutting. The spindle rotates the chuck and work piece.
3. Now check the work piece is turning properly. If it not set the work piece using dial gauge.
4. Now set the tool at desire feed by moving the tool post and carriage. The feed also play main role during cutting. Large feed may cause unwanted temperature increase.
5. After it tool is introduce between moving work piece at desire feed rate. It cut the metal from work piece. The feed rate is set at the cutting condition.
6. Now all unwanted metal is removed by moving the carriage form horizontally and vertically as desire according to the job requirement. After complete all process we got a well finished job.
7. Lathe can perform turning, boring, chamfering, shaping, facing, drilling knurling, grooving as shown in figure.

12.3 Casting

Casting is a manufacturing process in which a liquid material is usually poured into a mould, which contains a hollow cavity of the desired shape, and then allowed to solidify. The solidified part is also known as a casting, which is ejected or broken out of the mould to complete the process. Casting materials are usually metals or various cold setting materials that cure after mixing two or more components together; examples are epoxy, concrete, plaster and clay. Casting is most often used for making complex shapes that would be otherwise difficult or uneconomical to make by other methods.

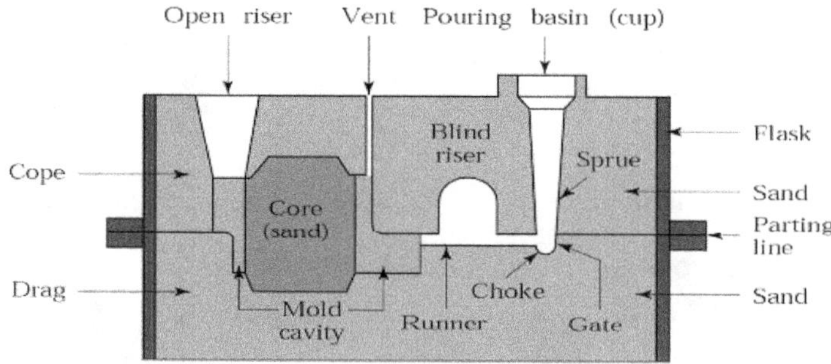

Fig. 12.11 Schematic diagram of casting process.

12.3.1 Classification of Casting Process

1. **Metal:** In metal working, metal is heated until it becomes liquid and is then poured into a mould. The mould is a hollow cavity that includes the desired shape, but the

mould also includes runners and risers that enable the metal to fill the mould. The mould and the metal are then cooled until the metal solidifies. The solidified part (the casting) is then recovered from the mould. Subsequent operations remove excess material caused by the casting process (such as the runners and risers).

2. **Plaster, Concrete, or Plastic Resin:** Plaster and other chemical curing materials such as concrete and plastic resin may be cast using single-use waste moulds, multiple-use 'piece' melds, or moulds made of small rigid pieces or of flexible material such as latex rubber. When casting plaster or concrete, the material surface is flat and lacks transparency. Often topical treatments are applied to the surface. For example, painting and etching can be used in a way that gives the appearance of metal or stone. Else, the material is altered in its initial casting process and may contain colored sand so as to give an appearance of stone. By casting concrete, rather than plaster, it is possible to create sculptures, fountains, or seating for outdoor use.

12.3.2 Sand Casting

Sand casting, also known as sand moulded casting, is a metal casting process characterized by using sand as the mould material. The term "sand casting" can also refer to an object produced via the sand casting process. Sand castings are produced in specialized factories called foundries. Over seventy percent of all metal castings are produced via sand casting process.

Moulds made of sand are relatively cheap, and sufficiently refractory for use in steel foundry. In addition to the sand, a suitable bonding agent (usually clay) is mixed with the sand. The mixture is moistened, typically with water, to develop the strength and plasticity of the clay and to make the aggregate suitable for moulding. The sand is typically contained in mould boxes known as flask. The mould cavities and gate system are created by compacting the sand around models, or patterns, or carved directly into the sand.

Moulding sands, also known as foundry sands, are defined by eight characteristics: refractoriness, chemical inertness, permeability, surface finish, cohesiveness, flowability, collapsibility, and availability/cost.

Steps in Sand Casting

There are six steps in this process:

1. Place a pattern in sand to create a mould.
2. Incorporate the pattern and sand in a gating system.
3. Remove the pattern.
4. Fill the mould cavity with molten metal.
5. Allow the metal to cool.
6. Break away the sand mould and remove the casting.

Pattern Making

Expendable moulds require some sort of pattern. The interior cavities of the mould, in which the molten metal will solidify, are formed by the impression of this pattern. Pattern design is crucial to success in manufacture by expendable mould metal casting. The pattern is a geometric replica of the metal casting to be produced. It is made slightly oversize to compensate for the shrinkage that will occur in the metal during the casting's solidification, and whatever amount of material that will be machined off the cast part afterwards. Although machining will add an extra process to the manufacture of a part, it improves surface finish and part dimensions considerably.

Pattern Material

The material from which the pattern is made is dependent upon the type of mold and metal casting process, the casting's geometry and size, the dimensional accuracy required, and the number of metal castings to be manufactured using the pattern. Patterns can be made from wood, like pine (softwood), or mahogany (hardwood), various plastics, or metal, like aluminum, cast iron, or steel. In most manufacturing operations, patterns will be coated with a parting agent to ease their removal from the mould.

Cores

For metal castings with internal geometry cores are used. A core is a replica, (actually an inverse), of the internal features of the part to be cast. Like a pattern, the size of the core is designed to accommodate for shrinkage during the metal casting operation. Unlike a pattern, a core remains in the mould while the metal is being poured. Hence, a core is usually made of a similar material as the mould. Once the metal casting has hardened, the core is broken up and removed much like the mould. Structural supports that hold the core in place are called chaplets. The chaplets are made of a material with a higher melting temperature than the casting material, and become assimilated into the part when it hardens.

Mould Making

When manufacturing by metal casting, consideration of the mould is essential. The pattern is placed in the mould and the mould material is packed around it. The mould contains two parts, the drag (bottom), and the cope (top). The parting line between the cope and drag allows for the mould to be opened and the pattern to be removed once the impression has been made.

There are many different metal casting processes used in the manufacture of parts. Those are expendable mould casting and permanent mould casting. As the name implies, expendable moulds are used for only one metal casting while permanent moulds are used for many. When considering manufacturing processes, there are advantages and disadvantages as discussed below:

Expendable Mould

- Can produce one metal casting only
- Made of sand, plaster, or other similar material
- Binders used to help material hold its form
- Mould that metal solidifies in must be destroyed to remove casting
- More intricate geometries are possible for casting

Permanent Mould

- Can manufacture many metal castings
- Usually made of metal or sometimes a refractory ceramic
- Mould has sections that can open/close, permitting removal of the casting
- Need to open mould limits part shapes

Elements of a Gating System

Pouring Basin

This is where the molten metal employed to manufacture the part enters the mould. The pouring basin should have a projection with a radius around it to reduce turbulence.

Down Sprue

From the pouring basin, the molten metal for the casting travels through the down sprue. This should be tapered so its cross-section is reduced as it goes downward.

Sprue Base

The down sprue ends at the sprue base. It is here that the casting's inner cavity begins.

Ingate/Choke Area

Once at the sprue base, the molten material must pass through the ingate in order to enter the inner area of the mould. The ingate is very important for flow regulation during the metal casting operation.

Runners

Runners are passages that distribute the liquid metal to the different areas inside the mould.

Main Cavity

The impression of the actual part to be cast is often referred to as the main cavity.

Vents

Vents help to assist in the escape of gases that are expelled from the molten metal during the solidification phase of the metal casting process.

Risers

Risers are reservoirs of molten material. They feed this material to sections of the mould to compensate for shrinkage as the casting solidifies. There are different classifications for risers.

***Top Risers*:** Risers that feed the metal casting from the top.

***Side Risers*:** Risers that feed the metal casting from the side.

***Blind Risers*:** Risers that are completely contained within the mould.

***Open Risers*:** Risers that are open at the top to the outside environment.

12.3.3 Die Casting

Die casting is a manufacturing process that can produce geometrically complex metal parts through the use of reusable moulds, called dies. The die casting process involves the use of a furnace, metal, die casting machine, and die. The metal, typically a non-ferrous alloy such as aluminum or zinc, is melted in the furnace and then injected into the dies in the die casting machine. There are two main types of die casting machines - hot chamber machines (used for alloys with low melting temperatures, such as zinc) and cold chamber machines (used for alloys with high melting temperatures, such as aluminum). However, in both machines, after the molten metal is injected into the dies, it rapidly cools and solidifies into the final part, called the casting.

Fig. 12.12 An engine block with aluminium and magnesium die castings.

Hot chamber die casting

Hot-chamber die casting, also known as gooseneck machines, relies upon a pool of molten metal to feed the die. At the beginning of the cycle the piston of the machine is retracted, which allows the molten metal to fill the "gooseneck". The pneumatic- or hydraulic-powered piston then forces this metal out of the gooseneck into the die. The advantages of this system include fast cycle times (approximately 15 cycles a minute) and the convenience of melting the metal in the casting machine. The disadvantages of this system are that it is limited to use with low-melting point metals and that aluminium cannot be used because it picks up some of the iron while in the molten pool. Therefore, hot-chamber machines are primarily used with zinc, tin, and lead-based alloys.

Cold chamber die casting

These are used when the casting alloy cannot be used in hot-chamber machines. These include aluminium, zinc alloys with a large composition of aluminium, magnesium and copper. The process for these machines starts with melting the metal in a separate furnace. Then a precise amount of molten metal is transported to the cold-chamber machine where it is fed into an unheated shot chamber (or injection cylinder). This shot is then driven into the die by a hydraulic or mechanical piston. The biggest disadvantage of this system is the slower cycle time due to the need to transfer the molten metal from the furnace to the cold-chamber machine.

12.3.4 Centrifugal Casting

It is a casting technique that is typically used to cast thin-walled cylinders. It is used to cast such materials as metal, glass, and concrete. It is noted for the high quality of the results attainable, particularly for precise control of their metallurgy and crystal structure. Unlike most other casting techniques, centrifugal casting is chiefly used to manufacture stock materials in standard sizes for further machining, rather than shaped parts tailored to a particular end-use.

Typical materials that can be cast with this process are iron, steel, stainless steels, glass, and alloys of aluminum, copper and nickel. Two materials can be cast together by introducing a second material during the process.

Typical parts made by this process are pipes, flywheels, cylinder liners and other parts that are axi-symmetric. It is notably used to cast cylinder liners and sleeve valves for piston engines, parts which could not be reliably manufactured otherwise.

The end result of centrifugal casting is a metal part that is:

- Clean
- Dense
- Uniform

12.3.5 Lost-Wax Precision Casting or Investment Casting

Lost-wax casting (precision casting) is the process by which a duplicate metal sculpture (often silver, gold, brass or bronze) is cast from an original sculpture. Depending on the sculptor's skills, intricate shapes can be manufactured by this method. Though the process today varies from foundry to foundry, the steps used in casting small bronze sculptures are fairly standardized. Variations of the process include: "lost mould", which recognizes that materials other than wax can be used (such as: tallow, resin, tar, and textile), and "waste wax process" (waste mould casting), because the mould is destroyed to remove the cast item.

Procedural steps for Investment Casting

Casts can be made of the wax model itself, the direct method, or of a wax copy of a model that need not be of wax, the indirect method. These are the steps for the indirect process:

1. **Model-making:** An artist or mould-maker creates an original model from wax, clay, or another material. Wax and oil-based clay are often preferred because these materials retain their softness.
2. **Mould making:** A mould is made of the original model or sculpture. The rigid outer moulds contain the softer inner mould, which is the exact negative of the original model. Inner moulds are usually made of latex, polyurethane rubber or silicone, which is supported by the outer mould. The outer mould can be made from plaster, but can also be made of fiber glass or other materials. Most moulds are made of at least two pieces, and a shim with keys is placed between the parts.
3. **Wax:** Once the mould is finished, molten wax is poured into it and swished around until an even coating, usually about $\frac{1}{8}$ inch (3 mm) thick, covers the inner surface of the mould. This is repeated until the desired thickness is reached. Another method is to fill the entire mould with molten wax and let it cool until a desired thickness has set on the surface of the mould. After this, the rest of the wax is poured out again, the mould is turned upside down and the wax layer is left to cool and harden.
4. **Removal of wax:** This hollow wax copy of the original model is removed from the mould. The model-maker may reuse the mould to make multiple copies, limited only by the durability of the mould.
5. **Chasing:** Each hollow wax copy is then "chased": a heated metal tool is used to rub out the marks that show the parting line or flashing where the pieces of the mould came together. The wax is dressed to hide any imperfections. The wax now looks like the finished piece. Wax pieces that were moulded separately can now be heated and attached; foundries often use registration marks to indicate exactly where they go.
6. **Spruing:** The wax copy is sprued with a treelike structure of wax that will eventually provide paths for the molten casting material to flow and for air to escape. The carefully planned spruing usually begins at the top with a wax "cup," which is attached by wax cylinders to various points on the wax copy. The spruing does not have to be hollow, as it will be melted out later in the process.
7. **Slurry:** A sprued wax copy is dipped into a slurry of silica, then into a sand-like stucco, or dry crystalline silica of a controlled grain size. The slurry and grit combination is called ceramic shell mould material, although it is not literally made of ceramic. This shell is allowed to dry, and the process is repeated until at least a half-inch coating covers the entire piece.

8. **Burnout:** The ceramic shell-coated piece is placed cup-down in a kiln, whose heat hardens the silica coatings into a shell, and the wax melts and runs out. The melted wax can be recovered and reused, although it is often simply burned up. Now all that remains of the original artwork is the negative space formerly occupied by the wax, inside the hardened ceramic shell.
9. **Testing:** The ceramic shell is allowed to cool, then is tested to see if water will flow freely through the feeder and vent tubes. Cracks or leaks can be patched with thick refractory paste. To test the thickness, holes can be drilled into the shell, then patched.
10. **Pouring:** The shell is reheated in the kiln to harden the patches and remove all traces of moisture, then placed cup-upwards into a tub filled with sand. Metal is melted in a crucible in a furnace, then poured carefully into the shell. The filled shells are then allowed to cool.
11. **Release:** The shell is hammered or sand-blasted away, releasing the rough casting. The sprues, which are also faithfully recreated in metal, are cut off, the material to be reused in another casting.
12. **Metal-chasing:** Just as the wax copies were chased, the casting is worked until the telltale signs of the casting process are removed, so that the casting now looks like the original model. Pits left by air bubbles in the casting and the stubs of the spruing are filed down and polished.

The methods used for small parts and jewellery vary somewhat from those used for sculpture. A wax model is obtained either from injection into a rubber mould or by being custom-made by carving. In dentistry, gold crowns, inlays and onlays are made by the lost-wax technique.

12.3.6 Special Casting Methods

Now a day's special casting process has been developed to effect a saving time and expense to produce a better quality casts. In comparing to sand casting, this process does not require drying or baking of moulds or cores. Rapid hardening action takes place due to chemical reactions in them.

Permanent Mould Casting

Permanent mould casting is a metal casting process that employs reusable moulds (permanent moulds), usually made from metal. The most common process uses gravity to fill the mould, however gas pressure or a vacuum are also used. A variation on the typical gravity casting process, called slush casting, produces hollow castings. It produces typical products or components such as gears, splines, wheels, gear housings, pipe fittings, fuel injection housings, and automotive engine pistons.

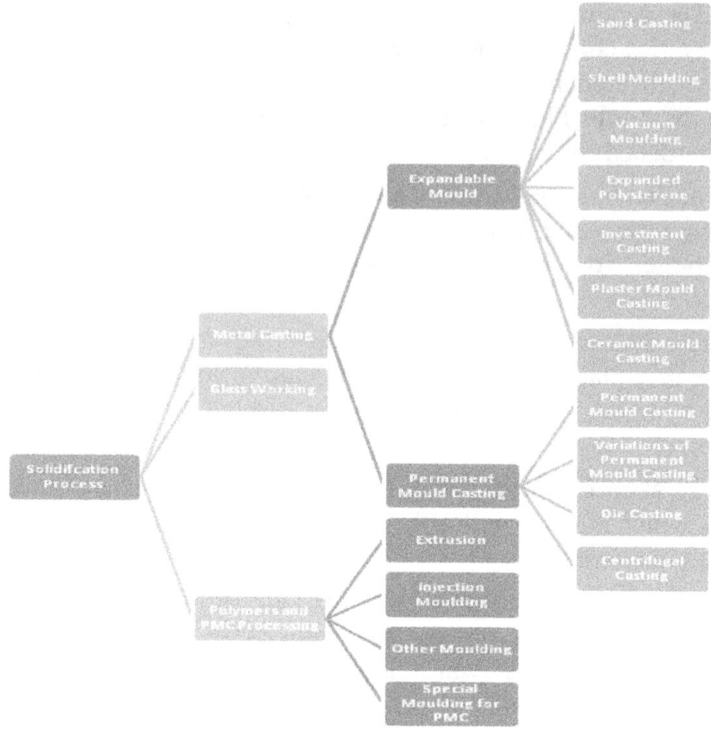

Fig. 12.13 Classification of casting processes.

Types of Permanent Mould Casting

There are four main types of permanent mould casting: gravity, slush, low-pressure, and vacuum. The same have been discussed below:

1. **Gravity Casting:** The gravity process begins by preheating the mould to 150–200 °C (300–400 °F) to ease the flow and reduce thermal damage to the casting. The mould cavity is then coated with a refractory material or a mould wash, which prevents the casting from sticking to the mould and prolongs the mould life. Any sand or metal cores are then installed and the mould is clamped shut. Molten metal is then poured into the mould. Soon after solidification the mould is opened and the casting removed to reduce chances of hot tears. The metal is poured at the lowest practical temperature in order to minimize cracks and porosity.
2. **Slush Casting:** It is to create a hollow casting or hollow cast. In the process the material is poured into the mould and allowed to cool until a shell of material forms in the mould. The remaining liquid is then poured out to leave a hollow shell. The process is usually used to cast ornamental products, from low-melting-point materials. It uses less material than solid casting, and results in a lighter and less expensive product.

Hollow cast shapes generally have a small hole where the excess liquid was poured out. Thus, slush moulding is used in automotive thermoplastic dashboard manufacture, where a liquid resin is poured into a hot, hollow mould and a viscous skin forms. Excess slush is drained off, the mould is cooled, and the moulded product is stripped out.

3. **Low-Pressure Casting:** Low-pressure permanent mould (LPPM) casting uses a gas at low pressure, usually between 3 and 15 psi (20 to 100 kPa) to push the molten metal into the mould cavity. The pressure is applied to the top of the pool of liquid, which forces the molten metal up a refractory pouring tube and finally into the bottom of the mould. The pouring tube extends to the bottom of the ladle so that the material being pushed into the mould is exceptionally clean. Majority of LPPM casting are from aluminium, magnesium, and copper alloys. Advantages include very little turbulence when filling the mould because of the constant pressure, which minimizes gas porosity and dross formation. Mechanical properties are about 5% better than gravity permanent mould castings. However, cycles times are longer than gravity permanent mould castings.
4. **Vacuum Casting:** Vacuum castings possess all the advantages of LPPM casting. Again, the dissolved gases in the molten metal are minimized and molten metal cleanliness is better. The process can handle thin-walled profiles and provides an excellent surface finish. Mechanical properties are usually 10 to 15% better than gravity permanent mould castings. The process is limited in weight of 0.2 to 5 kg (0.44 to 11.02 lb).

Advantages of Special Casting Methods

(a) Greater dimensional accuracy.
(b) Higher metallurgical quality.
(c) Lower production cost (in certain cases).
(d) Ability to cast extremely thin sections.
(e) High production rates.
(f) Better surface finish on the castings; therefore low labour and finishing costs. Minimum need for further machining of castings.
(g) Castings may possess a denser and finer grain structure.
(h) Castings are slightly stronger and more ductile than solid mould castings.

12.3.7 Shell Casting/Moulding

The 2-piece pattern is made of metal (aluminum or steel), it is heated to between 175°C-370° C, and coated with a lubricant, e.g. silicone spray. Each heated half-pattern is covered with a mixture of sand and a thermoset resin/epoxy binder. The binder glues a layer of sand to the pattern, forming a shell. The process may be repeated to get a thicker shell. The patterns are removed, and the two half-shells joined together to form the mould, metal is poured into the mould. When the metal solidifies, the shell is broken to get the part. Shell-mould casting yields better surface quality and tolerances.

Advantages of Shell Moulding

- Smoother cavity surface permits easier flow of molten metal and better surface finish on casting.
- Good dimensional accuracy Machining often not required.
- Mould collapsibility usually avoids cracks in casting Can be mechanized for mass production.

Disadvantages of Shell Moulding

- More expensive metal pattern.
- Difficult to justify for small quantities.

12.3.8 Casting Defects

A casting defect is an undesired irregularity in a metal casting process. Some defects can be tolerated while others can be repaired, otherwise they must be eliminated. They are broken down into five main categories:

(a) Gas porosity,
(b) Shrinkage defects,
(c) Mould material defects,
(d) Metal penetration,
(e) Pouring metal defects, and
(f) Metallurgical defects.

Gas Porosity

A condition existing in a casting caused by the trapping of gas in the molten metal or by mould gases evolved during the pouring of the casting. The defects in this category can be classified into blowholes and pinhole porosity. The above may be due to: metal containing gas, mould is too hot, poor mould burnout etc.

Shrinkage Cavities

These are caused by liquid shrinkage occurring during the solidification of the casting. To compensate for this, proper feeding of liquid metal is required. For this reason risers are placed at the appropriate places in the mould. Sprues may be too thin, too long or not attached in the proper location, causing shrinkage cavities. It is recommended to use thick sprues to avoid shrinkage cavities.

Moulding Material Defects

The defects in this category are cuts and washes, metal penetration, fusion, and swell. Cut and washes appear as rough spots and areas of excess metal, and are caused by erosion of moulding sand by the flowing metal. This is caused by the moulding sand not having enough strength and the molten metal flowing at high velocity.

Metal penetration

When molten metal enters into the gaps between sand grains, the result is a rough casting surface. This occurs because the sand is coarse or no mould wash was applied on the surface of the mould. The coarser the sand grains more the metal penetration.

***Fusion*:** The main reason for this is that the clay or the sand particles are of lower refractoriness or that the pouring temperature is too high.

***Swell*:** Under the influence of metallostatic forces, the mold wall may move back causing a swell in the dimension of the casting.

***Inclusions*:** Particles of slag, refractory materials, sand or deoxidation products are trapped in the casting during pouring solidification.

Pouring Metal Defects

The likely defects in this category are: mis-runs and cold shuts. A mis-run is caused when the metal is unable to fill the mould cavity completely and thus leaves unfilled cavities. A cold shut is caused when two streams while meeting in the mould cavity, do not fuse together properly thus forming a discontinuity in the casting.

Mould Shift

The mould shift defect occurs when cope and drag or moulding boxes have not been properly aligned.

12.3.9 Preventive Methods of Casting Defects

1. **Gas porosity** can be eliminated through good mould design or by introducing nitrogen into the aluminium metal before the liquid pour. Hipping can also be used to eliminate small amounts of porosity inside the casting. This involves placing the casting in a high temperature furnace, which is enclosed in a pressurised container. The casting is then heated in an inert gas, commonly argon, to eliminate the porous elements.
2. **Shrinkage** is addressed through the use of risers or chills. Pattern maker's shrinkage occurs due to the natural shrinkage of the molten metal as it cools to room temperature. A shrinkage allowance is therefore factored into the pattern design at the start of the process.
3. **Inclusions** can be addressed by implementing improved mould design and/or pouring techniques.
4. **Cavities** can be avoided by increasing the pouring temperature, faster material pouring, providing sufficient fluidity in the molten metal, and providing sufficient material to fill the mould.

5. **Dimensional inconsistencies** can be resolved through good pattern and mould design. Pressure from the molten metal lifting the top off the mould can also lead to dimensional anomalies. This can be addressed by applying metal weights to the mould during the casting process.
6. **Discontinuties** can be avoided by changing the chemical composition of the metal or by increasing the pouring temperature. There are several ways to prevent this problem, such as heating up the insert before the metal pour as well as using special coatings or shot blasting to improve adhesion.
7. **Misruns** are a common casting defect and they occur when the molten metal is unable to fill the extremities of the mould, which leaves unfilled cavities. All of these issues can be resolved through good process design.

12.4 Welding

Introduction

Welding is a process in which two or more parts are joined permanently at their touching surfaces by a suitable application of heat and/or pressure. Often a filler material is added to facilitate coalescence. The assembled parts that are joined by welding are called a weldment. Welding is primarily used in metal parts and their alloys.

12.4.1 Types of Grooves

Groove welding is a reference to the type of joint being prepared. There are different ways to prepare a groove for welding. "V-groove" is a common one with plates butted together forming a kind of "v" shape, this can be at multiple angles. There are also "U-groove", and "J-groove" (half a 'U-groove' and half 90 degree edge, or butt). In any case, no matter the type of joint, the type of groove, the cleaner, the better.

Types of Welding Positions

All welding can be classified according to the position of the workpiece or the position of the welded joint on the plates or sections being welded. Fillet, groove and surface welds may be made in all of the positions. There are four basic welding positions, as shown in figures:

Types of Welded Joints

Welds can be geometrically prepared in many different ways. The five basic types of weld joints are the butt joint, lap joint, corner joint, edge joint, and T-joint. Many welding processes require the use of a particular joint design, for example, resistance spot welding, laser beam welding, and electron beam welding are most frequently performed on lap joints. Other welding methods, like shielded metal arc welding, are extremely versatile and can weld virtually any type of joint. Some common welded joints are shown in the figure below:

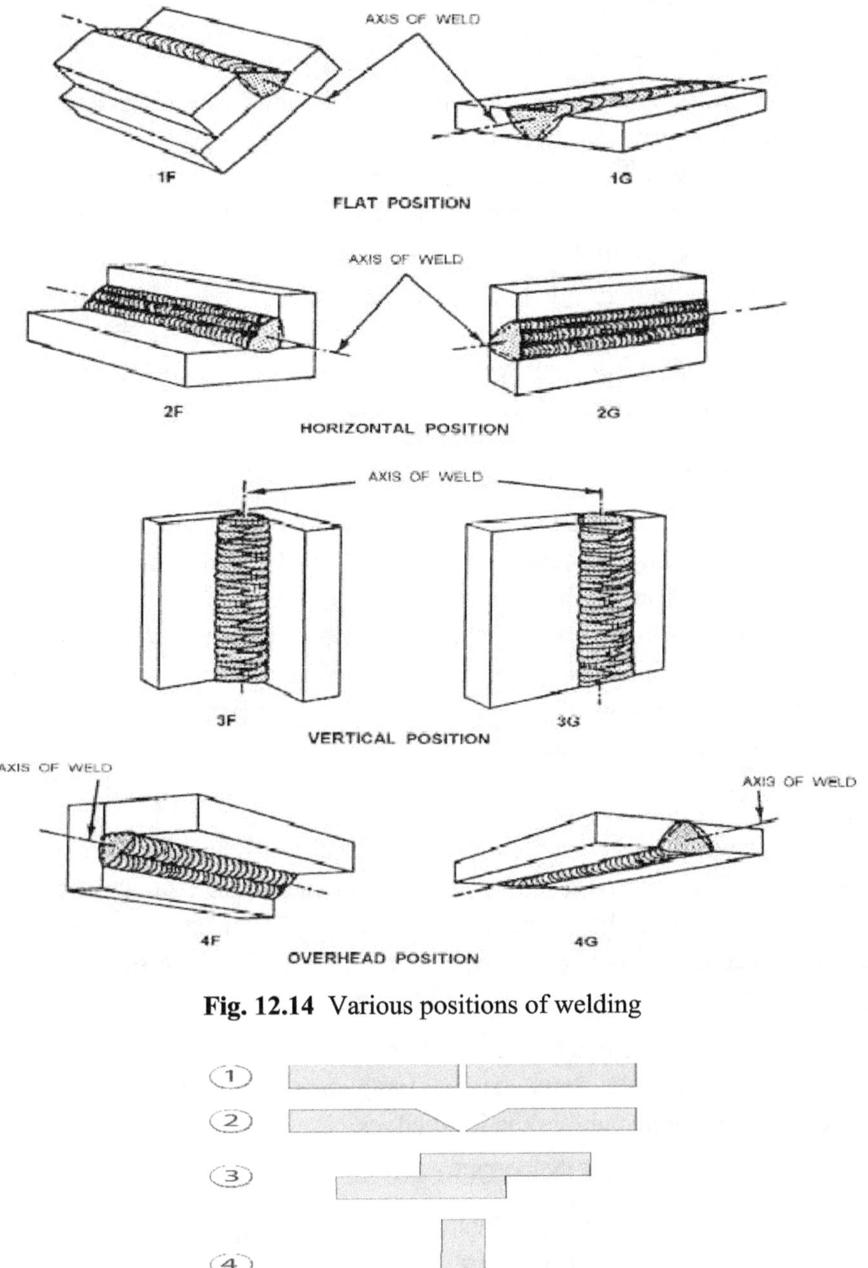

Fig. 12.14 Various positions of welding

1
2
3
4

Fig. 12.15 Common welding joint types – 1. Square butt joint, 2. V butt joint, 3. Lap joint, 4. T-joint.

12.4.2 Classification of Welding Processes

Welding processes are classified into two major groups:

1. **Fusion welding:** In this process, base metal is melted by means of heat. Often, in fusion welding operations, a filler metal is added to the molten pool to facilitate the process and provide bulk and strength to the joint. Commonly used fusion welding processes are: arc welding, resistance welding, oxyfuel welding, electron beam welding and laser beam welding.
2. **Solid-state welding:** In this process, joining of parts takes place by application of pressure alone or a combination of heat and pressure. No filler metal is used. Commonly used solid-state welding processes are: diffusion welding, friction welding, ultrasonic welding

12.4.3 Gas Welding

The most common gas welding process is oxyacetylene welding. It is one of the oldest and most versatile welding processes. It is widely used for welding pipes and tubes, as well as repair work. The equipment is relatively inexpensive and simple, generally employing the combustion of acetylene in oxygen to produce a high welding flame temperature. The flame, since it is less concentrated than an electric arc, causes slower weld cooling, which can lead to greater residual stresses and weld distortion, though it eases the welding of high alloy steels.

Oxyacetylene Weld

Oxyacetylene gas welding is commonly used to permanently join mild steel. A mixture of oxygen and acetylene, burns as an intense / focussed flame, at approximately 3,500 degrees centigrade. When the flame comes in contact with steel, it melts the surface forming a molten pool, allowing welding to take place. This type of welding is suitable for the prefabrication of steel sheet, tubes and plates.

The oxy-acetylene flame has two distinct zones. The inner zone (primary combustion zone) is the hottest part of the flame. The welding should be performed so as the point of the inner zone should be just above the joint edges.

$$C_2H_2 + O_2 - 2\ CO + H_2$$

The outer zone (secondary combustion zone) envelope performs two functions: (i) preheats the joint edges, (ii) prevents oxidation by using some of the surrounding oxygen from weld pool for combustion and gives off carbon dioxide and water vapour.

$$CO + H_2 + O_2 - CO_2 + H_2O$$

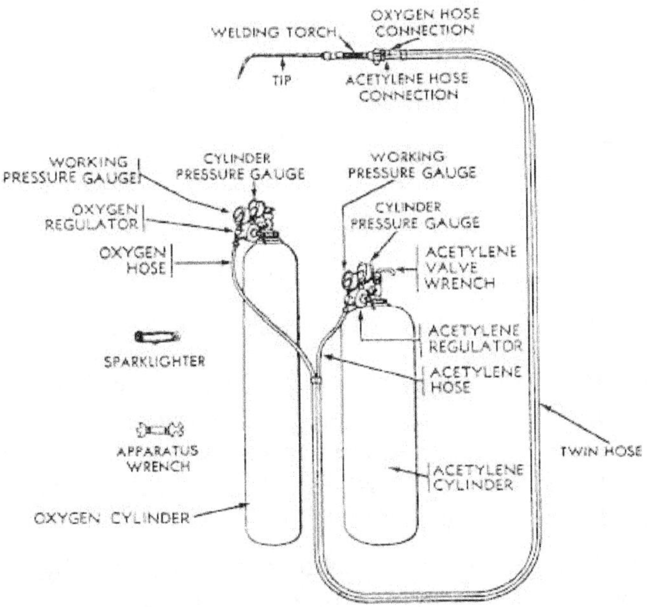

Fig. 12.16 Oxyacetylene welding.

Advantages of Oxyacetylene Welding

- It is easy to implement.
- The equipment is cheaper than most other types of welding rigs
- The equipment is more portable than most other types of welding rigs
- Oxyacetylene equipment can also be used to "flame-cut" large pieces of material.

Disadvantages of Oxyacetylene Welding:

- Weld lines are much rougher in appearance, and require more finishing operations.
- Welds have large heat affected zones

12.4.4 Electric Arc Welding

These processes use a welding power supply to create and maintain an electric arc between an electrode and the base material to melt metals at the welding point. They can use either direct (DC) or alternating (AC) current, and consumable or non-consumable electrodes. The welding region is sometimes protected by some type of inert or semi-inert gas, known as a shielding gas, and filler material is sometimes used as well.

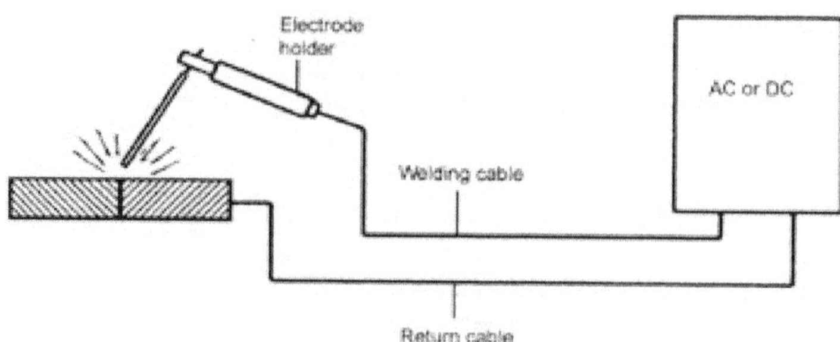

Fig. 12.17 Electric arc welding.

Direct current (DC) is the unidirectional flow of electric charge. DC is produced by sources such as batteries, power supplies, thermocouples, solar cells, or dynamos. It may flow in a wire, but can also flow through semiconductors, insulators, or even through a vacuum as in electron or ion beams. The electric current flows in a constant direction, distinguishing it from alternating current (AC).

DC may be obtained from an alternating current supply by use of a rectifier, which contains electronic elements (usually) or electromechanical elements (historically) that allow current to flow only in one direction. Direct current may be converted into alternating current with an inverter or a motor-generator set. Direct current is used to charge batteries and as power supply for electronic systems. Very large quantities of direct-current power are used in production of aluminum and other electrochemical processes. It is also used for some railways, especially in urban areas.

Types of Electric Arc Welding

Arc welding is a method of permanently joining two or more metal parts. It consists of combination of different welding processes wherein coalescence is produced by heating with an electric arc, with/without the application of pressure and with or without the use of filler metals depending upon the base plate thickness. A homogeneous joint is achieved by melting and fusing the adjacent portions of the separate parts. The final welded joint has unit strength approximately equal to that of the base material. The arc temperature is maintained approximately 4400°C. A flux material is used to prevent oxidation, which decomposes under the heat of welding and releases a gas that shields the arc and the hot metal. It employs an inert or nearly inert gas (helium, argon, and carbon dioxide) to form a protective envelope around the arc and the weld.

12.4.5 Shielded Metal Arc Welding (SMAW)

SMAW is also known as manual metal arc welding (MMA) or stick welding. Electric current is used to strike an arc between the base material and consumable electrode rod, which is made of filler material (typically steel) and is covered with a flux that protects the weld area from oxidation and contamination by producing carbon dioxide gas during the welding process. The electrode core itself acts as filler material.

The process is versatile and can be performed with relatively inexpensive equipment as shown in fig. Weld times are rather slow, since the consumable electrodes must be frequently replaced and because slag, the residue from the flux, must be chipped away after welding. Furthermore, the process is normally limited to welding ferrous materials, though special electrodes have made possible the welding of cast iron, nickel, aluminum, copper, and other metals.

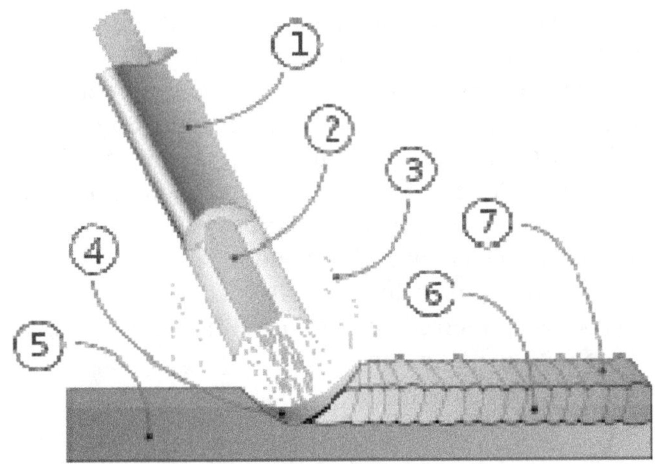

1. Coating Flow, 2. Rod, 3. Shield Gas, 4. Fusion, 5. Base metal, 6. Weld metal, 7. Solidified Slag

Fig. 12.18 Shielded metal arc welding.

Advantages

- The process can be carried out in any position with highest weld quality.
- It is the simplest of all the arc welding processes
- It finds innumerable applications, because of the availability of a wide variety of electrodes.
- Wide variety of metals and their alloys can be welded easily.
- The process can be very well employed for hard facing and metal resistance etc.
- Difficult to weld joints can be easily accomplished by this method.
- The welding equipment is portable and the cost is fairly low.

Disadvantages

- Due to flux coated electrodes, the chances of slag entrapment are more.
- Duo to fumes and particles of slag, welding control in this process is bit difficult.
- Due to limited length of electrode and brittle flux coating, mechanization is difficult.

- While welding long joints, if not properly cared, defects may occur at the place where welding is restarted with the new electrode
- The process uses stick electrodes and thus it is slower.

12.4.6 Metal Inert Gas Arc Welding (MIG)/Gas Metal Arc Welding (GMAW)

MIG welding is a semi-automatic or automatic process that uses a continuous wire feed as an electrode and an inert or semi-inert gas mixture to protect the weld from contamination. Since the electrode is continuous, welding speeds are greater. It is a commonly used high deposition rate welding process. Along with the wire electrode, a shielding gas feeds through the welding gun, which shields the process from contaminants in the air. A constant voltage, direct current power source is most commonly used with GMAW, but constant current systems, as well as alternating current, can be used. Originally developed for welding aluminium and other non-ferrous materials, GMAW was soon applied to steels also. Today, GMAW is the most common industrial welding process, preferred for its versatility, speed and the relative ease of adapting the process to robotic automation.

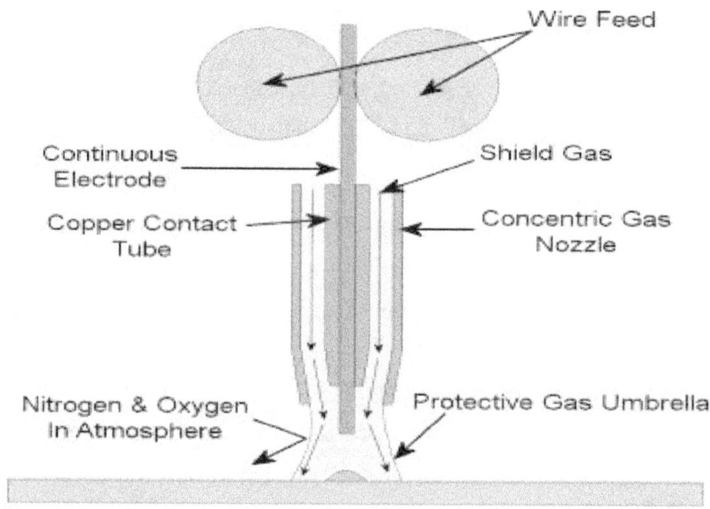

Fig. 12.19 Metal inert gas arc welding.

Advantages

1. Higher welding speeds
2. Greater deposition rates
3. Less post welding cleaning (e.g. no slag to chip off weld)
4. Better weld pool visibility

5. Low skill factor required to operate the welding torch
6. Positional welding offers no problems when compared to other processes
7. The process is easily automated
8. No fluxes required in most cases

Disadvantages

1. Higher initial setup cost
2. Process is limited to draught free conditions
3. Higher maintenance costs due to extra electronic components
4. The setting of plant variables requires a high skill level
5. Less efficient where high duty cycle requirements are necessary
6. Radiation effects are more severe

12.4.7 Tungsten Inert Gas Arc Welding (TIG)/Gas Tungsten Arc Welding (GTAW)

TIG welding is a manual welding process that uses a nonconsumable tungsten electrode, an inert or semi-inert gas mixture, and a separate filler material. Especially useful for welding thin materials, this method is characterized by a stable arc and high quality

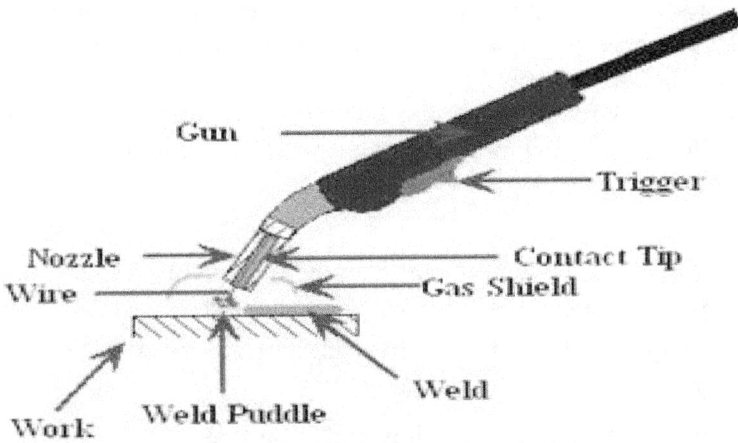

Fig. 12.20 Tungsten inert gas arc welding.

welds. The weld area is protected from atmospheric contamination by an inert shielding gas (argon or helium), and a filler metal is normally used. A constant-current welding power supply produces electrical energy, which is conducted across the arc through a column of highly ionized gas and metal vapors known as a plasma. GTAW can be used on nearly all weldable metals, though it is most often applied to stainless steel and light metals. It is used when quality welds are extremely important.

Advantages

1. Non-consumable electrode helps to provide flawless joints reducing downtime in production.
2. No flux is required because inert gas shields molten metal. So no slag and slag inclusion problems.
3. High quality and strong welding achieved.
4. Applicable to a very wide range of materials.
5. Especially good for welding thin sections and delicate workpieces
6. Capable of producing welds of high quality and appearance

Disadvantages

1. TIG is a time-consuming process, slower. Thus, low filler deposition rate.
2. More complicated, skilled and professional workers are needed.
3. Exposed to high intensity of light which can cause eye damage.
4. High initial cost.
5. Thicker sheets of metal are not welded.

12.4.8 Submerged Arc Welding (SAW)

SAW is a high-productivity welding method in which the arc is struck beneath a covering layer of flux. This increases arc quality, since contaminants in the atmosphere are blocked by the flux. The slag that forms on the weld generally comes off by itself, and combined with the use of a continuous wire feed, the weld deposition rate is high. This process uses AC or DC, high current 300-4000A. The process is commonly used in industry, especially for large products and in the manufacture of welded pressure vessels.

Advantages

- Lends itself to the production of consistently high quality welds with minimum operator skills.
- Minimum of welding fume and of arc visibility (radiation).
- Well suited to welding thick sections.
- Suitable for welding carbon, low alloy and alloy steels.
- Relatively high metal deposition rates

Disadvantages

- Flat or horizontal position welding only
- Care required to preserve correct electrode alignment, as electrode
- Tip and weld pool are underneath solid flux cover

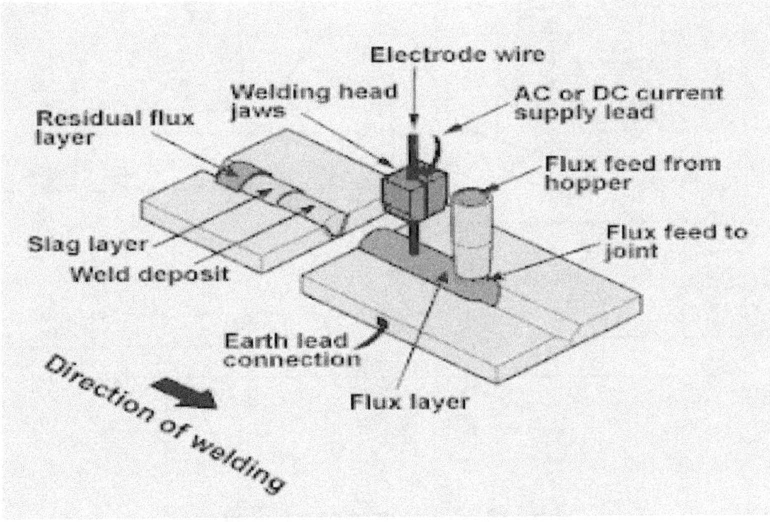

Fig. 12.21 Submerged arc welding

12.4.9 Electroslag Welding (ESW)

ESW is a highly productive, single pass welding process for thick materials in a vertical or close to vertical position. An electric arc is initially struck by wire that is fed into the desired weld location and then flux is added. Additional flux is added until the molten slag, reaching the tip of the electrode, extinguishes the arc. The wire is then continually fed through a consumable guide tube into the surfaces of the metal workpieces and the filler metal are then melted using the electrical resistance of the molten slag to cause coalescence. The wire and tube then move up along the workpiece while a copper retaining shoe that was put into place before starting is used to keep the weld between the plates that are being welded. This process uses a DC voltage usually ranging from about 600 A and 40-50 V, higher currents are needed for thicker materials.

Advantages

- Joint preparation is often much simpler than for other welding processes.
- Much thicker steels (450 mm) can be welded in single pass and more economically.
- It gives extremely high deposition rates.
- Residual stresses and distortion produced are low.
- Flux consumption as compared to that in submerged arc welding is very low.
- Since no arc exists, no spattering or intense arc flashing occurs.

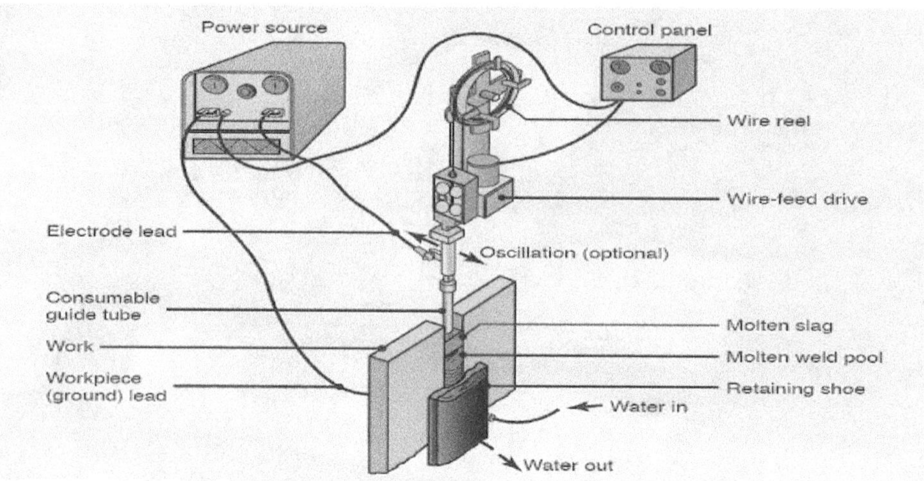

Fig. 12.22 Electroslag welding.

Disadvantages

- Uneconomical than other welding for joints below 60 mm.
- There is tendency toward hot cracking and notch sensitivity in the heat affected zone.

12.4.10 Plasma Arc Welding

This process is similar to TIG. A non-consumable electrode is used in this process. Arc plasma is a temporary state of gas. The gas gets ionized after the passage of electric current and becomes a conductor of electricity. The plasma consists of free electrons, positive ions, and neutral particles. Plasma arc welding differs from GTAW welding in the amount of ionized gas which is greatly increased in plasma arc welding, and it is this ionized gas that provides the heat of welding. The temperature of lasma gas may be as high as 33, 000^0 C, which is sufficient to melt any work piece.

Advantages

- Requires less operator skill due to good tolerance of arc to misalignments
- High welding rate
- High penetrating capability (keyhole effect)

Disadvantages

- Expensive equipment
- High distortions and wide welds as a result of high heat input

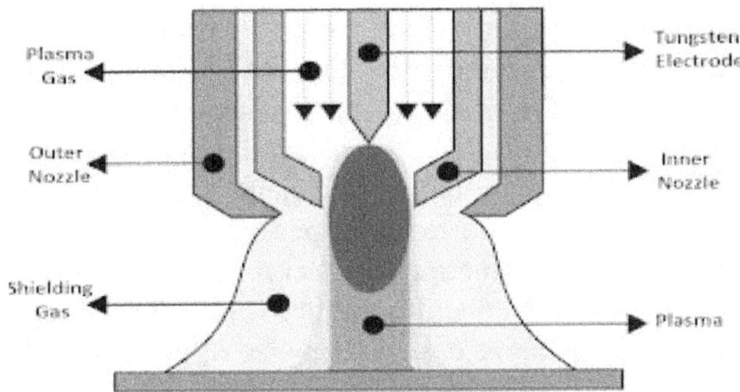

Fig. 12.23 Plasma arc welding.

12.4.11 Resistance Welding - Spot, Seam, Projection and Flash Welding

Resistance welding involves the generation of heat by passing current through the resistance caused by the contact between two or more metal surfaces. Small pools of molten metal are formed at the weld area as high current (1000–100,000 A) is passed through the metal. In general, resistance welding methods are efficient and cause little pollution, but the equipment cost is high.

All resistance welding like spot welding, seam welding, projection welding etc. are worked on same principle of heat generation due to electric resistance. When a current passes through electric resistance, it produces heat. This is same principle which is used in electric coil. The amount of heat produced depends on resistance of material, surface conditions, current supplied, time duration of current supplied etc. This heat generation takes place due to conversion of electric energy into thermal energy. The heat generation formula is:

$$H = I^2RT$$

where,

H = Heat generated in joule

I = Electric current in ampere

R = Electric resistance in Ohm

T = Time of current flow in second

This heat is used to melt the interface metal to form a strong weld joint by fusion. This process produces weld without application of any filler material, flux and shielding gases.

1. **Spot welding:** It is simplest type of resistance welding in which the work pieces are held together under pressu re of anvil face. The copper electrodes are brought

in contact with work piece and current start to flow through it. The work piece material applies some resistance in flow of current which cause local heat generation. At the interface surfaces the resistance is high due to air gap. The current starts to flow though it that melt down the interface surface. The amount of current supply and time should be sufficient for proper melting of interface surfaces. During the process, current stopped to flow but the pressure applied by electrode maintained for a fraction of second, while the weld rapidly cooled. Again, the electrode is removed and brought to contact at other spot. It will create a circular nugget. The nugget size depends on size of electrode, diameter of the order 4-7 mm.

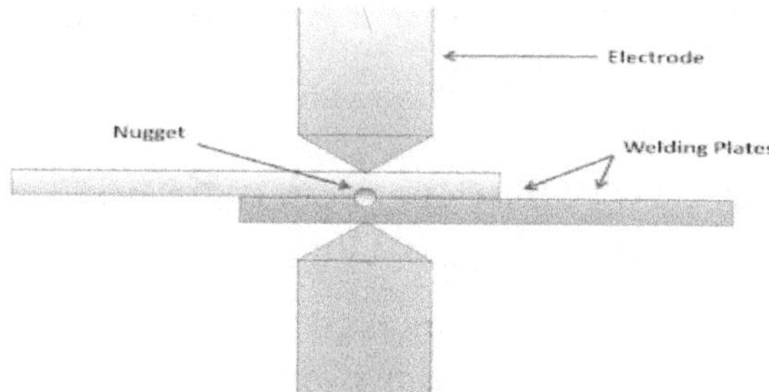

Fig. 12.24 Spot welding

2. **Seam Welding:** Seam welding is also called continuous spot welding in which a roller type electrode is used to flow current through work pieces. First the rollers are brought in contact with work piece. A high ampere current (2000-5000) is passed through these rollers. This will melt the interface surfaces and form a weld joint. Now the rollers start rolling at work plates. This will create a continuous weld joint. The timing of the weld and movement of electrode is controlled to assure that the weld overlap and work piece does not get too hot. The welding speed is about 60 in/min in seam welding. It is used to create air tight joints.

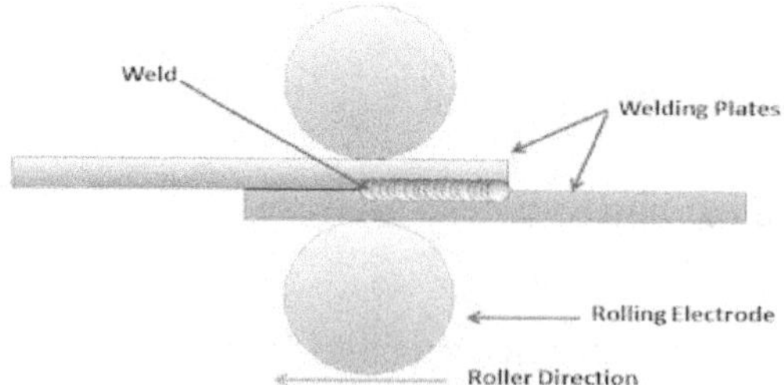

Fig. 12.25 Seam welding

3. **Projection Welding:** Projection welding is same as spot welding except a dimple is produced on work pieces at the location where weld is desired. Now the work pieces held between electrode and a large amount of current pass through it. A small amount of pressure is applied through electrode on welding plates. The current pass through dimple which melt down it and the pressure causes the dimple flatten and form a weld.

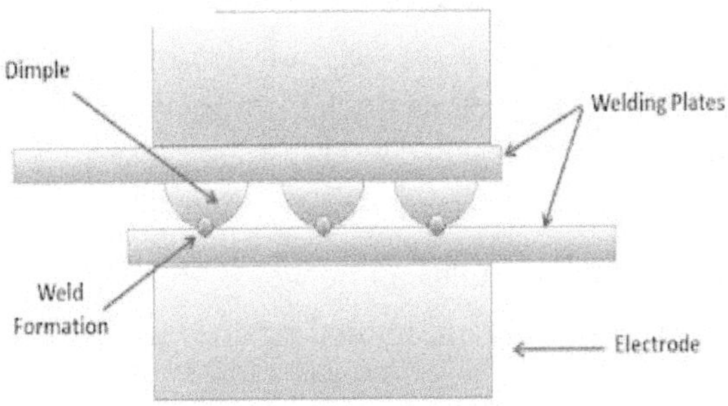

Fig. 12.26 Projection welding.

4. **Flash Welding: Flash welding** is a type of resistance welding that does not use any filler metals. The pieces of metal to be welded are set apart at a predetermined distance based on material thickness, material composition, and desired properties of the finished weld. Current in the range of 100000 ampere is applied to the metal, and the gap between the two pieces creates resistance and produces the arc required to melt the metal. Once the pieces of metal reach the proper temperature, they are pressed together, effectively forge welding them together.

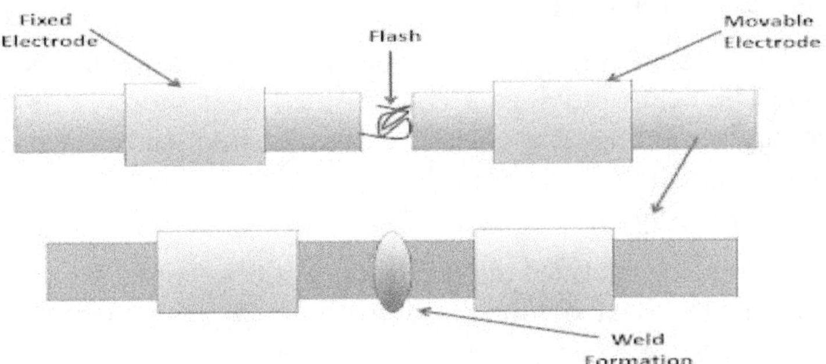

Fig. 12.27 Flash welding.

Applications of resistance welding

- Spot welding is widely used in automotive industries.
- Projection welding is widely used in production of nut and bolt.
- Seam welding is used to produce leak prove joint required in small tanks, boilers etc.
- Flash welding is used to welding pipes and tubes.

Advantages of resistance welding

- It can weld thin (0.1 mm) as well as thick (20mm) metals.
- High welding speed.
- Easily automated.
- Both similar and dissimilar metals can be weld.
- The process is simple and fully automated so does not required high skilled labor.
- High production rate.
- It is environment friendly process.
- It does not require any filler metal, flux and shielding gases.

Disadvantages of resistance welding

- High equipment cost.
- The thickness of work piece is limited due to current requirement.
- It is less efficient for high conductive materials.
- High electric power required.
- Weld joints have low tensile and fatigue strength.

12.4.12 Unconventional Welding Methods

12.4.13 Laser Beam Welding

Laser beam welding is a welding technique used to join multiple pieces of metal through the use of a laser. The beam provides a concentrated heat source, allowing for narrow, deep welds and high welding rates. The process is frequently used in high volume applications using automation, such as in the automotive industry. It is based on keyhole or penetration mode welding.

Laser beam welding is defined as a fusion welding process and coalescence is achieved by utilizing the heat obtained from a concentrated coherent light beam and impinging upon the surface to be joined. This process uses the energy in an extremely concentrated beam of coherent, mono-chromatic light to melt the weld metal. Lasers can interact with any material. It doesn't require a vacuum and it does not produce x-rays.

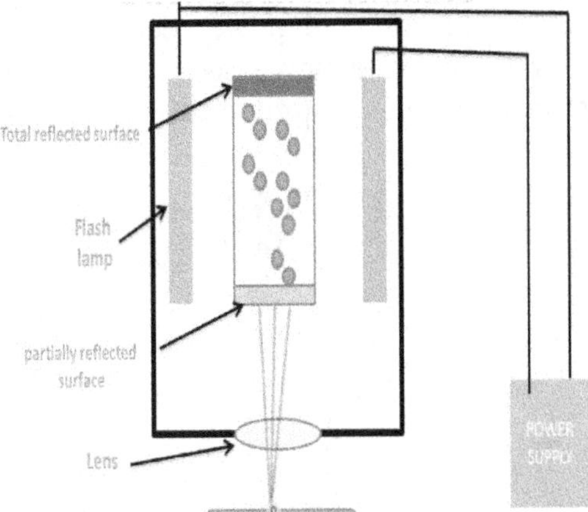

Fig. 12.28 laser beam welding.

Advantages

- Tool wear and breakage are not encountered
- Holes can be located accurately
- Even very tiny holes can be produced
- Machining is extremely rapid
- Holes can also be drilled at difficult entrance angles
- Operating cost is very low

Disadvantages

- The equipment cost is very high
- Tapers are normally encountered in direct drilling of holes
- A blind hole of precise depth is very difficult to achieve with a laser
- The thickness of material which has to be laser drilled is restricted to 50mm
- Adherent materials has to be removed for sure

12.4.14 Electron Beam Welding

Electron beam welding is a fusion welding process in which a beam of high-velocity electrons is applied to two materials to be joined. The work pieces melt and flow together as the kinetic energy of the electrons is transformed into heat upon impact. The process is often performed under vacuum conditions to prevent dissipation of the electron beam.

The electron beam generation takes place in a vacuum, and the process works best when the entire operation and the work piece are in a high vacuum. The beam is collimated and focussed by passing through the field of an electromagnet lens. The focussed area is 0.25-1 mm diameter and power density of 10 KE/mm and voltage 20-2000 KW. However, radiations namely ray, infrared and ultraviolet radiation generates and the welding operator must be protected.

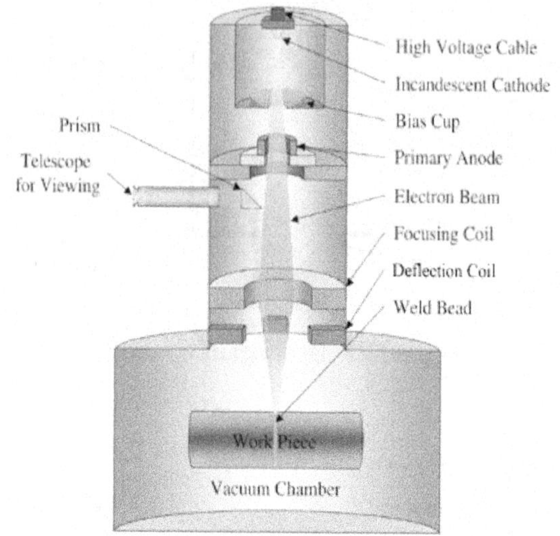

Fig. 12.29 Electron beam welding.

Advantages

- EBM is an excellent method for micro finishing.
- It can cut drill holes or cut slots which otherwise cannot be made.
- It is possible to cut any known material.
- No cutting tool pressure or wear occurred in this process.
- It is distortion free-machining so precise dimensions can be achieved.

Disadvantages

- High equipment cost.
- Highly skill operator required.
- Only small cuts are possible.

12.4.15 Ultrasonic Welding

Ultrasonic welding is used to connect thin sheets or wires made of metal or thermoplastic by vibrating them at high frequency and under high pressure. The equipment and methods involved are similar to that of resistance welding, but instead of electric current, vibration provides energy input. Welding metals with this process does not involve melting the materials; instead, the weld is formed by introducing mechanical vibrations horizontally under pressure. When welding plastics, the materials should have similar melting temperatures, and the vibrations are introduced vertically. Ultrasonic welding is commonly used for making electrical connections out of aluminum or copper, and it is also a very common polymer welding process.

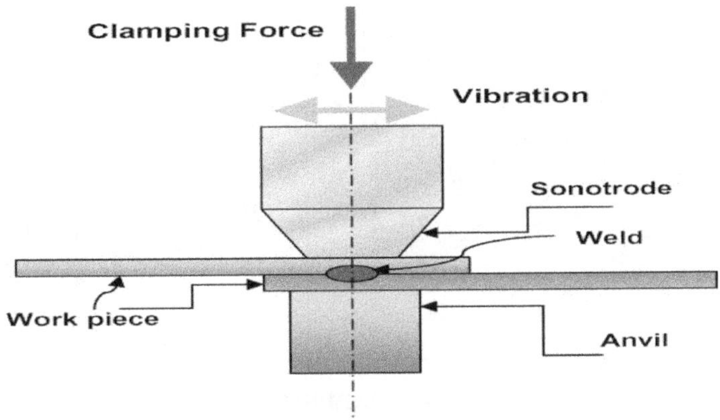

Fig. 12.30 Ultrasonic welding.

Advantages

- Fast, economical and easily automated.
- Mass production, upto 60 parts per minute is possible.
- Increased flexibility and versatility
- Possibility to join large structures.
- Used in health care industries due to clean welds.
- Produce the high strength joints consistently.

Disadvantages

- Large joints (>250 x 300 mm) cannot be welded in a single operation.
- Specifically designed joints are required.
- Ultrasonic vibrations can damage electric components.
- Tooling costs for fixtures are high.

12.4.16 Thermit Welding

Thermit welding employs molten metal to permanently join the conductors. The process employs an exothermic reaction of a thermite composition to heat the metal, and requires no external source of heat or current. The chemical reaction that produces the heat is an alumino-thermic reaction between aluminium powder and a metal oxide. It occurs without or with the application of pressure. The use of a filler metal is optional.

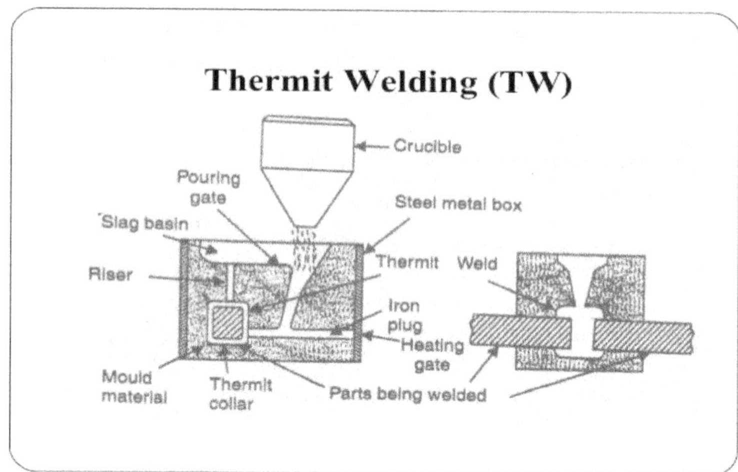

Fig. 12.31 Thermit welding.

In exothermic welding, aluminium dust reduces the oxide of another metal, most commonly iron oxide, because aluminium is highly reactive. Iron oxide is commonly used:

$$Fe_2O_3 + 2Al - 2\ Fe + Al_2O_3$$

The products are aluminium oxide, free elemental iron, and a large amount of heat. The reactants are commonly powdered and mixed with a binder to keep the material solid and prevent separation. Other metal oxides can be used, such as chromium oxide, to generate the given metal in its elemental form.

Applications

- It is mostly used to weld railroad at the site.
- It is used to weld thick plate before introduce electroslag welding.
- It is used to repair heavy castings.
- It is used to weld cable connectors of copper.
- It is used to make structure joints in large ships etc.
- It is used to joint pipe, thick plate etc. where power supply is not available.

Advantages

- It is simple and easy process.
- Low setup cost.
- Metal joining rate is high.
- Thermite welding can be done at site where casting is impossible.
- This can be used where power supply is not available.

Disadvantages

- It is used for limited metals like iron and copper.
- It is uneconomical for welding light parts.
- Highly depends on environmental condition like moisture contain, work piece alignment etc.

12.4.17 Welding Allied Processes

1. **Soldering:** Soldering is a process in which two or more items (usually metal) are joined together by melting and putting a filler metal (solder) into the joint, the filler metal having a lower melting point than the adjoining metal. Soldering differs from welding in that soldering does not involve melting the work pieces. In the past, nearly all solders contained lead, but environmental and health concerns have increasingly dictated use of lead-free alloys for electronics and plumbing purposes.

   Applications

   - It is used in plumbing, electronics, and metalwork from flashing to jewellery.
   - It provides reasonably permanent but reversible in sheet metal objects such as food cans, roof flashing, rain gutters and automobile radiators.
   - It is used to join lead came and copper foil in stained glass work.
   - Electronic soldering connects electrical wiring and electronic components to printed circuit boards (PCBs).

2. **Brazing:** Brazing is a metal-joining process in which two or more metal items are joined together by melting and flowing a filler metal into the joint, the filler metal having a lower melting point than the adjoining metal. In brazing, the filler metal melts at a higher temperature, but the work piece metal does not melt.

3. **Difference between Soldering and Brazing:** Brazing differs from welding in that it does not involve melting the work pieces and from soldering in using higher temperatures for a similar process, while also requiring much more closely fitted parts than when soldering. The filler metal flows into the gap between close-fitting parts by capillary action. The filler metal is brought slightly above its melting (liquidus) temperature while protected by a suitable atmosphere, usually a flux. It then flows over the base metal (known as wetting) and is then cooled to join the work pieces together. It is similar to soldering, except for the use of higher temperatures. A major advantage of brazing is the ability to join the same or different metals with considerable strength.

12.4.18 Welding Defects

The lack of training to the operator or careless application of welding technologies may cause discontinuities in welding. In aluminum joints obtained by fusion welding, the defects such as porosity, slag inclusion, solidification cracks etc., are observed and these defects deteriorates the weld quality and joint properties. Common weld defects may result in sudden failures which are unexpected as they give rise to stress intensities. The common weld defects are discussed below:

(i) Porosity
(ii) Lack of fusion
(iii) Inclusions
(iv) Cracking
(v) Undercut
(vi) Lamellar tearing

Porosity: Porosity occurs, when the solidifying weld metal has gases trapped in it. The presence of porosity in most of the welded joints is due to dirt on the surface of the metal to be welded or damp consumables. It is found in the shape of sphere or as elongated pockets. By storing all the consumables in dry conditions and degreasing and cleaning the surface before welding, porosity can be avoided.

Lack of Fusion: Due to too little input or too slow traverse of the welding torch, lack of fusion arises. By increasing the temperature, by properly cleaning the weld surface before welding and by selecting the appropriate joint design and electrodes, a better weld can be obtained. On extending the fusion zone to the thickness of the joints fully, a good quality joint can be obtained.

Inclusions: Due to the trapping of the oxides, fluxes and electrode coating materials in the weld zone the inclusions are occurred. It occurs while joining thick plates in several runs using flux cored or flux coated rods. By maintaining a clean surface before the run is started, providing sufficient space for the molten weld metal between the pieces to be joined, the inclusions can be prevented.

Cracking: Due to thermal shrinkage, strain at the time of phase change, cracks may occur in various directions and in various locations. Due to poor design cracking is observed. A stage-wise pre-heating process and stage-wise slow cooling will prevent such type of cracks. The cracking can be minimized by preferring fillers with low carbon and low impurity levels, by reducing the gaps and cleaning the surface before welding.

Undercutting: The undercut is caused due to incorrect settings or using improper procedure. Undercutting can be detected by a naked eye and the excess penetration can be visually detected.

Lamellar Tearing: Due to non metallic inclusions, the lamellar tearing occurs through the thickness direction. As the fusion boundary is parallel to the rolling plane in T and corner joints, the lamellar tearing occur. By redesigning the joint and by buttering the weld area with ductile material, the lamellar tearing can be minimized.

12.5 Forming

Forming processes are particular manufacturing processes which make use of suitable stresses (like compression, tension, shear or combined stresses) which cause plastic deformation of the materials to produce required shapes. During forming processes no material is added or removed, i.e. they are deformed and displaced. It is the metalworking process of fashioning metal parts and objects through mechanical deformation. Forming operates on the materials science principle of plastic deformation.

Characteristics

Metal forming tends to have more uniform characteristics across its subprocesses than its contemporary processes, cutting and joining. Forming is characterized by:

- Very high loads and stresses required, between 50 - 2500 N/mm^2 (7-360 ksi)
- Large, heavy, and expensive machinery in order to accommodate such high stresses and loads
- Production runs with many parts, to maximize the economy of production and compensate for the expense of the machine tools

Forming Processes

Forming processes tend to be categorised by differences in effective stresses. These categories and descriptions are highly simplified, since the stresses operating at a local level in any given process are very complex and may involve many varieties of stresses operating simultaneously.

Compressive forming involves those processes where the primary means of plastic deformation is uni- or multiaxial compressive loading. The details are as under:

- Rolling, where the material is passed through a pair of rollers
- Extrusion, where the material is pushed through an orifice
- Die forming, where the material is stamped by a press around or onto a die
- Forging, where the material is shaped by localized compressive forces
- Indenting, where a tool is pressed into the workpiece

Tensile forming involves those processes where the primary means of plastic deformation is uni- or multiaxial tensile stress.

- Stretching, where a tensile load is applied along the longitudinal axis of the workpiece
- Expanding, where the circumference of a hollow body is increased by tangential loading
- Recessing, where depressions and holes are formed through tensile loading

12.5.1 High Energy Rate Forming (HERF)

High energy rate forming is the forming of sheet metal by a high energy surge, delivered over a very short time. Since the forming of the metal occurs so quickly, desirable materials for HERF will be ductile at high deformation speeds. Explosive forming, electromagnetic forming, and electrohydraulic forming are the examples of high energy rate forming.

12.5.2 Explosive Forming

It is a metalworking technique in which an explosive charge is used instead of a punch or press. Typically the mould and work piece are submerged in water. The sheet metal is secured over the mould by a ring clamp. Air is drawn out, creating a vacuum in the die cavity. An explosive is placed between the die cavity and the work, a certain distance (standoff distance) from the work. Standoff distance depends on the size of the work, for larger parts it is usually about half the diameter of the blank. The explosive itself is also deeply submersed in water. Upon detonation, the shock wave travels through the water and delivers great energy to the work, forming it to the die cavity near instantaneously. This high energy rate forming process can be used to form big thick plates.

12.5.3 Electro Magnetic Forming (High Velocity Forming)

It is a high velocity, cold forming process for electrically conductive metals, most commonly copper and aluminium. The work piece is reshaped by high intensity pulsed magnetic fields that induce a current in the work piece and a corresponding repulsive magnetic field. In practice, the work piece is placed in proximity to a heavily constructed coil of wire (*work coil*). A huge pulse of current is forced through the work coil by rapidly discharging a high voltage capacitor bank using an ignitron or a spark gap as a switch. This creates a rapidly oscillating, ultrastrong electromagnetic field around the work coil. The high work coil current (tens or hundreds of thousands of amperes) creates ultrastrong magnetic forces, causing permanent deformation. The process occurs extremely quickly (tens of microseconds).

12.5.4 Electrohydraulic Forming (Electric Discharge Forming)

It is a unique high energy forming process for sheet metal that uses the energy from the combustion of a thin metal wire. Two electrodes, with a wire connecting them, are submersed in liquid. A sheet metal blank is secured on top of the mould with a ring clamp and a vacuum is created in the die cavity under the blank. Electrical energy is stored in a capacitor bank. The electricity is discharged through the electrodes and the wire, instantly vaporizing the wire, creating a shock wave that travels through the water. This shock wave forms the sheet metal to the mould cavity. It produces a shock wave of relatively low magnitude and is best suited for thinner work. The wire needs to be replaced after every operation.

Advantages of HERF

- Production rates are higher for High-Energy Rate Forming process
- Die costs are relatively lower compared to other processes.
- It is possible to form most metals including difficult to form metals.
- No or minimum spring back effect on the material.
- Production cost is low as power hammer (or press) is removed in the process.
- Complex shapes/profiles can be made much easily.
- The required final shape/dimensions of the product are obtained in a single step, thus eliminating intermediate forming steps and preforming dies.
- Suitable for a range of production volume such as small numbers, batches or mass production.

Limitations of HERF

- Process is not suitable for highly brittle materials.
- Source of energy (chemical explosive or electrical) must be handled carefully.
- Highly skilled personnel are required from design to execution.
- Dies need to be much bigger in size to withstand high energy rates and shocks and to prevent cracking.

Applications of HERF

- In ship building – to form large parts (up to 25 mm thick).
- Bending thick tubes or pipes (up to 25 mm thick).
- Elliptical domes used in space applications.
- Crimping of metal strips.
- The cladding of two large plates of dissimilar metals.

12.6 Drawing

12.6.1 Wire Drawing

In drawing, the cross section of a long rod or wire is reduced or changed by pulling (drawing) it through a die. Thus, the difference between drawing and extrusion is that in extrusion the material is pushed through a die, whereas in drawing it is pulled through it. Although the presence of tensile stresses is obvious in drawing, compression also plays a significant role because the metal is squeezed down as it passes through the die opening. For this reason, the deformation that occurs in drawing is sometimes referred to as indirect compression.

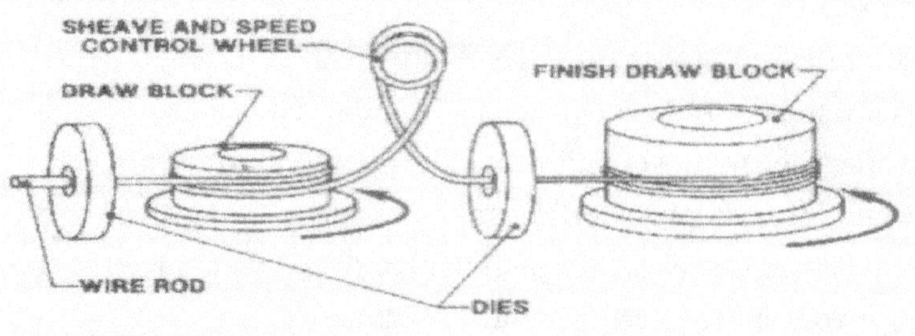

Fig. 12.32 Wire drawing process

The process of wire drawing involves the production of wire of close diameter and tolerance by drawing descaled hot-rolled rod through a die. The finished wire is coiled onto a motor driven block. The dies are manufactured from hard material e.g tungsten carbide. Diamond dies are used for small diameter wire. Soap is used for lubrication. If large diameter reductions are required the rod is progressively drawn through several stages of dies and blocks. Wire can be extruded at speeds of up to 2000m/min.

12.6.2 Bar Drawing

The basic difference between bar drawing and wire drawing is the stock size that is processed. Bar drawing is the term used for large diameter bar and rod stock, while wire drawing applies to small diameter stock. Wire sizes down to 0.03 mm (0.001 in) are possible in wire drawing. Bar drawing is generally accomplished as a single-draft operation, the stock is pulled through one die opening. Because the beginning stock has a large diameter, it is in the form of a straight cylindrical piece rather than coiled. This limits the length of the work that can be drawn. In contrast, wire is drawn from coils consisting of several hundred (or even several thousand) feet of wire and is passed through a series of draw dies. The number of dies varies typically between 4 and 12.

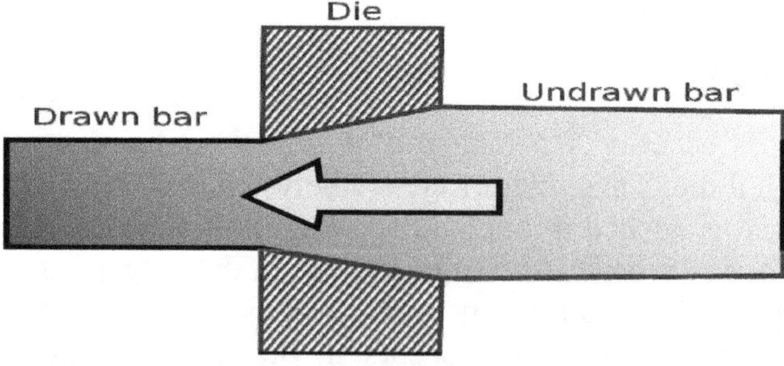

Fig. 12.33 Bar drawing process.

12.6.3 Tube Drawing

Tube drawing is very similar to bar drawing, except that the beginning stock is a tube. It is used to decrease the diameter, improve surface finish and improve dimensional accuracy. A mandrel may or may not be used depending on the specific process used. A draw bench is used to pull the tube through a drawing die. One end of the tube to be drawn is collapsed and fed through the die. This end is then clamped to a travelling carriage. The carriage is then move away from the die using either mechanical or hydraulically traction. In practice, several tubes may be drawn in parallel to increase productivity.

The die most commonly used in tube drawing is a sintered tungsten carbide insert encased in steel. The cobalt content is approximately 10 percent. Higher cobalt content provides more shock resistance, whereas a lower content provides better wear resistance. The basic tube drawing processes are sinking, rod (mandrel) drawing, and several types of plug drawing.

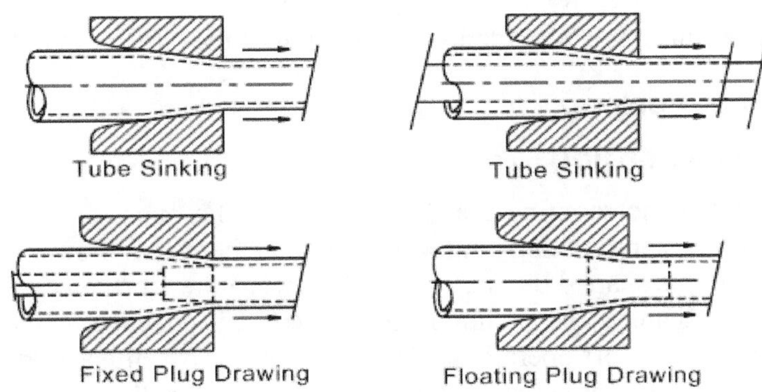

Fig. 12.34 Four types of tube drawing.

The diameter and wall thickness of tubes that have been produced by extrusion or other processes can be reduced by tube drawing process. The process of tube drawing is similar to wire or rod drawing except that it usually requires a mandrel of the requisite diameter to form the internal hole. Tubes as large as 0.3 m in diameter can be drawn using this process.

12.7 Forging

Forging is the process of forming and shaping metals through the use of hammering, pressing or rolling. The process begins with starting stock, usually a cast ingot (a "cogged" billet which has already been forged from a cast ingot), which is heated to its plastic deformation temperature, then upset or "kneaded" between dies to the desired shape and size. Forging can create a myriad of sizes and shapes with enhanced properties when compared to castings or assemblies.

Advantages of Forging Process

- Parts produced are stronger than casted parts or parts made in a machine.
- Internal grain changes its form and continuous forging strengthen the part characteristics.
- Closed dimensional tolerance.
- Smooth surface finish.
- Well-shaped metals are used for a wide range of applications.

Disadvantages of Forging Process

- Prevents work hardening and increases the difficulty of performing other machining operations on the part.
- Involves huge expenditure for the machinery, dies, tools and personnel.
- Some forging requires metal-forming dies, which are required to be precisely machined and heated to properly shape the piece.
- Poor surface finish due to oxidation and scaling of the surface in hot forging.

12.7.1 Types of Forging

12.7.2 Open Die Forging

Open die forging does not confine the flow of metal. The operator obtains the desired shape of forging by manipulating the work material between blows. Some specially shaped tools or a simple shaped die may be used between the work piece and the hammer or anvil to assist in shaping the required sections (round, concave, or convex), making holes, or performing cut – off operations. This process is most often used to make near – final shape of the part so that some further operation done on the job produces the final shape. It is the simplest forging process which is quite flexible but not suitable for large scale production. The resulting size and shape of the forging are dependent on the skill of the operator.

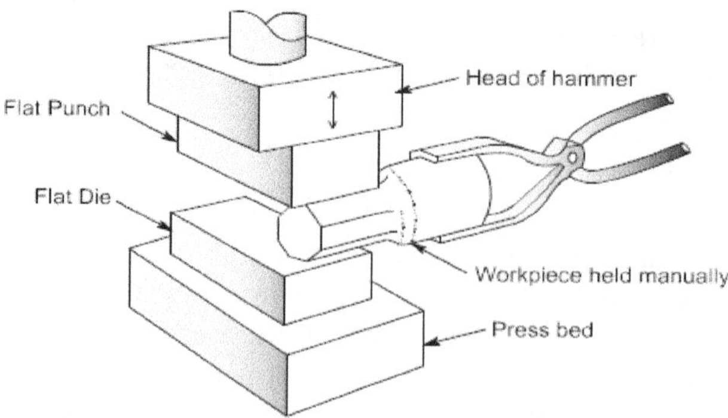

Fig.12.35 Open die forging.

Forging Force: In open die forging operation, the forging force F, to be applied on a solid cylindrical component can be determined from the relation:

$$F = \sigma_f . \frac{\pi}{4} d^2 \left(1 + \frac{\mu d}{3h}\right)$$

where, σ_f is the flow stress of the material, μ is the coefficient of friction, and d and h are the diameter and height of the work piece, respectively.

Advantages

- Reduced chance of voids
- Better fatigue resistance
- Improved microstructure
- Continuous grain flow
- Finer grain size
- Greater strength

Disadvantages

- Less control on mechanical properties and dimensions
- Simple shapes can be manufactured
- Process is slow
- Further machining may be required for finish

12.7.3 Closed Die Forging

The process uses shaped dies to control the flow of metal. The heated metal is positioned in the lower cavity and on it one or more blows are struck by the upper die. This hammering makes the metal to flow and fill the die cavity completely. Excess metal is squeezed out around the periphery of the cavity to form flash. On completion of forging, the flash is trimmed off with the help of a trimming die.

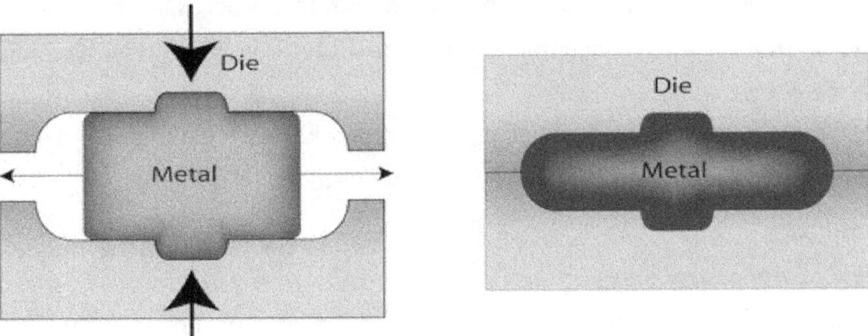

Fig.12.36 Closed die forging.

Most of the cases, die sets contain several cavities. The work material is given final desired shape in stages as it is deformed in successive cavities in the die set. The shape of the cavities causes the metal to flow in desired direction, thereby imparting desired structure to the component.

Advantages

- Generate internal grain flow and increase the tightness and strength of products.
- For large production runs, it is more cost-effective.
- Suitable for almost all the metals (steel, aluminum and brass).
- Surface finish is better.
- Less machining is required.
- Environment protected.

Disadvantages

- Size limitation
- Cost-effective for short production runs (less than 500pcs) for the high tooling cost.
- Shape limitation

12.7.4 Drop Forging

It is a forging process where a hammer is raised and then dropped onto the work piece to deform it according to the shape of the die. There are two types of drop forging: open-die drop forging and closed-die drop forging. As the names imply, the difference is in the shape of the die, with the former not fully enclosing the work piece, while the latter does.

Advantages

- The production efficiency is higher.
- Complex shaped metal parts can be forged with this process.
- Size of drop forgings is accurate, surface quality is better and less machining allowance
- Save metal material and reduces machining work. Less cost of spare parts.
- The production operation is easy, so less work force.

Disadvantages

- Prevent work hardening.
- Involve a lot of expenditure for the machinery, dies, tools and personnel.
- Require metal-forming dies, precisely machining to properly shape the piece.

12.7.5 Press Forging

Press forging, mostly used for forging of large sections of metal, uses hydraulic press to obtain slow and squeezing action instead of a series of blows as in drop forging. The continuous action of the hydraulic press helps to obtain uniform deformation throughout the entire depth of the work piece. So, the impressions obtained in press forging are cleaner.

Press forgings generally need smaller draft than drop forgings and have greater dimensional accuracy. Dies are generally heated during press forging to reduce heat loss, promote more uniform metal flow and production of finer details. Hydraulic presses are available in the capacity range of 5 MN to 100MN.

Advantages

- Greater accuracy in terms of tolerances within 0.01 to 0.02 inch.
- Less draft, so more complicated shapes can be forged with better dimensional accuracy.
- The speed, pressure and travel of the die are automatically controlled.
- Possibility of process automation through mechanism of blank feeding.
- The operation needs single squeezing action, thus saves the time.
- Provides uniform and simultaneous deformation throughout the metal.
- Number of working strokes per minute more. Thus, 40 to 50 parts produced per minute.
- Suitable for mass production of nuts, bolts, rivets, screws, break levers, bearing races, valves, etc.

Disadvantages

- The initial capital cost is higher compared with drop forging,
- The difficulty of descaling the blank is another short coming of this process.
- The process is economically suitable only when the equipment is efficiently utilized.

12.7.6 Upset Forging

Upset forging involves increasing the cross section of a material at the expense of its corresponding length. Parts can be upset – forged from bars or rods upto 200 mm in diameter in both hot and cold condition. In this case, the unsupported length (l) of the rod does not exceed $3d$ to prevent buckling, 'd' the rod diameter.

The process uses split dies with one or several cavities in the die. Upon separation of split die, the heated bar is moved from one cavity to the next. The split dies are then forced together to grip the part and a heading tool (or ram) advances axially against the bar, upsetting it to completely fill the die cavity. Upon completion of upsetting process, the heading tool comes back and the movable split die releases the stock. Examples of upset forged parts are fasteners, valves, nails, and couplings.

Advantages

- Greater strength and better quality.
- Contribute towards component weight reduction and eliminate machining operations.
- Require fewer operations, have lower scrap costs, so more cost efficient production
- Resist deformation and breakage during use.
- Widely used of all forging processes.

Disadvantages

- High tooling cost
- Suitable for forging of heavier job
- Unsymmetrical parts are difficult to manufacture.

12.7.7 Defects in Forging and its Remedies

A brief description of defects and their remedial methods is provided below:

- Incomplete forging penetration: Dendritic ingot structure at the interior of forging is not broken. Actual forging takes place only at the surface. Some sections of the die cavity are not completely filled by the flowing metal.

  Cause - Use of light rapid hammer blows

  Remedy - Forging press may be used for full penetration.
- Surface cracking: Surface cracking occurs due to excess working on surfaces at low temperature. Thus, so many cracks arise on work piece.

  Cause - Excessive working on the surface and too low temperature.

  Remedy - working temperature may be increased.
- Cracking at the flash: This crack penetrates into the interior after flash is trimmed off.

  Cause - Very thin flash

  Remedy - Increasing flash thickness, relocating the flash to a less critical region of the forging, hot trimming and stress relieving.
- Cold shut: Two surfaces of metal fold against each other without welding completely. Cause - Sharp corner (less fillet), excessive chilling, high friction

  Remedy - Increase fillet radius on the die.

- Unfilled Section (Unfilling/Underfilling): Some section of die cavity not completely filled by the flowing metal.

  Cause - Improper design of the forging die, less raw material, poor heating.

  Remedy - Proper die design, Proper raw material and Proper heating. \

- Die shift (Mismatch): Misalignment of forging at flash line. This is caused by the miss alignment of the die halve, making the two halve of the forging to be improper shape.

  Cause - Misalignment of the die halves.

  Remedy - Proper alignment of die halves.

- Scale Pits (Pit marks): Irregular depurations on the surface of forging.

  Cause - The oxide and scale gets embedded into the finish forging surface.

  Remedy - Proper cleaning of the stock prior to forging.

- Flakes: These are basically internal ruptures.

  Cause - Rapid cooling causes the exterior to cool quickly causing internal fractures.

  Remedy - Follow proper cooling practices.

- Residual stresses in forging: This defect occurs due to improper cooling of forged part. Too much rapid cooling is main causes of this type of defects.

  Cause - Inhomogeneous deformation and improper cooling (quenching) of forging.

  Remedy - Slow cooling of the forging in a furnace over a period of time

12.8 Extrusion

Extrusion is the process by which long straight metal parts can be produced. The cross-sections that can be produced vary from solid round, rectangular, to L shapes, T shapes. The process is done by squeezing metal in a closed cavity through a tool, known as die using either a mechanical or hydraulic press. Tubes and many other different shapes can be produced in this process.

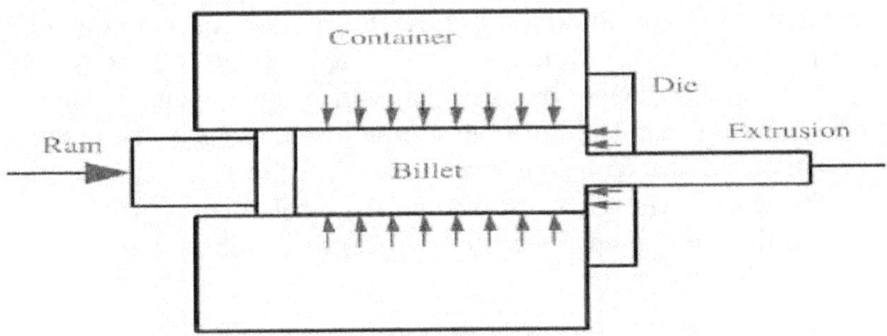

Fig. 12.37 Extrusion process.

Extrusion produces compressive and shear forces in the stock. No tensile force is produced, generating high deformation without tearing the metal. The cavity in which the raw material is contained is lined with a wear resistant material. This can withstand the high radial loads that are created when the material pushes the die.

1. **Hot Extrusion:** Hot extrusion is done at fairly high temperatures, approximately 50 to 75 % of the melting point of the metal. The pressures can range from 35-700 MPa (5076 - 101,525 psi). Due to the high temperatures and pressures and its detrimental effect on the die life as well as other components, good lubrication is necessary. Oil and graphite work at lower temperatures, whereas at higher temperatures glass powder is used.

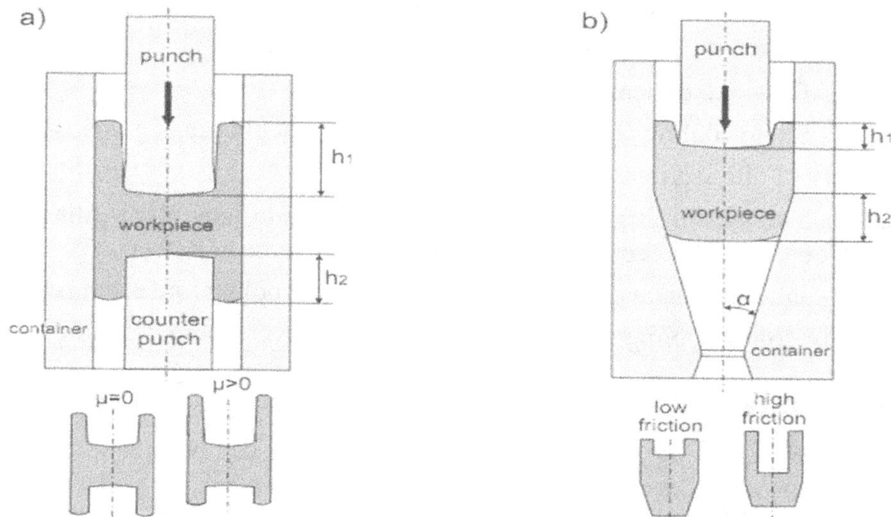

Fig. 12.38 (a) Backward extrusion, (b) Forward extrusion

In **forward extrusion** a metal billet is loaded into a container, and a ram compresses the material, forcing it to flow through one or more openings in a die at the opposite end of the container. As the ram approaches the die, a small portion of the billet remains that cannot be forced through the die opening. This extra portion, called the butt, is separated from the product by cutting it just beyond the exit of the die. One of the problems in direct extrusion is the significant friction that exists between the work surface and the walls of the container as the billet is forced to slide toward the die opening. This friction causes a substantial increase in the ram force required in direct extrusion.

In **backward extrusion,** the die is mounted to the ram rather than at the opposite end of the container. As the ram moves, the metal is forced to flow through the clearance in a direction opposite to the motion of the ram. Since the billet is not forced to move relative to the container, there is no friction at the container walls, and the ram force is therefore lower than in direct extrusion. Limitations of this extrusion are imposed by the lower rigidity of the hollow ram and the difficulty in supporting the extruded product as it exits the die.

Typical parts produced by extrusion are: trim parts used in automotive and construction applications, window frame members, railings, aircraft structural parts etc.

12.8.1 Cold Extrusion

Cold extrusion is the process carried out at room temperature or at slightly elevated temperatures. This process can be used for most materials-subject to designing robust enough tooling that can withstand the stresses created by extrusion. Examples of the metals that can be extruded are lead, tin, aluminum alloys, copper, titanium, molybdenum, vanadium, steel. Examples of parts that are cold extruded are collapsible tubes, aluminum cans, cylinders, gear blanks. The advantages of cold extrusion are listed as follows:

- No oxidation takes place.
- Good mechanical properties due to severe cold working
- Good surface finish

12.8.2 Impact Extrusion

Impact extrusion is performed at higher speeds and shorter strokes than conventional extrusion. In this process, the punch impacts the work part rather than simply applying pressure to it. Impacting can be carried out as forward extrusion, backward extrusion, or combinations of these. It is usually done cold on a variety of metals. Backward impact extrusion is most common. Products made by this process include toothpaste tubes and battery cases. The high-speed characteristics of impacting permit large reductions and high production rates, making this an important commercial process.

12.8.3 Advantages of Extrusion Process

- A variety of shapes/ cross-sections are possible, especially with hot extrusion.
- Grain structure and strength properties are enhanced in cold and warm extrusion.
- Good surface finish, fairly close tolerances are possible, in cold extrusion.
- Little or no waste material is produced.
- Extrusions often minimize the need for secondary machining.

Disadvantages of Extrusion Process

- Cross section of the extruded part must be uniform throughout its length.

Extrusion Defects

- Surface cracking - When the surface of an extrusion splits. This is often caused by the extrusion temperature, friction, or speed being too high. It can also happen at lower temperatures if the extruded product temporarily sticks to the die.
- Pipe - A flow pattern that draws the surface oxides and impurities to the centre of the product. Such a pattern is often caused by high friction or cooling of the outer regions of the billet.
- Internal cracking - When the centre of the extrusion develops cracks or voids. These cracks are attributed to a state of hydrostatic tensile stress at the centerline in the deformation zone in the die.
- Surface lines - When there are lines visible on the surface of the extruded profile. This depends heavily on the quality of the die production and how well the die is maintained, as some residues of the material extruded can stick to the die surface and produce the embossed lines.

Questions

1. Briefly describe the operations that can be performed on lathe machine.
2. Describe the various techniques to perform a taper turning on lathe.
3. What is casting? How it differs from other primary shaping processes?
4. Describe the advantages and disadvantages of sand casting?
5. Briefly describe the various die casting g processes.
6. Describe about the advantages of special casting techniques.
7. What is centrifugal casting? Why is it used?
8. Describe the various casting defects. Explain the causes and remedies for these defects.
9. Describe the working principle of arc welding. Explain the shielded arc welding.
10. Describe the basic fusion welding process. What is the role of flux in welding?
11. Explain the soldering vis-a-vis brazing process.
12. What is gas welding? Explain the various parameters in gas welding.
13. What are the welding defects? Explain the causes and remedies.
14. Describe high energy rate forming.
15. Describe the process of tube drawing.
16. Explain the difference between bar and wire drawing.
17. What is forging? Explain various forging processes.
18. Describe the various forging defects, causes and its remedies.
19. Differentiate between forward and backward extrusion.
20. Describe the process of impact extrusion.
21. Describe the advantages of extrusion process.
22. Mention the various defects and remedies in extrusion process.

CHAPTER 13

Coupling, Clutches and Brakes

13.1 Introduction

A machine is defined as the assemblage (collection) of number of links or elements which receives energy in some available form and uses it to do some particular type of work. For example:

(i) A lathe is a machine which receives mechanical energy from line shaft through the belt and uses this energy to remove the metal from a piece of work.

(ii) A crow bar with its fulcrum is as a machine which utilizes the muscular energy of a man in raising a heavy weight.

(iii) A heat engine is a machine which utilizes the heat energy derived from the combustion of fuel to produce mechanical work. This mechanical work can either be utilized to propel a vehicle or to run a generator or for some other purposes.

Each component or assembly of few components not having any relative motion between them is called a *machine element.*

Each machine element of the machine is always designed to perform the specific function for either transmitting the forces or the motion or both.

13.2 Coupling

A coupling is a device which is used to connect two co-axial shafts permanently so that the necessary torques and powers can be transmitted. For example a coupling is required to connect the driving shaft of a motor to the driven shaft of a centrifugal pump for transmission of power. However, a clutch permits the rapid connection and disconnection of the driving and driven shafts at wheel of the operator.

Important requirements of a good coupling

(i) It must transmit the full power of the shaft.

(ii) It should be easy to connect and disconnect a coupling.

(iii) It must keep the shafts in perfect alignment.

(iv) Projected parts of coupling such as bolt-heads, nuts etc., should be well protected for operator's safety.

Classification of Coupling

1. Sleeve or muff coupling
2. Clamp or compression or split muff coupling
3. Flange couplings

Flexible Coupling

(i) Bush pin type Flange coupling
(ii) Oldham coupling
(iii) Universal coupling

13.2.1 Rigid Coupling

Rigid couplings are used to connect two perfectly aligned shafts rigidly. Some of the couplings of this type are being discussed below.

1. ***Sleeve or muff coupling*:** Fig. 13.1 shows one form of muff coupling. It consists of a cast iron hollow cylinder called sleeve or muff. The sleeve is fitted over the ends of shafts with the help of a gib-head sunk key in the key ways of the shafts and the muff as shown in Fig. 13.1. The power is transmitted from the driving shaft to key, then to the muff and from the muff to the key and the driven shaft without any relative motion between them.

   Muff couplings are used for transmission of small powers in case of small shafts.

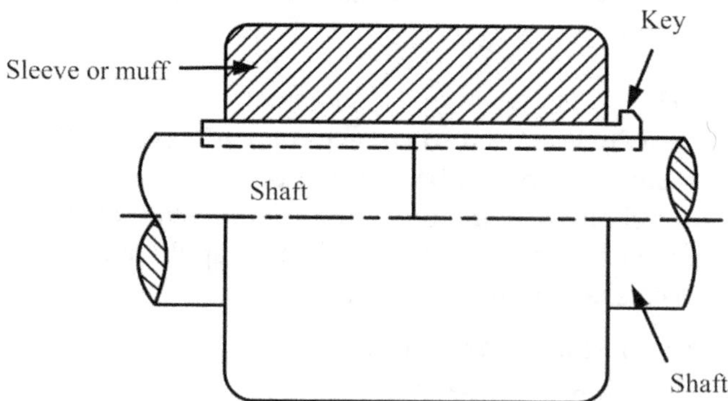

Fig. 13.1 Muff coupling.

2. ***Clamp or compression coupling*:** A variation of muff coupling called clamp or compression or split muff coupling is shown in Fig. 13.2. In this case the sleeve or muff is split into two sleeve one sleeve is fixed from the top and the other half from the bottom and these are clamped together by means of nuts and bolts, which housed in recesses formed in the sleeve.

   Such couplings are used for transmission of heavy torques as shown Fig. 13.2.

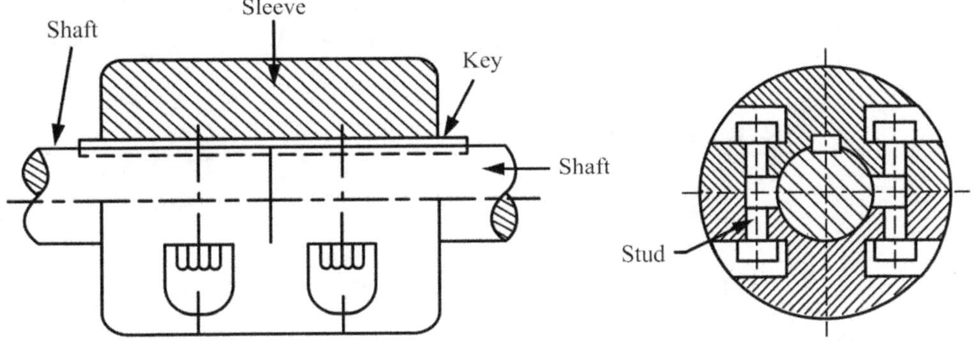

Fig. 13.2 Split muff coupling.

3. ***Flange coupling*:** A protected type of flange coupling is shown in Fig. 13.3. where each flange provided with a shroud for covering bolt-heads and nuts. This prevents bolts and nuts from catching the clothes of work men and ensure their safety.

   To ensure proper alignment of coupling one of the flanges has circular projection which fits into a corresponding recess in the other flange.

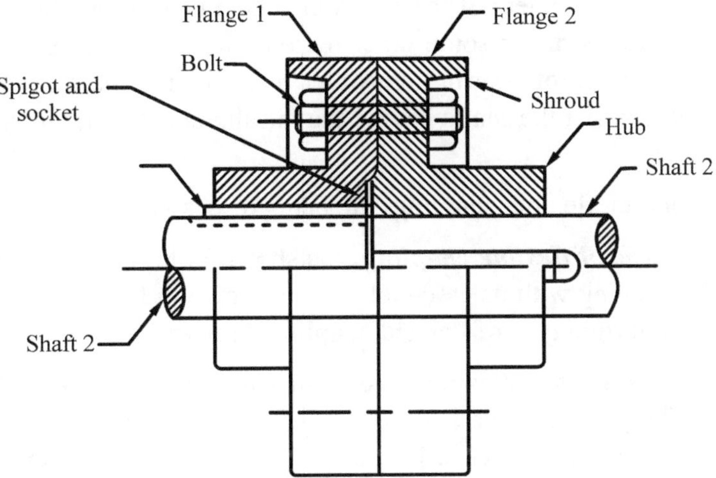

Fig. 13.3 Split muff coupling.

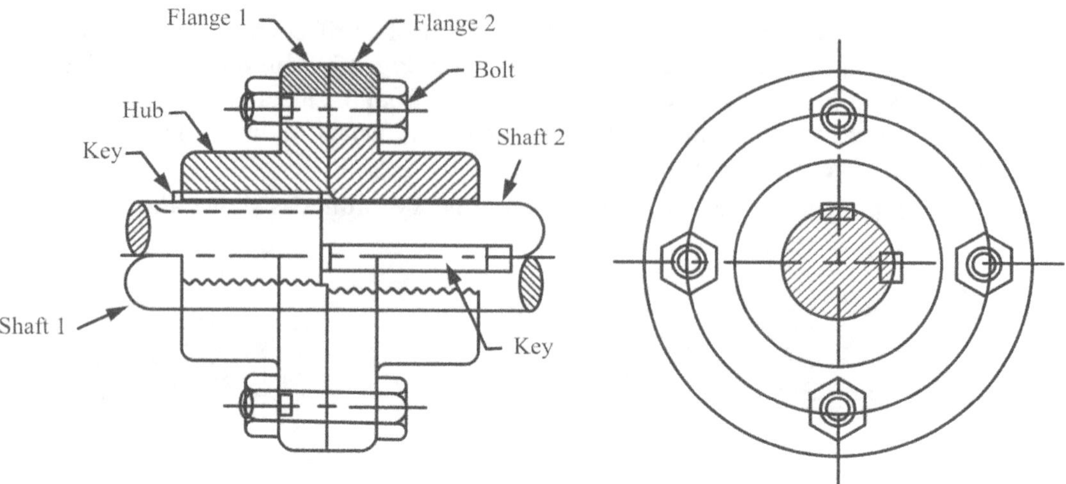

Fig. 13.4 Unprotected type rigid flange coupling.

4. As shown in Fig. 13.4. There are unprotected type of flange couplings. The flange coupling consists of two coupling halves, keyed to two shafts. These two halves are coupled together by means of suitable number of bolts fitted in the holes through two halves. In protected type flange coupling there is risk of accident of catching the cloths of work-men.

13.2.2 Flexible Couplings

In actual practice it is difficult to attain as well maintain the perfect alignment between the coupled shafts. Misalignment of the shafts may be due to installation errors or due to unequal wear of bearings or for some other reasons. It causes the stress reversal in the shafts and excessive wear of bearings. In order to overcome these problems a *flexible coupling* is used to connect the shafts having the misalignment either in lateral or angular direction.

Few of the important flexible couplings are being discussed below.

1. ***Bushed-pin type of flexible coupling*:** (as shown in Fig. 13.5). A bushed pin type of flexible coupling with its essential parts is shown in Fig. 13.5. It is basically the modification of rigid type of flanged coupling discussed above.

   The coupling bolts known as pins are of special shape are rigidly fixed by nuts to one of the flanges. The enlarged portion of the pin is fitted with brass bush and covered with a ring of rubber or leather bush. The power is transmitted through the flexible rubber bushes which can take care of misalignment of shafts from 1 mm to 12 mm depending upon the size of the coupling.

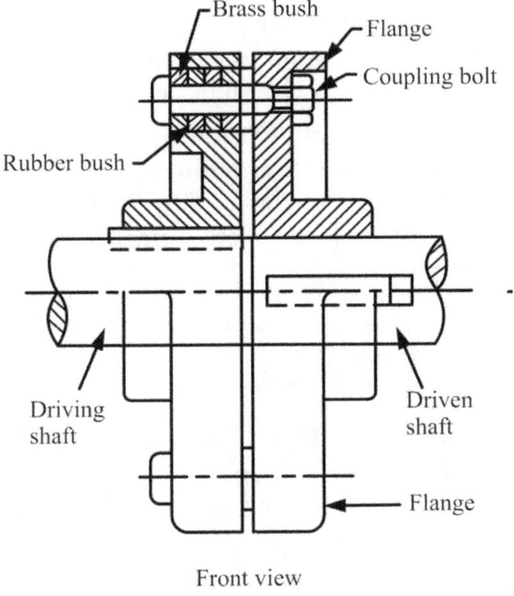

Front view

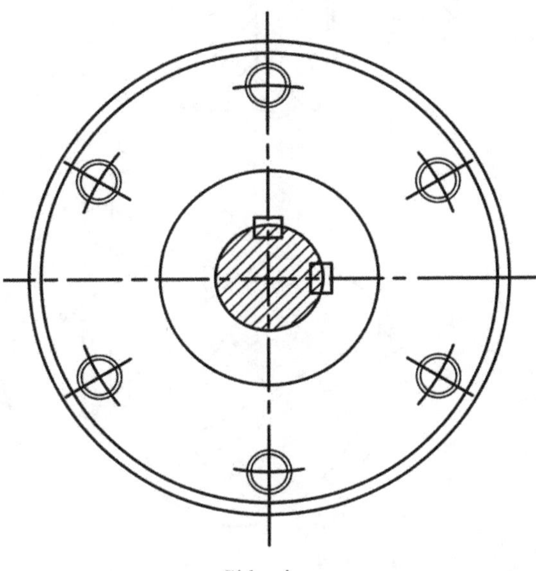

Side view

Fig. 13.5 Bush-pin type of flexible coupling.

These type of couplings can be directly connected from the prime mover (an electric generator or motor) to the machine to be driven like a centrifugal pump. It can transmit large torques and power.

2. ***Oldham's Coupling***: This type of coupling is used to connect two parallel shafts which are not co-axial and these are at a very short distance apart. The essential parts of an Oldham's coupling are shown in Fig. 13.6.

   The flanges A and B are rigidly fastened on the driving and driven shafts respectively. In between the flanges couplings has an intermediate piece C which has two projections x and y and these slide fit in the grooves provided in the flanges of the shafts. These projections are formed at right angles to each other but on opposite side of the piece C.

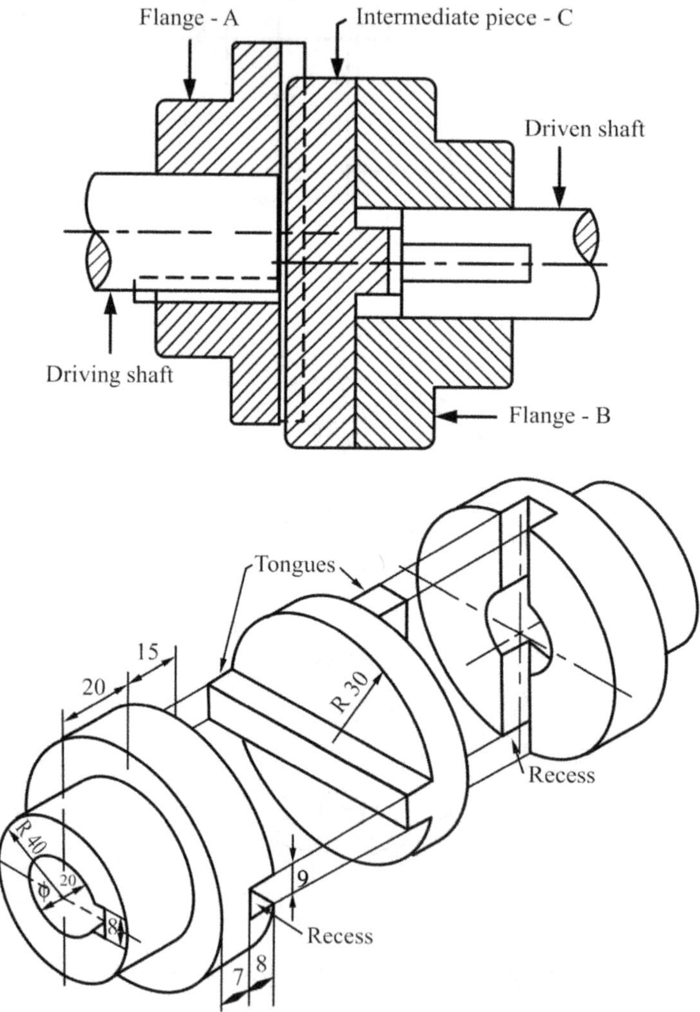

Fig. 13.6 Oldham's Coupling.

If the driving shaft is given a rotary motion, the flange A rotates at the same speed as it is rigidly fastened to the shaft. The flange A transmits the same motion to intermediate piece and from the intermediate piece to flange B and the driven shaft.

3. ***Universal coupling or Hooke's joint*:** A universal coupling is used to transmit motion and power to two non-parallel intersecting shafts, also in the cases of the shafts having angular misalignment. It's some of the applications are:

    (i) In an automobile to transmit power from the gear box of the engine to the rear axie.

    (ii) To transmit motion to adjustable shafts as spindles of multispindle drilling machines, rolling machines etc.

A single universal coupling is shown in Fig. 13.7

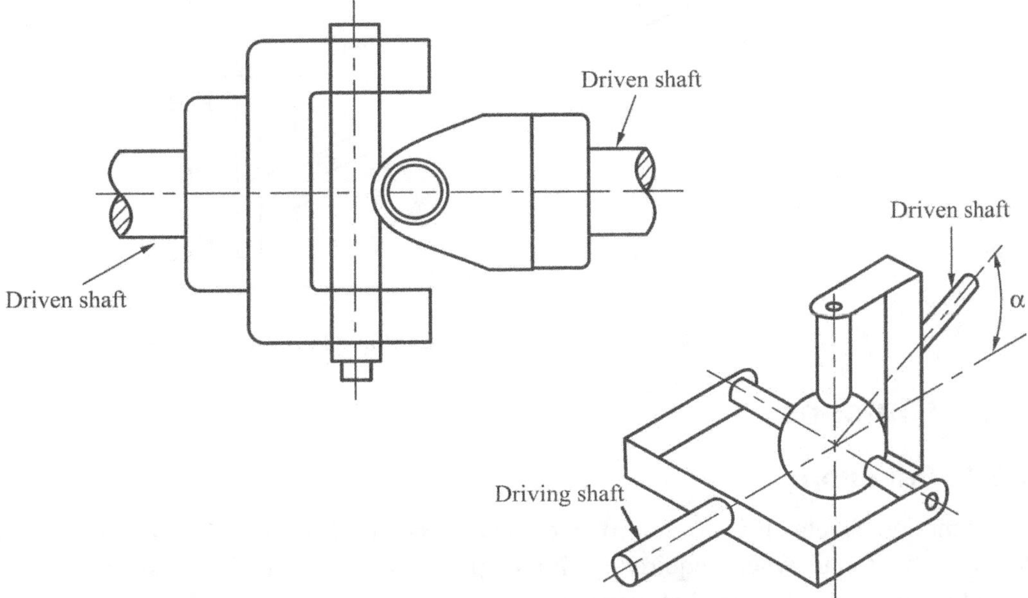

Fig. 13.7 Universal coupling.

The driving and driven shafts are inclined at an angle 'α' rotate in their fixed bearings. Each shaft has a fork at its ends. The four ends of the forks are connected by an intermediate block containing pins in the form of a cross. The cross pins rest on needle bearings in the fork ends. The four arms of the cross are at right angles to each other. The intermediate block can be in the shape of a cross, square or sphere.

In this coupling, a complete rotation of either shaft causes the other shaft to complete one rotation in the same time but with varying angular speeds.

In order to transit constant angular velocity from driving to driven shaft a double universal coupling is used shown in Fig. 13.8.

In order to have the constant velocity ratio, following conditions are required to be satisfied:

(a) The driving and driven shafts should make equal angles with the intermediate shaft.

(b) The forks of the intermediate shaft should lie in the same plane of rotation.

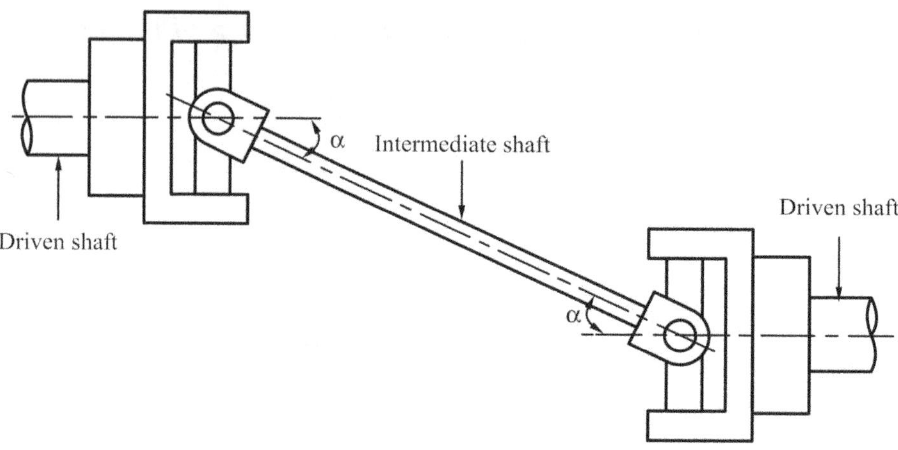

Fig. 13.8 Double universal coupling.

13.3 Clutches

13.3.1 Function

Main function of clutch is to control the flow of power from the driving shaft to the driven shaft as and when required. With help of clutch, the driven may be stopped without stopping the driving shaft of engine.

It is a mechanism by means of which engine is connected or disconnected from rest of transmission system.

In automobile where vehicle can be stopped for a (while) short time or to change the gear, the requirement is that driven shaft should be disconnected from driving shaft but engine should run under no load condition. This is achieved by using clutch mounted between engine shaft and gear-box (driven) shaft. Clutch is operated by a lever.

13.3.2 Classification of Clutches

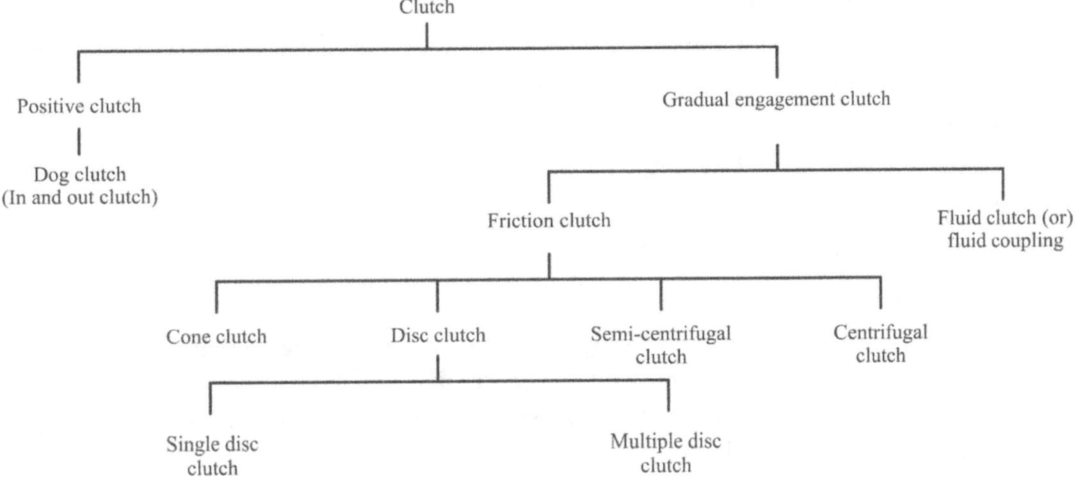

13.3.3 Positive Clutches (Dog clutch)

- Dog clutch is shown in Fig. 13.9.

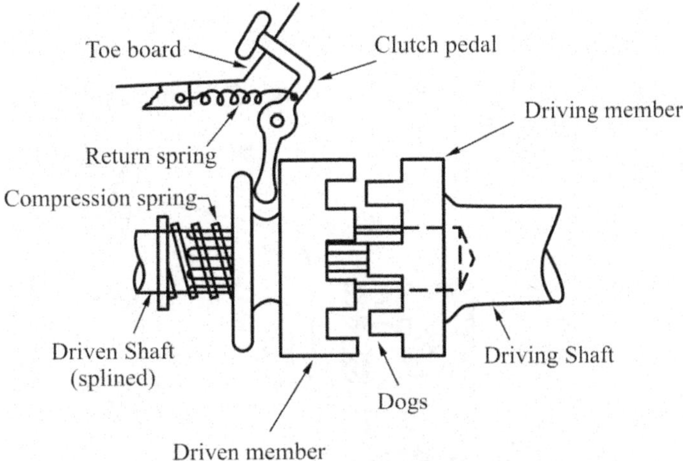

Fig. 13.9 Dog clutch (disengaged position)

- These clutches are also known as dog clutches (jaw clutch)
- This type of clutches use two members having dogs or notches at their facings. One member is fixed with the driving shaft and the other member is free to slide axially on the splined driven shaft. Driven member is moved to and from on the splined shaft by means of suitable lever.

In order to couple the shafts, the driven member is moved into contact with driving member through shifting device.

The shifting device is mounted on the hub of sliding flange which makes it possible to engaged with driving flange to transmit the motion and power to the driven shaft.

Advantages

It is easy to operate, do not slip and it is cheap.

They are of the positive type.

No heat is generated at clutch surface during engagement.

Disadvantages

1. It gives noise and shock during engagement as the dog member do not mesh together until speed of both the driving and driven shaft is brought to uniform.
2. They cannot be engaged at high speeds and some times cannot be engaged at rest unless the jaws are aligned.

Applications

They are used in following machines:

1. Power presses; 2. Punches; 3. Gear box and some machine tools

13.3.4 Cone Clutch

Cone clutch is as shown in Fig. 13.10. It consists of driving shaft, splined driven shaft, outer cone (cup), inner cone, helical spring.

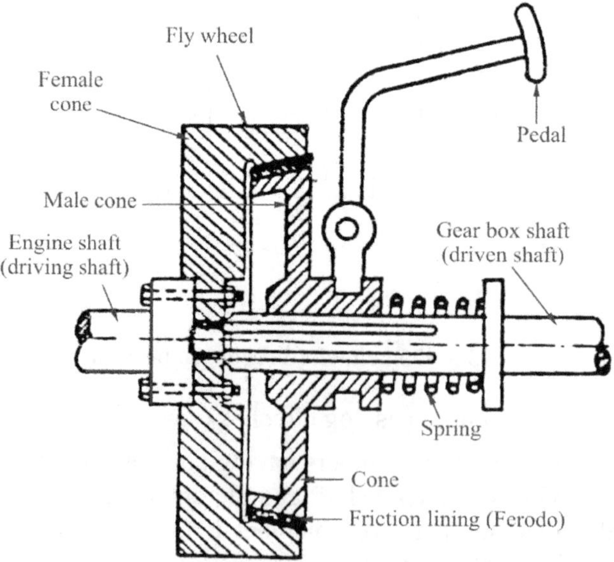

Fig. 13.10 Cone clutch

Outer cone known as the female cone is keyed to the driving shaft where as inner cone (male cone) is free to slide axially on the splined driven shaft. The spring force keeps the driven member in contact with the driving member and the clutch is considered to be engaged. Clutch is disengaged by means of fork fitted in groove by applying pedal force.

Two cones having leather facing. Cone angle kept is about 15° to 20°.

13.3.5 Friction Clutches (as shown in Fig. 13.11(a))

Operation of friction clutches is based on the friction force developed between surfaces of two clutch members. As soon as faces of driving and driven members are brought into contact, friction force starts acting between them resulting in the movement of driven shaft stationary member with driving member. When the driving and driven members are moving at same speed and there is no slip between them, the clutch is considered to be fully engaged.

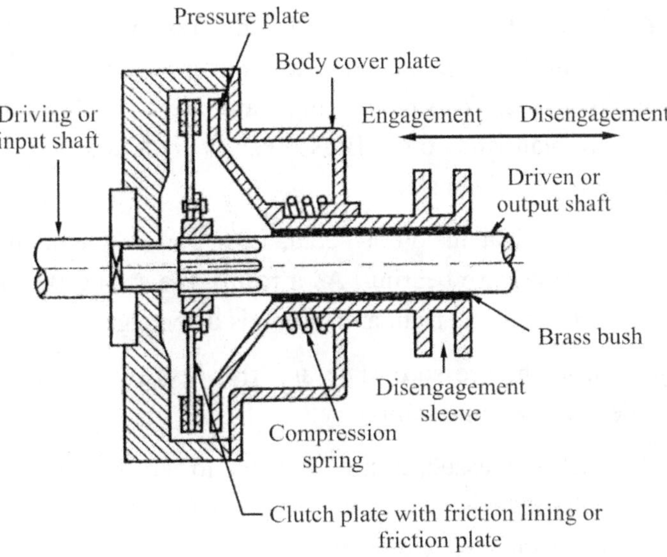

Fig. 13.11(a) Single plate clutch in disengaged position

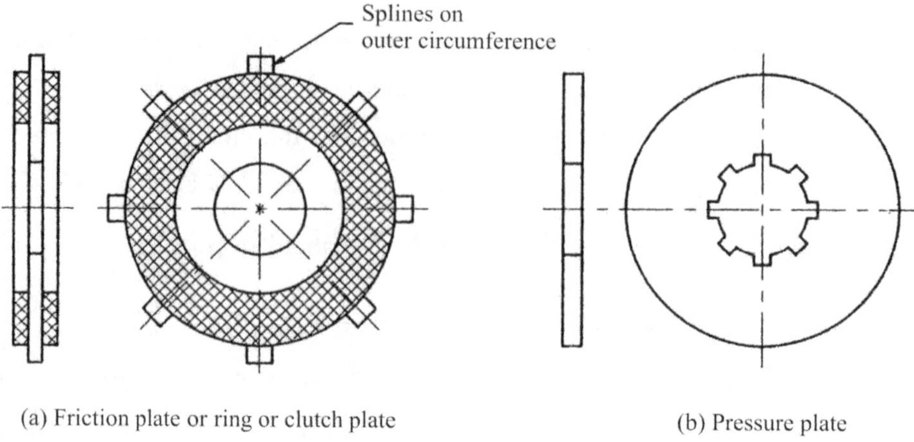

Fig. 13.11(b) Friction plate and pressure plate

1. *Single plate (disc) clutch*

Single plate clutch consists of friction plate, pressure plate driving shaft, splined driven shaft, splined hub, brass bush etc. Single plate clutch is shown in Fig. 13.11(a).

Driven plate, it is clutch plate which is held by the pressure plate with the flywheel. It is steel disc which is connected to splined hub through spring. Friction plate has a ring of friction material on either side. The linings (rings) are riveted to the clutch plate spring is placed between pressure plate and cover. Brass bush is pressed in side the pressure plate so that it can rotate freely on driven shaft.

When pressure is applied at the Foot-Pedal, it is transmitted to the levers through linkage, fork and pressure release bearing. As a result, the spring is compressed and the disengagement sleeve moves to the right and clutch is disengaged.

During disengagement, the pressure plate and flywheel rotate with the engine while the clutch plate inside them become stationary.

When the clutch pedal is released, it moves sleeve to left the clutch plate engages the driven shaft. The clutch plate is then held between the flywheel and pressure plate rotating as one unit. This is the engaged position.

13.3.6 Multiplate Clutch (wet clutch)

As shown in Fig. 13.12, it consists of driving shaft connected to engine, driven shaft connected to gear box, spring, two sets of frictional (driving) plates and pressure (driven) plates arranged alternatively driven shaft is splined.

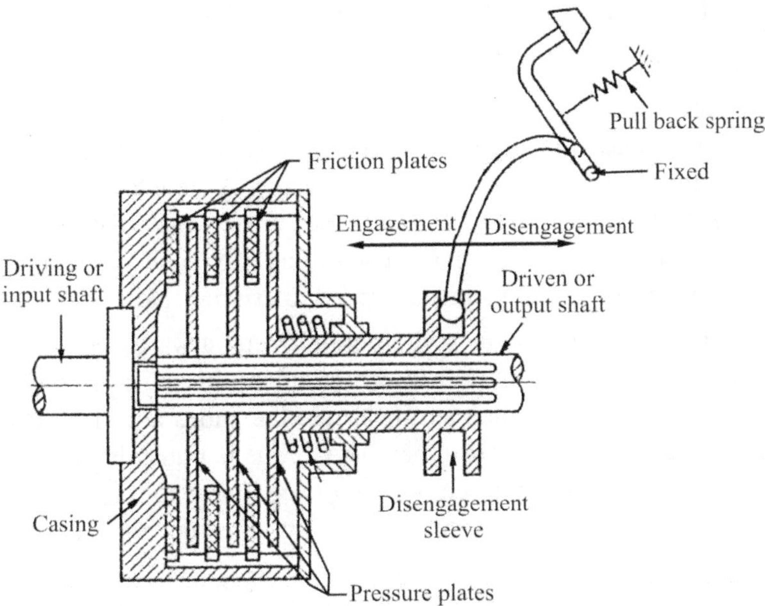

Fig. 13.12 Multiplate clutch at disengaged position

As shown in Fig.13.12, friction plates are splined on their outer circumference and engaged with corresponding splines on the casing. Friction plates having ring of friction lining on both sides, except the first plate which is adjacent to casing. Friction material used is generally ferodo. Pressure plates are splined on their inner circumference and engaged with splines on driven shaft. Both friction plates and pressure plates are free to move axially and when clutch pedal is pressed, clutch is disengaged, as the sleeve moves to the right, no axial force acts between friction plates and pressure plates. Driving shaft rotates no motion is transmitted to driven shaft.

When clutch pedal is released, the force due to spring will press the pressure plates to come in contact with friction plates and clutch is engaged. The power or torque is transmitted from driving shaft to casing, and from casing to friction plates and from friction to pressure plates and then to driven shaft.

Most multiple clutches run in oil bath.

Advantages

1. It is used where compact construction is required in case of scooters, mopeds, and motor cycles.
2. Oil is useful to remove friction material result due to clutch wear. These are drained with oil and trapped by oil filter.

13.3.7 Centrifugal Clutch (as shown in Fig. 13.13)

It works on principle of centrifugal force.

In this type the centrifugal clutch is in disengaged position when vehicle is not moving. It consists of spider having radial guide, sliding shoes, helical tension spring and drum. Spider having four equispaced radial guides, is keyed to driving shaft. Sliding shoes can slide radial in guides. The sliding shoes are covered with frictional material from outer side. These shoes are held against spider on driving shaft by helical spring. All these assembly is enclosed in drum which is keyed to driving shaft.

The shoes are pulled inward by the springs and forced outward due to centrifugal force caused by rotation of driving shaft. When the speed of driving goes on increasing, the centrifugal force will be greater than spring force, the shoes are pressed against inside surface of drum which is the driven member. The torque is transmitted from driving shaft to driven shaft because of friction force between shoe lining and inside surface of drum. With increasing speed the torque transmitting capacity of centrifugal clutch increases.

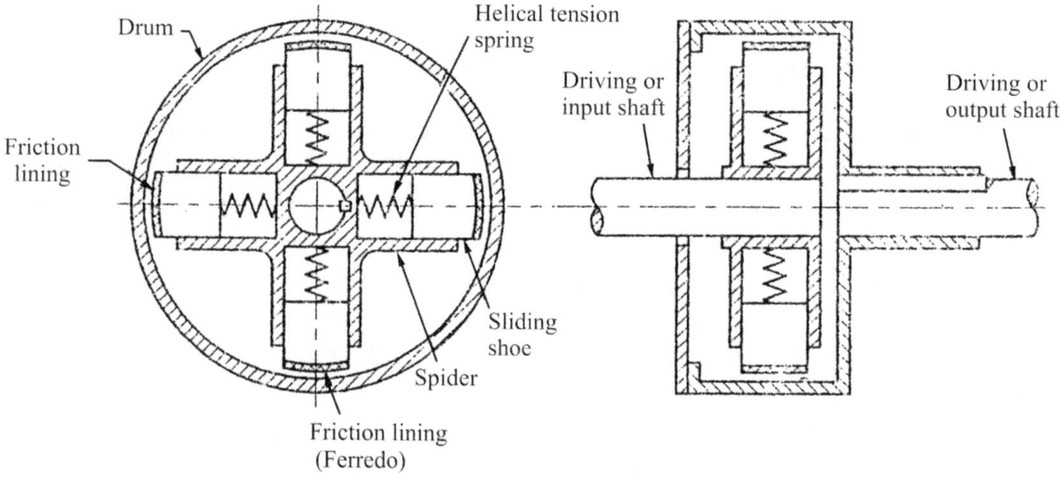

Fig. 13.13 Centrifugal clutch.

13.4 Definition of Brake and Materials Used

Power can be transmitted using friction from one shaft to another shaft by a **clutch.**

Sometimes frictional force is also used to absorb the energy possessed by a moving member. The device, used for this purpose is called as **Brake.** Therefore, a **brake is a device by means of which an artificial frictional resistance is applied to a moving body in order to retard or stop the motion of a body.**

Clutches and brakes work on the same principle of friction but the functional difference between clutch and brake is that the clutch connects one moving part to another moving part, whereas the brake connects one moving part to another stationary part.

During braking process, the brake absorbs either kinetic energy or potential energy or both by an object. In automobiles, brake absorbs kinetic energy of moving vehicles. In case of elevators and hoists brake absorb potential energy released by the objects during braking period.

The energy absorbed by the brake is converted in the form of heat which is dissipated to the surrounding air or water which is circulated through the passage in the brake drum.

The capacity of brake depends on unit pressure between braking surfaces, coefficient of friction, peripheral velocity of brake drum, projected area of friction surfaces and heat dissipating capacity of the brake.

The shoes are engaged with inner surface of cylindrical drum which is driven member.

13.4.1 Classification of Brakes

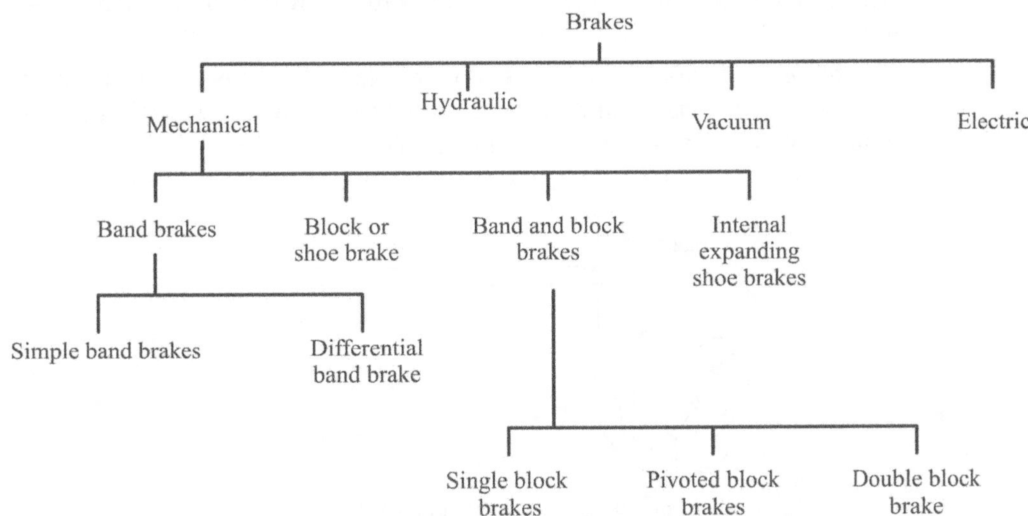

13.4.2 Material used for Braking Lining are

1. Wood on cast iron
2. Fiber on metal
3. Cork on metal
4. Asbestos
5. Leather on metal

13.4.3 Uses

Brakes are used in automobiles, train, hoists, elevators, presses etc.

13.5 Mechanical Brakes

13.5.1 Band Brakes

Introduction

- Band brake consists of rope, belt or flexible steel band lined with friction material which wrapped (embraces) party around the drum.
- When force P is applied on the lever, the band is pressed against the external surface of drum, the frictional force between drum and band will include torque on the drum.

This type of brakes is widely used in construction work and in automobile as hand brake.

There are two types of band brakes.

(i) Simple band brake (ii) Differential band brake

(i) *Simple band brake*

Simple band brake is shown in Fig. 13.14. In this type one end of the band is attached at fulcrum of the lever while the other end is at a distance 'b' form fulcrum.

When force 'p' is applied at free end of the level. It turns about the fulcrum 'O'. It tightens the band on drum and brakes are applied. The braking force is provided by the friction between band and the drum.

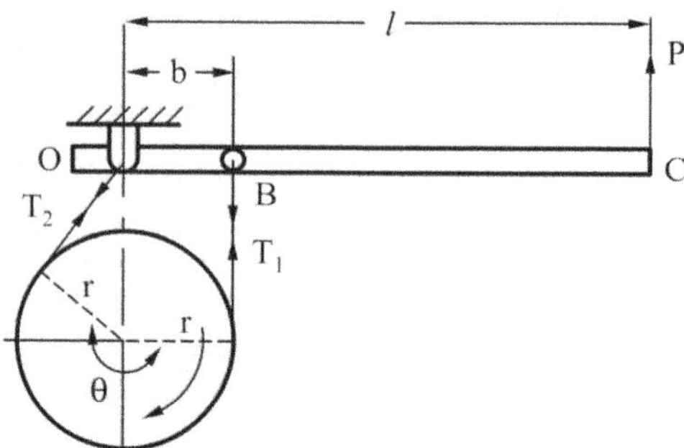

Fig. 13.14 Simple band brake.

(ii) *Differential band brake*

In differential band brake, the two ends of band are attached to the two point 'A' and 'B' on opposite side of the fulcrum as shown in Fig. 13.15.

The lever AOC is pivoted at fulcrum 'O'.

Differential band brake can provide higher braking torque compare to simple band brake, for same amount of effort 'P'.

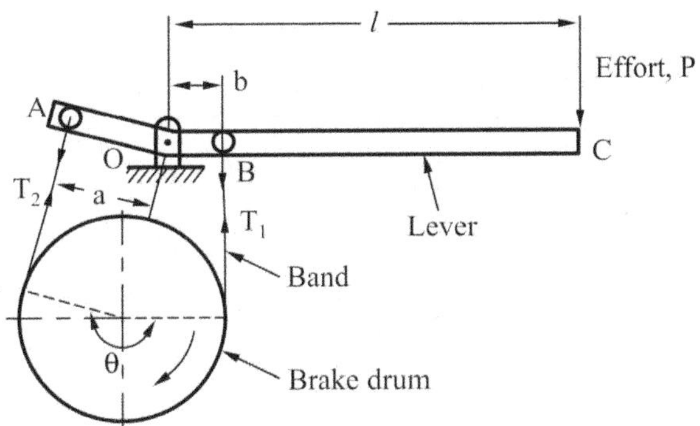

Fig. 13.15 Differential band brake.

13.5.2 Single Block or Shoe Brake

Single block or shoe brake consists of block or shoe which is pressed against a rotating drum by means of lever as shown in Fig. 13.16.

The block which is rigidly attached or pivoted to the lever is lined with friction material. The friction between friction lining on the block and drum retard the rotation of the drum.

The block or shoe is made up of softer material than the rim of the drum. The material of the block for light and slow vehicles is wood and rubber and for heavy and fast vehicles it is cast steel.

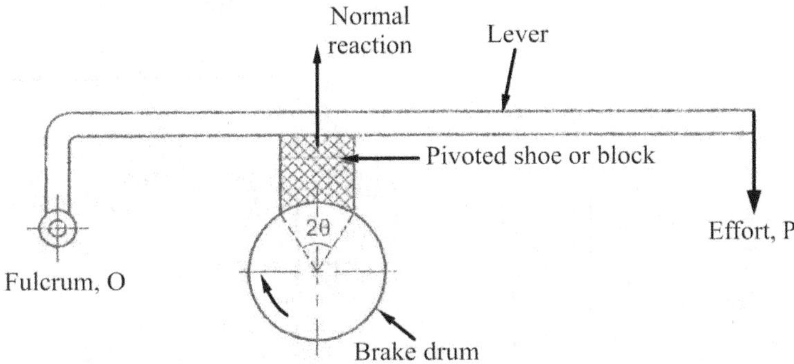

Fig. 13.16 Single block or shoe brake.

13.5.3 Pivoted Block Brake ($2\theta > 60°$)

A pivoted block brake is shown in Fig. 13.17. Unlike single block brake, in this the shoe is pivoted to lever to get uniform wear when 2θ is more than $60°$. These brakes have more life and provide higher braking torque.

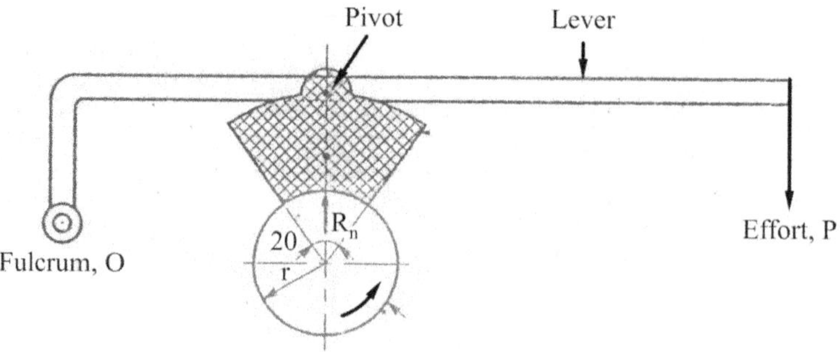

Fig. 13.17 Pivoted block or shoe brake.

13.5.4 Double Block or Shoe Brake

When a single block brake is pressed against the rotating drum, an unbalanced load on the bearing of the shaft supporting the drum will act. This load produces the bending of the shaft. It can be prevented by using a double block or shoe brake having two blocks on the two sides of the drum. Upper ends of brake arms are held together by using spring as shown in Fig. 13.18.

When effort P is applied in down ward direction to bell crank lever, then spring gets compressed and brake is released. When there is no force P on bell crank lever, the brake is engaged automatically. In this case braking torque becomes two times. Double block or shoe brake is used in electric cranes.

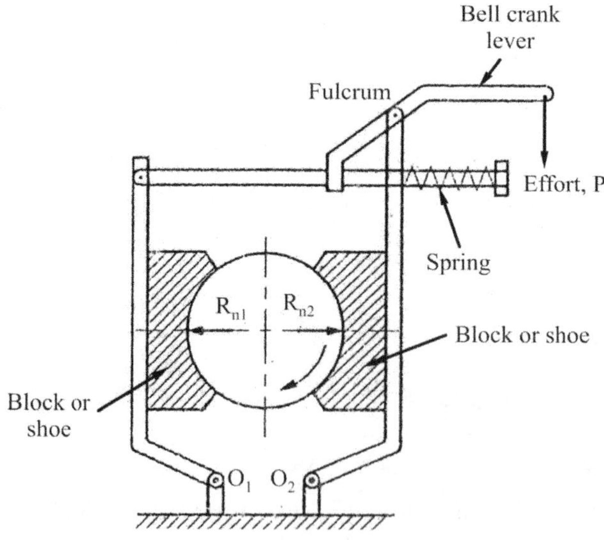

Fig. 13.18 Double block or shoe brake.

13.5.5 Band and Block Brake

A band and block brake consists of number of wooden blocks fixed inside a steel band. When the brake is applied, the blocks are pressed against the drum. The friction between blocks and drum provide the braking action as shown in Fig. 13.19.

The advantages of providing wooden blocks is that it provides higher co-efficient of friction and they can be easily replaced after being warn out.

Effectiveness of this brake increases.

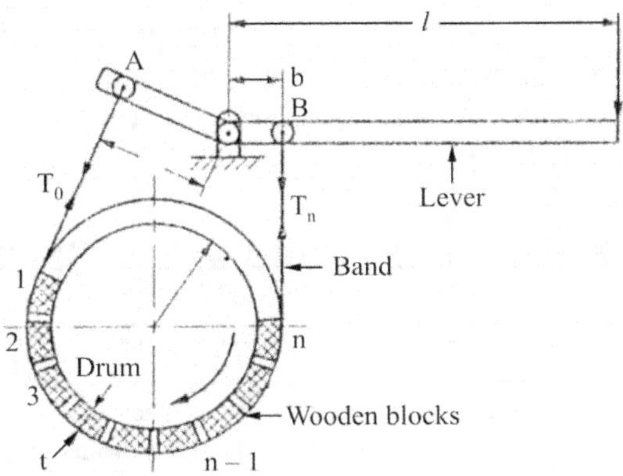

Fig. 13.19 Band and block brake.

13.5.6 Internal Expanding Shoe Brake

Fig. 13.20 shows the internal expanding shoe brake. It consists of two semi circular shoes R_1 and R_2. They are pivoted with back fixed plate at the fulcrum P_1 and P_2.

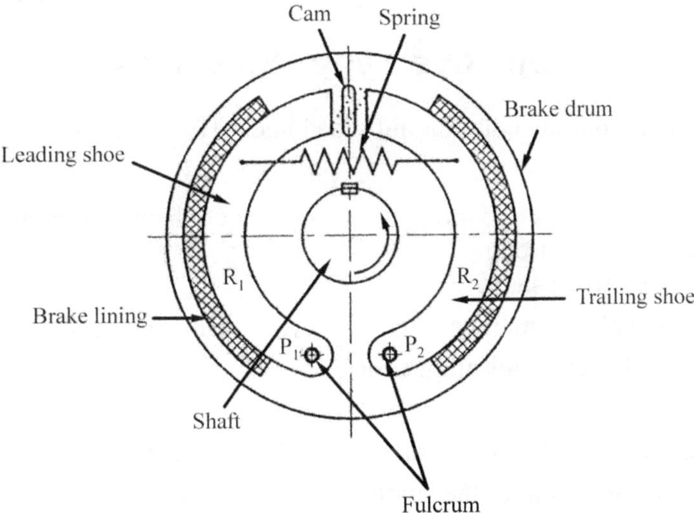

Fig. 13.20 Internal expanding shoe brake.

The outer surface of the shoes are lined with friction material (known as Ferodo).

One end of shoe is pivoted at fulcrum and other end is subjected to actuating force. The actuating force on both shoes is applied by cam mechanism or by Hydraulic wheel cylinder. When cam is operated mechanically, the shoes are pushed out-ward against the rim of the drum. The friction between shoes and the drum produces braking torque causing drum to slow down or stop completely. When cam is brought to vertical position, the spring brings the shoes to the original position and drum begins rotating.

Internal expanding shoe brake is used in motor cycle, scooter, cars and light trucks.

In hydraulically operated brake during braking the pressurized fluid enters into wheel cylinder fitted with two pistons, fluid exerts the pressure on piston, pistons move outwards and pushing free ends of the shoes causing braking of the drum. As the speed of drum is decreased or drum is stopped.

13.6 Difference between a Brake and a Clutch

Brake	Clutch
Used to slow down or stop the motion of a moving body.	Used to transmit the motion of moving body to another body, which may be moving or stationary.
During braking, the energy of the moving member is lost in friction	During working of clutches, the energy is not lost in friction except in case of slipping of clutch.
In a hoist, the brake normally remains in engaged condition.	In an automobile the clutch normally remains in engaged conditions.
Ordinarily the brake remains in disengaged condition.	In machine tools, presses etc, the clutch ordinarily remains in disengaged condition.

Objective Type Questions

1. Match A with appropriate coupling provided under B.

	A		B
a.	The coupling ensures the most accurate, rigid and strong connection between shaft.	(1)	Oldham's coupling
b.	The coupling allow small parallel misalignment angular misalignment between shaft.	(2)	Universal coupling
c.	The coupling is used when angular or offset misalignment of the shafts is more.	(3)	Pin type flexible coupling

d.	The coupling is used when the shafts are parallel but not in alignment.	(4)	Flange coupling

2. Make correct statement by joining A with B

	A		B
a.	They are fastening devices to transmit power from one rotating shaft to another	(1)	Clutch
b.	They are used to control or stop the speed of a machine.	(2)	Coupling
c.	It is used to transmit power from driver to driven shaft when it may be required to start or stop frequently.	(3)	Brake

Solution :

1. a. (4) b. (3) c. (2) d. (1)
2. a. (2) b. (3) c. (1)

Short Questions

1. Define a coupling. What are the requirements of a good coupling? How the couplings are classified?
2. What is the difference between muff and split muff coupling?
3. What is the difference between rigid coupling and flexible coupling? Describe bushed pin type flexible coupling with a neat sketch.
4. Explain oldham's coupling with a neat sketch.
5. Which coupling can be used to couple two shafts whose axis intersect? Give name of different parts of the coupling.
6. What is the function of a clutch in automobile vehicle? What are the various types of clutches? Name the clutch used in scooters and cars.
7. What is clutch? Mention broad classification of clutches?
8. What is function of brake? How are they classified? Explain with neat sketch working of any one type of make.
9. What is Clutch? How it differs from brake when multi plate clutches are used?
10. Draw neat diagram of

    (i) Block brake (ii) Internal expanding brake (iii) Disc Clutch (iv) Muff coupling

Answers

Correct 1, 2, 4, 5

In correct 3, 6

Match the Pair

1.

	Set A	Set B
A	Friction	1. irreversible flow
B	steady flow	2. Incompressible flow
C	low velocity gas flow	3. zero viscosity
D	Ideal fluid	4 Velocity at a point is constant

Answers

(1) A – 1, B – 4, C – 2, D – 3

2.

	Set A	Set B
A	Stream flow	1. Path of particle that passed a point
B	Streak flow	2. Path of a single particle that passed a point
C	Equipotential flow	3. equi potential line
D	Path line	4. Described by potential function

Answers

(2) A – 3, B – 2, C – 4, D – 2

CHAPTER 14

Power Transmission System

14.1 Introduction

The power produced by prime mover is used to drive the various machines in the work shop and manufacturing industries. The transmission of power from prime mover to machine can be obtained by two systems

1. Individual system
2. Group system

Individual Drive

In this type of drive separate motor is used for each machine. Machine shaft is directly connected with motor shaft or through belt, chain and pulleys. See the Fig. 14.1.

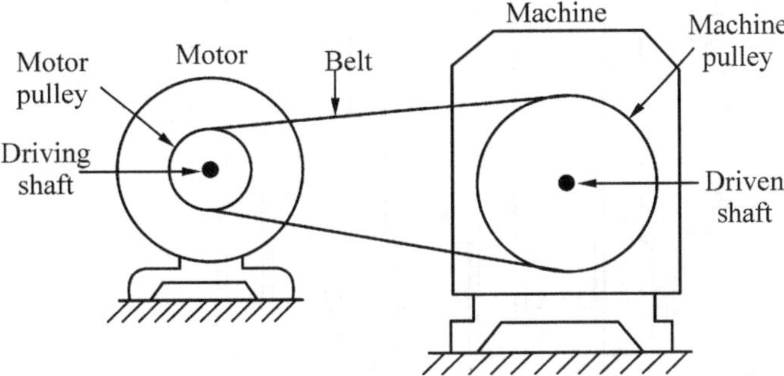

Fig. 14.1 Individual drive.

1. It is suitable when the machine is likely to be shifted at different places
2. Mobile or portable machines
3. Machine can be located at any place as per requirement

4. Shop view is clean and open belting is absent. So accidents are reduced.
5. In failure of motor, that particular machine is stopped but other machines continue to operate so idle time is minimized and production continuity is maintained.
6. Better control and speed range.

Disadvantages

1. Initial cost is more
2. All machines can not be simultaneously shut down by single control (switch)
3. It is not economical for large number of machines
4. Increase in maintenance and spare parts.

Group Drive

A single motor or prime-mover of large capacity drives a common overhead main shaft located from one end to the other end of workshop. This shaft has many pulleys, the motion is transmitted by this shaft to individual machines by flat belt and pulley systems.

The fast and loose pulleys are fitted on counter shaft near the machines to start and stop the machine without-stopping main motor. See the Fig. 14.2.

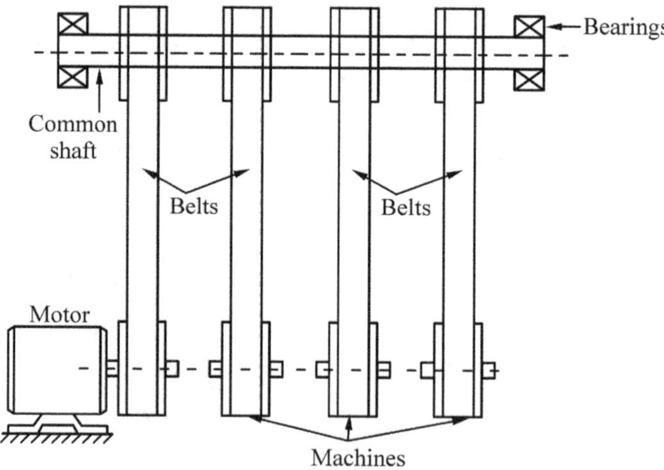

Fig. 14.2 Group drive.

Applications

1. Process industries
2. For operating machines which are arranged to perform sequential operations.

Advantages

1. Initial cost is low
2. Better control in stopping all machines simultaneously in case of process industries and for sequence operations.
3. Economical for large number of machines.

Disadvantages

1. Low efficiency due to power loss in shafting, belting and pulleys
2. Continuous power loss during idle running or running at low loads when few machines are in operation. Hence low efficiency
3. Uneconomical due to power loss
4. Failure of main motor causes total shutdown of the workshop
5. Replacement of belt and repair of motor increases time and expenses.
6. Chances of accidents are more because layout is difficult, presence of open belts, clumsy and not flexible and less illumination.
7. Shop view is not clean and overhead crane can not be used.

Comparison of individual and group drive

Individual Drive	Group Drive
No such power loss, therefore efficiency is comparatively more	Low efficiency due to power loss in shafting, belting and pulleys
Less power is wasted as each machine is run by individual motor	Continuous power loss during idle running or running at low loads when few machines are in operation
failure of motor does not cause total shut down of workshop	Failure of main motor causes total shut down of the work shop
Economical due to less power loss	Uneconomical due to comparatively more power loss
Easy of maintenance	maintenance is comparatively difficult
initial cost is less	initial cost is more
Shop is clean and over head crane can be used	shop view is not clean and over head crane can not be used
Machines can be located at convenient places	no flexibility in locating machines at convenient places
Better control in stopping all the machines simultaneously is not possible	Better control in stopping all the machines simultaneously in case of process industries and for sequence operation
Better illumination therefore less chances of accidents	Chances of accidents are more because of less illumination

14.2 Power Transmission Elements

Main component or elements of power transmission systems are as follows (Fig. 14.3)

1. Shaft
2. Bearings
3. Belt
4. Rope
5. Chain
6. Gear
7. Pulleys

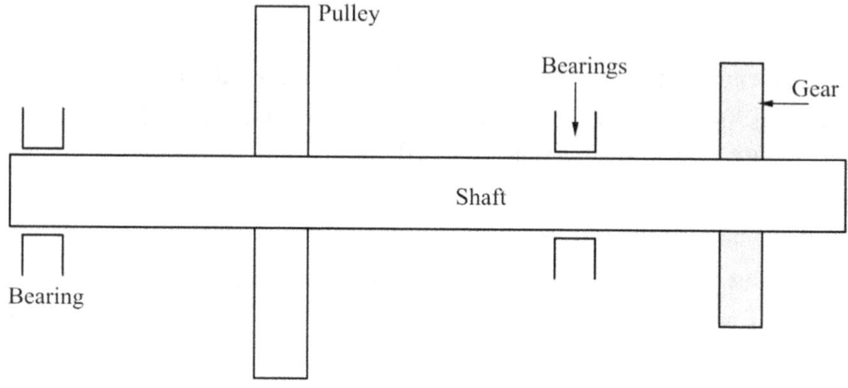

Fig. 14.3

Shaft is one of the main and important components for the power transmission system. It is supported and rotates inside bearings. Pulleys, gears are mounted or keyed to shaft for transmission of power. Part of shaft within bearing is known as journal.

Axle

It is stationary shaft on which other member such as pulley, gear etc rotates.

Spindle

It is machine shaft that drives or supports either cutting tool or the work piece on which machining is performed.

Bearings

1. It supports rotating shaft
2. It confines or guide in motion of shaft
3. They take up axial and radial thrust (load) imposed on shaft and transmit these loads to casing of machine or frame of the machine
4. It reduces the energy loss due to friction.

14.3 Power Transmission Method

The mechanical power from the prime mover (source) to the machine can be transmitted by the following methods.

1. Shaft drive
2. Friction drives
3. Coupling
4. Belt drive
5. Rope drive
6. Clutch drive
7. Gear drive
8. Chain drive

Selection of a particular drive depends on the power transmitted and distance between two shafts.

The important drives are discussed below

Belt drive

This is most common used method of power transmission in which the two shafts are rotated by an endless, thin belt running over two pulleys on the shaft.

The pulley on driving shaft is known as the 'driver' where as pulley on other shaft, to which the power is transmitted is known as driven pulley or follower pulley.

14.4 Types of Belt Drives

There are four types of belt drive namely:

1. Open belt drive
2. Cross belt drive
3. Quarter turn drive
4. Right angle drive

1. **Open belt drive**

   Two parallel shafts are connected by open belt drive such that the driver and follower both shafts rotate in same direction as shown in Fig. 14.4.

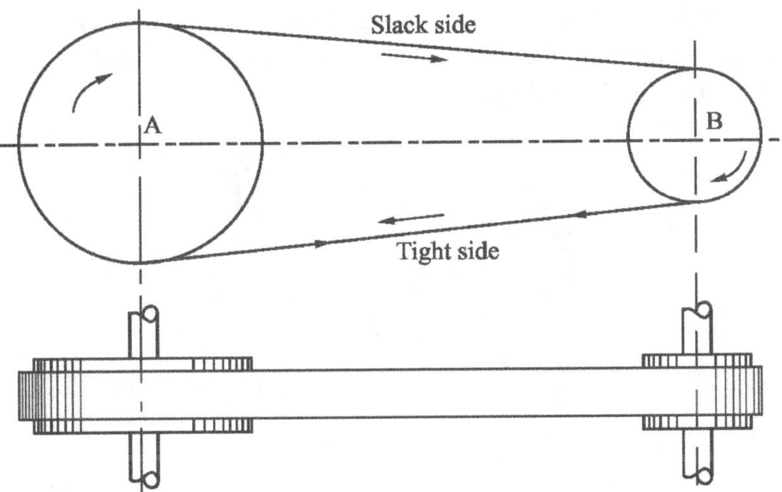

Fig. 14.4 Open belt drive.

2. Cross belt drive

Two parallel shafts are connected by belt-pulleys such that the follower pulley rotates in opposite direction to that of driver pulley.

Cross belts are subjected to greater wear. They are generally used for low speed and small power transmission. See the Fig. 14.5

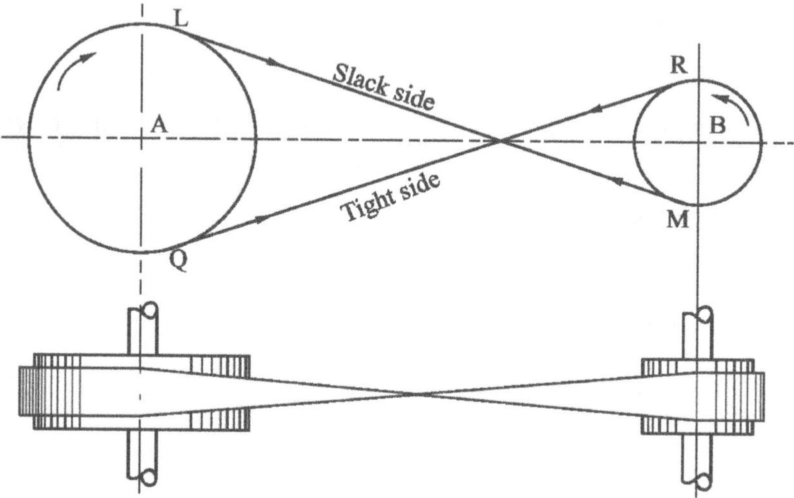

Fig. 14.5 Cross belt drive.

3. Quarter turn drive

Two non parallel and non intersecting shafts can be connected by this type of belt and pulleys as shown in Fig. 14.6. The centre line of belt near the pulley must lie in the central plane of that pulley.

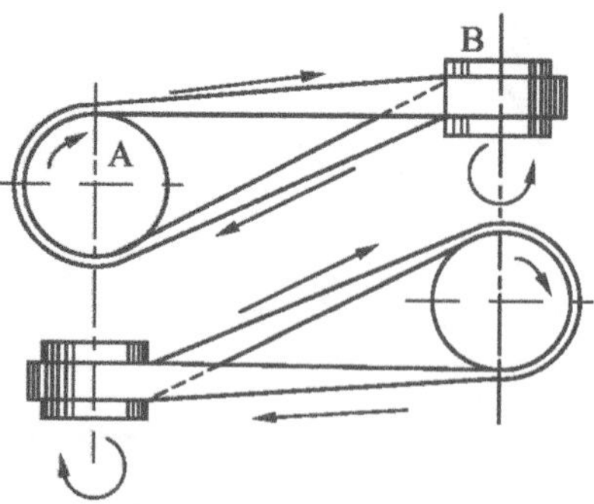

Fig. 14.6 Quater turn drive.

4. **Right angle drive**

In this method, additional guide pulley is used to direct the belt and to adjust tension of belt which connects the two non-parallel shafts as shown in Fig. 14.7. By following above principle, the other angle between two connected shaft can be obtained in the similar way

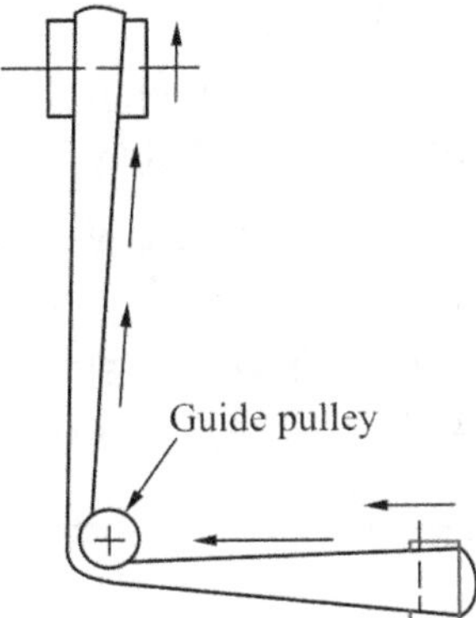

Fig. 14.7 Right angle drive.

14.5 Belt Materials

The following materials are used for belt constructions

(i) *Leather belt*: It is preferable used in dry as well wet conditions at normal temperatures.

(ii) *Cotton and canvas or woven belt*: It is woven by using cotton and camel hair. It is superior to leather belt in warm and humid places. They should be protected from oil and grease

(iii) *India rubber belt*: Vulcanized India rubber is cemented together with canvas piles to form belt. It is the best for damp conditions but it is costly and it is affected by oil and grease.

(iv) *Steel belt*: The thin steel bands are used as the belts in industries. It is not affected by oil or grease. It can be used in humid places.

(v) *Balata belt*: It is prepared by using cotton canvas and balata which is gum-like paste obtained from certain trees. It has wide applications in industries.

14.6 Belt Shapes

Cross section of belt is of three types:

(i) *Flat belt*: This belt is flat with certain thickness. The pulley surface is also flat and width of pulley is $1/4^{th}$ greater than the width of belt Fig. 14.8 (a).

(ii) *V belt*: V belt is trapezoidal in section is placed in pulley having 'V' shaped groove. It is made of fabric and VIR with cotton cord tension element.

It is used for large power transmission for short distance between the shafts. It is replaced by chain and gear drives for shorts. It is replaced by chain and gear drives for short-centre distance machine drives, due to positive drive and slip less-action Fig. 14.8 (b).

(iii) *Round belt (Rope)*: The cross-section of this belt is round and these ropes are passed in grooved pulleys. It is suitable for moderate distances.

(iv) *Timing belt*: It is similar to flat belts with teeth moulded on the under side face. They are known as positive drive belts. They are endless with standard length, width and pitch. They are expensive than V-belt and flat belts. They are capable to transmit moderate power for relatively short distance.

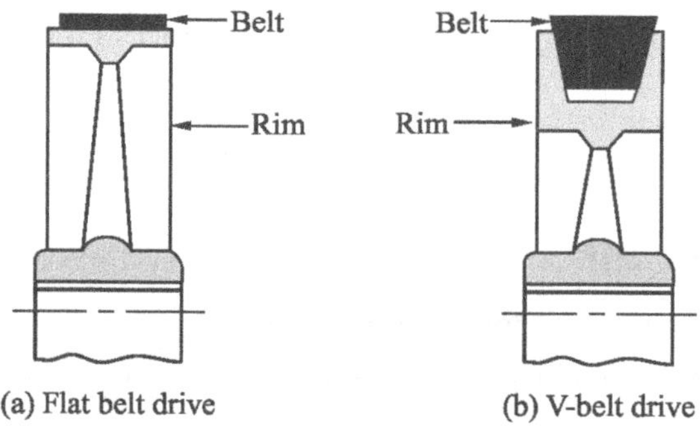

Fig. 14.8 Section of pulley rims.

14.7 Comparison Between Flat Belt Drive and V-belt Drive

Flat Belt Drive	V-Belt Drive
Belt is thin and flat	Belt section is trapezoidal
Pulley surface is flat	Pulley surface is V-grooved
Unsuitable for large transmission and very large power	Transmits high power
Possibility of slipping out from pulley	No possibility of coming out from pulley groove
No breakage on over load due to slip	There is possibility of breakage at over load due to no slippage of belt from pulley
It is not positive drive. Velocity ratio varies. Den-dense to slip.	Positive drive. Constant velocity ratio. Slip less drive.
Suitable for moderate distance	Suitable for short distance

Contd...

Flat Belt Drive	V-Belt Drive
Frequent shortening of belt is required due to elongation of belt	No elongation of belt due to usage. So no shortening of belt is required
Simple and cheap	Compact but costly
Used for low speeds	Used for high speeds
Absorbs shocks due to high elasticity	Can not absorbs shocks due to low elasticity
Small arc of contact is not permitted	Small arc of contact is permissible due to wedging action in groove

14.8 Belt Drive

Advantages

1. It is not costly method
2. It is used for parallel, non parallel, intersecting and non-intersecting shafts.
3. It is used for moderate or long distance shafts
4. It can not break at over loads due to slip
5. Absorbs shocks due to high elasticity
6. Belt is thin and flat.

Disadvantages

1. Complicated drive for non parallel shafts
2. Belt should be horizontal for open belt system
3. Frequent shortening is required due to elongation of belt during use
4. Unsuitable for short distance between two shafts
5. Not suitable for large loads (power) and high speeds
6. Slip increases as load increases
7. Constant velocity ratio is not maintained due to slip.
8. Possibility of falling out from the pulley
9. It is not positive drive
10. Power transmission efficiency is less due to slip.

Applications of belt drive

1. It is used in grinding and floor mills
2. For pump water from well, when pump and motor or engine are not on same level or away from each other
3. V-belt drive for compressors and radiators
4. For line shaft drive and group machines drive
5. Stopped pulley drive for lathes and drilling machines.

14.9 Tension in Belt Drive

Consider an open belt drive as shown in Fig. 14.11(a). When the belt is driven by driver pulley, there is more tension in the belt on one side of pulley i.e. the lower span of the belt is pulled due to friction between belt and pulley. The one side (lower span) is tight side. The upper side is known as slack side.

Let T_0 = Initial tension

T_1 = Tension on one side (lower span) of belt

T_2 = Tension on other side (upper span) of belt

∴ Increase in tension on tight side = $T_1 - T_0$

Decrease in slack side = $T_2 - T_0$

$$\therefore \quad (T_1 - T_0) = (T_2 - T_0)$$

$$\therefore \quad T_0 = \frac{T_1 - T_2}{2} \text{ or initial tension in belt} = \frac{T_1 - T_2}{2}$$

unequal tensions on pulley causes a torque on it.

14.10 Creep and Slip in Belt

1. When the belt passes from driver to follower, it is lengthened while it is shorten when it passes from follower towards the driver

   Due to these changes in length of the belt, there is relative motion between the belt and pulley surface. This relative motion is known as creep. The creep causes reduction in velocity ratio and loss of power. The loss of speed due to the creep is dependent on the elasticity of the belt material.

   Considering thickness of belt

$$V.R = \frac{(d_1 + t)}{(d_2 + t)} \times \frac{100 - S}{100}$$

t_1 = thickness of belt

S = slip

d_1 = diameter of driver pulley

d_2 = diameter of follower pulley

Slip

The difference between actual speed of the follower or driven pulley and its ideal speed in absence of creep is known as slip.

Slip is caused when force transmitted by the driver exceeds the friction force. Slip is also occurs because of air is trapped between pulley surface and belt. Due to slip velocity ratio does not remain constant. Its value increases with load. Due to slip positive drive is not obtained. But it is helpful in protecting the belt from breaking at over load.

Slip is actually mentioned as a percent of the ideal (calculated) speed.

$$\therefore \qquad \text{slip} = \frac{\text{calculated speed} - \text{actual speed}}{\text{calculated speed}} \times 100$$

let S_1 = slip between driver pulley and belt

S_2 = slip between follower pulley and belt

S = resultant slip

S = total % slip = $S_1 + S_2$

As we know surface speed of driver pulley = $\pi d_1 n_1$

where d_1 = diameter of driver pulley

n_1 = RPM of driver pulley

$$\therefore \qquad \text{Speed of belt} = \pi d_1 n_1 \left(\frac{100 - S_1}{100}\right)$$

The surface speed of follower = $\pi d_2 n_2$

where d_2 = diameter of follower

n_2 = RPM of follower

$$= \text{speed of belt} \times \frac{100 - S_2}{100}$$

$$\therefore \qquad \pi d_2 n_2 = \pi d_1 n_1 \times \frac{100 - S_1}{100} \times \frac{100 - S_2}{100}$$

$$= \pi d_1 n_1 \times \frac{(100 - S_1)(100 - S_2)}{100 \times 100}$$

$$= \pi d_1 n_1 \left[\frac{100 \times 100 - 100\,S_2 - 100\,S_1 + S_1 S_2}{100 \times 100}\right]$$

Neglecting S_1S_2 as they are very small we get

$$= \pi d_1 n_1 \left[\frac{100 \times (100 - (S_1 + S_2))}{100 \times 100}\right]$$

$$\therefore \text{ Speed of belt} = \pi\, d_1 n_1 \frac{100 - S_1}{100}$$

$$= \pi\, d_1 n_1 \frac{\pi d_2 n_2 (100 - S)}{100} \qquad \therefore [S = (S_1 + S_2)]$$

$$\frac{n_2}{n_1} = \frac{d_1}{d_2} \frac{(100 - S)}{100}$$

But velocity ratio $= \frac{n_2}{n_1}$

$$\therefore \qquad V.R = \frac{d_1}{d_2} \frac{(100 - S)}{100}$$

14.11 Length of Belt

It means that total length of belt required to connect driver and a followers. Here we will discuss the length of two types of belt drives.

1. Open belt drive
2. Cross belt drive

Length of open belt drive

In this case both pulleys rotate in same direction. Let the two pulleys are connected by open drive as shown in Fig. 14.9.

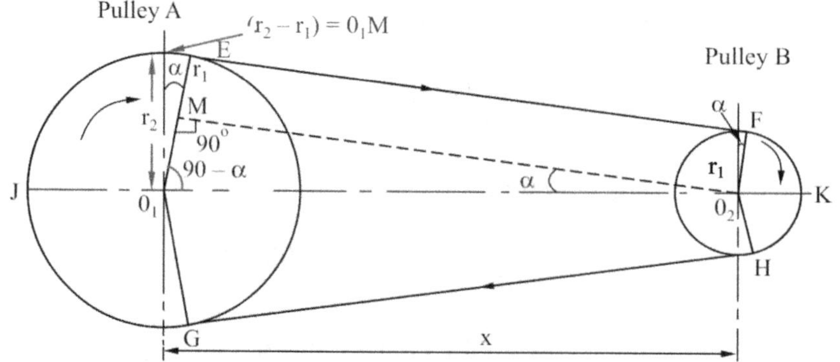

Fig. 14.9 Open belt.

Let $\quad L$ = length of open belt

r_1 = radius of larger pulley

r_2 = radius of smaller pulley

x = distance between centers of two pulleys i.e. O_1O_2

O_1 = centre of larger pulley and O_2 centre of smaller pulley. Belt leaves the larger pulley at E and G and smaller pulley at F and H.

Through O_2 draw O_2M parallel to EF. We find that O_2M will be perpendicular to O_1E because EF is tangent to larger pulley at E and EO_1 is radius of pulley make 90° at E. let angle $MO_2O_1 = \alpha$ radians.

Therefore length of belt L = (arc GJE + EF + arc FKH + HG)

or

$$L = 2\,[\text{arc JE} + EF + \text{arc FK}] \qquad \text{.....(i)}$$

From figure $$\sin\alpha = \frac{r_1 - r_2}{x}$$

Since α is very small, as central distance is very large and $(r_1 - r_2)$ is very small.

∴ $$\sin\alpha = \alpha.$$

So we can write $$\sin\alpha = \alpha = \frac{r_1 - r_2}{x} \qquad \text{.....(ii)}$$

$$\text{arc JE} = r_1\left(\frac{\pi}{2} + \alpha\right) \qquad \text{.....(iii)}$$

$$\text{arc FK} = r_2\left(\frac{\pi}{2} + \alpha\right) \qquad \text{.....(iii)}$$

$$EF = MO_2 = \sqrt{x^2 - \left(\frac{r_1 - r_2}{x}\right)^2} = x\sqrt{1 - \left(\frac{r_1 - r_2}{x}\right)^2}$$

Expanding this equation by binomial theorem

$$\sqrt{1 - \left(\frac{r_1 - r_2}{x}\right)^2} = \left[1 - \frac{1}{2}\left(\frac{r_1 - r_2}{x}\right)^2 + ...\right]$$

∴ $$EF = x\left[1 - \frac{1}{2}\left(\frac{r_1 - r_2}{x}\right)^2\right]$$

$$= x - \frac{(r_1 - r_2)^2}{2x^2}\,x$$

$$= \left[x - \frac{(r_1 - r_2)^2}{2x}\right]$$

Substituting the value of arc JE, arc FK and EF in equation (i) we get

$$L = 2\left[r_1\left(\frac{\pi}{2}+\alpha\right)+x-\frac{(r_1-r_2)^2}{2x}+r_2\left(\frac{\pi}{2}-\alpha\right)\right]$$

$$= 2\left[r_1\frac{\pi}{2}+\alpha\, r_1+x-\frac{(r_1-r_2)^2}{2x}+r_2\frac{\pi_2}{2}-r_2\,\alpha\right]$$

$$= 2\left[\frac{\pi}{2}(r_1+r_2)+\alpha\,(r_1-r_2)+x-\frac{(r_1-r_2)^2}{2x}\right]$$

$$= \left[\pi\,(r_1+r_2)+2\alpha\,(r_1-r_2)+2x-\frac{(r_1-r_2)^2}{x}\right]$$

Substituting $\alpha = \frac{r_1 - r_2}{x}$ in above equation we get

$$L = \pi\,(r_1+r_2)+2\frac{(r_1-r_2)}{x}\,(r_1-r_2)+2x-\frac{(r_1-r_2)^2}{x}$$

$$L = \pi\,(r_1+r_2)+\frac{(r_1-r_2)^2}{x}+2x \qquad \text{.....(v)}$$

From eq. (v) it is clear that in case of stepped pulleys with open belt the sum radius $(r_1 + r_2)$ is constant. But $(r_1 - r_2)$ is not constant hence, the open belt becomes perfectly fit for one pair of pulleys, where as it is slack for other pair of pulleys in the stepped pulleys.

Length of Cross Belt

In this case both pulleys rotate in opposite directions, larger pulley and small pulley are connected by cross belt Fig. 14.10.

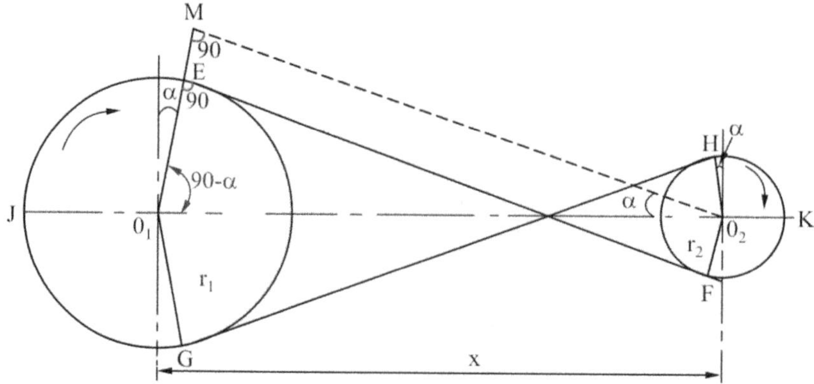

Fig. 14.10 In this case, (i.e., cross belt) both the pulleys rotate in the opposite directions.

Let r_1 = radius of larger pulley

r_2 = radius of smaller pulley

x = distance between the centers of two pulleys

L = total length of belt

Let belt leaves larger pulley at E and G and smaller pulley at F and H as shown in Fig. 14.10.

Through O_2 draw O_2M parallel to EF. We find that O_2M will be perpendicular to O_1E because angle between tangent EF and radius O_1E at E = 90°.

Let angle M $O_2 O_1$ = α radians.

Form figure length of the belt

$$L = [\text{arc GJE} + EF + \text{arc FKH} + HG]$$

$$= 2\,[\text{arc JE} + EF + \text{arc FK}] \quad .....(1)$$

α is very small, therefore putting

$$\sin\alpha = \alpha = \frac{r_1 - r_2}{x} \quad .....(2)$$

$$\text{arc JE} = r_1\left(\frac{\pi}{2} + \alpha\right) \quad .....(3)$$

$$\text{arc FK} = r_2\left(\frac{\pi}{2} + \alpha\right) \quad .....(4)$$

$$EF = MO_2 = \sqrt{x^2 - (r_1 + r_2)^2} \quad .....(5)$$

$$= x\sqrt{1 - \left(\frac{r_1 + r_2}{x}\right)^2} \qquad \text{By taking x common.}$$

Expanding the above equation by binomial theorem

$$= x\left[1 - \frac{1}{2}\left(\frac{r_1 + r_2}{x}\right)^2 + ...\right]$$

Substituting values of arc JE, arc FK and EF in equation (1) we get

$$L = 2\left[r_1\left(\frac{\pi}{2} + \alpha\right) + x - \frac{(r_1 + r_2)^2}{2x} + r_2\left(\frac{\pi}{2} + \alpha\right)\right]$$

$$= 2\left[r_1\frac{\pi}{2} + r_1\alpha + x - \frac{(r_1 + r_2)^2}{2x} + r_2\frac{\pi}{2} + r_2\alpha\right]$$

$$= 2\left[\frac{\pi}{2} + (r_1 + r_2) + \alpha\,(r_1 + r_2) + x - \frac{(r_1 + r_2)^2}{2x}\right]$$

$$= \left[\pi\,(r_1 + r_2) + 2\alpha\,(r_1 + r_2) + 2x - \frac{(r_1 + r_2)^2}{x}\right] \qquad .....(6)$$

Substituting $\alpha = \frac{r_1 + r_2}{x}$ in equation (6) we get

$$= \pi\,(r_1 + r_2) + \frac{2\,(r_1 + r_2)^2}{x} + 2x - \frac{(r_1 + r_2)^2}{x}$$

$$L = \pi\,(r_1 + r_2) + \frac{2\,(r_1 + r_2)^2}{x} + 2x$$

It may be noticed that above expression is function of $(r_1 + r_2)$ that is sum of two radii of two pulleys be constant, the length of belt required will also remain constant, provided the distance between centers of the pulleys remain unchanged.

14.12 Ratio of Driving Tension in Belt

1. When belt is not rotating, it has uniform initial tension to in its length. This is because of initial stretch.
2. When the driving pulley rotates, it pulls the belt from lower side and delivers it to upper side due to friction. Therefore side tension is increased from T_0 to T_1 and it is known as tight side. The upper side tension is decreased from T_0 to T_2 and it is known as slack side. See the Fig. 14.11.

   As belt length is constant, the mean tension is also constant

   $$\therefore \qquad T_1 - T_0 = T_0 - T_2$$

   $$\therefore \qquad T_0 = \frac{T_1 + T_2}{2} \qquad .....(1)$$

   let us consider a small length of belt subtending an angle at centre of pulley when belt begins to slip over the pulley against the friction, the various forces acting at an element arc xy of belt are

   1. Normal reaction between arc of belt and pulley = R
   2. Frictional force = μR
   3. Tension T acting at C
   4. Tension $T + \delta_T$ acting at D.

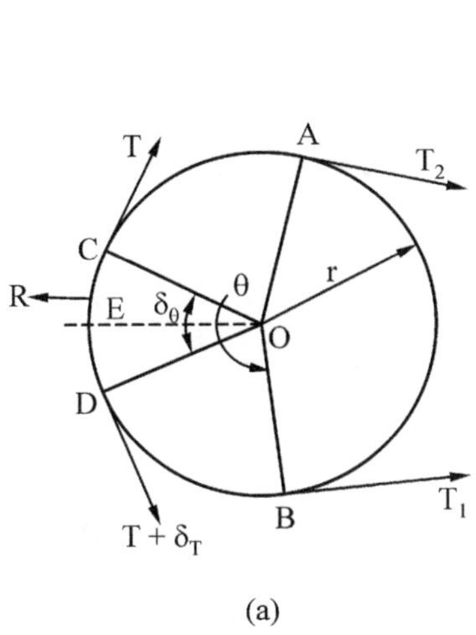

(a)

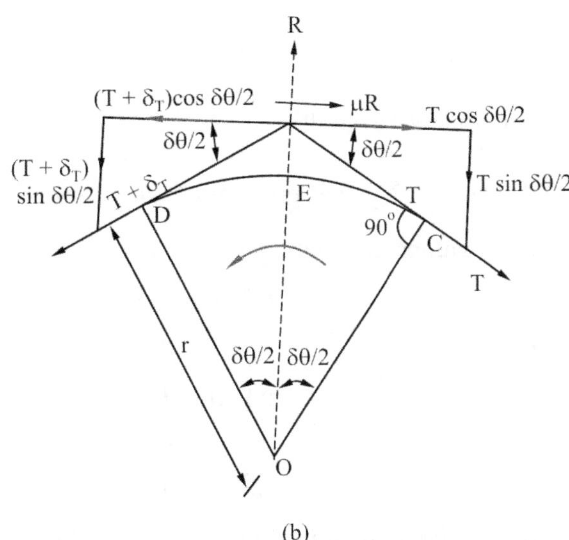

(b)

Fig. 14.11 Tensions in belt sides.

The tensions at free ends of the belt are T_1 and T_2 as shown Fig. 14.11(a).

For equilibrium, resolving the forces horizontally

$$R = T \sin\frac{\theta}{2} + (T + dT)\sin\frac{\theta}{2}$$

$$= \left(T\sin \theta/_2 + T\sin \theta/_2 + dT \sin d\,\theta/_2\right) \quad \text{... by simplifying}$$

$$= 2T\sin\frac{\theta}{2} + dT\sin\frac{\theta}{2} \qquad .....(1)$$

But $\delta\theta$ is very small, therefore $\sin\frac{\theta}{2} = \frac{\delta\theta}{2}$ and dT is also small and product of dT and $\sin\frac{\theta}{2}$ is very small and may be neglected. Therefore equation (1) becomes

$$R = 2T.\frac{\delta\theta}{2} \qquad .....(2)$$

Resolving vertically

$$\mu R = (T + \delta T)\cos\frac{\theta}{2} - T\cos\frac{\theta}{2}$$

$$\mu R = T\cos\frac{\theta}{2} + dT\cos\frac{\theta}{2} - T\cos\frac{\theta}{2} \qquad .....(3)$$

But δθ is very small

∴ $\cos\frac{\delta\theta}{2} = 1$ and product of dT cosδθ is very small therefore can be neglected. Put these values in equation 3 we get

$$\mu R = \delta T \qquad .....(5)$$

But $R = T.\,\delta\theta$ from equation (2)

∴ $\mu T\delta\theta = dT$

∴ $$\mu\,\delta\theta = \frac{dT}{T} \qquad .....(6)$$

Integrating equation between 0 to θ and tension T_2 to T_1

∴ $$\int_{T_2}^{T_1}\frac{\delta T}{T} = \int_{o}^{\theta}\mu d\theta$$

$$[\log_e T]_{T_2}^{T_1} = \mu\theta$$

∴ $$(\log_e T_1 - \log_e T_2) = \mu\theta$$

∴ $$\log_e\frac{T_1}{T_2} = \mu\theta$$

∴ $$\frac{T_1}{T_2} = e^{\mu\theta}$$

where θ is known as angle of lap and θ is taken in radian.

Similarly we can prove that for V belt or rope

$$\frac{T_1}{T_2} = e^{\mu\theta\operatorname{cosec}\alpha}$$

where α = semi grove angle of v-belt or rope.

14.13 Calculation for Angle of Lap

(i) ***For open belt***: Angle of contact or angle of lap is not same for two pulleys in open belt. It is smaller for both pulleys. For same co-efficient of friction "μ" for both pulleys, slipping first occurs on small pulley so angle of lap is considered for small pulley.

$$\frac{\theta}{2} = < DAC = \frac{\pi}{2} - \alpha = < NBA$$

$\therefore$ $$\cos NBA = \frac{BN}{AB} = \frac{BE - NE}{AB}$$

But from Fig. (a) $BE = r_2$ and $NE = CA = r_1$

$\therefore$ $$\cos NBA = \frac{r_2 - r_1}{D}$$

$\therefore$ $$\cos \theta/2 = \frac{r_2 - r_1}{D} = \frac{d_2 - d_1}{2D}$$

where D = central distance between two pulleys

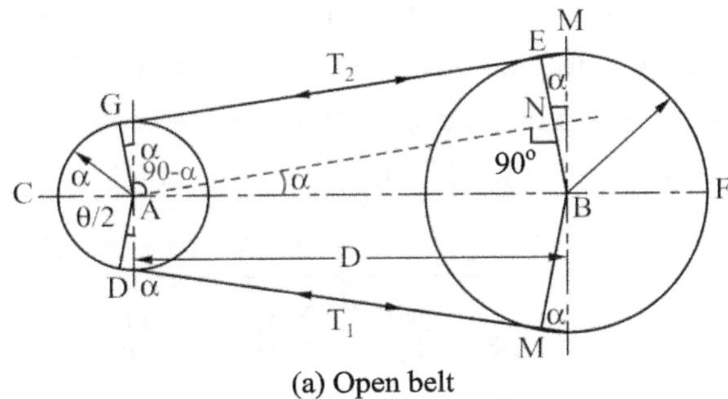

(a) Open belt

(ii) For cross belt: The angle of lap θ is same for both pulleys in this case.

$$\frac{\theta}{2} = \pi - < DAX = \pi - < ABN$$

$\therefore$ $$< ABN = \left(\pi - \frac{\theta}{2}\right)$$

$\therefore$ $$\cos ABN = \cos\left(\pi - \frac{\theta}{2}\right)$$

$\therefore$ $$\cos\left(\pi - \frac{\theta}{2}\right) = \cos ABN = \frac{BN}{AB} = \frac{BE + EN}{AB} = \frac{r_1 + r_2}{D} = \frac{d_1 + d_2}{2D}$$

$\therefore$ $$\cos\left(\pi - \frac{\theta}{2}\right) = \frac{d_1 + d_2}{D}$$

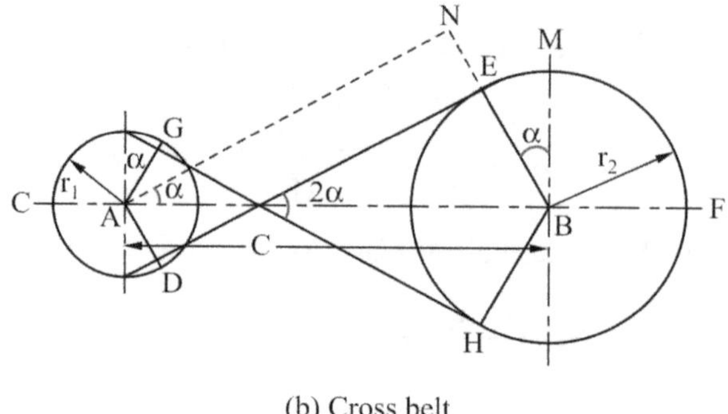

(b) Cross belt

14.14 Centrifugal Tension in Belt

When a belt moves on a straight path with uniform velocity there is no inertia force but when a belt reaches the pulley surface, it rotates with uniform speed along a circular path and belt is subjected to centrifugal force due to mass of belt. This centrifugal force tends to lift the belt from pulley and causes additional tension in belt. Centrifugal force produces equal tension on both side of belt. See the Fig. 14.12.

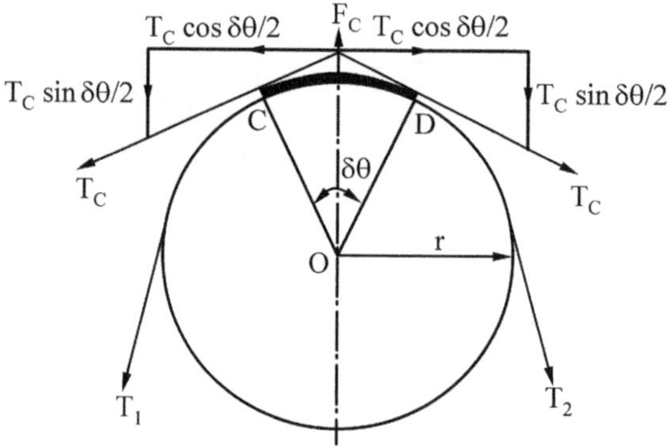

Fig. 14.12 Centrifugal stress in belt.

Let

T_c = centrifugal tension in belt in new ton

m = mass per unit length of belt

V = belt velocity in m/sec

$\therefore \qquad T_c = mV^2$

Power Transmitted by a Belt

A shown in Fig. 14.13, 'A' is driving pulley (driver) and 'B' is follower. Driving pulley pulls the belt from one side and delivers the same to the other. It is thus obvious that tension on former side (i.e. tight side) will be greater than latter side (slack side).

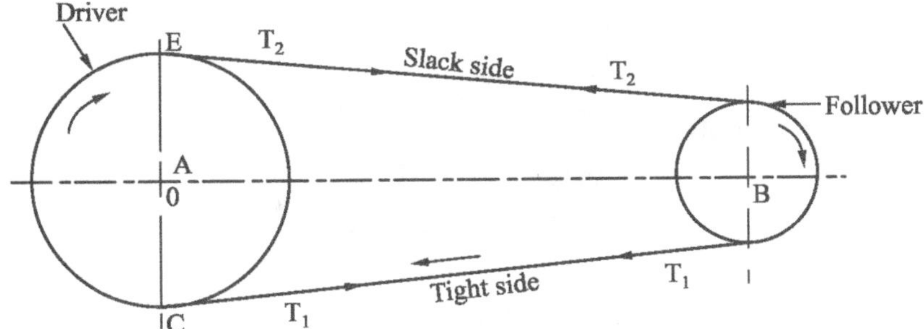

Fig. 14.13

T_1 = the tension on tight of pulley A acting at C

T_2 = the tension on the slack side of pulley 'A' acting at E.

r_1 = radius of the driver

r_2 = radius of the follower

V = velocity of belt in m/sec

The effective turning (driving) force at circumference of the follower is difference between two tensions i.e. $(T_1 - T_2)$

work done per second = force × velocity

$$= (T_1 - T_2)\,V \quad \text{watt}$$

14.15 Chain Drive

Introduction

1. In this drive, rigid links are hinged together to form a chain. The chain links provide flexibility for wrapping the chain over special toothed wheel known sprockets. They are very similar to spur gear. The sprockets have projected teeth, which fit into recesses in the links of chain Fig. 14.14.
2. Sufficient lubrication is required for this drive to avoid wear and tear. Lubricating oil also reduces noise.
3. There are two types of chains

   (i) roller chain for slow speed

   (ii) silent chain for high speed

4. Chain drives now-a-days are used to replace belt or rope drives as chain drive is positive drive.

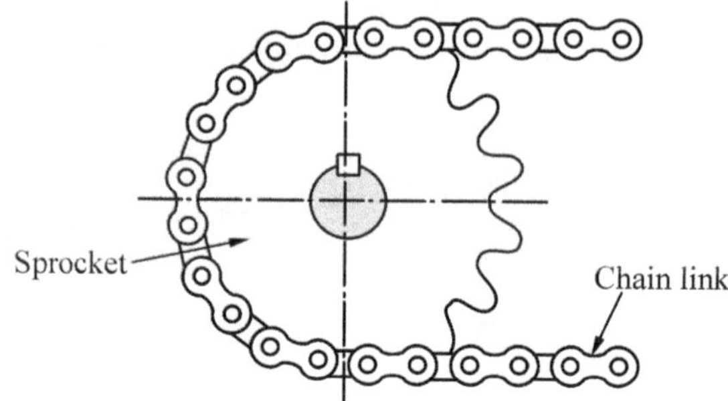

Fig. 14.14

Applications of chain drive

1. Used for short distance for transmitting power up to 4000 KW with large transmission numbers (12 to 16)
2. They are used in sugar machinery, agricultural and textile machineries where centre distance is about 7 to 8 m.
3. Chains are used in cycle, motor cycle and motor vehicles.
4. They are used in road rollers, work shop and factories

Advantages

1. Gives exact velocity ratio
2. No slip as it is positive drive
3. Highly efficient and effective drive
4. No initial tension in chain
5. Durable and strong drive
6. It is effective for large power transmission and for high speeds.

Disadvantages

1. It produces more noise
2. Lubrication is required
3. Its initial and maintenance cost is high
4. Power transmission for non parallel shafts is difficult.

14.16 Gear Drive

Introduction

Gears are toothed wheels. Teeth of driver gear engage in the teeth of follower gear for power and motion transmission. For obtaining desired speed and direction of rotation, gear trains of many gear wheels are used. Different types of gears used for connecting parallel, non parallel, intersecting and non intersecting shafts. The centre distance between the shafts should be very small.

Type of Gears

Following are the different types of gears generally used:

1. Spur gear
2. Spiral gear
3. Helical gear
4. Bevel gear
5. Worm and worm wheel
6. Rack and pinion

1. *Spur gear*: Gear whose axes are parallel and whose teeth are parallel to the centre line of gear are called spur gears (Fig. 14.15). They are used for parallel shafts to transmit motion and power at closed distance. These gears are used in variety of machines such as automobile vehicles, machine tools, watches and clock, cranes, precision measuring instruments, aero engines, marines, crushers, rolling mills etc.

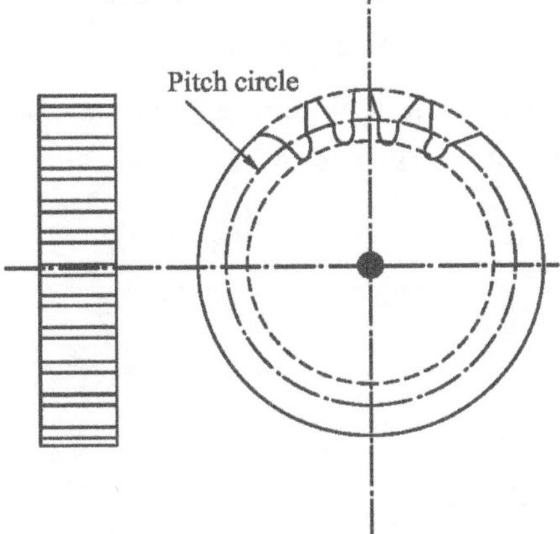

Fig. 14.15 Spur gear.

4. *Spiral gears*: The skew or spiral gear is used to connect non parallel and non intersecting shafts. Their teeth are like helical gears, cut on the wheel periphery (Fig. 14.16). The central distance between meting gear is the shortest distance

between the shafts. As the pitch surfaces are cylindrical and teeth has point contact. These gears are suitable for transmission of small power. They are not used to transmit heavy loads.

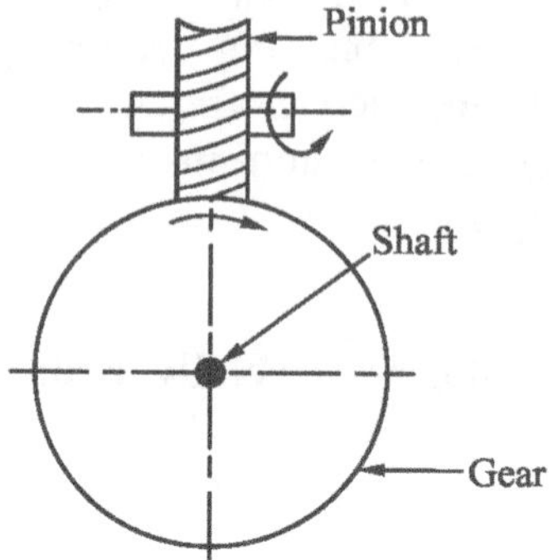

Fig. 14.16 Spiral gear.

3. *Helical Gears:*

   1. Like spur gears, helical gears connect shafts whose axes are parallel as well as non parallel and non-intersecting. But tooth spirals follow a helical path that is the pitch surfaces are cylindrical like screw threads, the teeth are helical Fig. 14.17 (a).
   2. There is line contact between the teeth. So friction loss and noise are low and transmission efficiency is high. So they are used for high power and high velocity ratio as compare to spur gears.
   3. The draw back of single helical gear is the axial thrust this is avoided by using double helical as shown in Fig. 14.17 (b).

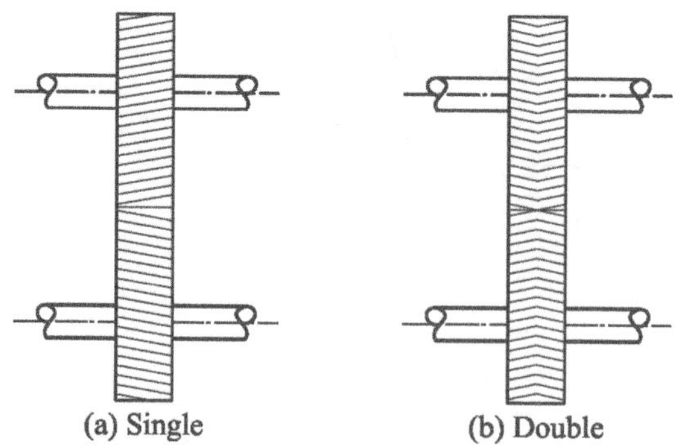

Fig. 14.17 Helical gear.

4. Helical gears produce an end thrust which is function of the helix angle. The larger the helix angle, the larger the axial thrust. Mounting assemblies must be built to absorb this thrust load.

4. *Bevel gear:*

   1. Bevel gears used for non parallel and intersecting shaft. Gears are cut on the conical surface. The angle between two shaft axes may be 90° or less than 90° (Fig. 14.18). The bevel gears having angle between shafts as 90° are known as mitre gears. If angle between shaft is less than 90° are called angular bevel gears.
   2. Bevel gears may be (i) straight tooth bevel gear (ii) spiral bevel gear (iii) skew bevel gear and (iv) hypoid gears are generally used in automobile differential gear box.

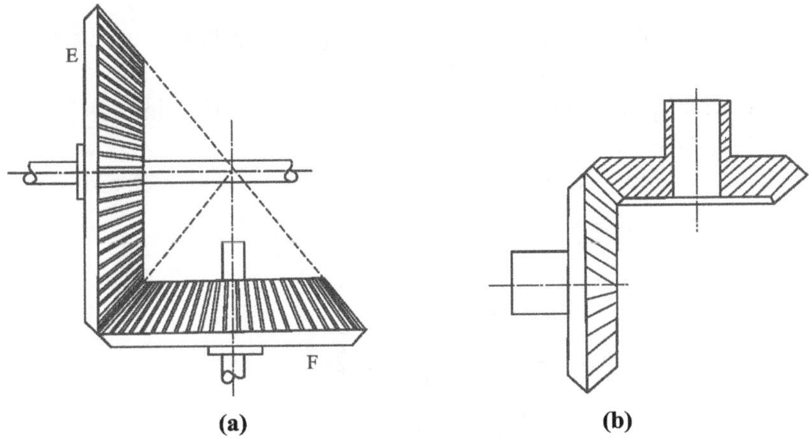

Fig. 14.18 Bevel grar.

6. *Worm and worm wheel*:

   1. Worm gear are used to connect shafts which lie in parallel planes but do not intersect at right angle.
   2. The small gear is known as worm gear and the larger as worm wheel (Fig. 14.19 (a) (b)).
   3. Because of screw action, worm gear drives are quite and vibration free.
   4. It is used for very high velocity ratio which depends on number of teeth on worm wheel and numbers of teeth on worm. It used in lathes, milling machine. They are also used for producing reduction gear box having large speed ratio.

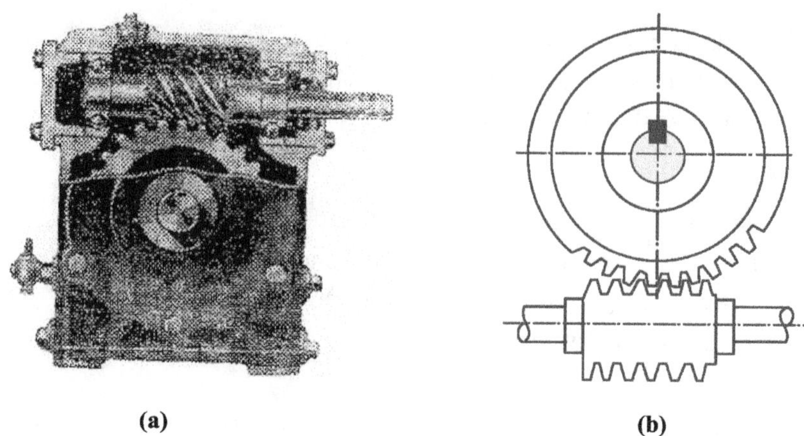

Fig. 14.19 Worm and worm wheel.

7. *Rack and Pinion*:
   1. The wheel is known as a pinion and straight strip have threads on one side and plane surface on other side is known as rack (Fig. 14.20).
   2. The rack and pinion assembly converts rotary motion into straight line motion and vice versa.
   3. It is used in lathes, drills and planner. It is also used in measuring instruments.

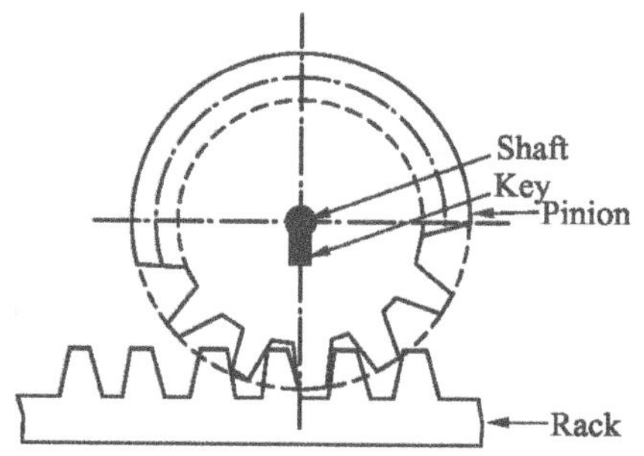

Fig. 14.20 Rack and pinion.

Advantages and disadvantages of gear drive

Advantages

1. It has on slip so its velocity ratio is constant i.e. it is positive drive.
2. Suitable precision and accurate speed control.
3. It gives long service without any attention.
4. It is strong and durable.
5. It gives smooth and noise less operation.

6. Power can be transmitted between the shafts which are parallel or non parallel, intersecting or non intersecting.
7. Useful for power transmission at short distance.
8. It is used in watches, clocks, measuring instruments, energy meter.
9. More compact device which can be enclosed in box for greater safety.

Disadvantages

1. It is very costly.
2. It should be well lubricated.
3. It can not be used for long distance between shafts.
4. Manufacturing is time consuming compare to chain drive or belt drive.

14.17 Comparison of Belt Drive, Chain Drive and Gear Drive

S.no.		Belt drive	Chain drive	Gear drive
1.	Slip	It occure during running	Slip does not take place during running. It is known as positive drive	No slip, it is known as positive drive
2.	Distance	Use for large distance between shafts	Suitable for moderate distance between shafts	Suitable for short distance between shafts
3.	Life	Less life	Moderate life	Long life
4.	Cost of installation	Low	Moderate (more than belt)	High installation cost compare to belt and chain
5.	Cost of manufacturing	Less	Moderate	High compare to chain and belt
6.	Time of manufacturing	Less	Moderate	High compare to chain and belt
7.	Velocity ratio	Generally gives low velocity ratio	Used for moderate ratio	Used high velocity ratio
8.	Lubrication	Lubrication does not required	Required sufficient lubrication	Requires proper lubrication
9.	Noise	Noisy operation	Noisy operation compare to belt	Give smooth and salient operation

Contd...

S.no.		Belt drive	Chain drive	Gear drive
10	Tension	Initial tension in belt is present	No initial tension	No initial tension
11	Space	Occupies large space	Occupies more space but less than belt drive	Occupies less space
12	Construction	Drive consists of 1. driving pulley 2. driven pulley 3. pulley shafts	Drive consists of 1. sprocket 2. chain 3. chain wheel 4. shafts	Consists of 1. driving gear 2. driven gear 3. shafts 4. box to project gears from dirt and dust
13.	Used	In lathers, drilling machines etc	Used in bicycle and automobiles etc	Used in gear box in automobile and machine tools

14.18 Gear Train

1. Gear train is defined as the combination of gear wheels used to transmit of power from one shaft to the other
2. There are two types of gear trains:
   (i) compound gear train as shown in Fig. 14.21 (a) and (c) for high velocity ratio.
   (ii) simple gear train as shown in Fig. 14.21 (b) for low velocity ratio

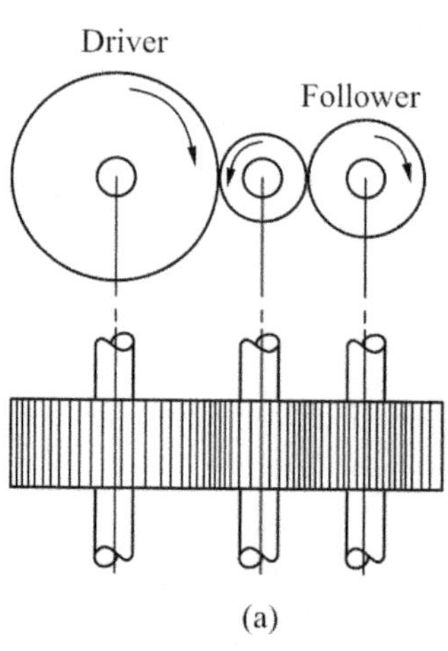

(a)

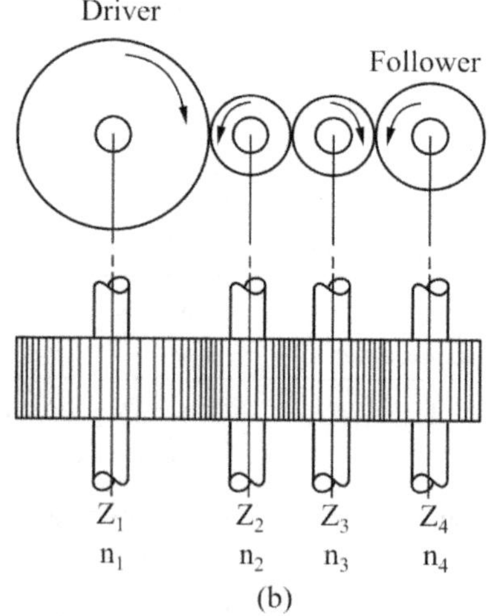

(b)

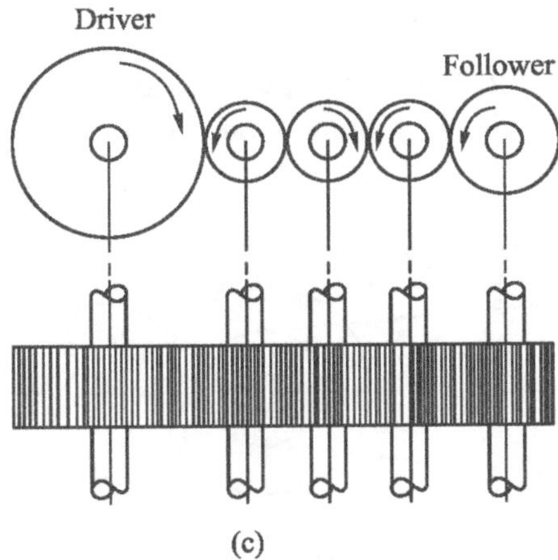

Fig. 14.21 (a) (b) & (c)

3. velocity ratio of gear train (V.R.): It is defined as ratio of speed of last gear wheel to speed of first wheel in RPM. Velocity ratio (V.R.) is positive when first and last wheel rotates in same direction and it is negative if they rotate in opposite direction.

Single gear train

In this type each shaft carries only one wheel. First and last wheels are driver and follower respectively. But intermediate wheels are drives as well as followers,

If letter "n" indicates speed in RPM and "z" indicates numbers of teeth's, then

$$\frac{n_1}{n_2} = \frac{z_2}{z_1}, \quad \frac{n_2}{n_3} = \frac{z_3}{z_2} \text{ and } \frac{n_3}{n_4} = \frac{z_4}{z_3}$$

$$\therefore \quad \frac{n_1}{n_2} = \frac{n_1}{n_2} \times \frac{n_2}{n_3} \times \frac{n_3}{n_4}$$

$$\frac{n_1}{n_2} = \frac{z_2}{z_1} \times \frac{z_3}{z_2} \times \frac{z_4}{z_3}$$

$$= \frac{z_4}{z_1}$$

$$\text{V. R.} = \frac{n_4}{n_1} = \frac{z_1}{z_4}$$

Thus velocity ratio is independent on the number of teethes of intermediate wheels. V.R is positive for odd number of shafts, otherwise it is negative.

Compound Gear Train

In this case except first and last shafts, all intermediate have two shafts, one acts as driver and other as follows. Two wheels mounted on the intermediate shafts have the same speed in RPM. From Fig. 14.22 (a), 14.22 (b),

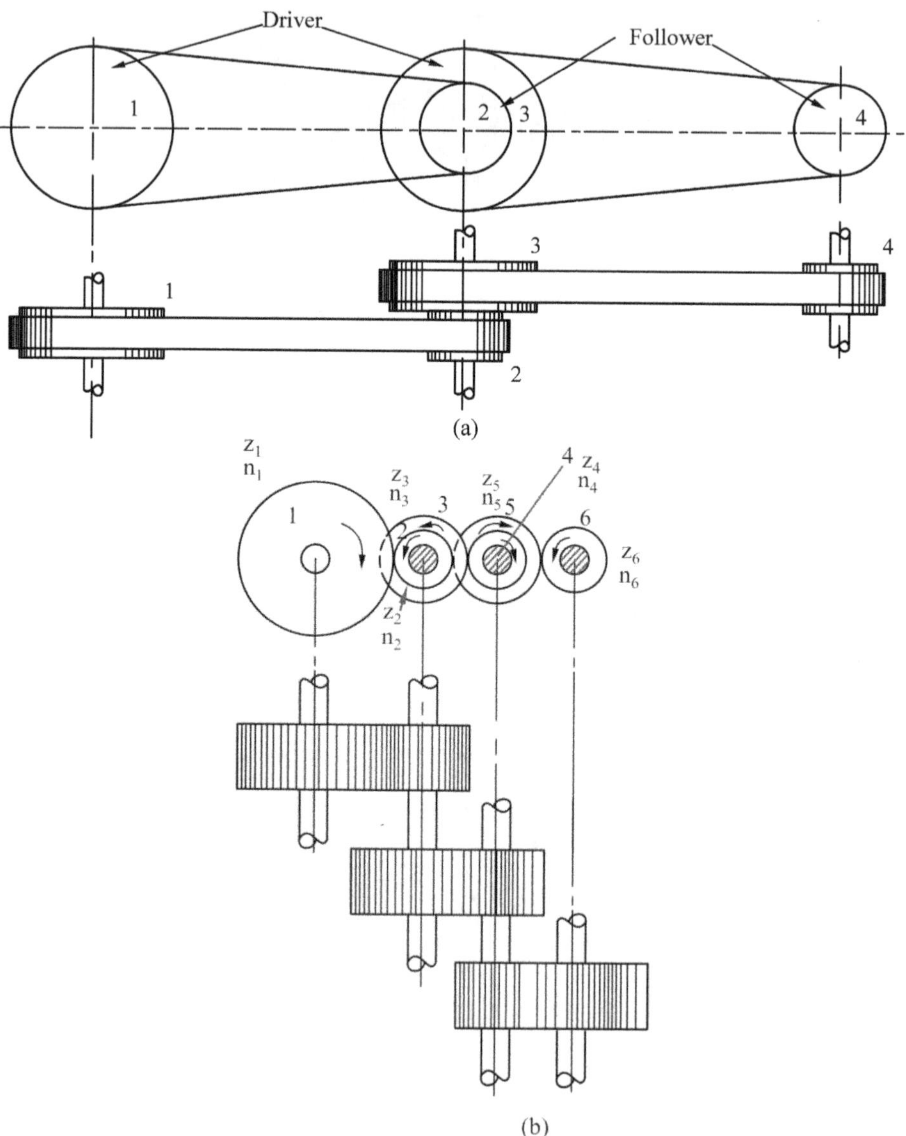

Fig. 14.22

$$\frac{N_2}{N_1} = \frac{Z_1}{Z_2} \qquad \text{.....(i)}$$

$$\frac{N_4}{N_3} = \frac{Z_3}{Z_4} \qquad \text{.....(ii)}$$

$$\frac{N_6}{N_5} = \frac{Z_5}{Z_6} \qquad \text{.....(iii)}$$

Multiply (i), (ii) and (iii)

$$\frac{N_2}{N_1} \times \frac{N_4}{N_3} \times \frac{N_6}{N_5} = \frac{Z_1}{Z_2} \times \frac{Z_3}{Z_4} \times \frac{Z_5}{Z_6}$$

But $N_2 = N_3$

$N_4 = N_5 \quad \therefore$

$$\frac{N_3}{N_1} \times \frac{N_5}{N_3} \times \frac{N_6}{N_5} = \frac{Z_1 \times Z_3 \times Z_5}{Z_2 \times Z_4 \times Z_6}$$

$$\therefore \qquad \frac{N_6}{N_1} = \frac{Z_1 \times Z_3 \times Z_5}{Z_2 \times Z_4 \times Z_6}$$

$$\frac{\text{speed of last follower}}{\text{speed of first driver}} = \frac{\text{product of teeths on the drivers}}{\text{Product of teeths on followers}}$$

14.19 Pulleys

Introduction: Pulleys are used to transmit power from one shaft to another by means of belt or rope running over them. Pulley is fixed on shaft by means of key (Fig. 14.23) as a result, when shaft rotates it also rotate along with shaft. The pulley are made up of cost iron, wrought iron, pressed steel, wood, plastic, aluminum alloys etc.

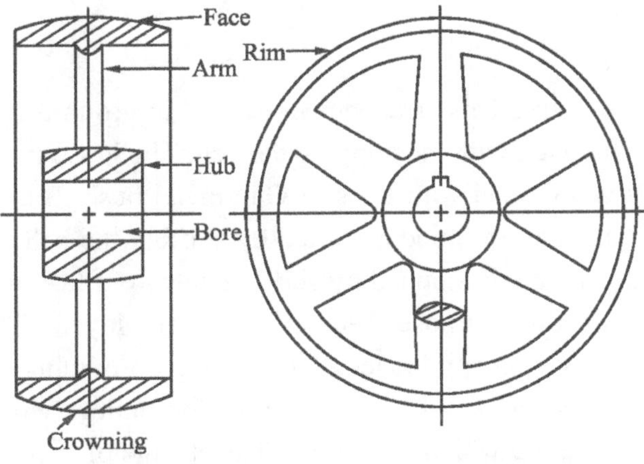

Fig. 14.23 Cast iron pulley. (construction)

Construction: Pulleys may be in one piece or halves to facilitate assembly on the shaft. The principal parts of pulley are (i) rim (ii) hub or boss (iii) arm or rib. See the Fig. 14.23.

(i) *Rim*: It is the outer portion of the pulley on which belt is mounted. In flat pulley rim are often given chamber or convex shape on its middle surface. The chamber is called crowning or draw. Due to the chamber the belt will not slip off the pulley when it rotates and gets a sound grip over the periphery of the pulley. In case of "V" belt or rope corresponding "V" or circular groove are cut on the rim face.

(ii) *Hub or Boss*: The central portion of pulley is called hub or boss. It has bore in its centre throughout its length. Bore has diameter equal to diameter of shaft. Shaft is mounted on the hub.

(iii) *Arm*: It connects the hub with rim. Arm of pulley may be straight or curved. The cross section of arm may be circular or elliptical. They add strength and stiffness to large pulleys.

Rib: It serves the same purpose as arm. The difference is in shape only. Rib is circular disc. Smaller size pulleys are provided with rib instead of arm.

Type of pulleys or classification of pulley

1. Fast and loose pulleys
2. Speed cones or stepped pulleys
3. Spit pulleys or built up pulley
4. Guide pulleys
5. Jockey pulleys
6. Grooved pulleys

1. *Fast and loose pulleys*: This arrangement is used in group drive where machine is to be started or stopped without stopping the belt. The loose pulley rotates freely on shaft as pulley is provided with Boss or Gun metal bush. But fast pulley is firmly keyed to the shaft. The machine is started by shifting the belt from loose pulley to fast pulley and machine is stopped by shifting belt from fast pulley to loose pulley (Fig. 14.24). For this purpose belt-shifter is employed. The loose pulley is prevented from moving along the length of the shaft providing collar on the shaft.

2. *Speed cones or stepped pulley*: They are cost-iron pulleys with two or more steps of different diameter on which belt is run. The change of speed of the machine can be obtained by simply shifting the belt from one step to the other step.

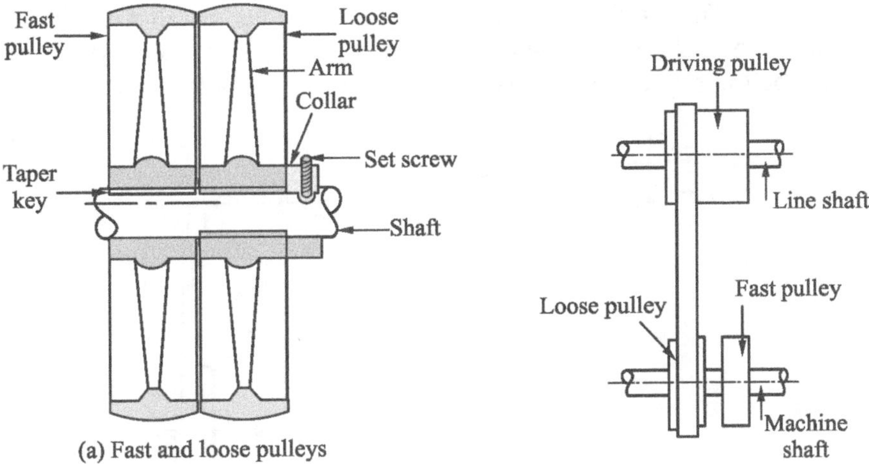

(a) Fast and loose pulleys

Fig. 14.24

The same belt is used for all steps. The driven pulley is placed just opposite to the driving pulley as shown in Fig. 14.25 (a), 14.25 (b).

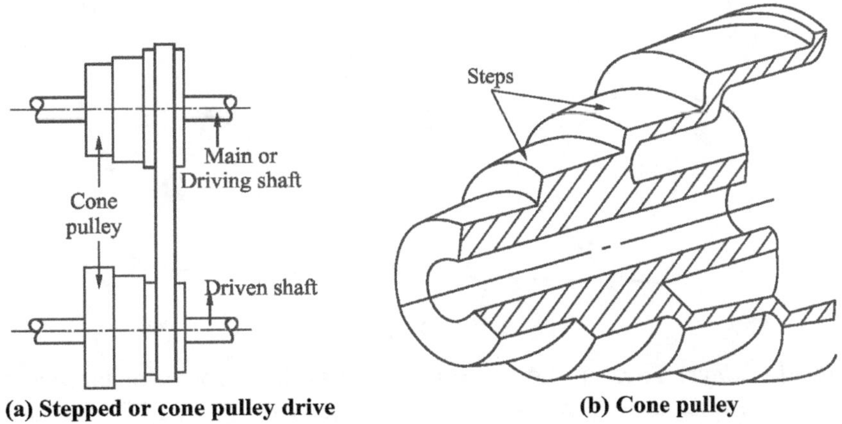

(a) Stepped or cone pulley drive **(b) Cone pulley**

Fig. 14.25

3. *Spit or built up pulley*: Large pulley is made in two halves and joint by bolt or rivet with a plate fixed inside of the rim. Arms of these pulleys are round and straight. One end of arm is screwed into hub and the other shrunk into the rim. They are generally made of wrought iron and mild steel. Split pulleys are lighter, stronger and more durable.

4. *Guide pulley*: These pulleys are used for connecting non parallel and intersecting or non intersecting shaft as shown in Fig. 14.26. The guide pulley guides the belt to keep the belt in proper plane. It is also used when two shafts are very close.

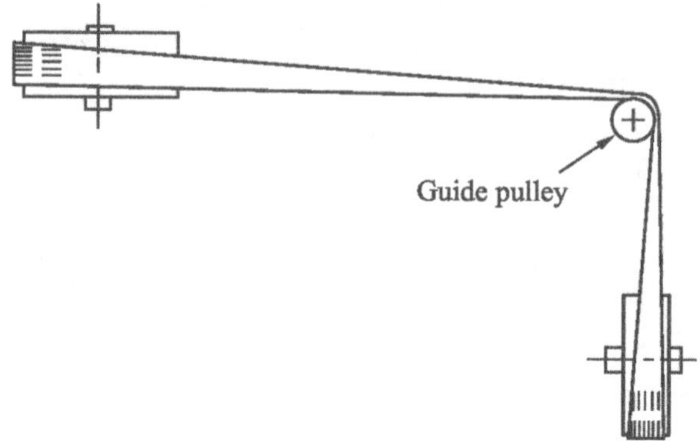

Fig. 14.26 Guide pulley.

5. *Jockey pulleys*: These pulleys are used to increase arc of contact and tension in belt as shown in Fig. 14.27. It is located near small pulley and on slack side of belt.

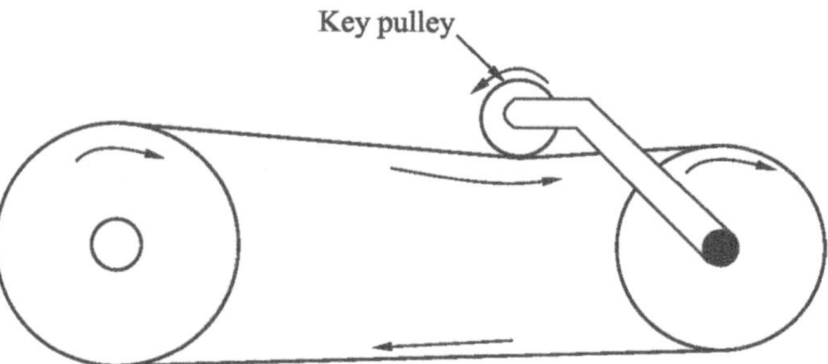

Fig. 14.27

6. *Grooved pulleys*: These pulleys are provided with v-groove or circular groove on the circumference to receive v-belt or rope as shown in Fig. 14.28 (a) and 14.28 (b). The angle between two faces of v-groove is 40° to 60°.

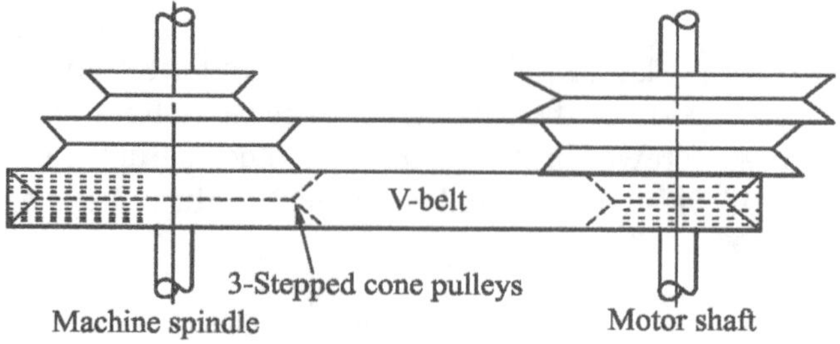

(a) V-belt cone pulley drive

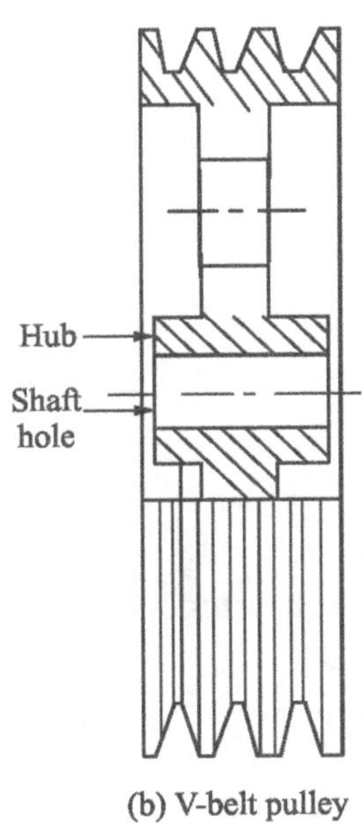

(b) V-belt pulley

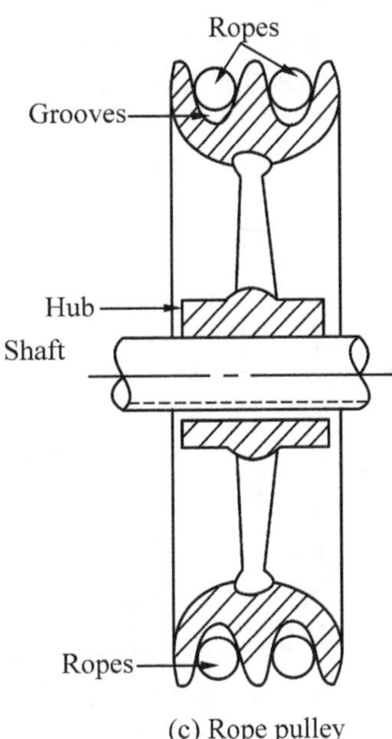

(c) Rope pulley

Fig. 14.28 Grooved pulleys.

These pulleys are used for short distance with v-belt and large distance with rope. The effect of grooves in pulley increases the friction between belt (or rope) and pulley as a result grip between belt and pulley is increased which reduces tendance to slip. In a rope pulley, the rope rests on two sides and not on bottom of groove.

Solved Problems

1. Power is transmitted from one shaft to another by means of cross belt. The diameter of larger pulley is 600 mm and that of smaller pulley is 300 mm. The distance between the centers of the two pulleys is 3 m. Find length of belt. The two shafts are parallel to each other and are in same plane.

Solution:

$D_1 = 600 \text{ mm} \quad \therefore \quad r_1 = 300 \text{ mm} = 0.3 \text{ m}$

$D_2 = 300 \text{ mm} \quad \therefore \quad r_2 = 150 \text{ mm} = 0.15 \text{ m}$

$x = 3\text{m}$

$$\text{length of cross belt} = \pi (r_1 + r_2) + \frac{r_1 + r_2}{x} + 2x$$

$$= \pi (0.3 + 0.15) + \frac{(0.3 + 0.15)}{3} + 2 \times 3$$

$$= 0.45\,\pi + 0.15 + 6$$

$$= 7.563 \text{ m}$$

2. Determine the rotational speed and direction of rotation of the follower in open and crossed belt drive as shown in figure.

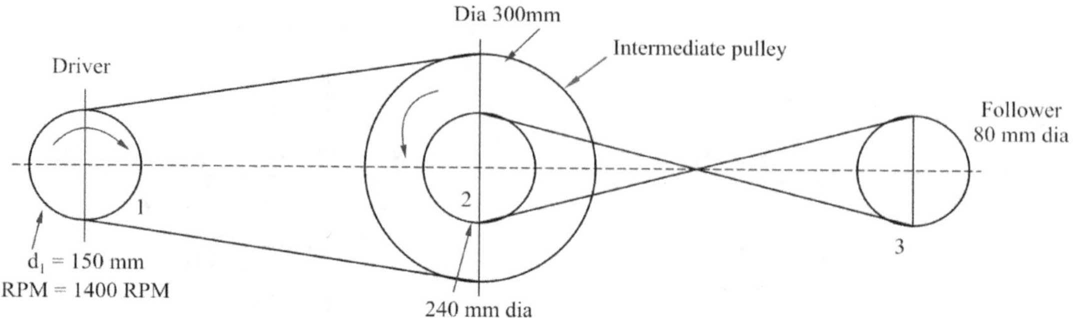

Solution: 300 mm and 240 mm diameter pulleys rotate together on intermediate shaft at same direction

Assume that drive rotate in clockwise direction the follower will rotate in anticlockwise direction

$$\frac{N_1}{N_2} = \frac{d_2}{d_1}$$

$$\frac{1400}{N_2} = \frac{300}{150}$$

N_2 = speed intermediate shaft

$$\frac{1400 \times 150}{300} = 700 \text{ RPM}$$

Small pulley on intermediate shaft will have same speed say N_2

$$\therefore \quad \frac{N_2}{N_3} = \frac{d_3}{d_2}$$

$$\therefore \quad \frac{700}{N_3} = \frac{80}{240}$$

$$N_3 = \frac{700 \times 240}{80} = 2100 \text{ RPM} \quad \text{(Anticlockwise direction)}$$

$$\text{Speed of follower} = \frac{\pi d_3 N_3}{60}$$

$$= \frac{\pi \times 80 \times 2100}{60}$$

$$= 8796 \text{ mm/sec}$$

$$= 8.8 \text{ m/sec}$$

3. A prime mover running at 300 RPM drives a D.C generator at 500 RPM by belt drive. Diameter of the pulley on out put shaft of prime mover is 60 cm. Assuming slip of 3% determine the diameter of generator pulley, if the belt running open it is 6 mm thick. Also find diameter of generator pulley when slip is neglected.

Solution: Driver speed $n_1 = 300$ RPM

$d_1 = 60$ cm

Driven speed $n_2 = 500$ RPM

Diameter of driven pulley = d_2 (is to be calculated)

Slip = 3%, t = 6 mm = 0.6 cm

$$\frac{N_2}{N_1} = \frac{d_1 + t}{d_2 + t} \times \frac{100 - S}{100}$$

$$\frac{500}{300} = \frac{60 + 0.6}{d_2 + 0.6} \times \frac{100 - 3}{100}$$

$\therefore$ $d_2 = 34.67$ cm diameter of generator pulley

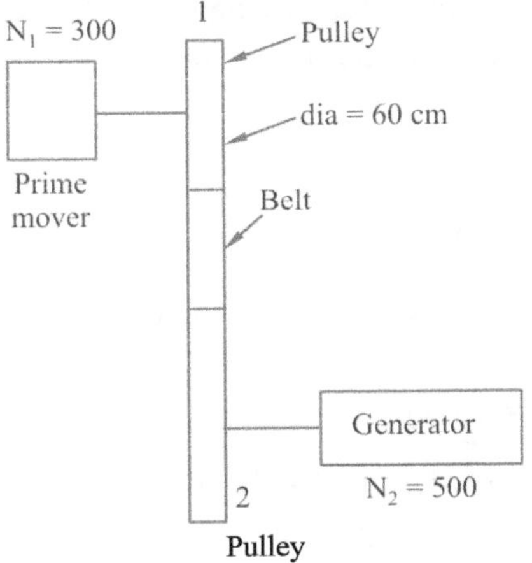

When slip is neglected

$$\frac{n_2}{n_1} = \frac{d_1 + t}{d_2 + t}$$

$$\frac{500}{300} = \frac{60 + 0.6}{d_2 + 0.6}$$

$$d_2 + 0.6 = \frac{60.6 + 300}{500}$$

$d_2 = 35.76$ cm

4. Two parallel shafts are connected by means of open belt. The shafts are in same plane. The distance between two centers is 4 m. The diameter of larger pulley is 600 mm and that of smaller pulley is 300 mm. Find length of belt.

Solution: For open belt $D_1 = 600$ mm $\therefore$ $r_1 = 300$ mm

$D_2 = 300$ mm $\therefore$ $r_2 = 150$ mm

$x = 4$ m $= 4000$ mm.

$\therefore$ length of open belt

$$= \pi (r_1 - r_2) + 2x + \frac{(r_1 - r_2)2}{x}$$

$$= \pi (300 - 150) + 2 \times 4000 + \frac{(300-150)^2}{4000}$$

$$= 471.238 + 8000 + 5.625$$

$$= 8476.625 \text{ mm}$$

$$= 8.47663 \text{ m}$$

5. The tension in two sides of belt are 600 N and 300 N respectively. If speed of driver is 250 RPM and its diameter is 1.25 m, find power transmitted by the belt.

Solution:

$T_1 = 600$ N

$T_2 = 300$ N

$$N = 250 \text{ RPM} = \frac{250}{60} \text{ RPS}$$

$D = 1.25$ m

$F = (T_1 - T_2) = (600 - 300) = 300$ N

$$P = \frac{FV}{1000 \times 60} \qquad \text{where } V = \frac{\pi DN}{60}$$

$$P = \frac{F \pi D N}{1000 \times 60} \qquad N = \text{in RPM}, \quad D = \text{in meter}, \quad F = \text{in N.}$$

$$P = \frac{300\,\pi \times 1.25 \times 250}{1000 \times 60} \qquad \therefore \text{ P we get in kw}$$

$$= 4.908 \text{ kW}$$

6. The driving pulley of motor has diameter 150 mm and 1200 RPM. The maximum tension is limited to 200 N per cm width of belt. To transmit 16 kW when ratio of tension is 2.5, determine the necessary width of belt.

Solution:

$$D = 150 \text{ mm}$$

$$N = 1200 \text{ RPM}$$

$$T_1 = 200 \text{ N/unit width of belt}$$

$$P = 16 \text{ kW}$$

$$\frac{T_1}{T_2} = 2.5$$

Width of belt is to be calculated

To calculate width of belt

(i) from equation of power transmitted force i.e. $(T_1 - T_2)$ can be found.

(ii) $\frac{T_1}{T_2}$ and $(T_1 - T_2)$ is known we can find T_1 and T_2.

(iii) From T_1 (max tension/cm), the width can be calculated

$$D = 150 \text{ mm} = 0.15 \text{ m}$$

$$N = 1200 \text{ RPM}$$

$$V = \pi DN$$

$$= \pi \times 0.15 \times 1200$$

$$= 565.2 \text{ m/min}$$

$$\text{Power} = \frac{F \times V}{1000 \times 60}$$

$$16 = \frac{F \times 565.2}{1000 \times 60}$$

$\therefore$ $F = (T_1 - T_2) = 1698.51 \text{ N}$ .....(1)

Now $\frac{T_1}{T_2} = 2.5$ $\therefore$ $T_1 = 2.5\ T_2$ put this value in equation (1)

we get $2.5\ T_2 - T_2 = 1698.51$

$$1.5\ T_2 = 1698.51$$

$$T_2 = 1132.343 \text{ N}$$

$\therefore$ $T_1 = 2.5\ T_2 = 2.5 \times 1131.77 = 2830.86 \text{ N}$

But $T = \text{maximum tension/cm} \times \text{width of belt in cm}$

$$2830.86 = 200 \times \text{width of belt in cm}$$

$$\therefore \quad \text{width of belt} = \frac{2830.86}{200} = 14.15 \text{ cm}$$

7. Find power transmitted by a belt running over a pulley of 60 cm diameter at 200 RPM. The co-efficient of friction between belt and pulley is 0.25 and angle of lap 160° and maximum tension in belt is 2453 N.

Solution:

$$D = 60 \text{ cm} = 0.6 \text{ m}$$

$$N = 200 \text{ RPM}$$

$$\mu = 0.25$$

$$\theta = 160^\circ = \frac{\pi}{180} \times 160 = 2.793 \text{ radian}$$

$$T_1 = 2453 \text{ N}$$

$$V = \pi DN = \pi \times 0.6 \times 200 = 120\,\pi \text{ m/min}$$

We know

$$\frac{T_1}{T_2} = e^{\mu\theta}$$

$$\frac{T_1}{T_2} = e0.2 \times 2.793 = e0.6982 = 2$$

$$\therefore \quad \frac{T_1}{T_2} = 2$$

$$\therefore \quad T_2 = \frac{T_1}{2} = \frac{2453}{2} = 1226.5 \text{ N}$$

But

$$T_1 - T_2 = 2453 - 1226.5 = 1226.5$$

$$P = \frac{F \times V}{1000 \times 60} = \frac{120\,\pi \times (T_1 - T_2)}{1000 \times 6} = 7.7 \text{ kW}$$

8. A leather belt transmits 30 kW from pulley of 750 mm diameter which runs at 500 RPM. The belt is in contact with pulley over an arc 160° and co-efficient of friction of 0.3. If permissible stress in belt is not to exceed $2 MN/m^2$, find minimum area of belt.

Solution:

$$D = 750 \text{ mm} = 0.75 \text{ m}$$

$$N = 500 \text{ RPM}$$

$$P = 30 \text{ kW}$$

$$\mu = 0.3$$

Stress $= \dfrac{2\ \text{MN}}{\text{m}^2} = \dfrac{2\times10^6}{10^4} = 200\dfrac{\text{N}}{\text{cm}^2}$

$$\theta = 160^\circ = \frac{\pi}{180}\times160 = 2.7925 \text{ Radians}$$

$$\frac{T_1}{T_2} = e^{\mu\theta}$$

$$= e^{0.3\times2.7925}$$

$$\frac{T_1}{T_2} = 2.311$$

$\therefore$ $T_1 = 2.311\ T_2$ .....(1)

$$P = \frac{F\times V}{1000\times60} = \frac{(T_1 - T_2)\,\pi DN}{1000\times60}$$

$$30 = \frac{(T_1 - T_2)\times\pi\times0.75\times500}{60\times1000}$$

$T_1 - T_2 = 1527.89$ N .....(2)

$2.311\ T_2 - T_2 = 1527.81$ $(\because T_1 = 2.311\ T_2)$

$\therefore$ $1.311\ T_2 = 1527.81$

$T_2 = 1165.38$ K

$T_1 = 2.311 \times 1165.38$ N

$T_1 = 2693.19$ N

Minimum cross section area of belt

$$= \frac{T_1}{\text{stress}}$$

$$= \frac{2693.19}{200} \qquad \left(\frac{\text{N}}{\text{N/cm}^2} = \text{cm}^2\right)$$

$$= 13.466\ \text{cm}^2$$

9. An open belt running over two pulleys 24 cm and 60 cm diameter connects two parallel shaft 3m apart and transmits 3.7 kW. Small pulley rotates at 300 RPM. The co-efficient of friction between belt and pulley is 0.3 and safe working tension is 98 N per cm width. Determine

(a) Minimum width of the belt

(b) Initial belt tension

(c) length of belt required

Solution: $d_2 = 24\text{ cm} = 0.24\text{ m} \quad \therefore\ r_1 = 0.12\text{ m}$

$d_1 = 60 = 0.60\text{ m} \quad \therefore\ r_2 = 0.3\text{ m}$

$x = 3\text{ m},\ N_2 = 300\text{ RPM},\quad \mu = 0.3,\quad P = 3.7\text{ kW}$

To find tension T_1 and T_2 apply equation of power that is

$$P = \frac{(T_1 - T_2)\,v}{1000 \times 60} \text{ and equation } \frac{T_1}{T_2} = e^{\mu\theta}.$$

Let T_1 = tension on the tight side

T_2 = tension on the slack side

$$P = \frac{(T_1 - T_2)\,v}{1000 \times 60} \qquad \text{.....(1)}$$

But $v = \pi\, d_2 N_2 = \pi \times 0.24 \times 300 = 226.1$ m/min.

Putting value of v in eq. (1) we get,

$$3.7 = \frac{(T_1 - T_2)\,226.1}{1000 \times 60}$$

$$\therefore \quad (T_1 - T_2) = \frac{3.7 \times 1000 \times 60}{226.1} = 981.95\text{ N} \qquad \text{.....(2)}$$

$$\frac{T_1}{T_2} = e^{\mu\theta}$$

is this equation θ is unknown that can be found out by applying

$$\sin\alpha = \frac{r_1 - r_2}{x} = \frac{30 - 12}{300} = 0.06 \quad (\because r_1, r_2 \text{ and x is taken in cm})$$

$$\therefore \quad \alpha = 3.45^\circ$$

$\therefore$ angle of lap on smaller pulley $= \theta = 180 - 2\alpha = 180 - 2 \times 3.45 = 173.1^0$

We know that $\dfrac{T_1}{T_2} = e^{\mu\theta}$ $\qquad \therefore\ \theta = \dfrac{\pi}{180} \times 173.1 = 3.02$

$$= e^{0.3 \times 3.02}$$

$$= e^{0.906}$$

$$= 2.475$$

$$\frac{T_1}{T_2} = 2.475$$

$$T_2 = 2.475\ T_2 \qquad \text{.....(3)}$$

From equation (2) and (3)

$$T_1 - T_2 = 981.95$$

$$2.475\ T_2 = 981.95$$

$$T_2 = \frac{981.95}{1.475} = 665.729\ N$$

$$\therefore \quad T_1 = 2.475 \times 665.729 = 1647.679\ N$$

(a) Minimum width of belt

since the safe working tension is 98 N per cm width

Therefore minimum width of belt

$$b = \frac{T_1}{98} = \frac{1647.679}{98} = 16.8\ cm$$

(b) Initial belt tension $= T_0 = \frac{T_1 + T_2}{2} = \frac{665.729 + 1647.679}{2} = 1156.7\ N$

$$= 1156.7\ N$$

(c) Length of the belt required

$$L = \pi(r_1 + r_2) + 2x + (r_1 - r_2)^2$$

$$= \pi\ (0.3 + 0.12) + 2 \times 3 + (0.3 - 0.12)^2$$

$$= 7.33\ cm$$

10. Rope drive is required to transmit 230 kW from pulley of 1 m diameter running at 450 RPM. The safe pull in each rope is 800 N and mass of rope is 0.46 kg per metre length. The angle of lap and groove angle is 160^0 and 45^0 respectively. The co-efficient of friction between rope and pulley is 0.3. Find number of ropes required.

Solution: Given

$$P = 230\ kW$$

$$D = 1\ m$$

$$N = 450\ RPM$$

$$T = 800\ N$$

$$m = 0.46\ kg/m$$

$$\mu = 0.3$$

$$2\alpha = 45^0 \text{ or } \alpha = 22.5^0$$

$$\theta = 160^0 = \frac{\pi}{180} \times 160 = 2.8 \text{ radian}$$

$v = \pi DN = \pi \times 1 \times 450 = 1413$ m/min or 23.55 m/sec

T_1 = Tension on tight side

T_2 = Tension on slack side

T_c = Centrifugal tension

$T_c = mv^2 = 0.46 \times (23.55)^2 = 255.4$ N

$\therefore \quad T_1 = T - T_c = 800 - 255.4 = 544.6$ N ..... (i)

$$\frac{T_1}{T_2} = e^{\mu\theta} \cos ec$$

$$= e^{0.3 \times 2.8 \text{ cosec} 22.5}$$

$$\left[\text{cosec } x = \frac{1}{\sin \alpha} = \frac{1}{\sin 22.5} = 2.613\right]$$

$$= e^{3 \times 2.8 \times 2.613}$$

$$= e^{2.195}$$

$$= 8.98$$

$$\therefore \quad T_2 = \frac{T_1}{8.98} = \frac{544.6}{8.98} = 60.64 \text{ N}$$

Power transmitted by one Rope

$$= \frac{(T_1 - T_2)\,v}{1000 \times 60} = \frac{(544.6 - 60.64)\,1413}{1000 \times 60}$$

$$= 11.397 = \frac{230}{11.397} = 20.18$$

$\therefore$ No. of Ropes

$$= \frac{\text{Total power transmitted}}{\text{Power transmitted by one Rope}} = \frac{230}{11.397} = 20.18$$

Say 21 Ropes

11. An engine running at 150 RPM drives the line shaft by means of belt. The engine pulley is 75 cm diameter and pulley on line shaft is 45 cm diameter. A 90 cm diameter pulley on the line shaft drives 15 cm diameter pulley keyed to Dynamometer shaft. Assuming slip of 2% at each drive, find speed of Dynamo shaft.

Solution: Engine (Drive) Pulley diameter = d_1 = 75 cm

Line shaft (follower) pulley diameter = d_2 = 45 cm

Line shaft (Driver) pulley diameter = d_3 = 90 cm

Dynamo follower pulley diameter = d_4 = 15 cm

Slip of each drive is 2%

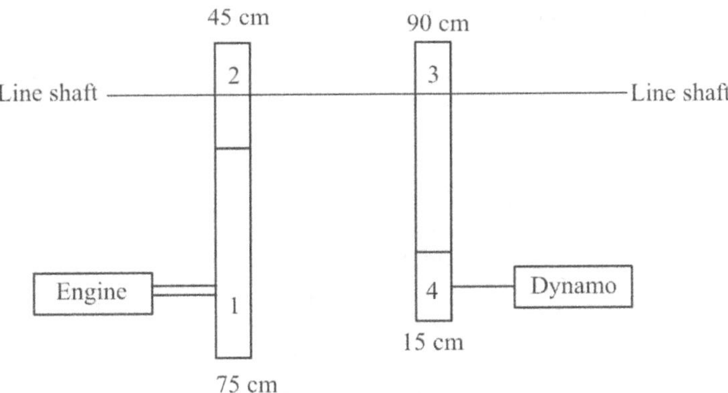

$$\frac{\text{Dynamo speed}}{\text{Engine speed}} = \frac{N_2}{N_1}$$

$$= \frac{d_2}{d_1} \times \frac{d_3}{d_4} \times \frac{(100 - S)}{100}$$

Total slip $= 2 + 2 = 4\%$

$$\therefore \quad \frac{\text{Dynamo shaft speed}}{\text{Engine speed}} = \frac{75 \times 90}{45 \times 15} \frac{(100 - 4)}{100} = 9.6$$

$\therefore$ Dynamo shaft speed

$= 9.6 \times$ Engine speed

$= 9.6 \times 150$

$= 1440$ RPM

12. A compound gear train consists of six gears. The number of teeth on the gear are as follows.

Motor shaft is connected to gear A and rotates at 975 RPM. The gear A drives gear B and gear C drives gear D and E drives F.

Gear	A	B	C	D	E	F
	20	50	25	75	26	65

***Solution*:** $N_A = 975$ RPM

$T_A = 20$

$T_B = 50$

$T_c = 25$

$T_D = 75$

$T_E = 26$

$T_F = 65$

Let N_F = Speed of shaft F

$$\frac{N_F}{N_A} = \frac{T_A \times T_C \times T_E}{T_B \;\; T_D \;\; T_F}$$

$$= \frac{20 \times 25 \times 26}{50 \times 75 \times 65} = \frac{4}{75}$$

$$\therefore \quad N_F = N_A \times \frac{4}{75}$$

$$= 975 \times \frac{4}{75}$$

$$= 52 \text{ RPM}$$

13. Two Parallel shafts are to connected by a gear drive. They are very nearly 1.5 m apart and their velocity ratio is 8: 2. If pitch of the gears is 90 mm, find the number of teeth in each of two wheels and distance between the shafts.

***Solution*:** Let d_1 and d_2 are the pitch diameter of the two gears

Centre distance between two gears $= \frac{d_1 + d_2}{2}$

$$1.5 = \frac{d_1 + d_2}{2} \qquad \text{..... (i)}$$

Velocity ratio $\frac{n_2}{n_1} = \frac{d_1}{d_2} = \frac{8}{2} = 4$

$\therefore$ $d_1 = 4\, d_2$, Put value in equation (i) we get

$$1.5 = \frac{4d_2 + d_2}{2}$$

$$\therefore \quad 5d_2 = 3$$

$$d_2 = \frac{3}{5} = 0.6 \text{ m} = 600 \text{ mm}$$

But $d_1 = 4d_2$

$\therefore$ $d_1 = 4 \times .6 = 2.4 \text{ m} = 2400 \text{ mm}$

Circular pitch $=P = \dfrac{\pi d_1}{z_1} = \dfrac{\pi d_2}{z_2} = 90$

$\therefore$ $z_1 = \dfrac{\pi d_1}{P} = \dfrac{\pi \times 2400}{90} = 83.77$ i.e., 84

z_1 = no. of teeth on gear '1'

$z_2 = \dfrac{\pi d_2}{P} = \dfrac{\pi \times 600}{90} = 20.94$ i.e., 21

z_2 = no. of teeth on gear '2'

As we have round up the teeth, the diameter d_1 and d_2 will change for same pitch. The actual diameters will be

$$d_1^1 = \frac{Pz_1}{\pi} = \frac{90 \times 84}{\pi} = 2406.42 \text{ mm}$$

$$d_2^1 = \frac{Pz_2}{\pi} = \frac{90 \times 21}{\pi} = 601.6 \text{ mm}$$

Actual diameter is found as above due to this centre distance given in problem will change therefore

Exact centre discharge $= \dfrac{d_1^1 + d_2^1}{2} = \dfrac{2406 + 601.6}{2} = 1504 \text{ mm}$

$= 1.504 \text{ m}$

14. Find out RPM of 100 teeth worm wheel when it is driven by worn shaft whose lead is equal to 3 pitch revolving at 450 RPM.

Solution: Velocity ratio of worm gearing $= \dfrac{\text{RPM of worm shaft}}{\text{RPM of Worm wheel}}$

$= \dfrac{\text{No. of teeth in worm wheel}}{\text{No.of teeth on worm shaft}}$

Lead of worm shaft being equal to 3 pitch. It is 3 start threads

Let n be the RPM of the worm wheel, therefore $\dfrac{450}{n} = \dfrac{100}{3}$

$\therefore$ $n = \dfrac{450 \times 3}{100} = 13.5 \text{ RPM}$

Questions

1. What is meant by individual and group drive.
2. Comparison between individual and group drive.
3. Advantage of individual and group drive.
4. Write short notes on
   (i) Individual drive
   (ii) Group drive
5. What are different type of belt drives? Describe with neat sketch any two types.
6. Obtain an expression for the length of belt in:
   (i) an open belt drive
   (ii) a cross belt drive
7. Explain clearly the terms "Slip" and "Creep" as applied to belt drive.
8. Compare Flat belt drive with "V" belt drive.
9. Derive the expression for power transmitted by the belt.
10. Write short notes on belt material.
11. Write short notes on various types of pulleys.
12. State advantages and disadvantages of:
    (i) Belt drive
    (ii) Gear drive
    (iii) Chain drive
13. Compare between Belt drive, chain drive and gear drive.
14. State the applications of:
    (i) Belt drive
    (ii) Gear drive
    (iii) Chain drive
15. Sketch the chain link and name the parts.
16. State the equation for:
    (i) Power transmitted by belt
    (ii) Ratio of two tensions
    (iii) Centrifugal tension in belt
    (iv) Angle of lap of belts
    (v) Initial tension
    (vi) Condition for maximum power transmission.

17. Discuss the effect of initial belt tension on maximum power transmitted by belt.
18. Derive expression for ratio of tensions for flat belt passing over a pulley, when it is just on the point of slipping.
19. What is effect of centrifugal tension on power transmitted by Flat belt? Derive expression for centrifugal tension.
20. Write short note on gear transmission system.
21. What are various types of gears used for power transmission? State the applications of each type.
22. Recommend drives for following applications giving reason for the same.
    (i) Bicycle
    (ii) Scooter
    (iii) Motor cycle
    (iv) Motor car
    (v) Flour Mill
    (vi) Compressor
    (vii) Water Pump
23. Belt drive is not used in Automobiles – Give reasons.
24. Name various types of chains
25. What is meant by gear train? What is its necessity?

Exercises

1. The driving pulley of motor has a diameter of 150 mm and runs at 1200 RPM. The maximum tension in belt is limited to 200 N per width of belt. Determine necessary width of belt to transmit 15 KW, when ratio of belt tension is 2.5.

   (width = 14.15 cm)

2. Suggest suitable drives for following and justify the answer.
   (i) Flour mill
   (ii) Automobile power transmission from engine shaft to propeller shaft
   (iii) Transmission of power from main spindle of lathe to its lead screw

   [(i) Flat belt drive (ii) chain drive (iii) gear drive]

3. A cross belt connects two pulleys of 500 mm diameter 2 meter apart. The initial tension in the belt is 500 N. If co-efficient of friction between belt and pulley is 0.3, find the power transmitted at 700 RPM. Also calculate the length of belt.

[Power transmitted = 9.1318 kw

Length of belt = 5.6965 m]

4. Two pulleys 400 mm diameter, 2m apart are connected by open belt. If belt tension ratio of 2.5 is required, what is the minimum co-efficient of friction? What tension ratio will be available, if the belt is used to drive the driven pulley in opposite direction?

[$\mu = 0.29166$; Tension ratio = 2.8116]

5. Two pulleys one 450 mm diameter and other 250 mm diameter are on parallel shaft 1.95 m apart and are connected by cross belt. Find length of belt required. What power can be transmitted by belt when larger pulley rotates at 200 RPM, if maximum permissible tension in belt is 1 KN and co-efficient of friction between belt and pulley is 0.25.

[length of belt = 4.975 m

Power transmitted = 2.736 kW]

6. Parallel shafts are to be connected by gear drive. They are very nearly 1 meter apart and their velocity ratio is 9:2. If pitch of gear is 57 mm, find the number of teethes in each of two wheels and distance between the shaft.

[$z_1 = 90$, $z_2 = 20$, Distance = 0.998 m]

7. Find out the RPM of a 50 teeth worm wheel when it is being driven by worm shaft whose lead is equal to 3 pitch revolving at speed of 300 RPM.

[18 RPM]

8. In simple train of gears A, B, C, D the number of teeth are $z_A = 30$, $z_B = 60$, $z_C = 60$, and $z_D = 40$. If A makes 36 RPM, find the RPM of the gear C and D.

[$n_c = 18$ RPM, $n_D = 27$ RPM]

CHAPTER 15

Important Engineering Materials

15.1 Introduction

All engineers of different disciplines – civil, mechanical, electrical, chemical, electronics, instrumentation or computer engineering, have to contribute for successful design of any engineering system. They have to use proper material to construct or manufacture of system-components.

Proper material will be selected with functional service and economic consideration. For selection of proper materials for the products, knowledge about the materials and their properties may be very helpful.

15.2 Properties of Materials

Property indicates the nature of material which help to predict behaviour of material under different conditions.

The mechanical properties of a material include Strength, Elasticity, Plasticity, Malleability, Ductility, Impact strength, Hardness, Toughness, Brittleness, Resilence, Creep, Fatigue etc. These properties are being discussed below.

1. ***Strength***
   - It represents the ability of a material to resist the breaking under the action of external loads.
   - The strength of a material varies according to the type of loading. It is possible to assess tensile, compressive, shearing or torsional strengths.
2. ***Elasticity***
   - It is a property of the material by virtue of which a body will deform when loaded and regains its original shape and size, after removal of load. Material possessing this property is known as elastic material.

     e.g: Springs used for reducing vibrations or shocks.

- Example of elastic materials are mild steel, copper, rubber etc.
- This property is desirable for materials used in tools and machines.

3. ***Plasticity***

- It is the property of a material due to which it undergoes permanent deformation without failure or rupture on application of load.
- Materials which possess this property are called plastic materials.

  This property is used for various processes of forging, rolling, forming, in stamping images on coins etc.

4. ***Malleability***

Malleability is the ability of a material to withstand deformation under compression without rupture.

e.g. Forging and rolling operations are easily carried out on malleable material.

Lead and gold are examples of good malleable materials.

5. ***Ductility***

It is a property of the material which enables it to be drawn into thin wires on application of tensile load. Examples of ductile materials are mild steel, copper, aluminium, nickel, zinc, tin and lead.

Usually ductility is associated with tensile and malleability with compressive property of a material.

Ductility can be measured by % Elongation $= \frac{\text{change in length}}{\text{original length}} \times 100$

6. ***Impact Strength***

Impact strength is defined as the capacity of a material to resist or absorb shock-energy, before it gets fracture. Ductile materials possess higher impact strength than brittle materials.

7. ***Hardness***

The ability of a material to resist indentation or scratching, wear, abrasion, cutting, penetration etc, is called hardness.

It is also means the ability of a metal to cut another metal. Comparative hardness of various material is determined by various tests like Brinell, Rockwell, and Vicker's hardness tests.

8. ***Toughness***

It is the ability of the material to absorb energy during entire elastic and plastic deformation, during sudden or impact loading.

Higher toughness is desirable property in materials used for gears, chain, crane hook etc.

9. ***Brittleness***

It is defined as a tendency to fracture without appreciable deformation under the action of tensile force. Brittleness is therefore opposite of ductility or malleability.

A brittle material is hard and has less ductility.

Examples of brittle materials are cast iron, glass, concrete, bricks, stones etc.

10. ***Resilence***

- It represents the capacity of a material to absorb energy when tensile and compressive loads are applied up to the elastic limit. On removal of the load, the material gives off the stored energy.
- Resilence of a material resists the shock and impact loads.
- This property is essential for spring materials.

11. ***Creep***

- The slow and progressive deformation of a material with time under constant external force, at high temperature is called creep.
- This property is useful in designing the machine components subjected to high temperatures like boilers, furnaces, internal combustion engines etc.

12. ***Fatigue***

It represents the property of material to determine its behaviour when subjected to fluctuation of load in which the maximum stress developed in each cycle, remains within elastic limit. Under these conditions a component may fail after certain number of load application cycles. Such type of failure is called fatigue-failure.

Factors affecting fatigue strength of a material:

- Nature of loading (fluctuating or reversing)
- Magnitude of maximum load
- Number of cycles up to failure.

Engineering Materials

Introduction

Materials are probably more deep-seated in our culture than most of us realize. Transportation, housing, clothing communication, food production and virtually every segment of our daily lives is influenced by some degree or another by materials. Materials have contributed to the advancement of a number of technologies, including medicine and health, information and communication, national security and space, transportation, structural materials, textiles, personal hygiene, agriculture and food science and the environment. The excitement of material science and engineering is amplified by its intimate connections with other disciplines and its impact on daily life. These inter-disciplinary interactions between the material sciences and other fields in the development of new materials and their applications also require coordination and communication between researchers working in diverse areas.

Over the years there have been changes in man's choice of materials for various engineering activities. The ages and times/period of man's activities on earth are sometimes referred to requirements when such materials were in vogue like the stone age, the iron age and the current silicon age. However, the challenges of current applications are constantly motivating the discovery and development of new kinds of materials with the desired properties and the right cost to meet the challenges of the current day world.

Mechanical Properties of Engineering Materials

Mechanical properties relate deformation to an applied load or force. The important mechanical properties affecting the selection of a material are listed below:

- ***Tensile Strength*:** This enables the material to resist the application of a tensile force. To withstand the tensile force, the internal structure of the material provides the internal resistance.
- ***Hardness:*** It is the degree of resistance to indentation or scratching, abrasion and wear. Alloying techniques and heat treatment help to achieve the same.
- ***Ductility*:** This is the property of a metal by virtue of which it can be drawn into wires or elongated before rupture takes place. It depends upon the grain size of the metal crystals.
- ***Impact Strength*:** It is the energy required per unit cross-sectional area to fracture a specimen, i.e., it is a measure of the response of a material to shock loading.
- ***Wear Resistance*:** The ability of a material to resist friction wear under particular conditions, i.e. to maintain its physical dimensions when in sliding or rolling contact with a second member.
- ***Corrosion Resistance*:** Those metals and alloys which can withstand the corrosive action of a medium, i.e. corrosion processes proceed in them at a relatively low rate.

- ***Density*:** This is an important factor of a material where weight and thus, the mass is critical, i.e. aircraft components.
- ***Shear strength*:** It is the ability of a material to resist the shear force applied on the material.
- ***Compressive strength*:** It is the ability of a material to withstand pressure acting on a given plane.
- ***Impact strength*:** It is the energy required per unit cross sectional area to fracture a specimen. The impact strength is a complex characteristic which takes into account both toughness and strength of a material.

15.3 Mechanical Testing of Engineering Materials

15.3.1 Tensile Test

Hooke's Law: Within the elastic limit of a solid material, the deformation (strain) produced by a force (stress) of any kind is proportional to the force. If the elastic limit is not exceeded, the material returns to it's original shape and size after the force is removed, otherwise it remains deformed or stretched. The force at which the material exceeds its elastic limit is called 'limit of proportionality'. Thus, it can be denoted as:

$$E = \sigma/\varepsilon$$

where, E is modulus of elasticity, σ is stress, and ε is strain.

Stress-Strain Diagram

Suppose that a metal specimen be placed in tension-compression-testing machine (UTM). As the axial load is gradually increased in increments, the total elongation over the gauge length is measured at each increment of the load and this is continued until failure of the specimen takes place. Knowing the original cross-sectional area and length of the specimen, the normal stress σ and the strain ε can be obtained. The graph of these quantities with the stress σ along the y-axis and the strain ε along the x-axis is called the stress-strain diagram.

Proportional Limit (Hooke's Law)

From the origin O to the point P called proportional limit, the stress-strain curve is a straight line. This linear relation between elongation and the axial force causing was first noticed by Sir Robert Hooke in 1678 and is called Hooke's law. Within proportional limit, stress is directly proportional to strain. i.e. $\sigma \propto \varepsilon$ or $\sigma = k\varepsilon$. The constant of proportionality is called the modulus of elasticity, E or Young'modulus.

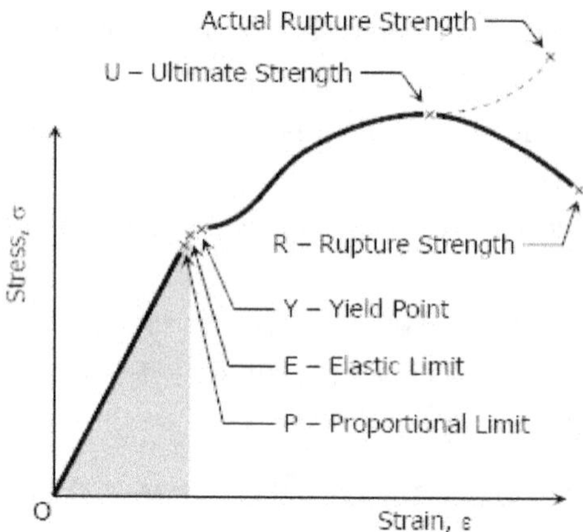

Fig. 15.1 Stress-strain diagram for medium carbon steel.

Elastic Limit

The elastic limit is the limit beyond which the material will no longer go back to its original shape when the load is removed, or it is the maximum stress that may be developed such that there is no permanent or residual deformation when the load is entirely removed.

Elastic and Plastic Ranges

The region in stress-strain diagram from O to P is called the elastic range. The region from P to R is called the plastic range.

Yield Point

Yield point is the point at which the material will have an appreciable elongation or yielding without any increase in load.

Ultimate Strength

The maximum ordinate in the stress-strain diagram is the ultimate strength or tensile strength.

Rapture Strength

Rapture strength is the strength of the material at rupture. This is also known as the breaking strength.

Modulus of Resilience

Modulus of resilience is the work done on a unit volume of material as the force is gradually increased from O to P, in $N \cdot m/m^3$. This may be calculated as the area under the stress-strain curve from the origin O to up to the elastic limit E (the shaded area in the figure). The resilience of the material is its ability to absorb energy without creating a permanent distortion.

Modulus of Toughness

Modulus of toughness is the work done on a unit volume of material as the force is gradually increased from O to R, in N·m/m^3. This may be calculated as the area under the entire stress-strain curve (from O to R). The toughness of a material is its ability to absorb energy without causing it to break.

Working Stress, Allowable Stress, and Factor of Safety

Working stress is defined as the actual stress of a material under a given loading. The maximum safe stress that a material can carry is termed as the allowable stress. The allowable stress should be limited to values not exceeding the proportional limit. However, since proportional limit is difficult to determine accurately, the allowable tress is taken as either the yield point or ultimate strength divided by a factor of safety. The ratio of this strength (ultimate or yield strength) to allowable strength is called the factor of safety.

Percentage Elongation

The ductility of a material in tension can be characterized by its elongation and by the reduction in area at the cross section where fracture occurs. It is the ratio of the extension in length of the specimen after fracture to its initial gauge length, expressed in percent.

$$\delta = \frac{(l_1 - l_g)}{l_1} \times 100$$

l_I = gauge length of specimen after fracture (the distance between the gage marks at fracture)

l_g= gauge length before fracture (initial gauge length)

For 50 mm gage length, steel may here a % elongation is of the order of 10% to 40%.

15.3.2 Hardness

Hardness is a measure of how resistant solid matter is to various kinds of permanent shape change when a compressive force is applied. It can be defined as resistance offered to penetration. Macroscopic hardness is generally characterized by strong intermolecular bonds, but the behaviour of solid materials under force is complex. Therefore, there are different measurements of hardness: *scratch hardness*, *indentation hardness*, and *rebound hardness*.

Scratch hardness is the measure of how resistant a sample is to fracture or permanent plastic deformation due to friction from a sharp object. The principle is that an object made of a harder material will scratch an object made of a softer material. When testing coatings, scratch hardness refers to the force necessary to cut through the film to the substrate. The most common test is Mohs scale. One tool to make this measurement is the sclerometer.

Indentation hardness measures the resistance of a sample to material deformation due to a constant compression load from a sharp object; they are primarily used in engineering and metallurgy fields. The tests work on the basic premise of measuring the critical dimensions of an indentation left by a specifically dimensioned and loaded indenter. Common indentation hardness scales are Rockwell, Vickers, Shore, and Brinell.

Rebound hardness, also known as dynamic hardness, measures the height of the "bounce" of a diamond-tipped hammer dropped from a fixed height onto a material. This type of hardness is related to elasticity. The device used to take this measurement is known as a scleroscope. Two scales that measures rebound hardness are the Leeb rebound hardness test and Bennett hardness scale.

Brinell Hardness Test

This method consists in pressing a hardened steel ball under a constant load P into a specially prepared flat surface on the test specimen as indicated in the figures below.

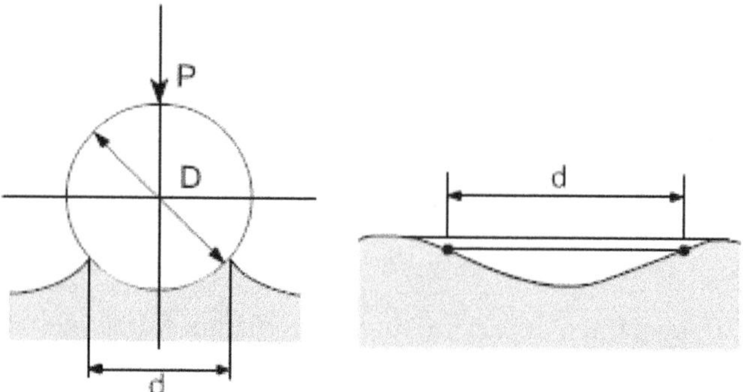

Fig. 15.2 Brinell hardness test.

After removing the load an indentation remains on the surface of the test specimen. If area of the spherical surface in the indentation is denoted as F sq. mm. Brinell Hardness number is defined as:

$$BHN = P / F$$

F is expressed in terms of D and d

P = applied load

D = ball diameter and

d = diametric of indentation

Thus, Brinell Hardness number may be expressed as:

$$Bhn = \frac{2P}{\pi D(D - \sqrt{D^2 - d^2})}$$

Vickers Hardness Test

The Vickers hardness test method consists of indenting the test material with a diamond indenter, in the form of a right pyramid with a square base and an angle of 136 degrees between opposite faces subjected to a load of 1 to 100 kgf. The full load is normally applied for 10 to 15 seconds. The two diagonals of the indentation left in the surface of the material after removal of the load are measured using a microscope and their average calculated. The area of the sloping surface of the indentation is calculated. The Vickers hardness is the quotient obtained by dividing the kgf load by the square mm area of indentation.

$$HV = 1.854 * (f / d^2)$$

where,

HV = Vickers Hardness

f = Load

d = Arithmetic Mean

Rockwell Hardness Test

The Rockwell hardness test method consists of indenting the test material with a diamond cone or hardened steel ball indenter. The indenter is forced into the test material under a preliminary minor load. When equilibrium has been reached, an indicating device, which follows the movements of the indenter and so responds to changes in depth of penetration of the indenter is set to a datum position. While the preliminary minor load is still applied an additional major load is applied with resulting increase in penetration. When equilibrium has again been reach, the additional major load is removed but the preliminary minor load is still maintained. Removal of the additional major load allows a partial recovery, so reducing the depth of penetration. The permanent increase in depth of penetration, resulting from the application and removal of the additional major load is used to calculate the Rockwell hardness number. The expression for RHN may be written as follows:

$$HR = N - (d/s)$$

where, d is the depth (from the zero load point), and

N and s are scale factors that depend on the scale of the test being used.

15.3.3 Impact Test (Izod and Charpy Test)

There are two types of test use for testing the impact resistance of any material: one is Izod Impact resistance test and other is Charpy Impact resistance test. Both of them use the same principle to test the impact resistance of any material but still there lies some difference in them.

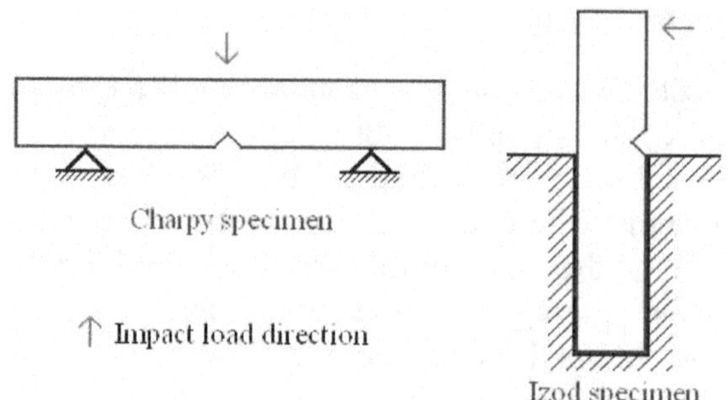

Fig. 15.3 Charpy and Izod test.

The *Izod test* involved the striker, the testing material, and the pendulum. The striker was fixed at the end of the pendulum. The test material was fastened at a vertical position at the bottom, and the notch was facing the striker. The striker swings downward, hitting the test material in the middle, at the bottom of it's swing, and is left free at the top. The notch is placed to concentrate the stress, and provoke delicate failure. It is used to evaluate the materials for overall hardiness. The notch is very important because it can affect the result of the test. It has an essential effect on the competence of the sample to absorb the collision. The blades in the notch can overheat the polymers, and deteriorate the materials surrounding the notch, which could lead to an inaccurate test result. The Izod method chose a short projection, supported at one end, to produce better steel tools for cutting metal.

The *Charpy test* includes striking an appropriate test material with a striker fastened at the end of a pendulum. The test material is secured horizontally in place at both ends, and the striker hits the center of the test material, behind a machined notch. The notch is positioned away from the striker, fastened in a pendulum. The test material usually measures 55x10x10 millimeters. The Charpy method has a machined notch across one of the larger faces. There are two types of charpy notch: a V-notch or a U-notch. The V-notch, measures 2 millimeters deep, with a 45 degree angle and 0.25 millimeter radius, parallel to the base. The U-notch, is 5 millimeters deep, with a 1 millimeter radius at the bottom of the notch. This method proved to be reliable, and gave qualitative collision data.

15.4 Certain other Important Materials

Various metals and their alloys are used in manufacture of various tools, equipments, instruments and machinery etc. Thus it is essential to study these materials and their properties.

Over and above certain non-metallic materials like plastic, rubber, leather, wood, glass and ceramics are also used in various applications.

Thus the engineering materials can be broadly classified as shown in Fig. 15.4.

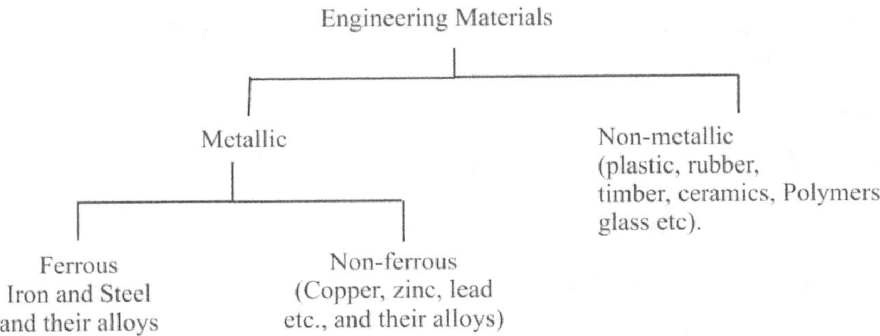

Fig 15.4 Classification of Engineering Materials.

Metals by virtue of composition are often grouped into two classes:

A. Ferrous materials

B. Non-ferrous materials

Ferrous materials include cast iron, and different categories of steel.

Non-ferrous materials have long list such as aluminium, copper, tin, lead etc and their alloys.

15.5 Ferrous Metals and Alloys

The ferrous metals commonly used in engineering practice are cast iron, wrought iron, Steels and alloy steels.

The principle raw material for all ferrous metal is pig iron.

Pig iron is obtained by melting the iron in a blast furnace with help of coke and lime-stone.

15.6 Cast Iron

The cast iron is obtained by re-melting pig iron with coke and lime-stone in a furnace called cupola.

Cast iron may be defined as an alloy of iron and carbon in which carbon content varies from 2.0% to 6.67%. In addition to carbon other elements also present are – silicon (Si) manganese (Mn), Phosphorous (P), Sulphur (S) etc.

Cast iron is the most important and widely used metal. Being brittle material it cannot be used in machine parts subjected to shocks.

Properties

- It has good casting characteristics, high compressive strength wear resistance, and it has low cost.

- Cast iron can be easily melted and casted with or without machining to the required final shape and size.

Applications

Machine tool beds, cylinder blocks, Piston ring, gears and many other parts are made of cast iron.

15.7 Types of Cast-Iron

Important types of cast iron are:

1. Grey cast iron
2. White cast iron
3. Malleable cast iron

1. *Grey Cast Iron*: It is an ordinary commercial iron. It is called grey cast iron since it is grayish in colour on fracture.

Its composition is:

C = 2 to 4.5%

Mn = 0.4 to 1%

Si = 1 to 1.25%

P = 0.15 to 1%

S = 0.02 to 0.15%

Properties

It has high compressive strength and low tensile strength, brittle, low impact toughness and no ductility.

Applications

Grey cast iron castings are used for machine tool bodies, automotive cylinder blocks and heads, Flywheels, Pipe fitting and agricultural applications etc.

2. *White Cast Iron*: It has a silvery-white fracture due to presence of carbon in form of carbide (Fe_3C) and also called as cementite.

Its composition is

C = 1.75 to 2.3%

S_i = 0.85 to 1.2%

Mn < 0.4%

P < 0.2%

and S < 0.2%

Properties

- Because of absence of graphite, it is called white C.I.
- It is extremely hard and brittle.
- It has excellent resistance to wear but has poor machinability.

Applications

- It is used as raw material in the production of malleable cast iron.
- It is not very fluid in molten state thus it is not suitable for foundry work.

3. ***Malleable Cast Iron***

It is produced by long heat treatment given to white cast iron at sufficiently high temperature (900 °C to 950 °C) and then allowed to cool slowly.

Properties

Good strength, ductility, machinability and wear resistance.

Applications

It is used in manufacture of connecting rod, gears, automobile parts, pipe fittings, valves etc.

15.8 Wrought Iron

It is the purest form of iron and it is obtained by eliminating most of the carbon and impurities.

Its composition is:

C = 0.02%

Si = 0.13%

Mn = 0.03%

P = 0.1%

S = 0.02%

Slag < 0.07% and remainder is iron (99.5% to 99.9%)

Properties

The fibrous structure of wrought iron improves strength, fatigue resistance, and corrosion resistance.

Applications

It is highly malleable and ductile and hence it is used for making hooks, chains, blast-plates, condenser tubes, heat exchangers, water and steam pipes etc.

15.9 Alloys of Cast Iron

A large variety of alloy cast irons have been developed by suitable addition of different alloying elements to overcome inherent drawbacks of cast iron.

15.9.1 Effect of various Elements on properties of Cast Iron are:

1. ***Carbon***: It varies from 2 to 4%. If Carbon is present in free state, then it makes the iron weak and brittle. If carbon is in combined form, it makes the iron hard and strong.
2. ***Silicon***: It is limited up to 4%. It helps in formation of free graphite by combining with iron and forming silicates.
3. ***Sulphur***: It is kept below 0.1%. It renders cast iron hard and brittle.
4. ***Manganese***: It is added to reduce the bad effects of sulphur. It makes C.I white and hard.
5. ***Phosphorous***: It improves fluidity and castability of molten metal.
6. ***Nickel (Ni)***: It's addition (0.25 to 2%) improves the machinability of cast iron.
7. ***Chromium (Cr)***: Addition of Cr from 0.15 to 0.19%, improves the wear resistance, tensile strength and hardness of C.I.
8. ***Vanadium***: It reduces graphite formation. It increases strength, hardness and machinability.

15.10 Plain Carbon Steels can be divided into three Classes

(a) Low Carbon Steel (up to 0.25% C)

(b) Medium Carbon steel (0.25% to 0.65% C)

(c) High Carbon steel (0.65% to 1.5% C)

Class of Steel	Properties	Applications
(a) Low carbon steel (0.15 to 0.3% C)	(i) Popularly known as "Mild steel" (ii) Relatively soft and ductile. (iii) Possess good formability and weldability	(i) Automobile and Refrigerator body material. (ii) Structural shape angles etc. (iii) Ship plates, Shaft, Wheels, nuts etc.
(b) Medium Carbon steel (0.30 to 0.8% C)	(i) Strength, ductility, toughness etc. can be obtained by heat treatment process. Adding alloying elements such as Cr, Ni and M_o, its mechanical properties increase.	(i) Railway wheel – tracks (ii) Gears, axle, Crank shaft, hand tools. (iii) Bolts, balls for ball-mill)

Contd...

Class of Steel	Properties	Applications
(c) High Carbon Steel (0.8 to 1.5% C)	(i) Steels are hardest, strongest and least ductile (ii) Possess good wear resistance (iii) Poor machinability	(i) Cutting tools (ii) dies, hacksaws (iii) Springs and Gauges (iv) Piston rings and chisels.

15.11 Alloy Steel

When elements other than carbon such as Cr, Va, Ni, Ti, Mo Mn, Si and tungsten (w) are added in required proportion in steels to improve its mechanical properties.

Properties

(i) More strength, hardness, ductility, toughness at low and high temperature.

(ii) Better resistance to wear and abrasion.

(iii) Better electric and magnetic properties.

Main types of alloy steels used in practice are:

(a) Stainless steel

(b) High speed steel (H.S.S)

(c) Spring steels

(d) Bearing steels

(a) Stainless Steel

Properties

(i) Most popular and widely used steels.

(ii) They have excellent resistance to corrosion and oxidation.

(iii) High corrosion resistance due to presence of chromium in steels.

Applications

(i) Numerous applications in power plants, nuclear plants, chemical and petrochemical industries.

(ii) Food processing units, and dairy industries.

(iii) Surgical instruments and house hold utensils and general engineering applications.

(b) High Speed Steels (H.S.S)

These steels are well suitable for manufacture of cutting tools, which can be operated at high speeds.

Alloying elements used are tungsten, molybdenum, chromium and vanadium

Carbon varies from 0.20 to 0.40%.

Applications

(i) High speed cutting tools

(ii) Milling machine-cutters

(iii) Tools for lathe and shaping machine

(iv) Press-working dies

(c) Spring Steels

These steels are used to produce various kinds of springs, operated under tension or compression.

Properties

(i) This spring material has a high resilience i.e., it will be capable of storing a large amount of elastic energy. So that is can spring back without permanent deformation.

(ii) It has high strength, high toughness and high fatigue strength.

Applications

(i) Used as leaf and helical springs.

(ii) Used as automobile and air craft valve-springs.

(iii) Used as laminated and coil springs for motor cars etc.

(b) Bearing Steels

Also known as anti-friction bearing steels.

Chemical composition

C	0.98 to 1.10%
Si	0.15 to 0.35%
Cr	1.30 to 1.40%
Mn	0.25 to 0.45%

and some traces of S and Ph

Applications

(i) Manufacturing of ball and roller bearings.

(ii) Manufacture of spindle, cold forming rollers, and hardened machine parts.

15.12 Non-Ferrous Metals and Alloys

They do not contain iron as a major element.

Important non-ferrous metals are – Aluminium, copper, lead, tin, nickel, zinc and titanium.

1. **Copper**

Properties

(i) Highest electrical and thermal conductivity.

(ii) Good ductile and malleable metal.

(iii) High resistance to various types of corrosion.

Applications

(i) Electrical Applications

(ii) Good heat conductor

(iii) Used for structural purposes

(iv) Used in telecommunications

(v) Automobile radiators, pressure vessels and water heaters.

(vi) Refrigerator and Heat exchangers.

2. **Copper Alloys**

(a) *Brasses*: Brass is an alloy of copper and zinc.

Properties

Addition of zinc in copper, improves colour, strength, machinability, but less expensive.

Applications

(i) Automobile radiators

(ii) Medals

(iii) Coins

(iv) Sockets

(v) Piston rods

(vi) Musical instruments etc.,

(b) *Bronzes*: Bronze is an alloy of copper and several other elements such as (Tin, Al, Si, Ni)

Properties

(i) Higher strength and hardness than brasses

(ii) Good bearing properties

(iii) Better corrosion resistance

(iv) Costlier than brass

Applications

They are used for bearings, bushes, springs, gears, dies, utensiles etc.

3. Aluminium and its Alloys

Important Properties

(i) Lighter than all engineering materials

(ii) Possess low density, hence low weight

(iii) High ductility and malleability

(iv) Excellent corrosion resistance

(v) Posses good electrical and thermal conductivity.

Applications

(i) Widely used for electrical conductors

(ii) Due to low weight, widely used in automobile, rail, road and marine applications.

(iii) Architectural and decorative items.

(iv) Used in air-craft frames, propellers, auto-body parts, missiles and space crafts.

4. Lead and its Alloys

Lead is the softest and heaviest among all common metals. In nature lead exists either in the form of sulphides or oxide.

Important Properties

(i) Very soft, ductile and malleable.

(ii) Low melting point.

(iii) It has low electrical conductivity and high co-efficient of thermal expansion.

(iv) Good lubricating properties.

(v) Its X-rays and Gamma rays absorbing power is high.

Applications

(i) Mainly used in manufacture of batteries, cables, flexible sheet and pipes.

(ii) Used for sheathing electric cables.

(iii) Used in manufacture of solders, printing metals and bearing.

(iv) Used in chemical industries components.

5. Nickel and its Alloys

Nickel is an important alloying element mostly used in chemical processing, space research and nuclear reactor.

Applications

(i) A great percentage of Ni is used for production of stainless steel.

(ii) Pure Ni is used as a catalyst in chemical reactions.

(iii) In chemical laboratory used as crucible material.

6. **Titanium and its Alloys**

Titanium is available in large quantity from earth crust. It is extracted mainly from titanium ore called rutile ($T_i\,O_2$).

Applications

(i) Pure titanium is used in form of sheets, tubes, rods, bars in chemical industries.

(ii) Minor addition of Ti, increases corrosion resistance in stainless steel.

(iii) Because of high strength-to-weight ratio, higher corrosion resistance, and high temperature strength, titanium-alloys are widely used in aircraft applications. Its melting point is 1675 °C.

15.13 Non-Metallic Materials

These are either natural materials, like rubber, wood, asbestos and ceramics or synthetic materials like nylon, bakelite, laminated sheets etc. These are used in engineering practice due to their low density, low cost, flexibility, resistant to heat and electricity.

Some of the non-metallic materials are being discussed below in brief.

1. ***Plastics***: Plastic can be defined as organic material moulded to shape by application of pressure with or without application of heat. These can be rolled and extruded, cast, laminated and machined. These have high resistance to corrosion. Generally plastics are strong, light weight, workable corrosion and chemical resistance and durable.

   Generally plastics are used for manufacture of household applications, such as bucket, plates, cups etc also used for manufacture of aero plane and automobile parts, laminated gears, pulleys, bearing, safety glasses etc.

2. ***Rubber***: It is a natural plastic material. It resists abrasion, heat, strong acids and alkalies.

   Soft rubber is used for electrical insulations, belts for power transmission, bushes etc. Hard rubber is used for piping and as lining material for pickling tanks.

3. ***Asbestos***: Its trade name is Ferrodo when asbestos is lined with lead oxide. It is used as friction lining for brakes and clutches. Asbestos is incombustible in nature and it is insulator of both heat and electricity. It is used as fire resisting curtains in theatres, partition walls, electrician gloves, clothing, clamps, wire gauge, piston and piston rod packing etc.

4. ***Timber***: The timber denotes wood which is suitable for engineering and building construction purposes, by sawing and converting into various commercial uses.

   Advantages of timber as a construction material:

   (i) High strength weight ratio.

   (ii) Ease of working

(iii) Good thermal insulations

(iv) Toughness and attractive appearance.

Drawbacks

(i) Susceptibility to decay and inflammability.

(ii) Variations in properties due to changes in moisture content.

(iii) Variation is strength along length and across length.

General Applications

It is used in framed structure, furniture, sound-proof construction, tool-handles, patterns for foundry work, railway sleepers, etc.

Indian Timber trees and their Applications:

There are more than 150 types of timbers are produced in India.

Following few which are used for engineering purposes:

Types of Timber	**Applications**
1. Babul	Agricultural instruments tool-handles etc.
2. Bamboo	Construction of wooden roof, fancy goods.
3. Deodar	Cheap and rough furniture, Railway sleeper, Packing boxes etc.
4. Mahogany	Furniture, Fancy goods etc.
5. Mango	Cheap furniture, toys, ship-building, doors and windows.
6. Sal	Railway sleeper, wagons, bridges structural work.
7. Shishom	Plywood, sport-goods etc.
8. Mulberry	Sport goods, Hockey sticks, cricket bats, Tennis rockets etc.

5. ***Abrasive Material*****:** An Abrasive is hard, mechanically resistant material used for cutting, grinding and polishing of components. Abrasive material are used in form of powder, emery paper, emery-cloth, solid discs, wheels or bars of various shapes and sizes.

Basically emery is an abrasive material and it is a mixture of oxides of Alumina and oxides of Iron.

6. ***Ceramic Materials*****:** Ceramics are inorganic non-metallic materials and in most cases, they are treated at high temperature, during manufacturing.

Ceramic materials include glass-products, cements and concrete, abrasive refractory lining for furnace.

Traditionally used in pottery, crockery and sanitary works.

A composite material of ceramic and metal is called cermet. Many ceramic materials are hard, porous and brittle.

Properties of Cermet

(i) Excellent hardness and temperature resistance.

(ii) Greater refractory properties.

Applications

(i) Used for high speed cutting tools

(ii) Components used in rocket and jet engines.

(iii) Refractories used as kiln linings, furnace linings etc.

7. ***Graphite*:** Graphite is an allotrope of carbon. It is an electric conductor, thus it can be used as electrodes of an arc-lamp. It is very stable material. It can be considered as highest grade of coal, but it cannot be easily ignited.

   It is also used in pencils for drawing and writing. It is commonly known as lead-pencil.

   Types

   (i) Crystalline flake graphite occurs as flat-plate.

   (ii) Amorphous graphite occurs as fine particles.

   (iii) *Lump Graphite*: It is used for making graphite crucibles in steel making, brake lining, as lubricants, pencils, zinc-carbon batteries etc.

8. ***Diamond*:** It is a naturally occurring material which is an allotrope of carbon.

   (i) It is a hardest abrasive material used for manufacture of drills.

   (ii) Used for dressing, grinding wheels.

   (iii) Used for polishing hard carbides-metals.

   It is the hardest and toughest material and acts as electrical insulators. Diamonds make superlative abrasive, due to which it is used in engraving tools. Some blue diamonds are semi-conductors. It has high thermal conductivity. It is yellow in colour due to presence of Nitrogen as impurity.

9. ***Glass*:** Glass is a ceramic material. It is a mixture of various insoluble silicates with excess of silica.

   These are of two types:

   (i) Glass composed of silica, alkali and lime used for crown and plate glass.

   (ii) Flint glass made of lead-silicates in large proportion with alkali. Lead makes it denser and lustrous.

Applications

(i) X-ray tubes.

(ii) Optical glasses in camera, microscope, telescope etc.

(iii) Domestic utensils.

(iv) As mirror, lenses, prism and Auto-head-lights etc.

10. ***Composite Materials*:** A composite material is, a microscopic combination of two or more distinct materials having a specific interfare between them.

Composite materials are different from alloys. The constituents, which form alloy, are soluble in each other and form a homogenous solid solution. Ideally the physical and chemical properties are same at every point in the alloy. In composite, the constituent's phases must be chemically dissimilar, separated by a distinct interface and essentially in soluble in one another. The properties are also different at different point and in different directions.

Wood is a beautiful example of a natural composite which consists of strong and flexible cellulose fibers surrounded by a stiff material called Lignin as shown in Fig 15.2.

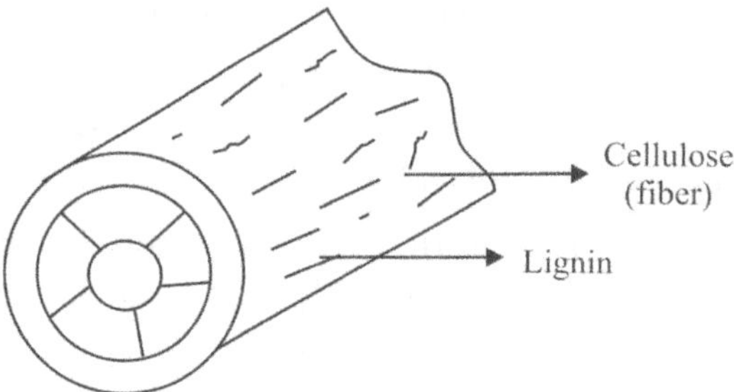

Fig. 15.5

11. ***Polymers*:** A polymer is an organic material composed of repeating structural units, connected by covalent bond, e.g., Plastics, adhesives, fibers etc.

Characteristics

(i) Low density

(ii) Low mechanical strength

(iii) Good corrosion resistance

(iv) Excellent surface finish and different colours can be obtained.

(v) Non-conductor of electricity and low thermal conductivity.

Applications

(i) Synthetic fibers and textile polymers make high structure fibers.
E.g: Silk and wool.

(ii) Used in optical industries, for making lenses, reflectors, transparent film package etc.

(iii) Used as packaging and coating material.

15.14 Classification of Engineering Materials

Engineering materials have been extensively tested for their properties and classified into broad groups. Basically engineering materials have been classified into two categories: Metals and Non-Metals. Further, non metals have been classified as: ceramics, organic polymers, composites, semiconductors, biomaterials, advanced materials.

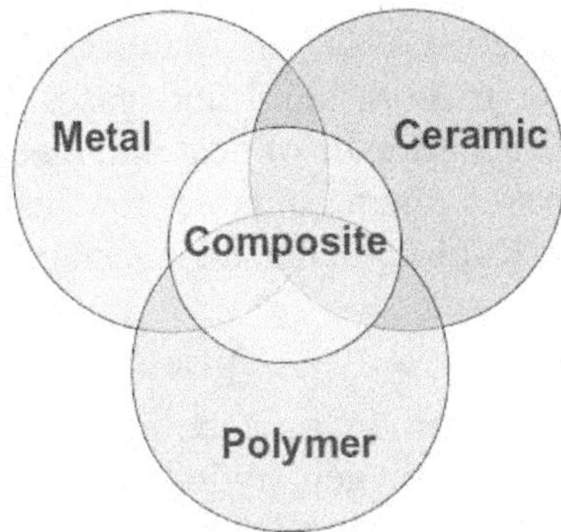

Fig. 15.6 Classification of engineering materials.

Metals

Metals are polycrystalline bodies which are having number of differentially oriented fine crystals. Normally major metals are in solid states at normal temperature. However, some metals such as mercury are also in liquid state at normal temperature. All metals are having high thermal and electrical conductivity, and positive temperature coefficient of resistance. Examples of metals: Silver, Copper, Gold, Aluminum, Iron, Zinc, Lead, Tin etc.

Metals can be further divided into two groups: Ferrous Metals, and Non-Ferrous Metals.

Ferrous Metals

All ferrous metals are having iron as common element, having very high permeability which makes these materials suitable for construction of core of electrical machines. **Examples:** Cast Iron, Wrought Iron, Steel, Silicon Steel, High Speed Steel, Spring Steel etc.

Non-Ferrous Metals

All non-ferrous metals are having very low permeability. Example: Silver, Copper, Gold, Aluminum etc.

Alloys

Alloys are the composition of two or more metals or metal and non-metals together. Pure metals are having very low mechanical strength, which sometimes does not match with the mechanical strength required for certain applications. To overcome this draw back alloys are used. Alloys are having good mechanical strength, low temperature coefficient of resistance.

Example: Steels, Brass, Bronze, Gunmetal, Invar. Super Alloys (Aluminium, copper, magnesium, titanium alloys) etc.

Non-Metals

Non-Metal materials are non-crystalline in nature. They exist in amorphic or mesomorphic forms. These are in both solid and gases forms at normal temperature. Normally all non-metals are bad conductor of heat and electricity. Non-metals are having very high resistivity which makes them suitable for insulation purpose in electrical machines. **Examples:** Plastics, Rubber, Leathers, Asbestos etc.

15.14.1 Organic Materials

All organic materials are having carbon as a common element. In organic materials, carbon is chemically combined with oxygen, hydrogen and other non-metallic substances. Generally organic materials are having complex chemical bonding.

Example: Plastics, PVC, Synthetic Rubbers etc.

15.14.2 Polymer

Polymers are usually long chain organic macro-molecules with covalent bonds. Each molecule is formed from a large number of unitary molecules known as *Monomer*. For this reason, the polymers can be bent and stretched easily. In certain polymers, some of the molecules cross link with each other thereby increasing the strength across the molecules. Characteristics of polymer are: Ductile/soft, Translucency, Low thermal stability, Lightweight (but high molecular weight), and Poor conductor of heat and electricity.

15.14.3 Composites

The history of composites dates back to ancient times for construction applications, straw was mixed with mud to form a building material known as adobe. The straw provided the structure and strength, while the mud acted as a binder, holding the straw together in place. Since the days of adobe, the use of composites has evolved to commonly incorporate a structural fibre and a plastic, this is known as fibre reinforced plastics (FRP). Like straw, the fibre provides the structure and strength to the composite, while a plastic polymer holds the fibre together. Common types of fibres used in FRP composites include:

- Fiber glass
- Carbon Fiber
- Aramid Fiber
- Boron Fiber
- Basalt Fiber
- Natural Fiber (Wood, Flax, Hemp, etc.)

In the case of fibre, hundreds of thousands of tiny glass fibres are compiled together and held rigidly in place by a plastic polymer resin. Common plastic resins used in composites include:

- Epoxy
- Vinyl Ester
- Polyester
- Polyurethane
- Polypropylene

There exists macro-physical combination of different phases where the aim is to combine beneficial properties of the constituent materials. Hence, very clear and abstract knowledge of properties of the constituent materials are needed in order to select compatible materials for composite fabrication. Typical engineered composite materials include:

- Mortars, concrete
- Reinforced plastics, such as fibre-reinforced polymer
- Metal composites
- Ceramic composites (composite ceramic and metal matrices)

15.14.4 Ceramic Materials

Ceramic materials are non-metallic solids. These are made of inorganic compounds such as Oxides, Nitrides, Silicides and Carbides. It possesses exceptional Structural, Electrical, Magnetic, Chemical and Thermal properties. These ceramic materials are now extensively used in different engineering fields. **Examples:** Silica, glass, cement, concrete, garnet, Mgo, Cds, Zno, SiC etc.

15.14.5 Plastics

Plastics are the most versatile materials ever invented. Indeed, the word "plastic," which derives from the Greek word *plastikos*, meaning to mold or form, has a general description for anything particularly adaptable or flexible. Most plastics are made from synthetic resins (polymers) through the industrial process of polymerization. Since the first plastic, celluloid, was developed as a replacement for elephant ivory in the 1860s, many different types of plastics, including nylon, polyethylene, and Teflon have revolutionized the manufacture of commercial goods. Two main types of plastics are thermoplastics and thermosets. Although the use of plastic continues to grow and revolutionary new plastics are constantly being developed, concerns for environmental effects promote invention of bioplastics.

15.15 Advanced Materials

The materials that are utilised in high-technology (or high-tech) applications are sometimes called advanced materials. High technology means a device or product that operates or functions using relatively intricate and sophisticated principles; for example, electronic equipment (VCRs, CD players, etc.), computers, fiber optic systems, spacecraft, aircraft and military rocketry etc. The advanced materials are typically either traditional material whose properties have been enhanced or newly developed high performance materials. Furthermore, advanced materials may be of all material types (e.g., metals, ceramics, and polymers) and are normally relatively expensive. Other advanced materials: materials that are used for lasers, ICs, magnetic information storage, liquid crystal display (LCDs), fiber optics, and the thermal projection system for the space shuttle orbiter.

15.16 Smart Materials (Materials of the Future)

Smart or intelligent materials form a group of new and state-of-the art materials now being developed that will have a significant influence on many of the technologies. The name-smart implies that these materials are able to sense changes in their environments and then respond to these changes in pre-determined manners-traits that are also found in living organisms. In addition, the concept of smart materials is being extended to rather sophisticated systems that consist of both smart and traditional materials. The field of smart materials attempts to combine the sensor (that detects an input signal), actuator (that performs a responsive and adaptive function) and the control circuit as one integrated unit. Actuators may be called upon to change shape, position, natural frequency, or mechanical characteristics in response to changes in temperature, electric fields, and/or magnetic fields. Usually, four types of materials are commonly used for actuators: shape memory alloys, piezoelectric ceramics, magnetostrictive materials, and electrorheological /magnetorheological fluids. Shape memory alloys are metals that, after deformed, revert back to their original shapes when temperature is changed. Piezoelectric ceramics expand and contract in response to an applied electric field (or voltage); conversely, these materials also generate an electric field when their dimensions are altered. The behaviour of magnetostrictive materials is analogous to that of the piezoelectric ceramic materials, except that they are responsive to magnetic fields. Also, electrorheological and magnetorheological fluids are liquids that experience dramatic changes in viscosity upon the application of electric and magnetic fields, respectively.

15.17 Biomaterials

A biomaterial is any material, natural or man-made, that comprises whole or part of a living structure or biomedical device which performs, augments, or replaces a natural function. These are employed in components implanted into the human body for replacement of diseased or damaged body parts. Biomaterials must not produce toxic substances and must be compatible with body tissues (i.e., these materials must not cause adverse biological reactions). All the above materials, i.e., metals, ceramics, polymers, composites, and semiconductors may be used as biomaterials. They are used in medicine for Dental applications, Surgery, Drug delivery, Joint replacements, Bone plates, Bone cement, Artificial ligaments and tendons, Dental implants for tooth fixation, Blood vessel prostheses, Heart valves, Skin repair devices, Cochlear replacements, Contact lenses, etc.

Questions

1. Describe the mechanical properties of engineering materials in brief.
2. Describe about the various testing methods for engineering materials.
3. Describe the various mechanical properties (hardness) of a material.
4. Describe the various classifications of engineering materials with their preparation and applications.
5. Write short notes on:
   (a) Composite materials (b) Plastics (c) Smart materials (d) Bio materials

Objective Type Questions

Choose the Correct Statements

1. Most malleable material is
   (a) Tin (b) Copper
   (c) Gold (d) Silver
2. Most important property of lead is
   (a) brittleness (b) malleability (c) ductility
3. Steel containing carbon between 0.15 to 0.30% is called as
   (a) mild steel (b) medium carbon steel
   (c) High carbon steel
4. Material commonly used for machine tool bodies is
   (a) copper (b) mild steel
   (c) Cast iron (d) wrought iron

5. Crank shaft of motor car is made of
   (a) Silicon steel (b) Medium Carbon steel
   (c) High speed steel (d) High Carbon steel
6. Material used for crane hook is
   (a) Cast iron (b) mild steel
   (c) Wrought Iron (d) Aluminium
7. Material used for automobile radiator is
   (a) Copper (b) Aluminium
   (c) Lead (d) Nickel

Answers

1. (c)	2. (b)	3. (a)	4. (c)
5. (b)	6. (c)	7. (a)	

Short Questions

1. Differentiate between the following:
   (a) Brass and Bronze (b) Glass and refractory
   (c) Diamond and Quartz (d) Cement and Concrete
2. Differentiate between the following mechanical properties.
   (a) Ductility and Malleability (b) Hardness and Toughness
   (c) Elasticity and Plasticity (d) Resilience and creep
3. Select suitable timber material used for the components shown in column A.

	A		B
(a)	Packing boxes	1	Mango
(b)	Cricket bats	2.	Shisham
(c)	Railway sleeper	3.	Mulberry
(d)	Door and windows	4.	Deodar
(e)	Plywood	5.	Sal

4. Define ferrous and non-ferrous materials
5. Differentiate between C.I and steel
6. State and classify steels. Give at least one example of their applications.
7. What is stainless steel? State some of its applications.
8. What do you understand by non-ferrous metals? Name any four and state their applications.

9. What do you understand by non-metallic materials? Give examples and state their practical importance.
10. What is cermet? State at least two its properties and practical applications.
11. What are the abrasive materials? State their applications.
12. Write any two applications of the following materials:

    (a) Plastic (b) Timber (c) Rubber
    (d) Ceramics (e) Asbestos (f) Glass
13. Match (A) with (B) as per proper combinations.

A	B
(a) Automobile body	1. Spring steel
(b) Household utensils	2. Bronze
(c) Helical Springs	3. Stainless steel
(d) Aircraft frame	4. Aluminium alloys
(e) Anti-friction bearings	5. Low Carbon steel

Steam Tables

PROPERTIES OF DRY AND SATURATED STEAM

(Pressure Table)

Pressure bar p	Sat. temp. °C t_s	Specific vol. Sat. steam m³/kg v_g	Enthalpy kJ/kg Sat. water h_f	Evap. h_{fg}	Sat. steam h_g	Entropy kJ/kg K Sat. water Φ_w	Sat. steam Φ_s
.006113	0.1	206.136	.01	2501.3	2501.4	.0000	9.1562
.007	1.89	181.255	7.91	2496.9	2504.4	.0288	9.1064
.008	3.77	159.675	15.81	2492.5	2508.3	.0575	9.0578
.009	5.45	142.789	22.89	2488.5	2511.4	.0829	9.0142
.010	6.98	129.208	29.30	2484.9	2514.2	.1059	8.9756
.011	8.37	118.042	35.17	2481.6	2516.8	.1268	8.9408
.012	9.66	108.696	40.58	2478.6	2519.1	.1460	8.9091
.013	10.86	100.755	45.60	2475.7	2521.3	.1637	8.8792
.014	11.98	93.922	50.31	2473.1	2523.4	.1802	8.8529
.015	13.03	87.980	54.71	2470.6	2525.3	.1957	8.8279
.016	14.02	82.763	58.87	2468.3	2527.1	.2102	8.8044
.017	14.95	78.146	62.80	2466 0	2528.8	.2238	8.7825
.018	15.84	74.030	66.54	2463.9	2530.5	.2368	8.7618
.019	16.69	70.337	70.10	2461.9	2532.0	.2401	8.7422
.020	17.50	67.004	73.48	2460.0	2533.5	.2607	8.7237
.021	18.28	63.981	76.74	2458.2	2534.9	.2719	8.7060
.022	19.02	61.226	79.85	2456.4	2536.3	.2826	8.6892
.023	19.73	58.705	82.83	2454.8	2537.6	.2928	8.6732
.024	20.42	56.389	85.72	2453.1	2538.8	.3026	8.6579
.025	21.08	54.254	88.49	2451.6	2540.0	.3120	8.6432
.026	21.72	52.279	91.17	2450.1	2541.2	.3211	8.6290
.027	22.34	50.446	93.75	2448.6	2542.3	.3299	8.6155
.028	22.94	48.742	96.27	2447.2	2543.4	.3384	8.6024
.029	23.52	47.152	98.7	2445.8	2544.5	.3466	8.5898
.030	24.08	45.665	101.5	2444.5	2545.5	.3545	8.5776
.032	25.16	42.964	105.57	2441.9	2547.5	.3697	8.5545
.034	26.19	40.572	109.84	2439.5	2549.3	.3840	8.5327
.036	27.16	38.440	113.90	2437.2	2551.1	.3975	8.5123
.038	28.08	36.527	117.77	2435.0	2552.8	.4104	8.4930
.040	28.96	34.800	121.46	2432.9	2554.4	.4226	8.4746

Contd...

PROPERTIES OF DRY AND SATURATED STEAM

(Pressure Table)

Pressure bar p	Sat. temp. °C t_s	Specific vol. Sat. steam m³/kg v_g	Enthalpy kJ/kg			Entropy kJ/kg K	
			Sat. water h_f	Evap. h_{fg}	Sat. steam h_g	Sat. water Φ_w	Sat. steam Φ_s
.42	29.81	33.234	125.00	2430.9	2555.9	.4343	8.4572
.44	30.62	31.806	128.39	2429.0	2557.4	.4455	8.4406
.46	31.40	30.500	131.65	2427.2	2558.8	.4562	8.4248
.48	32.15	29.299	134.79	2425.4	2560.2	.4665	8.4096
.50	32.88	28.192	137.82	2423.7	2561.5	.4764	8.3951
.55	35.58	25.769	144.95	2419.6	2564.5	.4997	8.3613
.60	36.16	23.739	151.53	2415.9	2567.4	.5210	8.3304
.65	37.63	22.014	157.67	2412.4	2570.0	.5408	8.3020
.70	39.00	20.530	163.40	2409.1	2572.5	.5592	8.2758
75	40.29	19.238	168.79	2406.0	2574.8	.5764	8.2515
.080	41.51	18.103	173.88	2403.1	2577.0	.5926	8.2287
.085	42.67	17.099	178.70	2400.3	2579.0	.6079	8.2073
.090	43.76	16.203	183.29	2397.7	2581.0	.6224	8.1872
.095	44.81	15.399	187.65	2395.2	2582.9	.6362	8.1632
.10	45.81	14.674	191.83	2392.8	2584.7	.6493	8.1502
.11	47.69	13.415	199.67	2388.3	2588.0	.6738	8.1168
.12	49.42	12.361	206.92	2384.1	2591.1	.6963	8.0863
.13	51.04	11.465	213.67	2380.2	2593.9	.7172	8.0584
.14	52.55	10.693	219.99	2376.6	2596.6	.7366	8.0325
.15	53.97	10.022	225.94	2373.1	2599.1	.7549	8.0085
.16	55.32	9.433	231.56	2369.9	2601.4	.7720	7.9860
.17	56.59	8.910	236.89	2366.8	2603.7	.7882	7.9649
.18	57.80	8.445	241.95	2363.8	2605.8	.8035	7.9451
.19	58.96	8.027	246.78	2361.0	2607.8	.8181	7.9263
.20	60.06	7.649	251.40	2358.3	2609.7	.8320	7.9085
.21	61.12	7.307	255.83	2355.7	2611.6	.8452	7.8916
.22	62.14	6.995	260.08	2353.2	2613.3	.8579	7.8756
.23	63.12	6.709	264.18	2350.8	2615.0	.8701	7.8602
.24	64.06	6.446	268.12	2348.5	2616.6	.8818	7.8455
.25	64.97	6.204	271.93	2346.3	2618.2	.8931	7.8314

Contd...

PROPERTIES OF DRY AND SATURATED STEAM

(Pressure Table)

Pressure bar p	Sat. temp. °C t_s	Specific vol. Sat. steam m³/kg v_g	Enthalpy kJ/kg Sat. water h_f	Evap. h_{fg}	Sat. steam h_g	Entropy kJ/kg K Sat. water Φ_f	Sat. steam Φ_s
.26	65.85	5.980	275.61	2344.1	2619.7	.9040	7.8179
.27	66.70	5.772	279.17	2342.0	2621.2	.9145	7.8049
.28	67.53	5.579	282.62	2340.0	2622.6	.9246	7.7924
.29	68.33	5.398	285.98	2338.0	2624.0	.9344	7.7803
.30	69.10	5.229	289.23	2336.1	2625.3	.9439	7.7686
.32	70.60	4.922	295.48	2332.4	2627.8	.9622	7.7465
.34	72.01	4.650	301.40	2328.8	2630.2	.9723	7.7257
.36	73.36	4.408	307.05	2325.5	2632.5	.9956	7.7061
.38	74.64	4.190	312.43	2322.3	2634.7	1.0159	7.6876
.40	75.87	3.993	317.58	2319.2	2636.8	1.0259	7.6700
.42	77.05	3.815	322.51	2316.2	2638.7	1.0400	7.6534
.44	78.18	3.652	327.26	2313.4	2640.6	1.0536	7.6375
.46	79.27	3.503	331.83	2310.6	2642.5	1.0666	7.6223
.48	80.32	3.367	336.23	2308.0	2644.2	1.0790	7.6078
.50	81.33	3.240	340.49	2305.4	2645.9	1.0910	7.5939
.55	83.72	2.964	350.54	2299.3	2649.8	1.1193	7.5615
.60	85.94	2.732	359.86	2293.6	2653.5	1.1453	7.5320
.65	88.01	2.535	368.54	2288.3	2656.3	1.1694	7.5048
.70	89.95	2.365	376.70	2283.3	2660.0	1.1919	7.4766
.75	91.78	2.217	384.39	2278.6	2663.0	1.2124	7.4564
.80	93.50	2.087	391.66	2274.1	2665.8	1.2329	7.4346
.85	95.14	1.972	398.57	2269.8	2668.4	1.2517	7.4141
.90	96.71	1.869	405.15	2265.7	2670.9	1.2695	7.3949
.95	98.20	1.777	411.83	2261.8	2673.2	1.2864	7.3766
1.00	99.63	1.694	417.46	2258.0	2675.5	1.3026	7.3594
1.01325	**100.00**	**1.6729**	**419.04**	**2257.0**	**2676.1**	**1.3069**	**7.3549**
1.05	101.00	1.6184	423.24	2254.4	2677.6	1.3184	7.3430
1.10	102.31	1.5495	428.79	2250.9	2679.7	1.3329	7.3273
1.15	103.58	1.4864	434.15	2247.5	2681.7	1.3471	7.3124
1.20	104.80	1.4284	439.32	2244.2	2683.5	1.3608	7.2981
1.25	105.99	1.3749	444.32	2241.0	2685.4	1.3740	7.2844

Contd...

PROPERTIES OF DRY AND SATURATED STEAM

(Pressure Table)

Pressure bar p	Sat. temp. °C t_s	Specific vol. Sat. steam m³/kg v_g	Enthalpy kJ/kg			Entropy kJ/kg K	
			Sat. water h_f	Evap. h_{fg}	Sat. steam h_g	Sat. water Φ_w	Sat. steam Φ_s
1.30	107.13	1.3254	449.15	2238.0	2687.1	1.3867	7.2712
1.35	108.24	1.2794	453.13	2235.0	2688.8	1.3990	7.2586
1.40	109.31	1.2366	458.39	2232.1	2690.4	1.4109	7.2464
1.45	110.36	1.1967	462.80	2229.1	2692.0	1.4224	7.2347
1.50	111.37	1.1593	467.11	2226.5	2693.6	1.4336	7.2233
1.55	112.36	1.1243	471.28	2223.8	2695.0	1.4444	7.2123
1.60	113.32	1.0914	475.36	2221.1	2696.5	1.4550	7.2017
1.65	114.26	1.0604	479.33	2218.5	2697.9	1.4652	7.1914
1.70	115.17	1.0312	483.20	2216.0	2699.2	1.4752	7.1814
1.75	116.06	1.0036	486.99	2213.6	2700.6	1.4849	7.1717
1.80	116.93	.9775	490.68	2211.2	2701.8	1.4944	7.1623
1.85	117.79	.9528	494.30	2208.8	2703.1	1.5036	7.1532
1.90	118.62	.9293	497.84	2206.5	2704.3	1.5127	7.1443
1.95	119.43	.9070	501.31	2204.2	2705.5	1.5225	7.1356
2.00	120.23	.8857	504.70	2201.9	2706.7	1.5301	7.1271
2.05	121.02	.8655	508.03	2199.8	2707.8	1.5386	7.1189
2.10	121.74	.8462	511.29	2197.6	2708.9	1.5468	7.1109
2.15	122.53	.8277	514.48	2195.5	2710.0	1.5549	7.1030
2.20	123.27	.8101	517.63	2193.4	2711.0	1.5628	7.0953
2.25	124.00	.7933	520.72	2191.3	2712.1	1.5706	7.0878
2.30	124.71	.7771	523.74	2189.3	2713.1	1.5782	7.0805
2.35	125.41	.7616	526.72	2187.3	2714.1	1.5856	7.0733
2.40	126.13	.7467	529.65	2185.4	2715.0	1.5930	7.0663
2.45	126.77	.7324	532.53	2183.5	2716.0	1.6002	7.0594
2.50	127.44	.7187	535.37	2181.5	2716.9	1.6072	7.0527
2.55	128.09	.7055	538.15	2179.7	2717.8	1.6142	7.0461
2.60	128.73	.6918	540.90	2177.8	2718.7	1.6210	7.0396
2.65	129.37	.6805	543.60	2176.0	2719.6	1.6277	7.0333
2.70	129.99	.6687	546.27	2174.2	2720.5	1.6343	7.0270
2.75	130.60	.6573	548.89	2172.4	2721.3	1.6408	7.0209

Contd...

PROPERTIES OF DRY AND SATURATED STEAM

(Pressure Table)

Pressure bar p	Sat. temp. °C t_s	Specific vol. Sat. steam m^3/kg v_g	Enthalpy kJ/kg			Entropy kJ/kg K	
			Sat. water h_f	Evap. h_{fg}	Sat. steam h_g	Sat. water Φ_w	Sat. steam Φ_s
2.80	131.21	.6463	551.48	2170.7	2722.1	1.6472	7.0149
2.85	131.81	.6357	554.02	2168.9	2723.0	1.6535	7.0090
2.90	132.39	.6254	556.54	2167.2	2723.8	1.6597	7.0032
2.95	132.97	.6154	559.02	2165.5	2724.5	1.6658	6.9975
3.00	133.55	.6058	561.47	2163.8	2725.3	1.6718	6.9919
3.05	134.11	.5965	563.88	2162.2	2726.1	1.6777	6.9864
3.10	134.67	.5875	566.27	2160.6	2726.8	1.6835	6.9810
3.15	135.22	.5787	568.62	2158.9	2727.6	1.6893	6.9756
3.20	135.76	.5702	570.95	2157.3	2728.3	1.6950	6.9704
3.25	136.30	.5620	573.25	2155.8	2729.0	1.7006	6.9652
3.30	136.83	.5540	575.52	2154.2	2729.7	1.7061	6.9601
3.35	137.35	.5463	577.76	2152.6	2730.4	1.7116	6.9551
3.40	137.87	.5387	579.97	2151.1	2731.1	1.7169	6.9502
3.45	138.38	.5314	582.16	2149.6	2731.8	1.7222	6.9453
3.50	138.88	.5243	584.33	2148.1	2732.4	1.7275	6.9405
3.55	139.38	.5173	586.47	2146.6	2733.1	1.7327	6.9358
3.60	139.87	.5106	588.59	2145.1	2733.7	1.7378	6.9311
3.65	140.36	.5040	590.68	2143.7	2734.4	1.7428	6.9265
3.70	140.84	.4926	592.75	2142.2	2735.0	1.7478	6.9219
3.75	141.32	.4914	594.81	2140.8	2735.6	1.7528	6.9175
3.80	141.79	.4853	596.83	2139.4	2736.2	1.7577	6.9130
3.85	142.26	.4794	598.84	2138.0	2736.8	1.7625	6.9087
3.90	142.72	.4736	600.83	2136.6	2737.4	1.7673	6.9044
3.95	143.18	.4680	602.80	2135.2	2738.0	1.7720	6.9001
4.00	143.63	.4625	604.64	2133.8	2738.6	1.7766	6.8959
4.1	144.53	.4519	608.59	2131.1	2739.7	1.7858	6.8877
4.2	145.40	.4417	612.36	2128.4	2740.8	1.7948	6.8796
4.3	146.26	.4321	616.06	2125.8	2741.9	1.8036	6.8717
4.4	147.10	.4228	619.68	2123.2	2742.9	1.8122	6.8641
4.5	147.93	.4140	623.25	2120.7	2743.9	1.8207	6.8565

Contd...

PROPERTIES OF DRY AND SATURATED STEAM

(Pressure Table)

Pressure bar p	Sat. temp. °C t_s	Specific vol. Sat. steam m^3/kg v_g	Enthalpy kJ/kg			Entropy kJ/kg K	
			Sat. water h_f	Evap. h_{fg}	Sat. steam h_g	Sat. water Φ_w	Sat. steam Φ_s
4.6	148.74	.4055	626.76	2118.2	2744.9	1.8290	6.8492
4.7	149.54	.3974	630.21	2115.7	2745.9	1.8371	6.8420
4.8	150.32	.3896	633.60	2113.2	2746.8	1.8451	6.8349
4.9	151.10	.3821	636.94	2110.8	2747.8	1.8530	6.8280
5.0	151.86	.3749	640.23	2108.5	2748.7	1.8607	6.8213
5.1	152.60	.3689	643.46	2106.1	2749.6	1.8682	6.8146
5.2	153.34	.3613	646.65	2104.8	2750.5	1.8757	6.8081
5.3	154.06	.3549	649.78	2101.5	2751.3	1.8830	6.8017
5.4	154.78	.3487	652.87	2099.3	2752.1	1.8902	6.7955
5.5	155.48	.3427	655.93	2097.0	2753.0	1.8973	6.7893
5.6	156.17	.3369	658.93	2094.8	2753.8	1.9043	6.7832
5.7	156.86	.3313	661.90	2092.7	2754.5	1.9112	6.7773
5.8	157.53	.3259	664.83	2090.5	2755.3	1.9180	6.7714
5.9	158.20	.3207	667.71	2088.4	2756.1	1.9246	6.7657
6.0	158.85	.3157	670.56	2086.3	2756.8	1.9312	6.7600
6.1	159.50	.3108	673.37	2084.2	2757.5	1.9377	6.7545
6.2	160.14	.3060	676.15	2082.1	2758.3	1.9441	6.7490
6.3	160.77	.3015	678.89	2080.1	2759.0	1.9504	6.7436
6.4	161.39	.2970	681.60	2078.0	2759.6	1.9566	6.7383
6.5	162.01	.2927	684.28	2076.0	2760.3	1.9627	6.7331
6.6	162.61	.2885	686.93	2074.0	2761.0	1.9688	6.7279
6.7	163.21	.2844	689.55	2072.1	2761.6	1.9748	6.7228
6.8	163.81	.2805	692.13	2070.1	2762.3	1.9807	6.7178
6.9	164.39	.2766	694.69	2068.2	2762.9	1.9865	6.7129
7.0	164.97	.2729	697.22	2066.3	2763.5	1.9922	6.7080
7.1	165.55	.2692	699.72	2064.4	2764.1	1.9979	6.7032
7.2	166.11	.2657	702.20	2062.5	2764.7	2.0035	6.6985
7.3	166.67	.2622	704.64	2060.7	2765.3	2.0091	6.6938
7.4	167.23	.2589	707.07	2058.8	2765.9	2.0146	6.6892
7.5	167.78	.2556	709.47	2057.0	2766.4	2.0200	6.6847

Contd...

PROPERTIES OF DRY AND SATURATED STEAM

(Pressure Table)

Pressure bar p	Sat. temp. °C t_s	Specific vol. Sat. steam m^3/kg v_g	Enthalpy kJ/kg			Entropy kJ/kg K	
			Sat. water h_f	Evap. h_{fg}	Sat. steam h_g	Sat. water Φ_w	Sat. steam Φ_s
7.6	168.32	.2524	711.85	2055.2	2767.0	2.0253	6.6802
7.7	168.86	.2493	714.20	2053.4	2767.5	2.0306	6.6758
7.8	169.39	.2463	716.52	2051.6	2768.1	2.0359	6.6714
7.9	169.91	.2433	718.83	2049.8	2768.6	2.0411	6.6670
8.0	170.43	.2404	721.11	2048.0	2769.1	2.0462	6.6628
8.1	170.95	.2376	723.38	2046.3	2769.7	2.0513	6.6585
8.2	171.46	.2349	725.62	2044.5	2770.2	2.0563	6.6544
8.3	171.97	.2322	727.83	2042.8	2770.7	2.0612	6.6502
8.4	172.47	.2295	730.04	2041.1	2771.2	2.0662	6.6462
8.5	172.96	.2270	732.22	2039.4	2771.6	2.0710	6.6421
8.6	173.46	.2245	734.38	2037.7	2772.1	2.0758	6.6381
8.7	173.94	.2220	736.52	2036.1	2772.6	2.0806	6.6342
8.8	174.43	.2196	738.64	2034.4	2773.0	2.0853	6.6303
8.9	174.90	.2173	740.75	2032.8	2773.6	2.0900	6.6264
9.0	175.38	.2150	742.82	2031.1	2773.9	2.0946	6.6226
9.1	175.85	.2127	744.91	2029.5	2774.4	2.0992	6.6188
9.2	176.31	.2105	746.96	2027.9	2774.8	2.1038	6.6151
9.3	176.78	.2084	749.00	2026.3	2775.3	2.1083	6.6114
9.4	177.24	.2063	751.02	2024.7	2775.7	2.1127	6.6077
9.5	177.69	.2042	753.02	2023.1	2776.1	2.1172	6.6041
9.6	178.14	.2022	755.01	2021.5	2776.5	2.1215	6.6005
9.7	178.59	.2002	756.98	2019.9	2776.9	2.1259	6.5969
9.8	179.03	.1982	758.94	2018.4	2777.3	2.1302	6.5934
9.9	179.47	.1963	760.88	2016.8	2777.7	2.1345	6.5899
10.0	179.91	.19444	762.81	2015.3	2778.1	2.1387	6.5865
10.2	180.71	.19080	766.63	2012.2	2778.9	2.1471	6.5796
10.4	181.62	.18730	770.38	2009.2	2779.6	2.1553	6.5729
10.6	182.46	.18392	774.08	2006.2	2780.3	2.1634	6.5664
10.8	183.28	.18067	777.74	2003.3	2781.0	2.1713	6.5599
11.0	184.09	.17753	781.34	2000.4	2781.7	2.1792	6.5536

Contd...

PROPERTIES OF DRY AND SATURATED STEAM

(Pressure Table)

Pressure bar p	Sat. temp. °C t_s	Specific vol. Sat. steam m^3/kg v_g	Enthalpy kJ/kg Sat. water h_f	Evap. h_{fg}	Sat. steam h_g	Entropy kJ/kg K Sat. water Φ_w	Sat. steam Φ_s
11.2	184.89	.17449	784.89	1997.5	2782.4	2.1869	6.5473
11.4	185.68	.17156	788.40	1994.6	2783.0	2.1945	6.5412
11.6	186.46	.16873	791.86	1991.8	2783.6	2.2020	6.5351
11.8	187.23	.16599	795.28	1989.0	2784.2	2.2093	6.5292
12.0	187.99	.16333	798.65	1986.2	2784.2	2.2166	6.5233
12.2	188.74	.16077	801.98	1983.4	2785.4	2.2238	6.5176
12.4	189.48	.15828	805.28	1980.7	2786.0	2.2309	6.5119
12.6	190.20	.15586	808.53	1978.0	2786.5	2.2378	6.5063
12.8	190.93	.15353	811.75	1975.3	2787.1	2.2447	6.5008
13.0	191.64	.15125	814.93	1972.7	2787.6	2.2515	6.4953
13.2	192.34	.14905	818.07	1970.0	2788.1	2.2582	6.4900
13.4	193.04	.14691	821.18	1967.4	2788.6	2.2648	6.4847
13.6	193.72	.14483	824.85	1964.8	2789.1	2.2714	6.4795
13.8	194.40	.14281	827.29	1962.3	2789.6	2.2778	6.4743
14.0	195.07	.14084	830.30	1959.7	2790.0	2.2842	6.4693
14.2	195.74	.13893	833.28	1957.2	2790.5	2.2905	6.4643
14.4	196.39	.13707	836.28	1954.7	2790.9	2.2968	6.4593
14.6	197.04	.13525	839.14	1952.2	2791.3	2.3029	6.4544
14.8	197.69	.13349	842.03	1949.7	2791.7	2.3090	6.4496
15.0	198.32	.13177	844.89	1947.3	2792.2	2.3150	6.4448
15.5	199.88	.12766	851.92	1941.2	2793.1	2.3298	6.4331
16.0	201.41	.12380	858.79	1935.2	2794.0	2.3442	6.4218
16.5	202.89	.12006	865.50	1929.4	2794.9	2.3582	6.4108
17.0	204.34	.11673	872.06	1923.6	2795.7	2.3718	6.4000
17.5	205.76	.11349	878.50	1917.9	2796.4	2.3851	6.3896
18.0	207.15	.11042	884.79	1912.4	2797.1	2.3981	6.3794
18.5	208.51	.10751	890.96	1906.8	2797.8	2.4109	6.3694
19.0	209.84	.10475	897.02	1901.4	2798.4	2.4233	6.3597
19.5	211.14	.10212	902.96	1896.0	2799.0	2.4354	6.3502
20.0	212.42	.09963	908.79	1890.7	2799.5	2.4474	6.3409

Contd...

PROPERTIES OF DRY AND SATURATED STEAM

(Pressure Table)

Pressure bar p	Sat. temp. °C t_s	Specific vol. Sat. steam m^3/kg v_g	Enthalpy kJ/kg			Entropy kJ/kg K	
			Sat. water h_f	Evap. h_{fg}	Sat. steam h_g	Sat. water Φ_w	Sat. steam Φ_s
20.5	213.67	.09725	914.52	1885.5	2800.0	2.4590	6.3318
21.0	214.90	.09498	920.15	1880.3	2800.5	2.4704	6.3229
21.5	216.10	.09281	925.69	1875.2	2800.9	2.4817	6.3141
22.0	217.29	.09073	931.14	1870.2	2801.3	2.4927	6.3056
22.5	218.45	.08875	936.49	1865.2	2801.7	2.5035	6.2972
23.0	219.60	.08685	941.77	1860.2	2802.0	2.5141	6.2890
23.5	220.72	.08502	946.97	1855.4	2802.3	2.5245	6.2809
24.0	221.83	.08327	952.09	1850.5	2802.6	2.5347	6.2729
24.5	222.92	.08159	957.13	1845.7	2802.9	2.5448	6.2651
25.0	223.99	.07998	962.11	1841.0	2803.1	2.5547	6.2575
26	226.09	.07692	971.85	1831.6	2803.6	2.5740	6.2425
27	228.12	.07409	981.34	1822.4	2803.8	2.5927	6.2280
28	230.10	.07145	990.59	1813.4	2804.0	2.6109	6.2139
29	232.02	.06899	999.61	1804.5	2804.1	2.6285	6.2002
30	233.90	.06668	1008.42	1795.7	2804.2	2.6457	6.1869
31	235.72	.06452	1017.04	1787.1	2804.1	2.6624	6.1740
32	237.51	.06249	1025.47	1778.6	2804.1	2.6787	6.1614
33	239.54	.06057	1033.72	1770.2	2803.9	2.6946	6.1491
34	240.94	.05877	1041.82	1761.9	2803.7	2.7101	6.1370
35	242.60	.05707	1049.75	1753.7	2803.4	2.7253	6.1253
36	244.23	.05545	1057.53	1745.6	2803.1	2.7401	6.1138
37	245.82	.05392	1065.17	1737.6	2802.8	2.7546	6.1025
38	247.38	.05247	1072.68	1729.7	2802.4	2.7688	6.0915
39	248.91	.05109	1080.05	1721.8	2801.9	2.7828	6.0807
40	250.40	.04978	1087.31	1714.1	2801.4	2.7964	6.0701
42	253.31	.04733	1101.48	1698.8	2800.3	2.8229	6.0495
44	256.12	.04510	1115.22	1683.8	2799.0	2.8485	6.0296
46	258.83	.04306	1128.58	1669.0	2797.6	2.8732	6.0103
48	261.45	.04118	1141.57	1654.5	2796.0	2.8970	5.9916
50	263.99	.03944	1154.23	1640.1	2794.3	2.9202	5.9734

Contd...

PROPERTIES OF DRY AND SATURATED STEAM

(Pressure Table)

Pressure bar p	Sat. temp. °C t_s	Specific vol. Sat. steam m³/kg v_g	Enthalpy kJ/kg			Entropy kJ/kg K	
			Sat. water h_f	Evap. h_{fg}	Sat. steam h_g	Sat. water Φ_w	Sat. steam Φ_s
52	266.45	.03783	1166.58	1626.0	2792.6	2.9427	5.9557
54	268.84	.03634	1178.66	1612.0	2790.7	2.9645	5.9385
56	271.17	.03495	1190.46	1598.2	2788.6	2.9858	5.9217
58	273.43	.03365	1202.02	1584.5	2786.5	3.0065	5.9052
60	275.64	.03244	1213.35	1571.0	2784.3	3.0267	5.8892
62	277.78	.03130	1224.46	1557.6	2782.1	3.0464	5.8734
64	279.88	.03023	1235.37	1544.3	2779.7	3.0657	5.8580
66	281.93	.02922	1246.09	1531.2	2777.2	3.0845	5.8428
68	283.93	.02827	1256.63	1518.1	2774.7	3.1030	5.8279
70	285.88	.02737	1267.00	1505.1	2772.1	3.1211	5.8133
72	287.79	.02652	1277.21	1492.2	2769.4	3.1389	5.7989
74	289.67	.02571	1287.28	1479.4	2766.7	3.1563	5.7847
76	291.50	.02494	1297.19	1466.6	2763.8	3.1734	5.7707
78	293.30	.02421	1306.98	1453.9	2760.9	3.1902	5.7569
80	295.06	.02352	1316.64	1441.3	2758.0	3.2068	5.7432
82	296.79	.02286	1326.18	1428.7	2754.9	3.2230	5.7297
84	298.49	.02222	1335.61	1416.2	2751.8	3.2391	5.7164
86	300.16	.02162	1344.93	1403.7	2748.7	3.2549	5.7032
88	301.80	.02104	1354.14	1391.3	2745.4	3.2704	5.6902
90	303.40	.02048	1363.26	1378.9	2742.1	3.2858	5.6772
92	304.99	.01995	1372.29	1366.5	2738.8	3.3009	5.6644
94	306.54	.01944	1381.23	1354.1	2735.4	3.3159	5.6517
96	308.07	.01895	1390.08	1341.8	2731.9	3.3306	5.6391
98	309.58	.01848	1398.86	1329.5	2728.3	3.3452	5.6265
100	311.06	.018026	1407.56	1317.1	2724.7	3.3596	5.6141
102	312.46	.017588	1416.19	1304.8	2721.0	3.3739	5.6017
104	313.96	.017167	1424.76	1292.5	2717.3	3.3880	5.5894
106	315.38	.016760	1433.26	1280.2	2713.5	3.4020	5.5771
108	316.77	.016367	1441.70	1267.9	2709.6	3.4158	5.5649
110	318.15	.015987	1450.10	1255.5	2705.6	3.4295	5.5527

Contd...

PROPERTIES OF DRY AND SATURATED STEAM

(Pressure Table)

Pressure bar p	Sat. temp. °C t_s	Specific vol. Sat. steam m³/kg v_g	Enthalpy kJ/kg			Entropy kJ/kg K	
			Sat. water h_f	Evap. h_{fg}	Sat. steam h_g	Sat. water Φ_w	Sat. steam Φ_s
112	319.50	.015620	1458.4	1243.2	2701.6	3.4430	5.5406
114	320.84	.015264	1466.7	1230.8	2697.5	3.4565	5.5285
116	322.16	.014920	1474.9	1218.4	2693.4	3.4698	5.5165
118	323.46	.014587	1483.1	1206.0	2689.2	3.4831	5.5044
120	324.75	.014263	1491.3	1193.6	2684.9	3.4962	5.4924
122	326.02	.013949	1499.4	1181.1	2680.5	3.5092	5.4804
124	327.27	.013644	1507.5	1168.6	2676.0	3.5222	5.4684
126	328.51	.013348	1515.5	1156.0	2671.5	3.5351	5.4564
128	329.73	.013060	1523.5	1143.4	2666.9	3.5478	5.4444
130	330.93	.012780	1531.5	1130.7	2662.2	3.5606	5.4323
132	332.12	.012508	1539.5	1118.0	2657.5	3.5732	5.4203
134	333.30	.012242	1547.4	1105.2	2652.6	3.5858	5.4082
136	334.46	.011983	1555.3	1092.4	2647.7	3.5983	5.3961
138	335.61	.011731	1563.2	1079.5	2642.7	3.6108	5.3839
140	336.75	.011485	1571.1	1066.5	2637.6	3.6232	5.3717
142	337.87	.011245	1579.0	1053.4	2632.4	3.6356	5.3595
144	338.98	.011010	1586.9	1040.2	2627.1	3.6479	5.3471
146	340.08	.010781	1594.8	1026.9	2621.7	3.6603	5.3348
148	341.16	.010557	1602.6	1013.5	2616.1	3.6726	5.3223
150	342.24	.010337	1610.5	1000.0	2610.5	3.6848	5.3098
152	343.30	.010123	1618.4	986.4	2604.8	3.6971	5.2971
154	344.35	.009912	1626.3	972.7	2598.9	3.7093	5.2844
156	345.39	.009706	1634.2	958.8	2593.0	3.7216	5.2716
158	346.42	.009504	1642.1	944.8	2586.9	3.7338	5.2586
160	347.44	.009306	1650.1	930.6	2580.6	3.7461	5.2455
162	348.44	.009111	1658.0	916.2	2574.3	3.7584	5.2323
164	349.44	.008920	1666.0	901.7	2567.7	3.7707	5.2189
166	350.43	.008732	1674.1	886.9	2561.0	3.7830	5.2053
168	351.40	.008547	1682.2	872.0	2554.2	3.7954	5.1916
170	352.37	.008364	1690.3	856.9	2547.2	3.8079	5.1777

Contd...

PROPERTIES OF DRY AND SATURATED STEAM

(Pressure Table)

Pressure bar p	Sat. temp. °C t_s	Specific vol. Sat. steam m^3/kg v_g	Enthalpy kJ/kg			Entropy kJ/kg K	
			Sat. water h_f	Evap. h_{fg}	Sat. steam h_g	Sat. water Φ_w	Sat. steam Φ_s
172	353.33	.008185	1698.5	841.5	2540.0	3.8204	5.1635
174	354.28	.008008	1706.8	825.8	2532.6	3.8330	5.1491
176	355.21	.007833	1715.1	809.9	2525.0	3.8457	5.1345
178	356.14	.007660	1723.5	793.6	2517.2	3.8586	5.1196
180	357.06	.007489	1732.0	777.1	2509.1	3.8715	5.1044
182	357.98	.007320	1740.7	760.1	2500.8	3.8846	5.0889
184	358.88	.007153	1749.4	742.8	2492.2	3.8978	5.0730
186	359.77	.006987	1758.3	725.1	2483.3	3.9112	5.0567
188	360.66	.006822	1767.3	706.1	2474.1	3.9249	5.0400
190	361.54	.006657	1776.5	688.0	2464.5	3.9388	5.0228
192	362.41	.006493	1785.9	668.7	2454.6	3.9530	5.0050
194	363.27	.006329	1795.6	648.6	2444.2	3.9676	4.9866
196	364.12	.006165	1805.5	627.8	2433.3	3.9825	4.9676
198	364.97	.006000	1815.7	606.1	2421.8	3.9979	4.9477
200	365.81	.005834	1826.3	583.4	2409.7	4.0139	4.9269
202	366.64	.005665	1837.3	559.6	2396.9	4.0305	4.9050
204	367.46	.005495	1848.9	534.3	2383.2	4.0479	4.8819
206	368.28	.005320	1861.1	507.3	2368.4	4.0663	4.8571
208	369.09	.005139	1874.2	478.1	2352.3	4.0861	4.8304
210	369.89	.004952	1888.4	446.2	2334.6	4.1075	4.8013
212	370.69	.004754	1904.2	410.6	2314.8	4.1313	4.7690
214	371.47	.004538	1922.2	369.6	2291.8	4.1585	4.7318
216	372.25	.004298	1943.8	320.3	2264.1	4.1914	4.6876
218	373.03	.004007	1972.6	255.1	2227.6	4.2351	4.6298
220	373.80	.003568	2025.0	140.6	2165.6	4.3154	4.5327
220.9	374.14	.003155	2099.3	0	2099.3	4.4298	4.4298

www.ingramcontent.com/pod-product-compliance
Lightning Source LLC
Chambersburg PA
CBHW080157170726
48282CB00005B/53